AF411151

Perspectives in Hadronic Physics

S. Boffi   C. Ciofi degli Atti   M.M. Giannini
(Eds.)

Perspectives in Hadronic Physics

4th International Conference
Held at ICTP, Trieste, Italy, 12–16 May 2003

 Springer

Editors:

Professor Sigfrido Boffi
Università di Pavia
Dipartimento di Fisica Nucleare e Teorica
Via Bassi 6
27100 Pavia, Italy
email: boffi@pv.infn.it

Professor Claudio Ciofi degli Atti
Università di Perugia
Dipartimento di Fisica
Via A. Pascoli 6/2
06100 Perugia, Italy
email: ciofi@pg.infn.it

Professor Mauro M. Giannini
Università di Genova
Dipartimento di Fisica
Via Dodecaneso 33
16146 Genova, Italy
email: giannini@ge.infn.it

ISBN 3-540-21064-4 Springer Berlin Heidelberg New York

Library of Congress Control Number: 2004103359

Springer is a part of Springer Science+Business Media.

springeronline.com

© Springer-Verlag Berlin Heidelberg 2004
Printed in Germany

Typesetting: Data prepared by authors and editors using a Springer TeX macro package
Cover design: *design & production* GmbH, Heidelberg

Printed an acid-free paper   10983188   55/3141/di   5 4 3 2 1 0

Preface

This volume contains the invited and contributed papers presented at the Fourth International Conference on Perspectives in Hadronic Physics and sent to the Editors within the deadline. The Conference was held at the Abdus Salam International Centre for Theoretical Physics (ICTP), Trieste, Italy, from May 12th to 16th, 2003, and was attended by about 100 scientists from 20 countries.

The series of Conferences on Perspectives on Hadronic Physics takes place every two years since 1997 and follows the seven Workshops on Perspectives in Nuclear Physics at Intermediate Energies, organized every two years at ICTP since 1983. The aim of these Conferences is to discuss the status-of-the-art concerning the experimental and theoretical investigations of hadronic systems, from nucleons to nuclei and dense nuclear matter, in terms of the relevant underlying degrees of freedom. For such a reason the Fourth Conference has been focused on those experimental and theoretical topics which have been in the last few years the object of intensive investigations, viz. the various approaches employed to describe the structure of hadrons in terms of QCD and QCD inspired models, the recent developments in the treatment of the properties and propagations of hadronic states in the medium, the relevant progress done in the solution of the few- and many- hadron problems, the recent results in the experimental investigation of dense hadronic matter and, last but not least, the physics programs of existing Laboratories and the suggested projects for new Facilities. All papers appearing in the Volume have been refereed and properly edited.

The conference was sponsored by the ICTP and the Istituto Nazionale di Fisica Nucleare (INFN). We would like to express our gratitude to Professor Katepalli R. Sreenivasan, Director of ICTP, and to Professor Enzo Iarocci, President of INFN, for their generous help.

We would also like to thank the members of the Advisory Committee for their contribution to make the conference a very successful one.

All the organizational work was taken care of by the efficient and highly professional staff of the ICTP. We are particularly indebted to Ms. Marina De Comelli, who was in charge for the preparation and the running of the conference, and to Ms. Anna Triolo of the public information office.

Sigfrido Boffi
Claudio Ciofi degli Atti
Mauro M. Giannini
The Editors

Contents

Laboratory physics programs and new facilities

The structure of hadrons

Eur Phys J A **19**, s01, 1–8 (2004)
Digital Object Identifier (DOI) 10.1140/epjad/s2004-03-001-5

Lattice calculations of hadron properties

Gunnar S. Bali

Department of Physics and Astronomy, The University of Glasgow, Glasgow G12 8QQ, Scotland

Received: 13 Aug 2003 / Accepted: 14 Nov 2003 /
Published Online: 6 Feb 2004 – © Società Italiana di Fisica / Springer-Verlag 2004

Abstract. Recent lattice studies of hadron properties, in particular of exotic states and charmonia are reviewed. Sea quark and quark mass effects are discussed as well as decays and mixing.

PACS. 12.39.Mk Glueball and nonstandard multi-quark/gluon states – 14.40.Cs Mesons with $S = C = 0$ – 14.40.Gx Mesons with $S = C = B = 0$ – 12.38.Gc Lattice QCD calculations

1 Introduction

While the simplicity and elegance of QCD is very appealing theoretically, the phenomenological observations of spontaneous chiral symmetry breaking and even more so of the confinement of colour charges turned it into a major calculational nightmare: it took almost twenty years after the discovery of asymptotic freedom to convincingly demonstrate that the QCD Lagrangian indeed implies these highly non-trivial collective phenomena. This was done by numerical simulations; an analytic proof is still lacking.

Fortunately, due to the property of asymptotic freedom, many short distance/high energy QCD problems can be approached by means of perturbation theory. This need not be so since the very basis of the perturbative expansion is shaky: confinement implies that quark and gluon fields never appear as asymptotic states. Fortunately, the success of jet phenomenology suggests that QCD is reasonably benign in the high energy region. This is very different in the low energy regime of *strong QCD*. Among the few analytical tools that exist are the strong coupling and the $1/N$ expansions, effective field theories (EFTs) and various QCD inspired or phenomenological models.

QCD can in principle be solved rigorously by means of lattice simulations on a computer. In practise however computational resources are finite. This leads to pion masses of typically more than 400 MeV or to simulations within the quenched approximation, where sea quark effects are neglected. While these systematic uncertainties reduce with faster computers, improved numerical algorithms and theoretical ingenuity, it will always remain desirable to combine lattice simulations with EFT methods or, where necessary, with QCD motivated models as only relatively simple questions can directly be addressed on the lattice. In many cases problems of phenomenological interest factorise naturally into a high energy electro-weak part and a low energy QCD part which can then be evaluated on the lattice, for instance electromagnetic form factors and weak decay matrix elements. In many cases QCD problems can also be factorised within the framework of EFTs into low and high energy parts. Lattice simulations turn out to be invaluable to gain insight into the dynamics of QCD as many parameters such as the number of active quark flavours, quark masses, number of colours, temperature etc. are not limited to their phenomenological values but can be varied.

Several good reviews of lattice calculations exist in the literature and I refer to them for details on theoretical aspects, calculational methods and wider phenomenological implications [1]. The topics covered in Chris Michael's review [2] "Exotics" have a non-vanishing overlap with this article, albeit written from a (slightly) different perspective. In my talk I covered the baryon spectrum and structure as well. Due to the page limit (and lack of new content) I refer to my recent review [3]. Subjectively selected highlights since then include two studies of generalised parton distributions [4,5] and a determination of the pion form factor [6]. Some progress has been made in the calculation of electromagnetic N to Δ transition form factors [7], a new study of the mass spectrum of excited nucleons has appeared [8] and a first step towards consolidation of previous results was performed [9]. I will not mention the Θ^+, because there are no lattice results. Instead I will concentrate on the spectroscopy of charmonia and exotics.

2 Glueballs and friends

It was realised as early as 1974 [10] that QCD offered the possibility of bound states composed only of energy. These closed-string-/glue-states/boxcitons became subsequently known as glueballs [11]. It is hard to imagine anything that demonstrates confinement more than the discovery of quarkless massive bound states. One can also fantasize that non-perturbative physics with particles similar to glueballs might play a rôle in future theories *beyond the Standard Model*. After all QCD is the only part of the Standard Model with a chance of a mathematically rigorous definition. So glueballs, the simplest possible colour-neutral states, are a challenge that certainly has to be addressed.

The situation is complicated theoretically as well as experimentally by the possibility of mixing with standard quark model states. Even worse QCD does not know what is *exotic* and what is not. The very meaning of this term is *"not quark model"* and it is not at all *a priori* clear what exactly we mean by *quark model*. For instance we can ask ourselves how closely the quark model describes something as profane as the QCD proton. While introducing the concept of constituent quarks eliminates the puzzle that more than 98 % of its rest mass is generated by spontaneous chiral symmetry breaking, i.e. by the glue, it is impossible to attribute the proton's spin and momentum exclusively to quark degrees of freedom. Gluodynamics plays an even greater rôle in properties of the η' and the π. So if the gluons already leave big footprints on *quark model states* how can we distinguish these from *glue states*? One possibility that immediately springs to mind would be to look for *spin-exotic* states: when coupling spin and angular momentum within mesons, only certain combinations of J^{PC} are allowed while for instance $0^{--}, 0^{+-}, 1^{-+}, 2^{+-}$ are forbidden. Again the situation is complicated by the possibility of $q\bar{q}q\bar{q}$ molecular states to couple to these same quantum numbers.

One toy model is the pure Yang Mills theory of gluodynamics. In this fictitious world no quark fields exist but many central features of QCD including confinement and asymptotic freedom are still reproduced. This well defined and self consistent theory yields a rich spectrum of glueballs, all of which are absolutely stable, with the exception of very heavy ones that decay into lighter glueballs.

One can go one step further towards QCD by including quark fields that propagate in the gluodynamic background but whose feedback onto the vacuum is neglected, the so-called quenched approximation. Quenched QCD (qQCD) is no quantum field theory since unitarity is violated, however, even in this approximation chiral symmetry is spontaneously broken. Neglecting the feedback of slowly moving heavy quark sources onto their environment is a very natural thing to do. Quenching is also justifiable in the limit of the number of colours $N \to \infty$, however, it is not always clear whether $1/3 \ll 1$. Simulations of pure $SU(N_c)$ gauge theories for [12,13] $N = 2, \ldots, 6$ seem to indicate this. As it should be ratios of light hadron masses from lattice simulations of qQCD have been found to be inconsistent with the observed spectrum [14] however the differences are typically smaller than 10 % suggesting that the quenched approximation has some predictive power if cautiously consumed. The consequences of violating unitarity at light quark mass can become dramatic in some channels and in particular in the scalar sector [15]: roughly

speaking as the axial anomaly does not exist in qQCD the η' will be a surplus light Goldstone boson. The impact of this can be investigated in quenched chiral perturbation theory where diagrams that include transitions from scalar to two πs yield unwanted contributions that explode as the πs, including the would-be η', become light.

One justification for quenching is that the computational effort is easily reduced by a factor of 10^3. So it makes complete sense to learn and to understand statistics and systematics from a quenched study, prior to doing the real thing. Moreover, most models used in hadron physics neglect quark pair creation and annihilation, so the model builders can still learn from qQCD. Also, "un-quenching" a model to compare it with lattice results is easier than un-quenching the lattice simulation. Last but not least, life is easier in quenched as lattice studies of strong decays and excited states are notoriously complicated: glueballs will not mix with quark model mesons, spin exotic meson-gluon hybrids are distinct from mesons, are stable and do not mix with four-quark molecules either.

2.1 Heavy glueballs and charmonia

In Fig. 1 we compile the glueball spectrum (filled boxes) of gluodynamics [13,16]. The scale $r_0^{-1} \approx 394$ is set from potential models. Since this is not the real world a systematic scale error of about 10 % should be assumed, for stable states that do not mix! While these newer results agree with the spectrum of [17], during the past decade the statistical errors have been reduced by factors varying between 1.5 and 2, depending on the channel. It has also been established that the 3^{++} glueball is lighter than the 1^{++}. This had only been suspected in the earlier reference, where the J assignment of this state remained ambiguous.

The lightest glueball is a scalar, followed by a tensor and a pseudoscalar. Possibly the lightest state with exotic quantum numbers is a 2^{+-} glueball around 4 GeV. Qualitative features can be understood in a bag model while the flux tube model seems inadequate [18]. Without numerical simulation it would have been completely impossible to come close even to a semi-quantitative understanding of these truly unconventional bound states.

We include experimental charmonium levels (lines) for comparison. The $^{2S+1}L_J$ notation refers to these states. We have only included confirmed resonances, with masses taken from the Particle Data Book. For the η_c' we took the Belle result from double charmonium production [19], 3630(8) MeV. One often thinks of the charm quark as heavy in the sense that its mass is much bigger than mesonic and baryonic QCD binding energies. However, the spectrum of states entirely made out of glue with no quarks at all covers a similar energy range! We also display recent quenched lattice results from two groups [20,21,22] (CP-PACS and Columbia). These have been obtained using relativistic charm quarks on anisotropic lattices with the ratio of spatial over temporal lattice spacings $\xi \approx 2$ (Columbia) and $\xi \approx 3$ (CP-PACS). A recent simulation of $J \leq 1$ S and P wave charmonia on isotropic ($\xi = 1$) lattices by QCD-TARO [23] (who also study charmonium

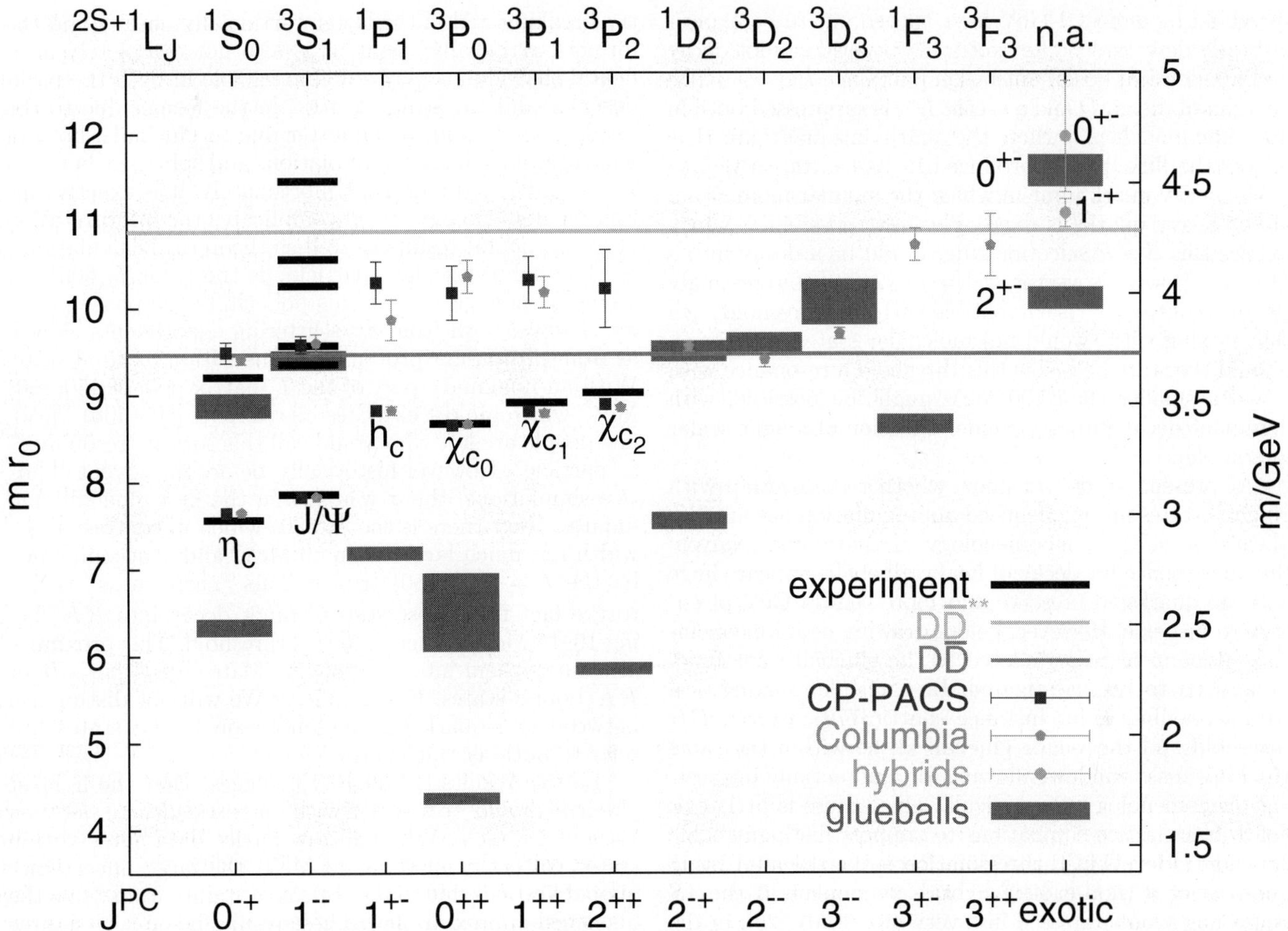

Fig. 1. The quenched charmonium spectrum (CP-PACS [20], Columbia [21,22]), glueballs [16,13] and spin-exotic $c\bar{c}$-glue hybrids [22], overlayed with the experimental spectrum

wave functions) confirms these findings. Note that in all these simulations the effect of diagrams with disconnected quark lines has been neglected. One might expect OZI violating contributions from these, in particular for states that lie close to glueballs with the same quantum numbers.

The charm quark mass has been adjusted such that the spin averaged $1S$ state reproduces experiment. This means that in this case the quenching errors of up to 10 % apply to spin-averaged level splittings with respect to this ground state, rather than with respect to zero energy. This systematic scale uncertainty caused by omitting sea quarks renders it irrelevant whether we use a "constructed" scale like r_0 or an experimental mass like the $1P - 1S$ gap as an input (which would increase the splittings by about 6 %). For the fine structure, potential models combined with lattice results [24] tell us that we should expect to undershoot the real world number by up to 40 %. For states above threshold or where mixing effects and strong decays play a big rôle radical changes might occur while for the $2P$ excitations finite volume effects could be an issue.

There are different ways of obtaining a given J^{PC}: for instance $J^{PC} = 1^{--}$ can be an S or a D wave. The corre-lation function associated with the respective D wave operator decays very fast in Euclidean time, into the same ground state as the respective S wave: it appears that the charm quark is too light to turn L into an (approximately) good quantum number. A similar behaviour has been observed for the 2^{++} and 1^{++} states (either P or F waves).

In addition to the standard charmonia and glueball states the figure contains the lightest two spin exotic $c\bar{c}$-gluon hybrids [22]. At least two other studies of hybrid charmonia with relativistic charm quarks exist to-date [25, 26]. Given the fact that not even L is a good quantum number it is not clear how one would distinguish non spin-exotic hybrids from conventional radial excitations. The lightest exotic hybrid turns out to be a vector, 1^{-+}. This is followed by 0^{+-} and 2^{+-} hybrids.

Once sea quarks are switched on, most of the states calculated on the lattice will decay strongly and mixing will occur too. In this context it is interesting to see that the spin-exotic 0^{+-} glueball and $c\bar{c}g$ hybrid have similar masses. We have included two experimental thresholds into the figure: 1^{--} charmonia can and will decay into a $D\overline{D}$ meson pair. Nonetheless resonances that exceed this threshold by almost 1 GeV have turned out to be experimentally detectable. The exotic 1^{-+} hybrid cannot decay into an identical boson-antiboson pair since $PC = -$. The next possibility, a $\overline{D}$ and a vector D^*, is suppressed both in flux tube models [27] where the quark-antiquark pair that breaks the flux tube is produced in its centre, as well as in the heavy quark limit in which the angular momentum of the heavy quarks is fixed. The next possibility, which follows this $S + P$ selection rule, would be a decay into a $\overline{D}$ and a P wave isovector $D^{**}[= D_1(2420)]$. Interestingly the mass of the 1^{-+} is compatible with this threshold such that mixing with (would be) molecular states is an issue. Should the state be below this threshold a resonance with a width smaller than 100 MeV might be possible, with dominant decay into a χ_c under emission of a light scalar meson [28].

At present we do not know whether the overlap with glueball states in the quenched approximation has any implications on the phenomenology of charmonia. Naïvely the phase space for decay of heavy glueballs appears huge with no quarks to preserve and more than 3 GeV of energy to disperse. However, before drawing definite conclusions dynamical issues related to the glueball wave functions need to be investigated. In particular mixing is a strong possibility: for instance the $\psi(3770)$, $\psi(2S)$, $D\overline{D}$ threshold and the vector glueball all lie within the same 100 MeV mass window, allowing for a potentially interesting phenomenology, in particular since there is little evidence from lattice simulations to support distinguishable $2S$ and $1D$ levels in this region. Recently, the interesting question of a (non-exotic) hybrid component in the $1S$ states has been addressed in a very nice study [29], in the framework of NRQCD. The conclusion is that this contribution is weak. The question becomes more exciting but also tremendously less approachable if the $2S$ charmonium state and glueball channels were considered in addition.

With data from Belle and Babar emerging, CLEO-c being online and the possibility of BES III starting to take data in 2006 the charmonium region is also exciting experimentally. Unfortunately, there is no way of producing say spin-exotic 1^{-+} states at any detectable rate in a decay starting from a vector resonance. For this we might have to wait for the proton-antiproton PANDA experiment at GSI. However, in the meantime there will definitely be progress in measuring spectrum and decay rates. Possibly the glue-richness of this mass region will leave its imprint, for instance by enhancing OZI suppressed processes.

2.2 Light scalars: Today

A lot of experimental attention has been devoted to the spectroscopy of light scalar mesons. The reasons are threefold: the scalar sector is intimately linked to chiral symmetry breaking which results in both light πs but at the same time also in heavy σs. Furthermore, the lightest glueball is predicted to be a scalar and, as an added extra, the lightest four-quark candidate state has scalar quantum numbers as well: pseudoscalar-pseudoscalar bound states can be lighter then quark model scalars since chiral symme-

try breaking makes the latter "artificially" heavy and the former "artificially" light. It is also not completely accidental that gluons play a dominant rôle in both the pseudoscalar and the scalar sectors, in the former due to the axial anomaly and in the latter due to glueballs, both of which result in large OZI violations and splittings between $I = 0$ and $I \neq 0$ mesons. Unfortunately, it is exactly this rich phenomenology which complicates the interpretation of experimental results as well as theoretical calculations.

The lightest scalar "particle" is the σ [or $f_0(600)$]. It seems by now clear that this pole shifts the phase of the $\pi\pi$ S wave, both from $\pi\pi$ scattering experiments as well as from proton-antiproton collisions with a 3π final state. With an imaginary part of the T matrix pole of 450(150) MeV, which almost exceedes its real part [800(400) MeV], few people are left who would call this sort of "resonance" a "particle". However historically, before the advent of lattice simulations, the σ was among the first glueball candidates. Next there is the $f_0(980)$ which in contrast has a width not much larger than 50 MeV, and is accompanied by the $I = 1$ $a_0(980)$ triplet. This "narrowness" is due to the fact that these states cannot decay into $K\overline{K}$, being 10–15 MeV lighter than this threshold. This proximity also turns them into natural candidates for either $s\overline{l}l\overline{s}$ or $K\overline{K}$ bound states. For simplicity We will not distinguish between four-quark and meson-meson bound states but refer to both as "molecules".

Lattice results [17,30,16,13] suggest that the lightest glueball should be a scalar with mass somewhere between 1.4 and 1.8 GeV. While all raw lattice data agree within statistical errors of some 40 MeV, the large uncertainty quoted above is due to the scale uncertainty from using the quenched approximation. These results lie on top of three experimental scalar $I = 0$ resonances, the broad $f_0(1370)$, the extremely well studied $f_0(1500)$ and the $f_0(1710)$. The standard picture is that these states are mixtures between a glueball and two $I = 0$ nonet mesons, one with dominantly $u\overline{u} + d\overline{d}$ and the other with $s\overline{s}$ quark content. The seven remaining $I \neq 0$ nonet members are most likely the four $K_0^*(1430)$ as well as three $a_0(1450)$ states. Whether there is an excess of experimental states over quark model states or not critically depends on how resonances are organised into nonets. It has to be said that the $a_0(1450)$ is quite broad and that the experimental evidence for these states is not rock hard. On the other hand, with a splitting of 450 MeV, it is extremely hard to reconcile the K_0^* and the $a_0(980)$ into the same nonet, in particular the near degeneracy of the a_0 with the $f_0(980)$ suggests very tiny OZI violating effects. [This in turn also means that there is little chance for a significant glueball component in the $f_0(980)$.] The remaining puzzle is the absence of $S \neq 0$ partners of the $a_0/f_0(980)$. Most likely these states have a large molecular component and are part of an inverted nonet [31], together with the $f_0(980)$, a would-be-$\pi\pi$ σ state and would-be-πK κs, which just happen to be extremely fragile, due to the light π involved.

2.3 Light scalars: Future

To support the above picture an unambiguous experimental identification of the $a_0(1450)$ and a theoretical clarification of the molecular nature of the $a_0/f_0(980)$ states rank high on the wish list. Different experimental and theoretical inputs have led to various mixing models [32, 33, 34, 35]. Amsler and Close [32] suggest that the light quark state mainly goes into the $f_0(1370)$, with a subleading component mixing into the $f_0(1710)$ while the $s\bar{s}$ and the glueball are mainly distributed between the $f_0(1500)$ and $f_0(1710)$. This view has been superceeded by Close and Kirk's more recent analysis of decays and production rates [35] which suggests the $f_0(1500)$ to be dominantly glueball in character with light and strange quark components mixing destructively and additively into the $f_0(1370)$ and $f_0(1710)$, respectively. In contrast [33, 34] predict the $f_0(1710)$ to be the (dominant) glueball-state with the two lighter resonances being composed primarily of the quark model mesons.

All these predictions crucially depend on the input parameters used, in particular on the ordering of "unmixed" states and on assumptions on the mixing matrix. One might expect the unmixed $d\bar{d} + u\bar{u}$ state to have a mass close to that of the $a_0(1450)$ but unfortunately the latter resonance is very broad experimentally and not overly well established. While in QCD with light sea quarks all states will automatically be "mixed", the quenched approximation is ideal to address the question of ordering of "unmixed" states and even for estimating off-diagonal mass-matrix elements. The main problem however is that in this case correlation functions will loose positivity at small quark masses as the π and the η' are degenerate [15]. In practise this means that at best $s\bar{s}$ scalars can be approached using traditional methods.

In [34] significant finite size effects and a strong lattice spacing dependence are observed. After extrapolations the quenched $s\bar{s}$ scalar appears to be about 300 MeV lighter than the pure glueball. RBC [36] quote an even lower mass of 1.04(7) GeV for the light quark a_0 and 0.9(1) GeV for the $I = 0$ f_0/σ. The leading chiral correction from the $\pi\eta'$ loop was included into this study. One would however hope that systematic uncertainties will be identified and investigated in the future. Should this result be confirmed, then the $a_0(980)$ could be explained as a quark model state but what is the $f_0(980)$ and why are the K_0^*s, that are experimentally well established, so heavy? In a very sophisticated study Bardeen $et\ al.$ [37, 15] determine the relevant low energy parameters of the quenched chiral Lagrangian and remove the leading order quenched chiral power and log. After testing the approach in the pseudoscalar sector they are able to determine an a_0 mass of 1.33(5) GeV, expressed in units of the ρ mass, which is by about 200 ± 100 MeV lighter than the glueball, extrapolated to the continuum limit and converted into the same units. The unmixed $s\bar{s}$ state might then be expected to be fairly equal in mass to the glueball. The main difficulty is that with only two lattice spacings at hand the extrapolation to the continuum limit is still a bit shaky.

The situation has also been studied by UKQCD [38] in QCD with sea quarks at lattice spacings $a \approx 0.1$ fm and $a \approx 0.13$ fm. In this case the a_0 is significantly lighter but the same holds true for the flavour singlet meson/glueball (see Fig. 2 below).

In addition to the studies of quark model scalars some results exist on four quark molecules, but with degenerate masses, with the aim of addressing the possibility of a $\pi\pi$ bound state. Results of quenched studies [39] are inconclusive due to the requirement of large volumes to avoid squeezing the πs on top of each other when they intend to be elsewhere: without a careful finite size study a residual interaction energy can hardly be disentangled from the $\pi\pi$ scattering phase [40]. Diagrams with disconnected quark lines have also been neglected (with one exception [41]). In the absence of definite results in the quenched approximation, the SCALAR collaboration has started the very ambitious program of determining the mass of an $I = 0\ \pi\pi$ state incorporating sea quarks [42]. In view of the above it is highly attractive to study simplified cases, with heavier quarks and $I = 1$, like $K\bar{K}$ or DK (see also Sect. 2.4 below) molecules, both with at least equally important phenomenological implications.

Few exploratory studies of strong decays exist. Since we are working in Euclidean time there is no concept of asymptotic states. Neither can we calculate the imaginary part of a forward amplitude but there are bug-fixes: the method of choice is the computationally very challenging exploitation of finite size effects introduced by Lüscher [40]. The poor man's approximation has been developed by Gottlieb $et\ al.$ [43] and Michael [44]. In the latter case an on-shell transition matrix element is evaluated by adjusting the rest mass of the decaying particle to the energy of the outgoing state. In QCD with sea quarks this method can only be approximate as the operator used to create the initial state will in general also have a non-vanishing overlap with the final state. The transition matrix element is related to a coupling which in turn, assuming momentum independence, can be normalised to phase space, predicting the partial decay width. The on-shell condition implies that for a given mass of the final state the relative momentum has to be adjusted to guarantee energy conservation. As momentum is discretized on a lattice this imposes some constraints. The initial state can be boosted such that at a fixed mass several relative momenta can be (approximately) realised, on sufficiently large volumes, and the momentum dependence of the coupling checked. Chiral perturbation theory can also help to connect results obtained at different quark mass, once these are sufficiently light.

Promising results using this approach have been obtained recently for the $\rho \to \pi\pi$ decay in $n_f = 2$ QCD by McNeile and Michael [45]. Decay couplings of a glueball to two pseudoscalars have so far only been computed by GF11 [30] some 9 years ago: a mass dependence has been observed with a stronger coupling to heavier mesons, which then has to be folded with phase space.

Elements of the glueball-scalar mass matrix have been calculated by two groups, again GF11 in qQCD [34] and UKQCD for $n_f = 2$ [46], for quark masses around the strange quark. Both collaborations obtain mixing energies of about 300 MeV and 500 MeV on relatively coarse lattices, respectively. Starting from two degenerate unmixed states this would imply level splittings between the mixed states of $0.5 - 1$ GeV. An extrapolation to the continuum limit by GF11 who also simulated three finer lattice spacings resulted in 61(45) MeV, a very reasonable value but with a 100 % uncertainty.

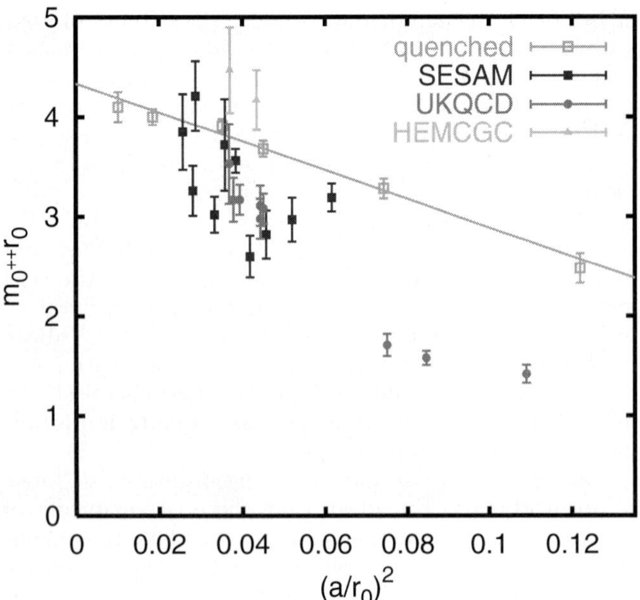

Fig. 2. The scalar "glueball": qQCD vs. $n_f = 2$

We are still in the position that the combined "world data" on the scalar $n_f = 2$ "glueball", plotted as a function of the squared lattice spacing a^2, fits into Fig. 2. The quenched case [13,17] is included for reference. The un-quenched results have been obtained by use of three different lattice discretizations of the Dirac action: staggered (HEMCGC [47]), Wilson (SESAM [48]) and clover (UKQCD [46,38]). The quarks are all not much lighter than the strange quark, the scalar meson is still stable and the wave function turns out to be very close to that of the quenched glueball [48,46]. Most $n_f = 2$ points clearly lie below the quenched line, however, there is certainly a slope in the results, such that the mass in the physical $a = 0$ limit does not contradict the quenched result. Within the SESAM data set there is an apparent discontinuity because different points have been obtained at different quark masses; the "glueball" becomes lighter as the quark mass is reduced. Clearly additional studies at lighter quark masses and different lattice spacings are required.

2.4 The $D_{sJ}^+(2317)$

Recently a narrow $D_{sJ}^+(2317)$ state has been detected by BaBar [49], dominantly decaying into $D_s\pi$. This finding was confirmed by CLEO [50] and by Belle [51]. The latter two collaborations also reported a narrow resonance around 2537 MeV, decaying into $D_s^* + \pi$. Both states lie by about 40 MeV below the respective DK and D^*K thresholds. In contrast potential model calculations suggest [52] much heavier masses for the missing D_{s0} and D_{s1}' P wave states, rendering these into broad resonances. These expectations have very recently received experimental support from Belle's observation of the missing D_0^* and D_1' states at $2308(17)(15)(28)$ MeV and at 2427 $(26)(20)(15)$ MeV, respectively [53]. These states indeed

strongly decay into $D\pi$ and $D^*\pi$, respectively, with widths of order 300 MeV.

If the new D_{sJ} mesons were P wave $c\bar{s}$ mesons, why are they so light but the corresponding $c\bar{d}$ mesons are not? Naturally this question invites speculation that the new scalar state might be of a DK (or 4-quark) molecular nature, somewhat resembling the $f_0/a_0(980)$ system [54]. Of course such ideas eventually have to be substantiated by a QCD calculation and indeed lattice results exist: simulations in the static limit within the quenched approximation [55] and with sea quarks [56], simulations including NRQCD/HQET $1/m$ corrections [57] and simulations with a relativistic charm quark [58,59]. The interpretation of these results is controversial: I [56] observed that both effects, including sea quarks and including relativistic corrections to the static limit, increase the mass of a quark model $c\bar{s}$ P wave scalar state, pushing it above the DK threshold and into agreement with potential model predictions. I then concluded that lattice results are incompatible with a pure quark model nature of the new D_{sJ} states. On the other hand, while confirming an increase in the predicted mass when incorporating sea quarks, the authors of [59] conclude that their results are consistent with the states observed by Babar and CLEO, however, within errors no disagreement is seen with the potential model predictions either. Interestingly, while $1/m$ corrections to the static limit are substantial for charm quarks and increase the $0^+ - 0^-$ splitting by about 25 % [56], the $1^+ - 1^-$ splitting is found to agree with the $0^+ - 0^-$ splitting within statistical and systematic errors of about 15 %, in quenched as well as with sea quarks [59], as suggested by chiral symmetry in the heavy quark limit [60].

The same chiral symmetry argument would also apply to molecular states, such that the mere discovery of similar B_{sJ} mesons would not help in discriminating between meson and molecule. Dynamical issues need to be addressed and decays investigated. In particular electromagnetic decays, where the theoretical understanding is much better than for strong decays, would reveal information about the internal structure of these states. The narrowness of these resonances suggests that this need not be a completely hopeless enterprise. If the lightest D_{sJ} state was dominantly a molecule, then a quark model $J = 0$ resonance should exist in addition. The recent Belle [53] evidence of the $D_0^*(2308)$ suggests that another broad resonance, centred around 2.4 to 2.5 GeV with dominant decay into DK, might be detectable.

Better lattice studies are needed to clarify these important issues, and also with respect to the possibility of similar states in the B meson system. Of course at physical light and strange quark masses lattice QCD should reproduce the experimental spectrum. So to gain insight into the nature of these states a high precision quenched benchmark study is required: in this approximation molecules and mesons are clearly distinct. In addition, the light sea quark mass dependence of both, a possible $c\bar{l}l\bar{s}$ molecule and a $c\bar{s}$ meson has to be traced down into the region where mixing can set in. Fortunately, the DK system is more user friendly than its $K\bar{K}$ counterpart: heavier

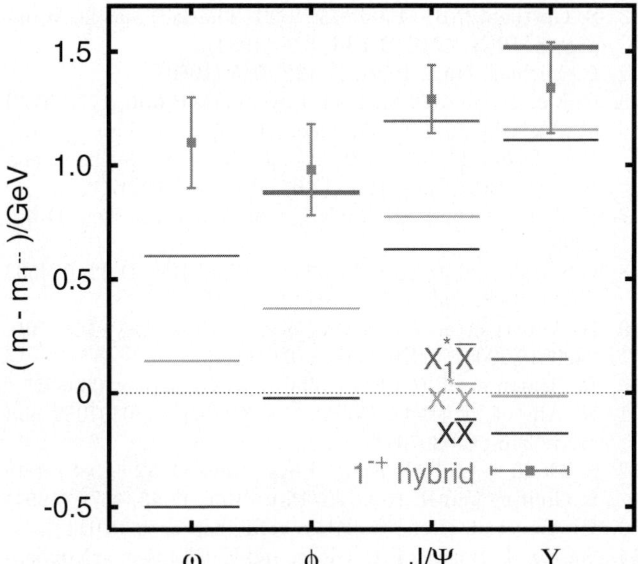

Fig. 3. Splitting of hybrid meson masses with respect to the respective triplet S wave ground states for the light, strange, charm and bottom cases. X denotes π, K, D and B mesons. X^* denotes the ρ, K^*, D^* and B^* vector mesons while X_1^* stands for the (likely) 1^+ states $b_1(1235)$, $K_1(1270)$, $D_1(2400)$ and $B_J^*(5732)$, respectively

particles can be accommodated in smaller volumes and the binding energy that is suggested by phenomenology is comfortably large with about 40 MeV, rather than a mere 10–15 MeV. It is also conceivable to calculate matrix elements that are related to electromagnetic decay rates.

2.5 More hybrids

The spectrum of $c\bar{c}$ glue hybrids has already been discussed in Sect. 2.1 above. Light hybrids have been studied to some extent as well, with [61,62] and without [63, 25,26] sea quarks. All these results yield the same ordering as in the charmonium case (and in fact also the bottomonium case) with 1^{-+} being the lightest exotic, followed by 0^{+-} and 2^{+-}. The 1^{-+} is consistent with a mass of 1.9(2) MeV, with the corresponding strange quark exotic about 200 MeV heavier, in quenched as well as in un-quenched simulations. This is heavier than the $\pi_1(1600)$ candidate. Mixing with molecular states is a possible explanation and the feasibility of studying this has been demonstrated by MILC [26]. However, at present the quarks are still so heavy that for instance the combined mass of a b_1 and a π is around 1.9 GeV too. Clearly the lattice calculations are incomplete. Molecules have to be included and the quark mass dependence of the mixing matrix has to be studied carefully.

Heavy hybrids can be studied using NRQCD [64, 65,66,67,68,29,18] or the Born-Oppenheimer approximation [69,64,67,28,18]. In the charmonium case simulations with relativistic quarks on anisotropic lattices have been pursued by several groups [22,23,25] as well as with isotropic lattices [26,62]. Even a simulation of bottomo-

nium with relativistic quarks on an anisotropic lattice exists [70]. With the statistical errors of present simulations differences between quenched and un-quenched data cannot clearly be resolved. In Fig. 3 We summarise the present estimates of the splitting of the 1^{-+} hybrid with respect to the respective vector meson state: the flavour dependence is tiny. In addition three decay thresholds are displayed. As detailed in Sect. 2.1 a strong decay into two pseudoscalars is forbidden while the decay into vector and pseudoscalar is suppressed, in particular in the heavy quark limit. Theoretically the $b\bar{b}$ hybrid is most clean-cut but experimentally hard to produce.

On the lattice electromagnetic matrix elements can be calculated [23,29] but strong decays are very challenging. To this end McNeile *et al.* [28] predict that if the lightest bottomonium hybrid is indeed below the $B^{**}\overline{B}$ threshold the dominant decay channel should be deexcitation by emission of a scalar: $H_b \to \chi_b\pi\pi$, with a width of about 100 MeV. The same argument should also be valid in the charmonium case, where phase space would reduce the width even further, but not necessarily for light hybrids.

3 Conclusions: From fiction to fact

A combination of new theoretical methods and computing technology has allowed us to arrive at the boundary between qualitative test of principle and quantitative prediction in the complicated area of flavour singlet physics, strong decays and mixing. A few years ago at least three major technical challenges had to be overcome: light quarks, sea quarks and disconnected quark lines. We are about half way through by now. With enough effort devoted onto the topics covered in this article, quantitative predictions are possible within the time scale that is relevant for experiments like glueX at JLAB or PANDA at GSI.

Acknowledgements. I warmly thank Sigfrido Boffi, Claudio Ciofi degli Atti and Mauro Giannini for organizing this stimulating meeting. This work has been supported by PPARC grants PPA/A/S/2000/00271 and PPA/G/0/2002/0463.

References

1. See e.g. C. McNeile: arXiv:hep-lat/0307027; C. Davies: arXiv:hep-ph/0205181; G.S. Bali: Phys. Rept. **343**, 1 (2001); M. Di Pierro: arXiv:hep-lat/0009001
2. C. Michael: arXiv:hep-lat/0302001
3. G.S. Bali: arXiv:nucl-th/0302039
4. P. Hagler et al. (LHPC Collab.): arXiv:hep-lat/0304018
5. M. Göckeler et al. (QCDSF Collab.): arXiv:hep-ph/0304249
6. J. van der Heide, M. Lutterot, J.H. Koch, and E. Laermann: Phys. Lett. B **566**, 131 (2003)
7. C. Alexandrou et al.: arXiv:hep-lat/0307018
8. D. Brommel et al. (BGR Collab.): arXiv:hep-ph/0307073
9. S.J. Dong et al.: arXiv:hep-ph/0306199

10. K.G. Wilson: Phys. Rev. D **10**, 2445 (1974); P.G. Freund and Y. Nambu: Phys. Rev. Lett. **34**, 1645 (1975); H. Fritzsch and P. Minkowski: Nuovo Cim. A **30**, 393 (1975)

11. D. Robson: Phys. Lett. B **66**, 267 (1977) and Nucl. Phys. B **130**, 328 (1977)

12. B. Lucini, M. Teper, and U. Wenger: Phys. Lett. B **545**, 197 (2002).

13. B. Lucini and M. Teper: JHEP **0106**, 050 (2001)

14. S. Aoki et al.: arXiv:hep-lat/0206009

15. W.A. Bardeen et al.: Phys. Rev. D **65**, 014509 (2002)

16. C.J. Morningstar and M.J. Peardon: Phys. Rev. D **60**, 034509 (1999)

17. G.S. Bali et al. (UKQCD Collab.): Phys. Lett. B **309**, 378 (1993)

18. See e.g. K.J. Juge, J. Kuti, and C. Morningstar: arXiv:nucl-th/0307116

19. K. Abe et al. (Belle Collab.): arXiv:hep-ex/0306015 and Phys. Rev. Lett. **89**, 142001 (2002)

20. M. Okamoto et al. (CP-PACS Collab.): Phys. Rev. D **65**, 094508 (2002)

21. P. Chen: Phys. Rev. D **64**, 034509 (2001)

22. X. Liao and T. Manke: arXiv:hep-lat/0210030

23. S. Choe et al. (QCD-TARO Collab.): arXiv:hep-lat/0307004

24. G.S. Bali and P. Boyle: Phys. Rev. D **59**, 114504 (1999)

25. Z.H. Mei and X.Q. Luo: arXiv:hep-lat/0206012

26. C.W. Bernard et al. (MILC Collab.): Phys. Rev. D **56**, 7039 (1997)

27. P.R. Page, E.S. Swanson, and A.P. Szczepaniak: Phys. Rev. D **59**, 034016 (1999)

28. C. McNeile, C. Michael, and P. Pennanen (UKQCD Collab.): Phys. Rev. D **65**, 094505 (2002)

29. T. Burch and D. Toussaint: arXiv:hep-lat/0305008

30. J. Sexton, A. Vaccarino, and D. Weingarten: Phys. Rev. Lett. **75**, 4563 (1995)

31. R.L. Jaffe: Phys. Rev. D **15**, 267 (1977)

32. C. Amsler and F.E. Close: Phys. Rev. D **53**, 295 (1996)

33. L. Burakovsky and P.R. Page: Phys. Rev. D **59**, 014022 (1999) [Erratum-ibid. D **59**, 079902 (1999)]

34. W.J. Lee and D. Weingarten: Phys. Rev. D **61**, 014015 (2000)

35. F.E. Close and A. Kirk: Eur. Phys. J. C **21**, 531 (2001)

36. S. Prelovsek and K. Orginos (RBC Collab.): arXiv:hep-lat/0209132

37. W. Bardeen, E. Eichten, and H. Thacker: arXiv:hep-lat/0307023

38. A. Hart and M. Teper (UKQCD Collab.): Phys. Rev. D **65**, 034502 (2002); A. Hart, C. McNeile, and C. Michael: arXiv:hep-lat/0209063

39. M.G. Alford and R.L. Jaffe: Nucl. Phys. B **578**, 367 (2000) and references therein

40. M. Lüscher: Commun. Math. Phys. **104**, 177 (1986); Commun. Math. Phys. **105**, 153 (1986); Nucl. Phys. B **354**, 531 (1991); Nucl. Phys. B **364**, 237 (1991)

41. M. Fukugita et al.: Phys. Rev. D **52**, 3003 (1995)

42. T. Kunihiro et al. (SCALAR Collab.): arXiv:hep-lat/0210012

43. S. Gottlieb, P.B. Mackenzie, H.B. Thacker, and D. Weingarten: Phys. Lett. B **134**, 346 (1984)

44. C. Michael: Nucl. Phys. B **327**, 515 (1989)

45. C. McNeile and C. Michael: Phys. Lett. B **556**, 177 (2003)

46. C. McNeile and C. Michael (UKQCD Collab.): Phys. Rev. D **63**, 114503 (2001); C. McNeile, C. Michael, and K.J. Sharkey: Phys. Rev. D **65**, 014508 (2002)

47. K.M. Bitar et al. (HEMCGC Collab.): Phys. Rev. D **44**, 2090 (1991)

48. G.S. Bali et al. (SESAM Collab.): Phys. Rev. D **62**, 054503 (2000); G.S. Bali: arXiv:hep-ph/0110254

49. B. Aubert et al. (BABAR Collab.): Phys. Rev. Lett. **90**, 242001 (2003) [arXiv:hep-ex/0304021]

50. D. Besson et al. (CLEO Collab.): arXiv:hep-ex/0305017

51. K. Abe et al. (Belle Collab.): arXiv:hep-ex/0307052 and arXiv:hep-ex/0307041

52. S. Godfrey and N. Isgur: Phys. Rev. D **32** (1985) 189; S. Godfrey and R. Kokoski: Phys. Rev. D **43**, 1679 (1991)

53. K. Abe et al. (Belle Collab.): arXiv:hep-ex/0307021

54. See e.g. T. Barnes, F.E. Close, and H.J. Lipkin: arXiv:hep-ph/0305025, and references thereof

55. C. Michael and J. Peisa (UKQCD Collab.): Phys. Rev. D **58**, 034506 (1998); C. Alexandrou et al.: Nucl. Phys. B **414**, 815 (1994); A. Duncan et al.: Phys. Rev. D **51**, 5101 (1995); A.K. Ewing et al. (UKQCD Collab.): Phys. Rev. D **54**, 3526 (1996)

56. G.S. Bali: arXiv:hep-ph/0305209

57. J. Hein et al.: Phys. Rev. D **62**, 074503 (2000); R. Lewis and R.M. Woloshyn: Phys. Rev. D **62**, 114507 (2000)

58. P. Boyle: Nucl. Phys. Proc. Suppl. **63**, 314 (1998) and Nucl. Phys. Proc. Suppl. **53** (1997) 398

59. A. Dougall, R.D. Kenway, C.M. Maynard, and C. McNeile (UKQCD Collab.): arXiv:hep-lat/0307001

60. M.A. Nowak, M. Rho, and I. Zahed: Phys. Rev. D **48**, 4370 (1993); W.A. Bardeen and C.T. Hill: Phys. Rev. D **49**, 409 (1994); D. Ebert, T. Feldmann, and H. Reinhardt: Phys. Lett. B **388**, 154 (1996); W.A. Bardeen, E.J. Eichten, and C.T. Hill: arXiv:hep-ph/0305049

61. P. Lacock and K. Schilling (SESAM Collab.): Nucl. Phys. Proc. Suppl. **73**, 261 (1999)

62. C. Bernard et al. (MILC Collab.): arXiv:hep-lat/0301024

63. P. Lacock et al. (UKQCD Collab.): Phys. Lett. B **401**, 308 (1997) and Phys. Rev. D **54**, 6997 (1996)

64. S. Collins, G. Bali, and C. Davies (UKQCD Collab.): Nucl. Phys. Proc. Suppl. **63**, 335 (1998)

65. T. Manke et al. (UKQCD Collab.): Phys. Rev. D **57**, 3829 (1998)

66. T. Manke et al. (CP-PACS Collab.): Phys. Rev. Lett. **82**, 4396 (1999)

67. K.J. Juge, J. Kuti, and C.J. Morningstar: Phys. Rev. Lett. **82**, 4400 (1999)

68. T. Manke et al. (CP-PACS Collab.): Phys. Rev. D **64**, 097505 (2001)

69. S. Perantonis and C. Michael: Nucl. Phys. B **347**, 854 (1990)

70. X. Liao and T. Manke: Phys. Rev. D **65**, 074508 (2002)

Eur Phys J A (2004) **19**, s01, 9–14
Digital Object Identifier (DOI) 10.1140/epjad/s2004-03-002-4

EPJ A direct
electronic only

Nucleon electromagnetic form factors from lattice QCD

J.D. Ashley, D.B. Leinweber, A.W. Thomas, and R.D. Young

Special Research Centre for the Subatomic Structure of Matter, and
Department of Physics, University of Adelaide,
Adelaide SA 5005 Australia

Received: 22 Aug 2003 / Accepted: 14 Nov 2003 /
Published Online: 6 Feb 2004 – © Società Italiana di Fisica / Springer-Verlag 2004

Abstract. It is imperative that lattice QCD serve to develop our understanding of hadron structure and, where possible, to guide the interpretation of experimental data. There is now a great deal of effort directed at the calculation of the electroweak form factors of the nucleon, where for example, measurements at Jefferson Laboratory have recently revealed surprising behaviour in the ratio G_E/G_M. While, for the present, calculations within the framework of lattice QCD are limited to relatively large quark mass, there has been considerable progress in our understanding of how to extrapolate to the chiral limit. Here we report the results of the application of these techniques to the most recent form factor data from the QCDSF Collaboration. The level of agreement with all of the form factors, for Q^2 below 1 GeV^2, is already impressive.

PACS. 12.38.Gc Lattice QCD calculations – 11.30.Rd Chiral symmetries – 13.40.Gp Electromagnetic form factors – 14.20.Dh Protons and neutrons

1 Introduction

The electromagnetic form factors of the nucleon provide a fundamental constraint for any theoretical description of the structure of the nucleon [1]. Even though they have been studied experimentally for more than 50 years, recent experiments at Jefferson Lab [2] have revealed surprising new behaviour in the ratio G_E/G_M for the proton. In addition, there are a range of other new results on the neutron electric and magnetic form factors [3] – this is an exciting and rapidly developing field.

In this context it is vital that lattice QCD, our only rigorous method of solving non-perturbative QCD, be used to inform our understanding of this data and the various models of hadron structure that are used to describe it. There was a relatively long hiatus in calculations of nucleon form factors in lattice QCD after the pioneering work of Leinweber and collaborators in the early 90's [4, 5,6]. However, the activity has intensified in the last few years [7,8,9]. In spite of this activity, limitations in computer speed mean that we are currently limited to lattice simulations at quark masses a factor 5-10 higher than the physical values and these calculations are currently made in quenched approximation (QQCD).

If one is to compare these state of the art simulations with experiment, it is necessary to make an extrapolation as a function of quark mass to the physical region [10, 11]. This extrapolation is made non-trivial by the non-analytic behaviour as a function of m_q which follows from the fact that chiral symmetry is dynamically broken in

QCD. Until recently the absence of data has meant that efforts at chiral extrapolation of form factors has been focussed on baryon magnetic moments [12,13,14,15] and charge radii [16,17]. The focus on non-analyticity [18] of hadron electromagnetic properties has inspired considerable investigation of the formal constraints in both full QCD and QQCD [19,20,21,22,23].

The practical issue is then how to incorporate these formal chiral constraints into a practical chiral extrapolation, given that the radius of convergence of the formal expansion dictated by chiral perturbation theory is rather small [24]. In the case of the nucleon mass, where there is extremely accurate data in full QCD from the CP-PACS Collaboration [25], one can formally demonstrate the model independence of the extrapolation procedure [24,26,27]. For the magnetic moments the data is only now improving to the point where the model independence of the choice of finite range regulator [28] can be examined [12]. In the case of the form factors this is not yet possible.

The procedure employed in this paper is to draw on the earlier phenomenological experience with chiral extrapolations of the magnetic moments and charge radii, taking simple phenomenological functional forms which build in the correct non-analytic behaviour and any other asymptotic constraints that are known [29]. For the present the accuracy of the lattice data is not such that it could discriminate between different functional forms for the Q^2-dependence of the nucleon form factors. In particular, it cannot yet address the JLab issue of whether G_E/G_M decreases as Q^2 increases. We therefore parametrize separately the isoscalar and isovector lattice data, at a given value of m_q, or m_π, as a dipole and then extrapolate the dipole mass as a function of m_π, building in the appropriate chiral constraints. Clearly this simple approach can be systematically improved as the lattice simulations become capable of making a better discrimination between possible functional forms. For the present we shall see that our relatively simple approach already produces quite impressive results.

2 Extrapolations

It is well known [1] that the Q^2-dependence of the nucleon Sachs electromagnetic form factors (with the exception of the neutron electric form factor G_E^n) are described in first approximation by a dipole form

$$G(Q^2) = \frac{G(0)}{(1 + Q^2/\Lambda^2)^2}. \tag{1}$$

Here Λ is the dipole mass and the charge form factor of the proton satisfies $G_E^p(0) = 1$, while $G_M^p(0) = \mu_p$ and $G_M^n(0) = \mu_n$ are the proton and neutron magnetic moments. As explained in the Introduction, we will use this phenomenological fact to construct a simple but effective extrapolation formula.

To isolate the chiral behaviour of the form factors we rearrange them into isovector and isoscalar combinations:

$$G^v = G^p - G^n \tag{2}$$

and

$$G^s = G^p + G^n, \tag{3}$$

respectively. These also display a dipole-like Q^2-dependence but with different dipole masses and magnetic moments.

To extrapolate lattice QCD results for the electromagnetic form factors from the large pion masses at which they are calculated to the physical regime, we extract dipole masses and magnetic moments from the lattice data and then extrapolate these as a function of m_π.

Following [13,14], one can use a Padé approximant which builds in both the correct chiral non-analytic behaviour as $m_\pi \to 0$ and the correct asymptotic behaviour as $m_q \to \infty$ to extrapolate the neutron and proton magnetic moments

$$\mu_i(m_\pi) = \frac{\mu_0}{1 - \frac{\chi_i}{\mu_0} m_\pi + c m_\pi^2}. \tag{4}$$

The chiral coefficients for the isovector and isoscalar moments are $\chi_v = -8.82$ and $\chi_s = 0$ respectively and μ_0 and c are fitting parameters, to be determined by the lattice data.

In order to build a suitable extrapolating function for the Q^2-dependence of the form factors, we use the connection between the mass parameter in a dipole form factor and the corresponding mean-square radius. For the isovector magnetic form factor the mean square radius is

$$\langle r^2 \rangle_M^v = -\frac{6}{G_M^v(0)} \frac{dG_M^v}{dQ^2} \Big|_{Q^2=0}. \tag{5}$$

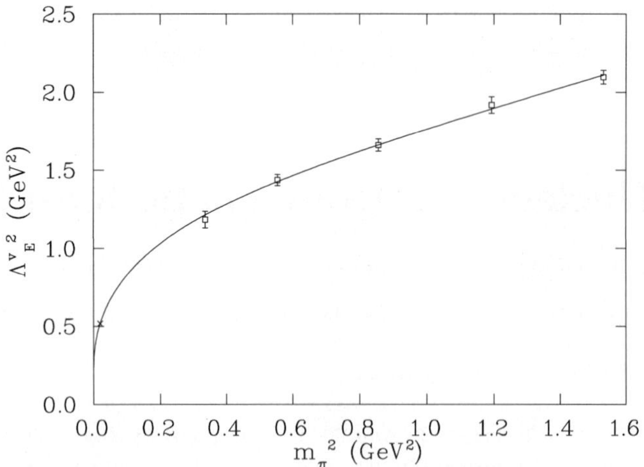

Fig. 1. Fit to values of the isovector electric form factor dipole mass extracted from lattice data with lattice spacing $a = 0.051$ fm. The physical value predicted by the fit is also indicated

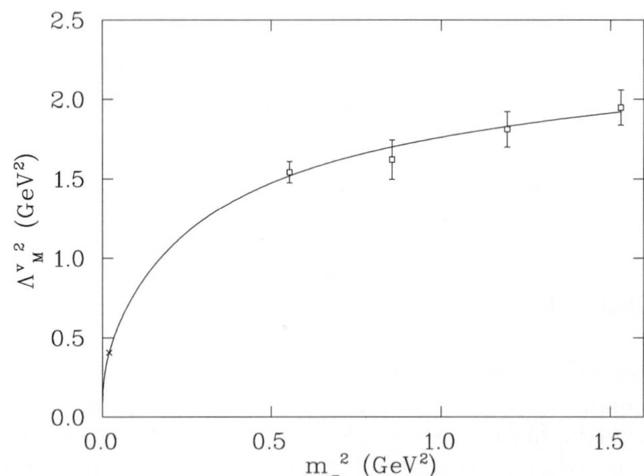

Fig. 2. Fit to values of the isovector magnetic form factor dipole mass extracted from lattice data with lattice spacing $a = 0.051$ fm. The physical value predicted by the fit is also indicated

Comparing this to the dipole of (1), we find an expression relating the dipole mass to $\langle r^2 \rangle_M^v$,

$$(\Lambda_M^v)^2 = \frac{12}{\langle r^2 \rangle_M^v}. \tag{6}$$

The chiral behaviour of the magnetic mean squared radius is known from chiral perturbation theory [30]

$$\langle r^2 \rangle_M^v \sim \frac{\chi_1}{m_\pi} + \chi_2 \ln(\frac{m_\pi}{\mu}). \tag{7}$$

The constants χ_1 and χ_2 are given by

$$\chi_1 = \frac{g_A^2 m_N}{8\pi f_\pi^2 \kappa_v}, \tag{8}$$

$$\chi_2 = -\frac{5g_A^2 + 1}{8\pi^2 f_\pi^2}, \tag{9}$$

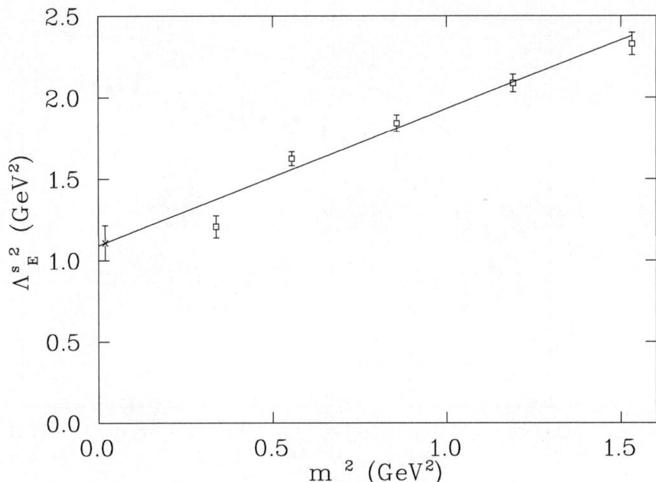

Fig. 3. Linear fit to values of the isoscalar electric form factor dipole mass extracted from lattice data with lattice spacing $a = 0.051$ fm. The physical value predicted by the fit is also indicated

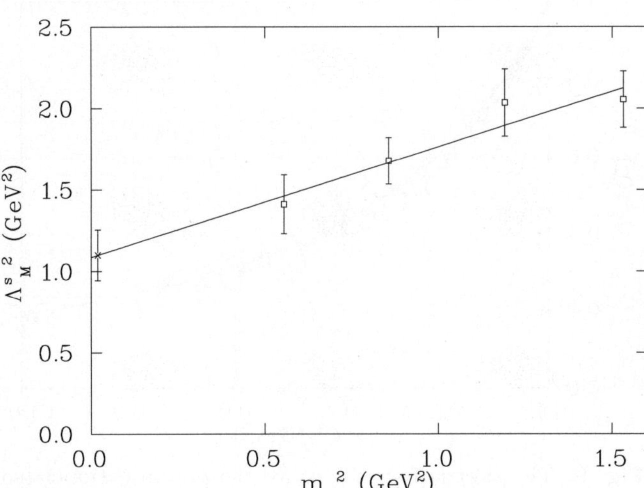

Fig. 4. Linear fit to values of the isoscalar magnetic form factor dipole mass extracted from lattice data with lattice spacing $a = 0.051$ fm. The physical value predicted by the fit is also indicated

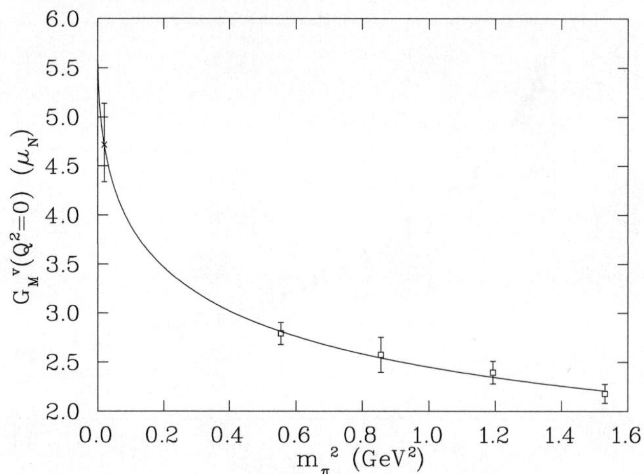

Fig. 5. Fit to values of the isovector magnetic moment extracted from lattice data with lattice spacing $a = 0.051$ fm. The physical value predicted by the fit is also indicated

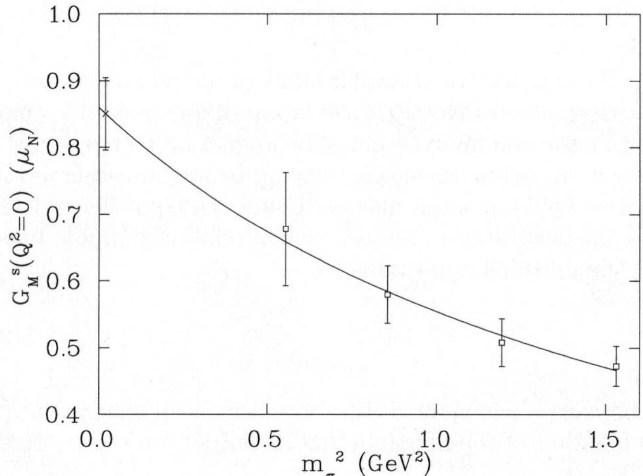

Fig. 6. Fit to values of the isoscalar magnetic moment extracted from lattice data with lattice spacing $a = 0.051$ fm. The physical value predicted by the fit is also indicated

where $g_A = 1.27$ is the axial coupling constant and $f_\pi = 93$ MeV is the pion decay constant, $m_N = 940$ MeV is the nucleon mass and $\kappa_v \approx 4.2$ is the isovector anomalous magnetic moment of the nucleon (in the chiral limit).

Earlier experience with nucleon properties as a function of quark mass has suggested that, as a consequence of the finite size of the source of the pion field, chiral loops are strongly suppressed for $m_\pi > 0.4$ GeV [31]. Thus in order to build an effective chiral extrapolation formula one needs to modify the non-analytic terms given above so that they are suppressed above this mass. For guidance as to an appropriate parametrization of this suppression we have evaluated the pion loops that give rise to the non-analytic chiral terms using the Cloudy Bag Model (CBM) [32,33,

34]. The results led us to replace (7) with the expression

$$\langle r^2 \rangle_M^v \sim \frac{\chi_1}{m_\pi} \frac{2}{\pi} \arctan(\mu/m_\pi) + \frac{\chi_2}{2} \ln\left(\frac{m_\pi^2}{m_\pi^2 + \mu^2}\right), \quad (10)$$

which ensures the correct chiral behaviour at low-m_π but suppresses it at values larger than μ.

Substituting this functional form into (6) and introducing a linear dependence on m_π^2 to give the expected behaviour at large quark mass, we get the following expression for the dipole mass associated with the isovector magnetic form factor

$$(\Lambda_M^v)^2 = \frac{12(1 + A_1 m_\pi^2)}{A_0 + \frac{\chi_1}{m_\pi} \frac{2}{\pi} \arctan(\mu/m_\pi) + \frac{\chi_2}{2} \ln\left(\frac{m_\pi^2}{m_\pi^2 + \mu^2}\right)}.$$
$$(11)$$

Here A_0 and A_1 are unknown parameters, adjusted to fit the lattice data.

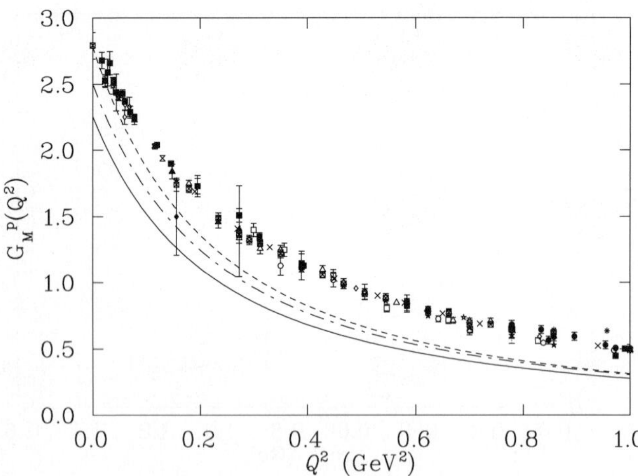

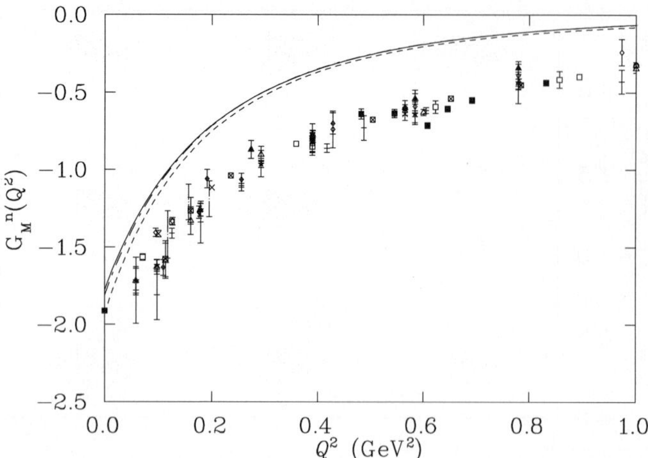

Fig. 7. The magnetic form factor for the proton extrapolated from lattice data with lattice spacings $a = 0.093$ fm (single line), $a = 0.068$ fm (dash-dot line) and $a = 0.051$ fm (dotted line) and compared to experimental results

Fig. 8. The magnetic form factor for the neutron extrapolated from lattice data with lattice spacings $a = 0.093$ fm (single line), $a = 0.068$ fm (dash-dot line) and $a = 0.051$ fm (dotted line) and compared to experimental results

We can perform a similar analysis to find an expression for the isovector electric form factor dipole mass Λ_E^v, combining the non-analytic chiral behaviour predicted by χPT at low m_π^2 with the slowly varying behaviour observed in lattice QCD at large masses. Using the equivalent of (5) for the electric form factor, we can relate the dipole mass to the mean charge radius by

$$(\Lambda_E^v)^2 = \frac{12}{\langle r^2 \rangle_E^v}. \tag{12}$$

The chiral behaviour of the isovector mean charge radius is like that of the magnetic radius but with no $1/m_\pi$ term [30],

$$\langle r^2 \rangle_E^v \sim \chi_2 \ln\left(\frac{m_\pi}{\mu}\right). \tag{13}$$

Combining this chiral behavior with a linear dependence on m_π^2 produces the following predicted form for the isovector electric form factor dipole mass,

$$(\Lambda_E^v)^2 = \frac{12(1 + B_1 m_\pi^2)}{B_0 + \frac{\chi_2}{2} \ln\left(\frac{m_\pi^2}{m_\pi^2 + \mu^2}\right)}, \tag{14}$$

where B_0 and B_1 are general fitting parameters.

Having isolated the chiral non-analytic behaviour to the isovector form factors, we expect that the isoscalar dipole masses should be roughly linear in m_π^2 – as observed in CBM results.

The forms chosen to represent the dipole masses as a function of m_π are able to reproduce the predictions of the CBM [34] very well.

3 Results

The QCDSF Collaboration [7] has recently reported QQCD results for the isovector and isoscalar nucleon electric and magnetic form factors calculated at different Q^2

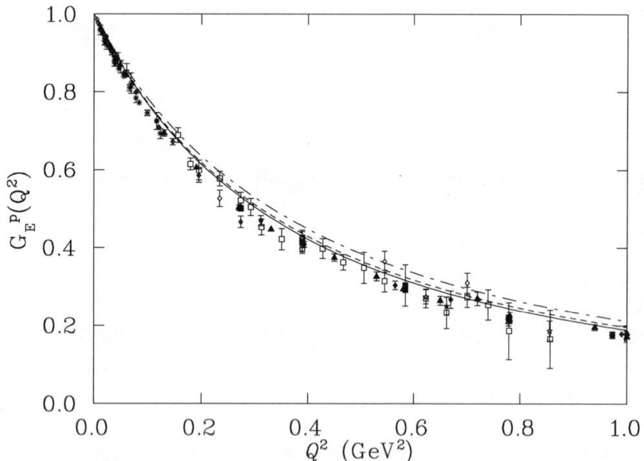

Fig. 9. The electric form factor for the proton extrapolated from lattice data with lattice spacings $a = 0.093$ fm (single line), $a = 0.068$ fm (dash-dot line) and $a = 0.051$ fm (dotted line) and compared to experimental results

values and pion masses and for three different lattice spacings ($\beta = 6.0, 6.2$ and 6.4). Using this data, with the scale set using the Sommer method ($r_0 = 0.5$fm [35]), we have plotted the form factors G_M^v, G_M^s, G_E^v and G_E^s, calculated on the lattice as a function of Q^2, at each pion mass and lattice spacing, and fit each graph with a dipole form, finding a best-fit dipole mass and magnetic moment. Plots of these best-fit dipole masses and magnetic moments against m_π^2 for the lattice spacing $a = 0.051$ fm are shown in Figs. 1, 2, 3 and 4.

The electric and magnetic isovector dipole masses are fitted with functions of the form (11) and (14), respectively. The scale $\mu = 0.41$ GeV was chosen to give the best simultaneous fit to the lattice data for both the electric and magnetic isovector form factors at all three lattice spacings. We observe that the curves fit the data well and

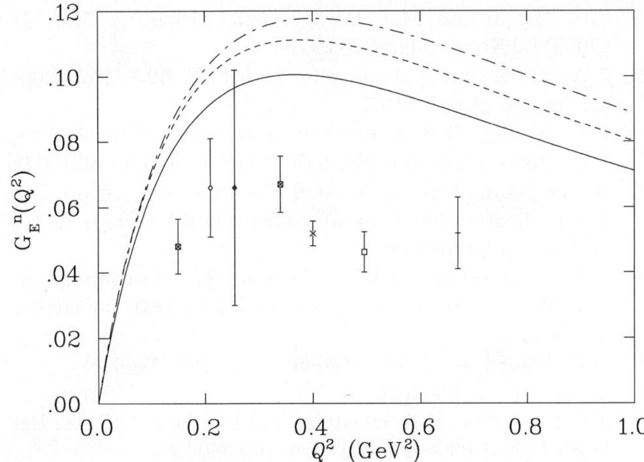

Fig. 10. The electric form factor for the neutron extrapolated from lattice data with lattice spacings $a = 0.093$ fm (single line), $a = 0.068$ fm (dash-dot line) and $a = 0.051$ fm (dotted line) compared to experimental results

predict physical results significantly different from those obtained by a naive linear extrapolation. In both cases, the form of the fitting functions greatly restricts the physical value, resulting in a small error. Finally, we note that the electric and magnetic isoscalar dipole mass plots are fitted with linear functions in m_π^2, because of the absence of any leading non-analytic behaviour in this case.

The magnetic moment plots are extrapolated using the Padé approximant, (4). The resulting errors in the physical magnetic moments dominate over errors in the dipole mass and are the main source of error in our results for the physical magnetic form factors. Similar fits were produced for lattice results at lattice spacings $a = 0.068$ fm and $a = 0.093$ fm.

Using the extrapolated physical values for the dipole masses and magnetic moments, we reconstruct the electric and magnetic form factors for the proton and neutron as a sum of two dipoles in Q^2. Figures 7, 8, 9 and 10 show the electric and magnetic form factors for the proton and neutron extrapolated from that lattice data at each of the three lattice spacings ($a = 0.093$ fm, $a = 0.068$ fm and $a = 0.051$ fm) compared to the experimental data from [36].

Except for the case of the neutron electric form factor (which is very sensitive because it is the result of a cancellation between G_E^v and G_E^s which yields a relatively small result) our extrapolated curves match the experimental points very well, particularly for G_E^p and G_M^p. Although our curve for G_E^n is almost twice as big as the experimental values, it does describe the correct shape but peaks at slightly too high a value of Q^2.

4 Conclusion

We have proposed a relatively simple approach to the extrapolation of lattice QCD data for the nucleon electromagnetic form factors. The data from QCDSF has been parametrized by a simple dipole form, with the dipole mass parameter taken to be a function of m_π which incorporates the leading non-analytic behaviour of chiral perturbation theory. For the isoscalar case, where there is no leading non-analytic behaviour, the extrapolation is a simple linear function of m_π^2, while for the isovector case we have used a functional form suggested by studies based on the cloudy bag model, which suppress the chiral behaviour for pion masses larger than 400 MeV.

The level of agreement between the empirical Q^2 dependence of the proton electric form factor and the proton and neutron magnetic form factors is impressive. This is quite remarkable when one considers that the data is based on quenched approximation and is extrapolated from a light quark mass a factor of twenty above the physical mass. This is only possible because of the remarkable fact that pion loops are suppressed by finite size effects once the pion Compton wavelength is smaller than the size of the source – empirically, $m_\pi > 0.4$ GeV. The extrapolation of the data has, of course, been performed using chiral coefficients appropriate to full QCD.

One of the main open questions in the analysis concerns the neutron electric form factor, which will always be a challenge given that it vanishes at $Q^2 = 0$ – as a result of the cancellation between the isoscalar and isovector contributions. In addition, we observe that there appears to be some residual β dependence in the value of the nucleon magnetic moments and this is the major ambiguity in the current reconstruction of the proton and neutron magnetic form factors. It is encouraging that the results for $\beta = 6.4$, corresponding to the smallest lattice spacing, is in the best agreement with the experimental data. Future lattice simulations will undoubtedly clarify this issue and also give us data at lower values of the quark mass. It would also be valuable to have full QCD simulations (rather than QQCD) to ensure that there is no unanticipated systematic difference. Most importantly, as the accuracy of the data and the range of Q^2 covered increases, we can extend the present analysis by employing more complicated fitting functions which should allow us to test the behaviour of properties such as G_E/G_M against experiment.

Acknowledgements. We would like to thank G. Schierholz for helpful communications concerning the lattice data from QCDSF and J. Kelly, R. Madey and A. Semenov for providing a compilation of the experimental data on the nucleon form factors. This work was supported by the Australian Research Council and the University of Adelaide.

References

1. A.W. Thomas and W. Weise: "The Structure Of The Nucleon," Wiley-VCH (Berlin, 2001) pp. 389
2. M.K. Jones et al. (Jefferson Lab Hall A Collaboration): Phys. Rev. Lett. **84** (2000) 1398 [arXiv:nucl-ex/9910005]
3. H.y. Gao: Int. J. Mod. Phys. E **12** (2003) 1 [arXiv:nucl-ex/0301002]

4. D.B. Leinweber, R.M. Woloshyn, and T. Draper: Phys. Rev. D **43** (1991) 1659

5. D.B. Leinweber: Phys. Rev. D **45** (1992) 252

6. T. Draper, K.F. Liu, D.B. Leinweber, and R.M. Woloshyn: Nucl. Phys. A **527** (1991) 531C

7. M. Gockeler, T.R. Hemmert, R. Horsley, D. Pleiter, P.E. Rakow, A. Schafer, and G. Schierholz (QCDSF Collaboration): arXiv:hep-lat/0303019

8. Proceedings of the XXth Int. Symposium on Lattice Field Theory, eds. R. Edwards, J. Negele, and D. Richards: Nucl. Phys. B(Proc. Suppl.)119 (2003)

9. S.J. Dong, K.F. Liu, and A.G. Williams: *Workshop on Future Directions in Quark Nuclear Physics*, eds. A.W. Thomas and K. Tsushima (World Scientific, Singapore, 1999) p. 131.

10. A.W. Thomas: Nucl. Phys. B (Proc. Suppl.) **119**, 50 (2003) arXiv:hep-lat/0208023

11. C. Bernard et al.: Nucl. Phys. B (Proc. Suppl.) **119**, 170 (2003) arXiv:hep-lat/0209086

12. I.C. Cloet, D.B. Leinweber, and A.W. Thomas: Phys. Lett. B **563**, 157 (2003) [arXiv:hep-lat/0302008]

13. E.J. Hackett-Jones, D.B. Leinweber, and A.W. Thomas: Phys. Lett. B **489**, 143 (2000) [arXiv:hep-lat/0004006]

14. D.B. Leinweber, D.H. Lu, and A.W. Thomas: Phys. Rev. D **60**, 034014 (1999) [arXiv:hep-lat/9810005]

15. T.R. Hemmert and W. Weise: Eur. Phys. J. A **15**, 487 (2002) [arXiv:hep-lat/0204005]

16. E.J. Hackett-Jones, D.B. Leinweber, and A.W. Thomas: Phys. Lett. B **494**, 89 (2000) [arXiv:hep-lat/0008018]

17. G.V. Dunne, A.W. Thomas, and S.V. Wright: Phys. Lett. B **531**, 77 (2002) [arXiv:hep-th/0110155]

18. D.B. Leinweber, A.W. Thomas, and R.D. Young: Phys. Rev. Lett. **86**, 5011 (2001) [arXiv:hep-ph/0101211]

19. D. Arndt and B.C. Tiburzi: arXiv:hep-lat/0307003

20. D. Arndt and M.J. Savage: Nucl. Phys. A **697**, 429 (2002) [arXiv:nucl-th/0105045]

21. M.J. Savage: Nucl. Phys. A **700**, 359 (2002) [arXiv:nucl-th/0107038]

22. S.R. Beane and M.J. Savage: Nucl. Phys. A **709**, 319 (2002) [arXiv:hep-lat/0203003]

23. J.W. Chen and X.d. Ji: Phys. Lett. B **523**, 107 (2001) [arXiv:hep-ph/0105197]

24. R.D. Young, D.B. Leinweber, and A.W. Thomas: Prog. Part. Nucl. Phys. **50**, 399 (2003) [arXiv:hep-lat/0212031]

25. A. Ali Khan et al. (CP-PACS Collaboration): Phys. Rev. D **65**, 054505 (2002) [Erratum-ibid. D **67**, 059901 (2003)] [arXiv:hep-lat/0105015]

26. D.B. Leinweber, A.W. Thomas, K. Tsushima, and S.V. Wright: Phys. Rev. D **61**, 074502 (2000) [arXiv:hep-lat/9906027]

27. D.B. Leinweber, A.W. Thomas, and R.D. Young: arXiv:hep-lat/0302020

28. J.F. Donoghue, B.R. Holstein, and B. Borasoy: Phys. Rev. D **59** (1999) 036002 [arXiv:hep-ph/9804281]

29. A.W. Thomas, J.D. Ashley, W. Detmold, D.B. Leinweber, W. Melnitchouk, and R.D. Young: Nucl. Phys. A **721**, 915 (2003)

30. V. Bernard, N. Kaiser and U.G. Meissner: Int. J. Mod. Phys. E **4**, 193 (1995) [arXiv:hep-ph/9501384]

31. W. Detmold, D.B. Leinweber, W. Melnitchouk, A.W. Thomas, and S.V. Wright: Pramana **57**, 251 (2001) [arXiv:nucl-th/0104043]

32. A.W. Thomas: Adv. Nucl. Phys. **13**, 1 (1984)

33. S. Theberge and A.W. Thomas: Phys. Rev. D **25**, 284 (1982)

34. D.H. Lu, A.W. Thomas, and A.G. Williams: Phys. Rev. C **57**, 2628 (1998) [arXiv:nucl-th/9706019]

35. M. Guagnelli, R. Sommer, and H. Wittig (ALPHA collaboration): Nucl. Phys. B **535**, 389 (1998) [arXiv:hep-lat/9806005]

36. H. Zhu et al. (E93026 Collaboration): Phys. Rev. Lett. **87**, 081801 (2001) [arXiv:nucl-ex/0105001]; W. Xu et al.: Phys. Rev. Lett. **85**, 2900 (2000) [arXiv:nucl-ex/0008003]; and a compilation of earlier data provided by J. Kelly, R. Madey, and A. Semenov

Eur Phys J A (2004) **19**, s01, 15–22

Digital Object Identifier (DOI) 10.1140/epjad/s2004-03-003-3

EPJ A direct

electronic only

Successful renormalization of a QCD-inspired Hamiltonian

Hans-Christian Pauli

Max-Planck-Institut für Kernphysik, D-69029 Heidelberg, e-mail: `pauli@mpi-hd.mpg.de`

Received: 18 Sept 2003 / Accepted: 14 Nov 2003 /
Published Online: 6 Feb 2004 – © Società Italiana di Fisica / Springer-Verlag 2004

Abstract. The long standing problem of non perturbative renormalization of a gauge field theoretical Hamiltonian is addressed and explicitly carried out within an (effective) light-cone Hamiltonian approach to QCD. The procedure is in line with the conventional ideas: The Hamiltonian is first regulated by suitable cut-off functions, and subsequently renormalized by suitable counter terms to make it cut-off independent. Emphasized is the considerable freedom in the cut-off function which eventually can modify the Coulomb potential of two charges at sufficiently small distances. The approach provides new physical insight into nature of gauge theory and the potential energy of QCD and QED near the origin. The so obtained formalism is applied to physical mesons with a different flavor of quark and anti-quark. The excitation spectrum of the ρ-meson with its excellent agreement between theory and experiment is discussed as a pedagogical example.

PACS. 11.10.Ef – 12.38.Aw – 12.38.Lg – 12.39.-x

1 Introduction

When starting in 1984 with Discretized Light-Cone Quantization (DLCQ) [1] and with a revival of Dirac's Hamiltonian front form dynamics [2], all challenges of a gauge field Hamiltonian theory were essentially open questions, particularly the non perturbative bound state problem, the many-body aspects, regularization, renormalization, confinement, chirality, vacuum structure and condensates, just to name a few. The step from the gauge field QCD Lagrangian down to a non relativistic Schrödinger equation was completely mysterious. Now we know better [3]. We have learned how to partition the problem and how to shape our thinking in four major steps:

$$\mathcal{L}_{\mathrm{QCD}} ,$$
$$\begin{aligned}
\hookrightarrow \quad H_{\mathrm{LC}}|\Psi\rangle &= M^2|\Psi\rangle, \\
\hookrightarrow \quad H_{\mathrm{eLC}}|\Psi_{q\bar{q}}\rangle &= M^2|\Psi_{q\bar{q}}\rangle, \\
\hookrightarrow \quad H_{\mathrm{eff}}|\varphi\rangle &= E|\varphi\rangle, \\
\hookrightarrow \left(\tfrac{\mathbf{p}^2}{2m_r} + V(\mathbf{r})\right)\psi(\mathbf{r}) &= E\psi(\mathbf{r}).
\end{aligned} \tag{1}$$

We have understood, for example, that the chiral phase transition, in which the quarks are supposed to get their mass, is not the major challenge. The challenge is to understand what happens *after* the phase transition, at zero temperature. The challenge is to understand the spectrum of physical hadrons and to get the corresponding eigenfunctions, the light cone wave functions.

The light-cone wave functions Ψ for a hadron with mass M encode all possible quark and gluon momentum, helicity and flavor correlations and, in principle, are obtained by diagonalizing the QCD light-cone Hamiltonian

$$H_{\mathrm{LC}} = P^+P^- - P_\perp^2, \text{ where } P^\pm = P^0 \pm P^z,$$

$$H_{\mathrm{LC}}|\Psi_i\rangle = M_i^2|\Psi_i\rangle \tag{2}$$

in a complete basis of Fock states with increasing complexity. For example, the positive pion has the Fock expansion:

$$|\Psi_{\pi^+}\rangle = \sum_n \langle n|\pi^+\rangle|n\rangle$$
$$= \Psi_{u\bar{d}/\pi}^{(\Lambda)}(x_i, \mathbf{k}_{\perp i})|u\bar{d}\rangle + \Psi_{u\bar{d}g/\pi}^{(\Lambda)}(x_i, \mathbf{k}_{\perp i})|u\bar{d}g\rangle + \dots ,$$

representing the expansion of the exact QCD eigenstate at scale Λ in terms of non-interacting quarks and gluons. The i_n particles in a Fock state (n) have longitudinal light-cone momentum fractions x_i and relative transverse momenta $\mathbf{k}_{\perp i}$, with

$$x_i = \frac{k_i^+}{P^+} = \frac{k_i^0 + k_i^z}{P^0 + P^z} , \quad \sum_{i=1}^{i_n} x_i = 1 , \quad \sum_{i=1}^{i_n} \mathbf{k}_{\perp i} = \mathbf{0}_\perp .$$

The form of $\Psi_{n/H}(x_i, \mathbf{k}_{\perp i})$ is invariant under longitudinal and transverse boosts; i.e., the light-cone wave functions expressed in the relative coordinates x_i and $k_{\perp i}$ are independent of the total momentum $(P^+, \mathbf{P}_\perp)$ of the hadron. The first term in the expansion is referred to as the valence Fock state, as it relates to the hadronic description in the constituent quark model. The higher terms are related to the sea components of the hadronic structure. It has been shown that the rest of the light-cone wave function is determined once the valence Fock state is known [4,5], with explicit expressions given in [5].

The key issue is to overcome the problem of any gauge theory, that the unregulated theory exposes logarithmic

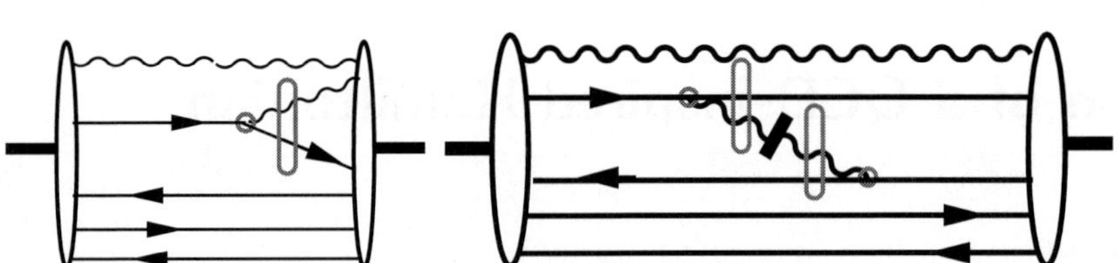

Fig. 1. Regularization of the interaction by vertex regularization. In a matrix element, as illustrated on the left for a vertex, a quark changes its four-momentum from k_1 to k_2, i.e. $Q^2 = -(k_1 - k_2)^2$. The vertex interaction is regulated by multiplying with a form factor $F(Q^2)$, as indicated by the circle. – Instantaneous interactions are treated correspondingly, as illustration on the right for a seagull

singularities. The problem of regularization and renormalization has been solved in the perturbative context of scattering theory, but not in the non perturbative context of a Hamiltonian. It is addressed to in the first two sections and applied in the remainder of this paper.

2 Regularization

Canonical field theory with the conventional QCD Lagrangian allows to derive the components of the total canonical four-momentum P^μ. Its front form version [3] rests on two assumptions, the light cone gauge $A^+ = 0$ [6] and the suppression of all zero modes [3,7]. The front form vacuum is then trivial.

I find it helpful to discuss the problem in terms of DLCQ [1,3]. In the back of my mind I visualize an explicit finite dimensional matrix representation of the Light-Cone Hamiltonian as it occurs for finite harmonic resolution. Such one is schematically displayed in Fig. 2 of [3]. All of its matrix elements are finite for any finite x and $k_\perp$.

The problem arises for ever increasing harmonic resolution, on the way to the continuum limit: The numerical eigenvalues are numerically unstable and diverge logarithmically [15,16], contrary to the calculations in 1+1 dimension [1]; see also actual DLCQ calculations in 3+1 by Hiller [8].

The reason is inherent to Dirac's relativistic vertex interaction $\langle k_1, h_1 | V | k_2, h_2; k_3, h_3 \rangle$, in which some particle '1' is scattered into two particles '2' and '3' with their respective four-momenta k and helicities h, see Fig. 1. The matrix element for bremsstrahlung, for example, is proportional to $k_\perp$, $\langle k_1, \uparrow | V | k_2, \uparrow; k_1, \uparrow \rangle \propto |\mathbf{k}_\perp|$, see Table 9 in [3], when the quark maintains its helicity while irradiating a gluon with four-momentum $k_3^\mu = (xP^+, \mathbf{k}_\perp, k_3^-)$. Singularities arise typically when squares of such matrix elements are integrated over all $\mathbf{k}_\perp$ as in the integrations of perturbation theory.

The singularities are avoided *a priori* by *vertex regularization*, by multiplying each (typically off-diagonal) matrix element with a regulating *form factor* F:

$$\langle k_1, h_1 | V | k_2, h_2; k_3, h_3 \rangle \Longrightarrow$$
$$\langle k_1, h_1 | V | k_2, h_2; k_3, h_3 \rangle \, F(Q) . \tag{3}$$

It took several years to realize that it is the Feynman four-momentum transfer across a vertex, $Q^2 = -(k_1 - k_2)^2$, which governs any effective interaction. The minimal requirement for such a form factor is

$$F(Q; \Lambda) = \begin{cases} 1 , & \text{for } Q^2 \to 0 , \\ 0 , & \text{for } Q^2 \to \infty . \end{cases} \tag{4}$$

The job would be done by a step function, $F(Q) = \Theta(Q^2 - \Lambda^2)$. The limit $\Lambda \to 0$ suppresses the interaction all together, the limit $\Lambda \to \infty$ restores the interaction and its problems. Any finite value of Λ^2 restricts Q^2 to be finite and eliminates the singularities. But the sharp cut-off generates problems in an other corner of the theory and $F(Q)$ must be an analytic function of Q, as to be seen below.

Vertex regularization takes thus care of the ultraviolet divergences. The (light-cone) infrared singularities are taken care of as usual by a kinematical gluon mass.

As usual, regularization is not unique and many ways can do that. Dimensional regularization, for example, is not applicable in a matrix approach which is stuck with the precisely 3+1 dimensions of the physical world. *Vertex regularization* should be confronted with the *Fock space regularization* of Lepage and Brodsky [6], see also [3], which has blocked the renormalization aspects for many years. It also should be confronted with [9] and [10]. After applauding the light-cone approach [9], Wilson and collaborators [10] have attempted to base their considerations almost entirely on a renormalization group analysis, but no concrete technology has emerged thus far.

3 Renormalization

The non perturbative renormalization of the Hamiltonian was stuck for many years by the fact that the coupling constant g and the regulator function $F(Q)$ multiply each other in (3). It was always clear that one may add *non local counter terms* [10], but is was not clear how they could be constructed. Progress has come from recent work on a particular model [19], which did allow to formulate a paradigmatic example in modern renormalization theory.

Here is the general but abstract procedure.

Suppose to have solved (2) for a fixed value of the 7 'bare' parameters in the Lagrangian, for the coupling constant $g = g_0$ and the 6 flavor quark masses $m_f = (m_f)_0$, and for a fixed value of exterior cut-off scale $\Lambda = \Lambda_0$. Suppose further that these 7+1 parameters are chosen such, that the calculated M_i^2 agree with the corresponding experimental values. Next, suppose to change the cut-off by a small amount $\delta\Lambda$. Every calculated eigenvalue will then change by δM_i^2. Renormalization theory is then the attempt to reformulate the Hamiltonian, such, that all changes δM_i^2 vanish identically.

The fundamental renormalization group equation is therefore:

$$dM_i^2\big|_0 = dM_i^2\big|_{g=g_0, m_f=m_{f_0}, \Lambda=\Lambda_0} = 0 \,, \qquad (5)$$

for all eigenstates i. Equivalently one requires that *the Hamiltonian is stationary* with respect to small $\delta\Lambda$:

$$\delta H_{\rm LC}\big|_0 = 0 \,. \qquad (6)$$

Hence forward reference to $(g_0, m_{f_0}, \Lambda_0)$, to the 'renormalization point', will be suppressed.

The Hamiltonian can be made stationary by making g and the m_f functions of Λ, by introducing *physical* coupling constants and masses, $\overline{g}$ and $\overline{m}_f$, respectively, which themselves are functions of the bare g and m_f, and which are functionals of the regulator $\overline{F} = F$. The variation of $H_{\rm LC}$ reads then

$$\delta H_{\rm LC} = \delta\overline{g}\frac{\delta H_{\rm LC}}{\delta\overline{\alpha}} + \sum_f \delta\overline{m}_f \frac{\delta H_{\rm LC}}{\delta\overline{m}_f} + \delta\overline{F}\frac{\delta H_{\rm LC}}{\delta\overline{F}} = 0 \,,$$

with the familiar variational derivatives. However, since $\overline{g}$ and $\overline{m}_f$ are themselves functionals of $\overline{F}$, this reduces to

$$\delta H_{\rm LC} = \delta\overline{F}\frac{\delta H_{\rm LC}}{\delta\overline{F}} = 0 \,.$$

Equation (5) as the fundamental equation of renormalization theory is then replaced by

$$\delta\overline{F} = \delta\Lambda\frac{\partial\overline{F}}{\partial\Lambda} = 0 \,, \qquad (7)$$

since the variational derivative of the Hamiltonian with respect to the regulator is unlikely to vanish.

It can be solved by counter term technology, as follows. A counter term is added to the Hamiltonian, whose interaction has exactly the same structure except that the regulator $F(Q)$ is replaced by $C(Q)$. This defines

$$\overline{F}(Q, \Lambda) = F(Q, \Lambda) + C(Q, \Lambda) \,, \qquad (8)$$

subject to the constraint that the counter term vanishes at the renormalization point,

$$C(Q, \Lambda)\bigg|_{\Lambda=\Lambda_0} = 0 \,. \qquad (9)$$

The fundamental (7) defines then a differential equation

$$\frac{dC(Q; \Lambda)}{d\Lambda} = -\frac{dF(Q; \Lambda)}{d\Lambda} \,, \qquad (10)$$

which, in its integral form, includes the initial condition

$$C(Q, \Lambda) = -\int_{\Lambda_0}^{\Lambda} ds\, \frac{dF(Q, s)}{ds} = F(Q, \Lambda_0) - F(Q, \Lambda) \,. \qquad (11)$$

The renormalized regulator function, $\overline{F} = F + C$,

$$\overline{F}(Q, \Lambda) = F(Q, \Lambda_0) \,, \qquad (12)$$

is *manifestly independent of Λ.* By construction, the value of Λ_0 is determined by experiment.

One should emphasize an important point: In deriving (12), use was made of assuming the regulator function has well defined derivatives with respect to Λ. The theta function of the sharp cut-off, however, is a distribution with only ill defined derivatives.

This raises an other important point: If $F(Q, \Lambda)$ is a function of Q/Λ other than a theta function, one must specify how the function approaches the limiting values of (4). The case of the 'soft' regulator

$$F(Q, \Lambda) = \frac{\Lambda^2}{\Lambda^2 + Q^2} \qquad (13)$$

is only a very special example. In a more general approach the soft regulator plays the role of a generating function

$$F(Q, \Lambda) = \left[1 + \sum_{n=1}^{N}(-1)^n s_n \Lambda^n \frac{\partial^n}{\partial\Lambda^n}\right]\frac{\Lambda^2}{\Lambda^2 + Q^2} \,. \qquad (14)$$

The partials $\Lambda^n\,\partial^n/\partial\Lambda^n$ are dimensionless and independent of a change in Λ. The arbitrarily many coefficients $s_1, \ldots, s_N$ are renormalization group invariants and, as such, subject to be determined by experiment.

4 The effective (light-cone) Hamiltonian

In a field theory, one is confronted with a many-body problem of the worst kind: Not even the particle number is conserved. For to formulate effective Hamiltonians more systematically, a novel many-body technique had to be developed, the *method of iterated resolvents* [5,12], whose details are not important here.

Important is that the *effective light-cone Hamiltonian* $H_{\rm eLC}$ has the same eigenvalue as the *full light-cone Hamiltonian* $H_{\rm LC}$ and that it generates the bound state wave function of valence quarks by an one-body integral equation in $(x, \mathbf{k}_\perp)$:

$$M^2\psi_{h_1 h_2}(x, \mathbf{k}_\perp) = \left[\frac{\overline{m}_1^2 + \mathbf{k}_\perp^2}{x} + \frac{\overline{m}_2^2 + \mathbf{k}_\perp^2}{1-x}\right]\psi_{h_1 h_2}(x, \mathbf{k}_\perp)$$

$$-\frac{1}{3\pi^2}\sum_{h'_q, h'_{\bar{q}}}\int \frac{dx'd^2\mathbf{k}'_\perp}{\sqrt{x(1-x)x'(1-x')}}\,\psi_{h'_1 h'_2}(x', \mathbf{k}'_\perp)$$

$$\times F(Q_q)F(Q_{\bar{q}})\left(\frac{\overline{\alpha}(Q_q)}{2Q_q^2} + \frac{\overline{\alpha}(Q_{\bar{q}})}{2Q_{\bar{q}}^2}\right)$$

$$\times [\overline{u}(k_1, h_1)\gamma^\mu u(k'_1, h'_2)]\,[\overline{v}(k'_2, h'_2)\gamma_\mu v(k_2, h_2)] \,. \qquad (15)$$

One has achieved step 2 of (1): $H_{\rm eLC}|\Psi_{q\bar{q}}\rangle = M^2|\Psi_{q\bar{q}}\rangle$. Here, M^2 is the eigenvalue of the invariant-mass squared. The associated eigenfunction $\psi_{h_1 h_2}(x, \mathbf{k}_\perp)$ is the probability amplitude $\langle x, \mathbf{k}_\perp, h_1; 1-x, -\mathbf{k}_\perp, h_2|\Psi_{q\bar{q}}\rangle$ for finding the quark with momentum fraction x, transversal momentum $\mathbf{k}_\perp$ and helicity h_1, and correspondingly the anti-quark.

Expressions for the (effective) quark masses $\overline{m}_1$ and $\overline{m}_2$ and the (effective) coupling function $\overline{\alpha}(Q)$ are given in [12]. Q_q and $Q_{\bar{q}}$ are the Feynman momentum transfers of quark and anti-quark, respectively, and $u(k_1, h_1)$ and $v(k_2, h_2)$ are their Dirac spinors in Lepage Brodsky convention [6], given explicitly in [3]. They arrange themselves in the Lorenz scalar spinor matrix

$$\langle h_1, h_2|S|h_1', h_2'\rangle = [\overline{u}(k_1, h_1)\gamma^\mu u(k_1', h_1')]$$
$$\times [\overline{v}(k_2', h_2')\gamma_\mu v(k_2, h_2)]$$

which is a rather complicated (matrix) function of its six arguments $x, x', \mathbf{k}_\perp, \mathbf{k}_\perp'$, as tabulated in [11]. Finally, the form factors $F(Q)$ restrict the range of integration and regulate the interaction. Note that the equation is fully relativistic and covariant.

It should be emphasized that (15) is valid only for quark and anti-quark having different flavors [5,12]. The additional annihilation term for identical flavors is omitted. At present, it is investigated by [13]. It should also be emphasized that the same structure was obtained with a completely different method, with Wegner's Hamiltonian flow equations [14]. In [14] is also shown why the concept of a 'mean momentum transfer', $Q^2 = \frac{1}{2}\left(Q_q^2 + Q_{\bar{q}}^2\right)$ is a meaningful simplification. It allows to replace (15) by

$$M^2\psi_{h_1 h_2}(x, \mathbf{k}_\perp) = \left[\frac{\overline{m}_1^2 + \mathbf{k}_\perp^2}{x} + \frac{\overline{m}_2^2 + \mathbf{k}_\perp^2}{1-x}\right]\psi_{h_1 h_2}(x, \mathbf{k}_\perp)$$
$$- \frac{1}{3\pi^2}\sum_{h_q', h_{\bar{q}}'}\int \frac{dx' d^2\mathbf{k}_\perp'\, \psi_{h_1' h_2'}(x', \mathbf{k}_\perp')}{\sqrt{x(1-x)x'(1-x')}}\frac{\overline{\alpha}(Q)}{Q^2}\overline{R}(Q)$$
$$\times [\overline{u}(k_1, h_1)\gamma^\mu u(k_1', h_2')][\overline{v}(k_2', h_2')\gamma_\mu v(k_2, h_2)] \ . \ (16)$$

The form factors $F(Q)$ have made their way into the regulator function $\overline{R}(Q) = F^2(Q)$. Krautgärtner et al. [15] and Trittmann et al. [16] have shown how to solve numerically such an equation with a high precision. But since the numerical effort is so considerable, it is reasonable to work first with (over-)simplified models, as specified next.

The Singlet-Triplet model. Quarks are at relative rest when $\mathbf{k}_\perp = 0$ and $x = \overline{x}$, with $\overline{x} \equiv \overline{m}_1/(\overline{m}_1 + \overline{m}_2)$. An inspection of (33) in [11] reveals that for very small deviations from the equilibrium values, the spinor matrix $\langle h_1, h_2|S|h_1', h_2'\rangle$ is proportional to the unit matrix,

$$\langle h_1, h_2|S|h_1' h_2'\rangle \simeq 4\overline{m}_1\overline{m}_2\, \delta_{h_1, h_1'}\, \delta_{h_2, h_2'} \ . \ (17)$$

For very large deviations, particularly for $\mathbf{k}_\perp'^2 \gg \mathbf{k}_\perp^2$, holds

$$Q^2 \simeq \mathbf{k}_\perp'^2 \ , \quad \text{and} \quad \langle \uparrow\downarrow |S| \uparrow\downarrow\rangle \simeq 2\mathbf{k}_\perp'^2 \ . \ (18)$$

The *Singlet-Triplet (ST) model* combines these aspects:

$$\langle h_1, h_2|S|h_1', h_2'\rangle = \delta_{h_1, h_1'}\, \delta_{h_2, h_2'}\, \langle h_1, h_2|S|h_1, h_2\rangle \ , (19)$$

$$\frac{\langle h_1, h_2|S|h_1, h_2\rangle}{Q^2} = \begin{cases} \frac{4\overline{m}_1\overline{m}_2}{Q^2} + 2, & \text{for } h_1 = -h_2, \\ \frac{4\overline{m}_1\overline{m}_2}{Q^2}, & \text{for } h_1 = h_2. \end{cases} \ (20)$$

For anti parallel helicities $h_1 = -h_2$ (singlets) the model interpolates between two extremes: For small momentum

transfer Q, the '2' in (18) is unimportant and the Coulomb aspects of the first term prevail. For large Q, the Coulomb aspects are unimportant and the hyperfine interaction is dominant. The '2' carries the *singlet triplet mass difference*: Its value is understood by $g_s\left(\frac{1}{4} - \left(-\frac{3}{4}\right)\right) = g_s$, with the spin-g factor $g_s = 2$. For parallel helicities $h_1 = h_2$ (triplets) the model reduces to the Coulomb kernel. The model over emphasizes many aspects but its simplicity has proven useful for fast and analytical calculations. Most importantly, the model allows to drop the helicity summations which technically simplifies the problem enormously. A more detailed investigation of the spinor matrix can be found in [20].

The model can not be justified in the sense of an approximation, but it emphasizes the point that the '2', or any other constant in the kernel of an integral equation, leads to numerically undefined equations and thus singularities. Replacing the function $\overline{\alpha}(Q)$ by the strong coupling constant $\alpha_s = g^2/4\pi$ completes the model assumptions. Hence forward, the overline bars for the effective quantitites will be suppressed.

5 The potential energy

It is possible to subtract a c-number from H_{eLC} and to define an effective Hamiltonian H_{eff} implicitly by

$$H_{\text{eLC}} \equiv (m_1 + m_2)^2 + 2(m_1 + m_2)H_{\text{eff}} \ . \ (21)$$

Its eigenvalues have the dimension of an energy

$$H_{\text{eff}}|\varphi\rangle = E|\varphi\rangle \ ,$$

achieving this way step 3 of (1). Note that mass and energy in the front form, on the light cone, are related by

$$M^2 = (m_1 + m_2)^2 + 2(m_1 + m_2)E \ , \quad (22)$$

and **not** by $M^2 = (m_1 + m_2)^2 + 2(m_1 + m_2)E + E^2$, as usual. Only if the energy is negligible as compared to the quark masses, *i.e.* only if $(E/(m_1 + m_2))^2 \ll 1$, the two relations coincide.

A rather drastic technical simplification is achieved by a transformation of the integration variable. One can substitute the integration variable x by the integration variable k_z, which, for all practical purposes, can be interpreted [3] as the z-component of a 3-momentum vector $\mathbf{p} = (k_z, \mathbf{k}_\perp)$. For equal masses $m_1 = m_2 = m$, the transformation is, together with its inverse,

$$x(k_z) = \frac{1}{2}\left[1 + \frac{k_z}{\sqrt{m^2 + \mathbf{k}_\perp^2 + k_z^2}}\right] \ . \ (23)$$

$$k_z^2(x) = (m^2 + \mathbf{k}_\perp^2)\frac{\left(x - \frac{1}{2}\right)^2}{x(1-x)} \ . \ (24)$$

Inserting these substitutions into (16) and defining the reduced wave function φ by

$$\psi_{h_1 h_2}(x, \mathbf{k}_\perp) = \frac{\sqrt{A(k_z, \mathbf{k}_\perp)}}{\sqrt{x(1-x)}}\varphi_{h_1 h_2}(k_z, \mathbf{k}_\perp), \ (25)$$

$$A(\mathbf{p}) = \sqrt{1 + \frac{\mathbf{p}^2}{m^2}} \ , \ (26)$$

leads to an integral equation in the components of $\mathbf{p}$, in which all reference to light-cone variables has disappeared. Using in addition the ST-model of (19), (16) translates for singlets identically into

$$M^2\varphi(\mathbf{p}) = 4\left[m^2 + \mathbf{p}^2\right]\varphi(\mathbf{p}) \tag{27}$$
$$- \frac{\alpha_c}{2\pi^2}\int\frac{d^3p'}{\sqrt{A(p)A(p')}}\left(\frac{4m^2}{Q^2}+2\right)\frac{R(Q)}{m}\,\varphi(\mathbf{p'})\,,$$

with $\alpha_c = \frac{4}{3}\alpha_s$. The equation for the triplets is obtained by dropping the '2'. In the ST-model, the helicity arguments in the wave functions can be suppressed. Applying the relation between mass and energy, as given in (22), the equation is converted to

$$E\varphi(\mathbf{p}) = \frac{\mathbf{p}^2}{2m_r}\varphi(\mathbf{p}) \tag{28}$$
$$- \frac{\alpha_c}{2\pi^2}\int\frac{d^3p'}{\sqrt{A(p)A(p')}}\left(\frac{4m^2}{Q^2}+2\right)\frac{R(Q)}{4m^2}\,\varphi(\mathbf{p'})\,,$$

since the reduced mass for $m_1 = m_2 = m$ is $m_r = m/2$.

The first term in this equation, $\mathbf{p}^2/2m_r$, coincides with the kinetic energy in a conventional non-relativistic Hamiltonian. This is remarkable in view of the fact that no approximation to this extent has been made. The fully relativistic and covariant light-cone approach has no relativistic corrections in the kinetic energy!

Since the first term in (28) is a kinetic energy, the second must be a potential energy — in a momentum representation. In principle, it could be Fourier transformed with $e^{-i\mathbf{pr}}$ to a configuration space with the variable $\mathbf{r}$. But due to the factor $A(p)A(p')$ in the kernel, the resulting potential energy would be non-local, see f.e. [21].

The non-locality of the potential is certainly mathematically exact. But I do not expect this to generate aspects of leading importance, and avoid it by the simplification $A(p) \equiv 1$, both in (25) and (28).

With $A(p) = 1$, the mean four momentum transfer Q^2 reduces to the three momentum transfer $q^2 = (\mathbf{p} - \mathbf{p'})^2$. In consequence, the kernel of (28),

$$U(\mathbf{q}) = -\frac{\alpha}{2\pi^2}\left(\frac{4m^2}{q^2}+2\right)\frac{R(q)}{4m^2}\,, \tag{29}$$

depends only on $\mathbf{q} = \mathbf{p} - \mathbf{p'}$. Its Fourier transform is a local function,

$$V(\mathbf{r}) = \int d^3q\, e^{-i\mathbf{qr}}\, U(\mathbf{q})\,, \tag{30}$$

which plays the role of a conventional *potential energy* in the Fourier transform of (28), *i.e.* in

$$E\,\psi(\mathbf{r}) = \left[\frac{\mathbf{p}^2}{2m_r} + V(\mathbf{r})\right]\psi(\mathbf{r})\,. \tag{31}$$

Here is the Schrödinger equation from (1)! Despite its conventional structure it is a front form equation, designed to calculate the light-cone wave function $\psi(\mathbf{r}) \to \varphi(\mathbf{p}) \to \psi_{q\bar{q}}(x, \mathbf{k}_\perp)$.

I conclude this section with a subtle point, which needs clarification in the future. The *simplification* $A(p) = 1$ is different from a *non-relativistic approximation*. The approach is certainly valid also for relativistic momenta $p^2 \gg m^2$, particularly (28) and (30). The reason is that $A(p)$ occurs only under the integral. There, the large momenta are suppressed by the regulator, anyway.

6 The renormalized Coulomb potential

Hence forward, I restrict consideration to the triplet case, *i.e.* to Coulomb kernels like $U(\mathbf{q}) \sim R(q)/q^2$. The renormalized Coulomb potential is *always finite at the origin*, as opposed to the conventional $\frac{1}{r}$–singularity. It is instructive to verify this explicitly for two regulators:

$$U(q) = \begin{cases} -\dfrac{\alpha_c}{2\pi^2 q^2}\dfrac{\lambda^2}{q^2+\lambda^2}, & \text{for the soft cut-off,} \\ -\dfrac{\alpha_c}{2\pi^2 q^2}\,\Theta(q^2 - \lambda^2), & \text{for the sharp cut-off.} \end{cases} \tag{32}$$

The Fourier transform according to (29) gives

$$V(r) = \begin{cases} -\dfrac{\alpha_c}{r}\left(1 - e^{-\lambda r}\right), & \text{for the soft cut-off,} \\ -\dfrac{\alpha_c}{r}\dfrac{2}{\pi}\mathrm{Si}(\lambda r), & \text{for the sharp cut-off,} \end{cases} \tag{33}$$

where Si is the Integral Sine. Asymptotically holds:

$$\lim_{r\to\infty} V(r) = \begin{cases} -\dfrac{\alpha_c}{r}, & \text{for the soft cut-off,} \\ -\dfrac{\alpha_c}{r}, & \text{for the sharp cut-off.} \end{cases} \tag{34}$$

Both cut-offs produce the conventional Coulomb potential. Near the origin, however, holds:

$$\lim_{r\to 0} V(r) = \alpha_c\lambda \cdot \begin{cases} -1 + \dfrac{(\lambda r)}{2}, & \text{for the soft cut-off,} \\ -\dfrac{2}{\pi} + \dfrac{(\lambda r)^2}{9\pi}, & \text{for the sharp cut-off.} \end{cases} \tag{35}$$

The renormalized Coulomb potential is *finite* but the constant is cut-off dependent. Even the r-dependence differs: The soft cut-off gives a linear and the sharp cut-off a quadratic dependence.

The cut-off dependence near the origin is one of the most important aspects of the present work and has a deep physical reason to be discussed below. Recalling the discussion in Sect. 3 and replacing the soft cut-off in analogy to (14) with

$$R(q) = \left[1 + \sum_{n=1}^{N}(-1)^n s_n\lambda^n\frac{\partial^n}{\partial\lambda^n}\right]\frac{\lambda^2}{\lambda^2 + q^2}\,, \tag{36}$$

gives straightforwardly the generalized Coulomb potential

$$V(r) = -\frac{\alpha_c}{r}\left[1 + \sum_{n=1}^{N}(-1)^n s_n\lambda^n\frac{\partial^n}{\partial\lambda^n}\right]\left(1 - e^{-\lambda r}\right)$$
$$= \frac{\alpha_c}{r}\left[-1 + e^{-\lambda r}\sum_{n=0}^{N} s_n(r\lambda)^n\right], \tag{37}$$

with $s_0 \equiv 1$. This result illustrates an other important point: The Laguerre polynomials are a complete set of

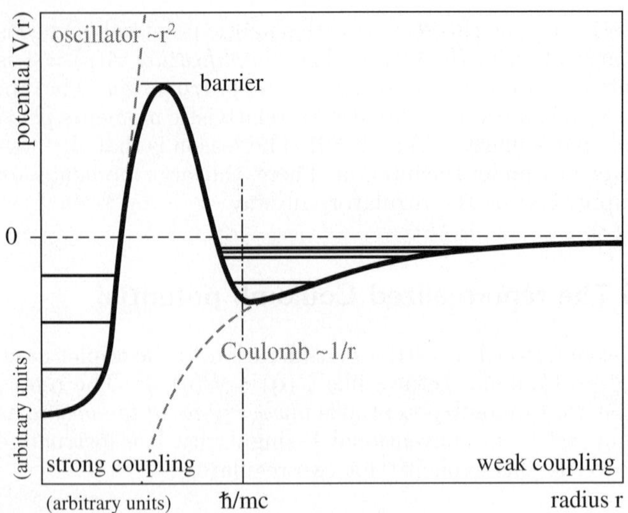

Fig. 2. Schematic behavior of the renormalized Coulomb potential, see also the discussion in the text

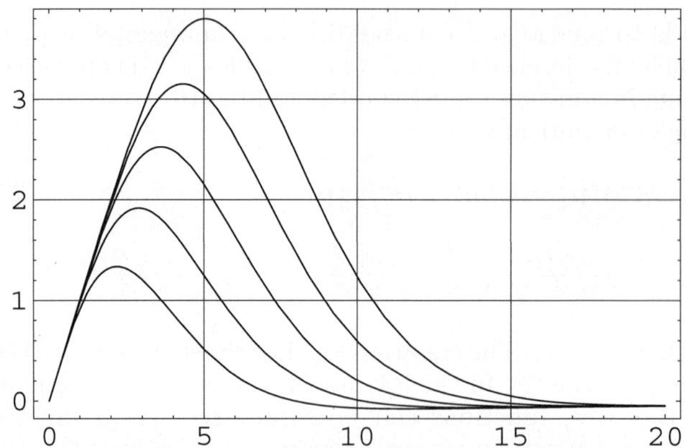

Fig. 3. The dimensionless Coulomb potential $W_N(y; 0, 1, 0)$ is plotted versus the radius parameter $y = \lambda r$ for different N, *i.e.* from bottom to top for $N = 4, 5, 6, 7, 8$

functions. The term added to the -1 in (37) is thus potentially able to reproduce an arbitrary function of r. The description in terms of a generating function, as in (36) or (37), is therefore *complete*.

The physical picture which develops is illustrated in Fig. 2. In the far zone, for sufficiently large r, the potential energy coincides with the conventional Coulomb potential $-\frac{\alpha_c}{r}$. Since the potential is attractive, it can host bound states which are probably those realized in weak binding. In the near zone, for sufficiently small r, the potential behaves like a *power series* $c_0 + c_1 r + c_2 r^2$ which potentially can host the bound states of strong coupling, provided the actual parameter values allow for that. In the intermediate zone, the actual potential must interpolate between these two extremes, since (37) is an analytic function of r. Most likely this is done by developing a *barrier of finite* height, depending on the actual parameter values. The onset of the near and intermediate regimes must occur for relative distances of the quarks, which are comparable to the Compton wave length associated with their reduced mass. If the distance is smaller, one expects deviations from the classical regime by elementary considerations on quantum mechanics, indeed.

The large number of parameter in (37) can be controlled by the following construction: The coefficients s_n in (37) are expressed in terms of only three parameters a, b, and c, by

$$s_n = \frac{1}{n!} + \frac{a}{(n-1)!} + \frac{b}{(n-2)!} + \frac{c}{(n-3)!} . \quad (38)$$

The first few coefficients are then explicitly

$$
\begin{aligned}
s_0 &= 1 , \\
s_1 &= 1 + a , \\
s_2 &= \tfrac{1}{2} + a + b , \\
s_3 &= \tfrac{1}{6} + \tfrac{a}{2} + b + c , \\
s_4 &= \tfrac{1}{24} + \tfrac{a}{6} + \tfrac{b}{2} + c .
\end{aligned}
\quad (39)
$$

As a consequence, the dimensionless Coulomb potential,

$$W_N(y; a, b, c) = \frac{V(r)}{\alpha_c \lambda} = \frac{1}{y}\left(-1 + e^{-y}\sum_{n=0}^{N} s_n y^n\right), \quad (40)$$

which depends on r only through the dimensionless combination $y = \lambda r$, is at most a quadratic function of y,

$$W_N(y; a, b, c) = a + by + cy^2 , \quad (41)$$

in the near zone, and thus independent of N. The remainder starts at most with power y^{N+1}. A value of $a = c = 0$ and $b = 1$ should therefore yield a linear set of functions $W_N(y; a, b, c) = y$ in the near zone. As shown in Fig. 3 this happens to be true for surprisingly large values of y, *i.e.* not only for $y \ll 1$. The value of N essentially controls the height of the barrier. Similarly, $W_N(y; 0, 0, 1) = y^2$ generates a set of functions which are strictly quadratic in the near zone. Again, N controls the height of the barrier, as to be seen below in Fig. 5.

7 Determining the parameters by experiment

The QCD-inspired model developed thus far has a considerable number of renormalization group invariant parameters, which must be determined once and for all by experiment.

In doing this [23], we have been inspired by the work of Anisovich et al. [25]. Enumerating the excited states of a hadron by a counting index $n = 0, 1, 2, \ldots$, these authors have found the linear relation $M_n^2 = M_0^2 + n\chi$ for practically all hadrons. As an example, I present in Fig. 4 the spectrum of the π- and the ρ-meson.

The linear relation between mass–squared and energy on the light cone, (22), allows then to conclude that the potential energy in the near zone must be a pure oscillator,

$$V(r) = -c_t + \frac{1}{2} f_t r^2 , \quad (42)$$

at least to first approximation, and that $b = 0$ in (41). If one addresses to reproduce the spectra of all flavor off-diagonal triplet mesons (pseudo-vector mesons), except

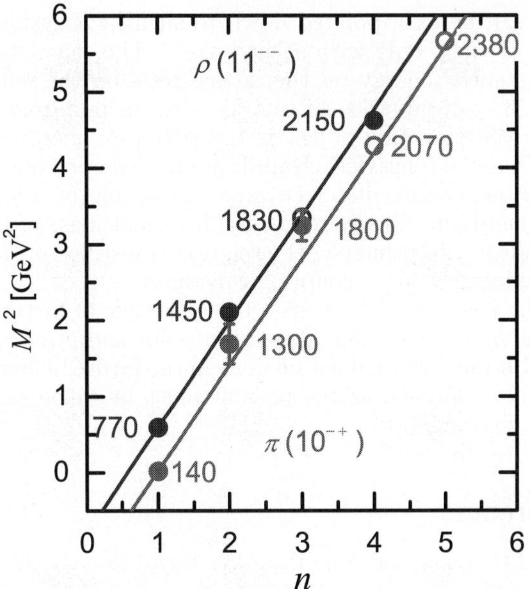

Fig. 4. The invariant mass-squares of all available π^{+-} and ρ^{+}–states are plotted versus a counting index $n = 0, 1, 2, \ldots$. The straight lines correspond to $M_n^2 = M_0^2 + n\chi$, with the value $\chi = 1.39$ GeV^2, taken from Anisovich et al. [25]. The filled circles correspond to states which have been empirically [24] seen, the empty ones correspond to the predictions [25]. — Plot courtesy of Shan-Gui Zhou

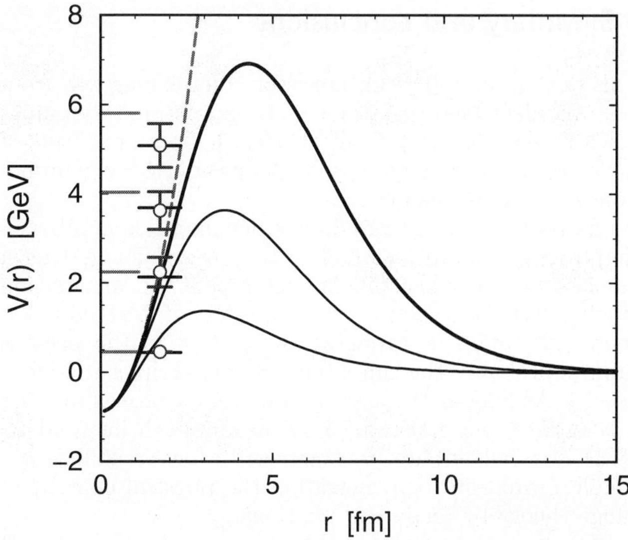

Fig. 5. The continuous lines display the generalized Coulomb potential $V(r) = \alpha_c \lambda W_N(\lambda r; a, 0, c)$ in physical units as function of r, for the values $N = 4, 5, 6$ from bottom to top. The circles indicate the experimental eigenvalues E_n for the ρ^+. They agree with the calculated eigenvalues for $N = 6$, shown by the horizontal lines. — The dashed line displays the harmonic approximation; the horizontal lines on the left indicate the oscillator states. — See the discussion in the text

the topped ones, one has to determine 6 parameters: The 2 constants from the oscillator model, c_t and f_t, and the 4 effective flavor quark masses $m_u = m_d$, m_s, m_c, and m_b. To determine them, one needs 6 experimental numbers, and I take from [24]:

$$M_{u\bar{d},t0}=0.768, \quad M_{u\bar{s},t0}=0.892, \quad M_{u\bar{c},t0}=2.010,$$
$$M_{u\bar{d},t1}=1.465, \quad M_{u\bar{s},t1}=1.680, \quad M_{u\bar{b},t0}=5.325, \tag{43}$$

all in GeV. The notation should be self-explanatory. For example, $M_{u\bar{d},t1}$ refers to the first excited state of the ρ^+. The so obtained parameter values are:

m_u	m_s	m_c	m_b	c_t	f_t	
0.218	0.438	1.749	5.068	0.880	0.0869	(44)
GeV	GeV	GeV	GeV	GeV	GeV^3	

The numbers differ slightly from those in [23], due to choosing the empirical data set different from (43), but yield about the same overall agreement with all available experimental states of pseudo-vector mesons.

Reverting the argument, one concludes as in [23] that the oscillator model in (42) explains quite naturally the systematics found by Anisovich et al. [25]. But one can do even better.

8 Relating the oscillator model to QCD

The oscillator model in (42) is only the harmonic approximation to the QCD–inspired, generalized Coulomb poten-

tial in (37). Their parameters are related obviously by

$$c_t = -\alpha_c \lambda a, \qquad b = 0, \qquad f_t = 2\alpha_c \lambda^3 c. \tag{45}$$

One needs more experimental information to pin down the value of a, c and N. Choosing λ as the QCD scale, i.e.

$$\lambda = 200 \text{ MeV}, \tag{46}$$

one can use the expressions for $\overline{\alpha}(Q)$ in [12] to calculate $\alpha_s \equiv \overline{\alpha}(0)$ from the measured value of the coupling constant at the Z-mass $M_Z = 91.2$ GeV,

$$\overline{\alpha}(M_Z) = 0.118, \quad \text{thus} \quad \alpha_s \equiv \overline{\alpha}(0) = 0.1695, \tag{47}$$

as to be shown in greater detail in [22]. Having fixed $\alpha_c = \frac{4}{3}\alpha_s$ and λ allows to calculate a and c from c_t and f_t, i.e.

$$a = -19.5, \qquad c = 24.0. \tag{48}$$

We are thus able to draw the generalized Coulomb potential $V(r) = \alpha_c \lambda W_N(\lambda r; a, 0, c)$ for different N as done in Fig. 5. The 'experimental' eigenvalues E_0—E_3 for the ρ–meson, obtained by means of $E_n = (M_n^2 - 4m_u^2)/(4m_u)$, see (22), are also inserted, including the empirical limits of error. The experimental error $\delta E_{\rho,3} \sim \pm 0.5$ GeV (thus $\delta M_{\rho,3} \sim \pm 0.1$ GeV) is hypothetical, since $M_{\rho,3}$ is not confirmed. Taking it for granted, the lowest possible value for N is thus

$$N = 6. \tag{49}$$

This completes the determination of all parameters. They are universal within the model. I thank Harun Omer[26] for giving me the exact eigenvalues prior to publication.

9 Summary and conclusions

This work is an important mile stone on the long way from the canonical Lagrangian for quantum chromo dynamics down to the composition of physical hadrons in terms of their constituting quarks and gluons, by the eigenfunctions of a Hamiltonian.

As part of a on-going effort, a denumerable number of simplifying assumptions had to be phrased for getting a manageable formalism [5]. Among them is the formulation of an effective interaction by the method of iterated resolvents [12], but the strongest assumption in the present work is probably the simplifying Singlet-Triplet model in Sect. 4. As long as the assumption are not proven at least *a posteriori*, one must speak of an approach inspired by QCD. It is advantageous, however, to have a sufficiently simple formalism for penetrating the physical content of gauge theory by analytical relations.

The biggest progress of the present work can be found in Sects. 2 and 3. It is related to a consistent regularization and renormalization of a gauge theory. The ultraviolet divergences in gauge theory are caused less by the possibly large momenta of the constituent particles, but by the large momentum *transfers* in the interaction. In a Hamiltonian approach, such as the present, one has not much choice else than to chop them off by a regulating form factor in the elementary vertex interaction.

The form factor makes its way into a regulator function which suppresses the large momentum transfers in the Fourier transform of the Coulomb interaction, see Sect. 6. The arbitrariness in chopping off the *large momentum transfers* is reflected in the arbitrariness of the potential at *small relative distances*. It is this arbitrariness which allows for a pocket in the potential which binds the quarks in a hadron.

The problem is then how to fix this function with its many parameters, by experiment. In practice this is less difficult than anti-cipated, see Sect. 7. It suffices to determine only three parameters, two continuous ones and one counting index.

The potential energy of the present work vanishes at an infinite separation of the quarks. This seems be be in conflict with the potential energies of phenomenological models [27] which rise forever. It also seems to be in conflict with lattice gauge calculations [28,29]. Is a finite ionization limit in conflict also with 'confinement', *i.e.* with the empirical fact that free quarks have not been observed? — The present model prohibits free quarks as a stable solution, since the sum of the constituent quark masses is always larger than the mass of the corresponding hadron and a pion. Free constituent quarks would hadronize very quickly into bound states. This is different from atomic physics with its free constituents, where the binding energy is always much smaller than the mass of positronium proper.

The most disturbing aspect of the present work is its obvious conflict with lattice gauge calculations [28,29] and their successes. Several points however should be made: I have not checked to which extent a linear term in the potential is consistent with the excellent agreement between theory and experiment presented in this work. — Even with present day computers lattice gauge calcula-

tions can be extrapolated down to such light systems as the π or the ρ only with a head ake. – The calculation of the potential energy on the lattice rests on the assumptions of static quarks, of quarks with an infinitely large mass. Whether this object is the potential energy to be used in a non relativistic Hamiltonian is an open question, as well as whether its eigenvalue can simply be added to the constituent masses to get the invariant mass of physical hadrons. In principle, the relation is justified only only for sufficiently small coupling constants.

The present work opens a broad avenue of further applications, among them also the baryons and physical nuclei. But much work must be done in the future before such a simple approach as the present must be taken serious. It is a first step only.

References

1. H.C. Pauli and S.J. Brodsky: Phys. Rev. D **32**, 1993 (1985)
2. P.A.M. Dirac: Rev. Mod. Phys. **21**, 392 (1949)
3. S.J. Brodsky, H.C. Pauli, and S.S. Pinsky: Phys. Rep. **301**, 299–486 (1998)
4. A.H. Mueller: Nucl. Phys. B **415**, 373 (1994)
5. H.C. Pauli: in: *New directions in Quantum Chromodynamics*, C.R. Ji and D.P. Min, Eds., American Institute of Physics, 1999, pp. 80–139; hep-ph/9910203
6. G.P. Lepage and S.J. Brodsky: Phys.Rev. D **22**, 2157 (1980)
7. A.C. Kalloniatis: Phys. Rev. D **54**, 2876 (1996)
8. J. Hiller: Nucl. Phys. B (Proc.Suppl.) **90**, 170 (2000)
9. K. Wilson: in *Lattice '89*, R. Petronzio, Ed.; Nucl. Phys. B (Proc. Suppl.) **17**, 82. (1990)
10. K.G. Wilson, T. Walhout, A. Harindranath, W.M. Zhang, R.J. Perry, and S.D. Glazek: Phys. Rev. D **49**, 6720 (1994)
11. H.C. Pauli: "Compendium of Light-Cone Quantization", Nucl. Phys. B (Proc. Suppl.) **90**, 259 (2000)
12. H.C. Pauli: Eur. Phys. J. C **7**, 289 (1998)
13. H.C. Krahl: master thesis (University of Heidelberg, 2004)
14. F. Wegner: Nucl. Phys. B (Proc. Suppl.) **90**, 141 (2000); H.C. Pauli: Nucl. Phys. B (Proc. Suppl.) **90**, 147 (2000)
15. M. Krautgärtner, H.C. Pauli, and F. Wölz: Phys. Rev. D **45**, 3755 (1992)
16. U. Trittmann and H.C. Pauli: Nucl. Phys. B (Proc. Suppl.) **90**, 161 (2000)
17. H.C. Pauli: Nucl. Phys. B (Proc. Suppl.) **90**, 154 (2000)
18. H.C. Pauli: Nucl. Phys. B (Proc. Suppl.) **108**, 273 (2002)
19. M. Frewer, T. Frederico, and H.C. Pauli: Nucl. Phys. B (Proc. Supp.) **108**, 234 (2002)
20. A. Krassnigg and H.C. Pauli: Nucl. Phys. B (Proc. Supp.) **108**, 251 (2002)
21. H.C. Pauli and J. Merkel: Phys. Rev. D **55**, 2486 (1997)
22. H.C. Pauli: 2003, in preparation
23. T. Frederico, H.C. Pauli, and S.G. Zhou: Phys. Rev. D **62**, 116011–8 (2002)
24. C. Caso et al: Eur. Phys. J. C **3**, 1 (1998)
25. A.V. Anisovich, V.V. Anisovich, and A.V. Sarantsev: Phys. Rev. D **66**, 051502–5 (2002)
26. H. Omer: master thesis (University of Heidelberg 2004)
27. S. Godfrey and N. Isgur: Phys. Rev. D **32**, 189 (1985)
28. K. Schilling: Nucl. Phys. B (Proc.Suppl.) **83**, 140 (2000)
29. G. Schierholz: Nucl. Phys. B (Proc.Suppl.) **90**, 207 (2000)

Eur Phys J A (2004) **19**, s01, 23–27

Digital Object Identifier (DOI) 10.1140/epjad/s2004-03-004-2

EPJ A direct

electronic only

Viewing the proton through "color" filters

Xiangdong Ji

Department of Physics, University of Maryland, College Park, MD20742, USA

Received: 31 July 2003 / Accepted: 14 Nov 2003 /
Published Online: 6 Feb 2004 – © Società Italiana di Fisica / Springer-Verlag 2004

Abstract. While the form factors and parton distributions provide separately the shape of the proton in coordinate and momentum spaces, a more powerful imaging of the proton structure can be obtained through quantum phase-space distributions. Here we introduce the Wigner-type quark and gluon distributions which depict a full-3D proton at every fixed Feynman momentum, like what is seen through momentum("color")-filters. After appropriate reductions, the phase-space distributions are related to the generalized parton distributions (GPDs) and transverse-momentum dependent parton distributions measurable in high-energy experiments.

PACS. 12.90.+b proton structure – 13.60.Fz high-energy scattering

1 Introduction

In exploring the microscopic structure of matter, there are two frequently-used approaches. First, the spatial distribution of matter (or charge) in a system can be probed through elastic scattering of electrons, or photons, or neutrons, etc. The physical quantity that one measures is the elastic form (structure) factors which depend on three-momentum transfer to the system. The Fourier transformation of the form factors provides direct information on the spatial distributions. The well-known examples include the study of charge distribution in an atom and the atomic structure of a crystal. The second approach is designed to measure the population of the constituents as a function of momentum, or the momentum distribution, through knock-out scattering. Here the well-known examples include the nucleon distributions in nuclei measured through quasi-elastic electron scattering, and the distribution of atoms in a quantum liquid probed through neutron scattering. The scattering cross section sometimes depends on the reaction dynamics which must be understood before the momentum distribution can be extracted.

Both approaches are complementary, but bear similar drawbacks. The form factor measurements do not yield any information about the underlying dynamics of the system such as the speed of the constituents, whereas the momentum distribution does not give any information on the spatial location of the constituents. More complete information about the microscopic structure lies in the correlation between the momentum and coordinate spaces, i.e., to know where a particle is located and, at the same time, with what velocity it travels. This information is certainly attainable for a classical system for which one can define and study the phase-space distribution of the constituents. For a quantum mechanical particle, however,

the notion of a phase-space distribution seems less useful because of the uncertainty principle. Nonetheless, the first phase-space distribution in quantum mechanics was introduced by Wigner in 1932 [1], and many similar distributions have been studied thereafter. These distributions have been used for various purposes in very diverse areas such heavy-ion collisions, quantum molecular dynamics, signal analysis, quantum information, optics, image processing, non-linear dynamics, etc.[2].

In this talk, we explore to what extent one can construct physically interesting and experimentally measurable phase-space distributions in quantum chromodynamics (QCD), and what information it contains about the QCD parton dynamics [3]. To facilitate the construction, we examine the uncertainty in the traditional interpretation of electromagnetic form factors due to relativity, and analyze the physical content of the Feynman parton distributions in the rest frame of the proton. We then introduce the phase-space Wigner distributions for the quarks and gluons in the proton, which contain most general one-body information of partons, corresponding to the full one-body density matrix in technical terms. After integrating over the spatial coordinates, one recovers the familiar transverse-momentum dependent parton distributions [4]. On the other hand, some reduced version of the distributions is related, through a specific Fourier transformation, to the generalized parton distributions (GPDs) which have been studied extensively in the literature in recent years [5]. Roughly speaking, a GPD is a one-body matrix element which combines the kinematics of both elastic form factors and Feynman parton distributions, and is measurable in hard exclusive processes. Therefore, the notion of phase-space distribution provides a new 3D interpretation of the GPDs in the rest frame of the proton.

Other interpretations of the GPD in the literature have been made in the IMF and impact parameter space [6].

2 Relativity constraint on interpretation of form factors and parton distributions

The electromagnetic form factors are among the first measured and mostly studied observables of the proton. They are defined as the matrix elements of the electromagnetic current between the proton states of different four-momenta. Because the proton is a spin one-half particle, the matrix element defines two form factors,

$$\langle p_2 | j_\mu(0) | p_1 \rangle = \bar{U}(p_2) \left\{ F_1(q^2)\gamma_\mu + F_2(q^2)\frac{i\sigma_{\mu\nu}q_\nu}{2M_N} \right\} U(p_1),$$
$$(1)$$

where F_1 and F_2 are the well-known Dirac and Pauli form factors, respectively, depending on the momentum transfer $q = p_2 - p_1$, and $U(p)$ is proton spinor normalized as $\bar{U}(p)U(p) = 2M_N$.

Since the beginning, it has been known that the physical interpretation of the proton form factors is complicated by relativistic effects [7]. Consider a system of size R and mass M. In relativistic quantum theory, the system cannot be localized to a precision better than its Compton wavelength $1/M$. Any attempt to do this with an external potential will result in creation of particle-antiparticle pairs. As a consequence, the static size of the system cannot be defined to a precision better than $1/M$. If $R \gg 1/M$, which is the case for all non-relativistic systems, the above is not a significant constraint. One can probe the internal structure of the system with a wavelength $(1/|\mathbf{q}|)$ comparable to or even much smaller than R, but still large enough compared to $1/M$ so that the probe does not induce an appreciable recoil. A familiar example is the hydrogen atom for which $RM_H \sim M_H/(m_e\alpha_{\rm em}) \sim 10^5$, and the form factor can be measured through electron scattering with momentum transfer $|\mathbf{q}| \ll M_H$.

When the probing wavelength is comparable to $1/M$, the form factors are no longer determined by the internal structure alone. They contain also the dynamical effects of Lorentz boosts because the initial and final protons have different momenta. In relativistic quantum theory, the boost operators involve nontrivial dynamical effects which result in the proton wave function being different in different frame (in the usual instant form of quantization). Therefore in the region $|\mathbf{q}| \sim M$, the physical interpretation of the form factors is complicated because of the entanglement of the internal and the center-of-mass motions in relativistic dynamics. In the limit $|\mathbf{q}| \gg M$, the former factors depend almost entirely on the physical mechanism producing the overall change of the proton momentum. The structural effect involved is a very small part of the proton wave function (usually the minimal Fock component only).

For the proton, $M_N R_N \sim 4$. Although much less certain than in the case of the hydrogen atom, it seems still sensible to have a rest-frame picture in terms of the electromagnetic form factors, so long as one keeps in mind that

equally justified definitions of the proton sizes can differ by $\sim 1/M_N(R_N M_N)$. For example, the traditional definition of the proton charge radius in terms of the slope of the Sachs form factor $G_E(q^2)$ is 0.86 fm [8]. On the other hand, if one uses the slope of the Dirac form factor F_1 to define the charge radius, one gets 0.79 fm, about 10% smaller.

Since relativity makes the interpretation of the electromagnetic form factors non-unique, the best one can do is to choose one particular interpretation and work consistently. For example, when extracting the proton charge radius from the Lamb shift measurements, one shall use the same definition as from the electric form factor. The most frequently-used definition is that of Sachs [8], but other schemes are equally good and the scheme dependence disappears in the limit $MR \to \infty$. This is very much like the renormalization scheme dependence of parton densities due to radiative corrections at finite strong coupling constant $\alpha_{\rm s}$. In this paper, we adopt the Sachs interpretation of the form factors, which means that *the spatial distributions are defined to be the Fourier transformation of the Breit frame matrix elements.*

Parton distributions were introduced by Feynman to describe deep-inelastic scattering. They have the simplest interpretation in the IMF as the densities of partons in the longitudinal momentum x. To construct the quantum phase-space distributions for the quarks, we need an interpretation of the Feynman densities in the rest frame. This is because the IMF involves a Lorentz boost along the z-direction which destroys the rotational symmetry of the 3D space.

The physics of the Feynman quark distribution in the rest frame is made more clear through the notion of the *spectral function* (gauge-link omitted)

$$S(k) = \frac{1}{2p^+} \int d^4\xi \, e^{ik\cdot\xi} \langle p|\bar{\Psi}(0)\gamma^+\Psi(\xi)|p\rangle \ . \qquad (2)$$

which is the dispersive part of the single-quark Green's function in the proton. The physical meaning of $S(k)$ can be seen from its spectral representation,

$$S(k) = \sum_n (2\pi)^4\delta^{(4)}(p-k-p_n)\langle p|\bar{\Psi}_k|n\rangle\gamma^+\langle n|\Psi(0)|p\rangle/2p^+$$
$$\sim \sum_n (2\pi)^4\delta^{(4)}(p-k-p_n)|\langle n|\Psi_{k+}|p\rangle|^2 \qquad (3)$$

where Ψ_k is a Fourier transformation of quark field $\Psi(\xi)$: It is the probability of annihilating a quark (or creating an antiquark) of *four*-momentum k (three-momentum $\mathbf{k}$ and the off-shell energy $E = k^0$) in the proton, leading to an "on-shell" state of energy-momentum $p_n = p - k$. The quark here is off-shell because if p_n and p are both "on-shell", $k^2 \neq m_q^2$ in general. [That the partons are off-shell are in fact also true in the IMF calculations.] Of course, in QCD $|n\rangle$ is not in the Hilbert space, but the spectral function itself is still a meaningful quantity.

Since the quarks are ultra-relativistic, Ψ_k contains both quark and antiquark Fock operators. One cannot in general separate quark and anti-quark contributions, unlike in the non-relativistic systems in which only the particle or antiparticle contribute. In fact, if one expands the above expression, one finds pair creations and annihilation terms. However, this is also true for the usual charge density. Therefore we can speak of $S(k)$ as a distribution of vector charges and currents, but not a particle density. In nuclear physics where the non-relativistic dynamics dominates, the proton spectral function in the nucleus is positive definite and can be regarded as a particle density. The nuclear spectral function is directly measurable through pick-up and knock-out experiments, in which E and $\mathbf{k}$ are called the missing energy and missing momentum, respectively (see for example [9]).

It is now easy to see that in the rest frame of the proton, the Feynman quark distribution is

$$q(x) = \sqrt{2} \int \frac{d^4k}{(2\pi)^4} \delta(k^0 + k^z - x M_N) S(k) . \qquad (4)$$

The x variable is simply a special combination of the off-shell energy k^0 and momentum k^z. The parton distribution is the spectral function of quarks projected along a special direction in the four-dimensional energy-momentum space. The quarks with different k^0 and k^z can have the same x, and moreover, the both $x > 0$ and $x < 0$ distributions contain contributions from quarks and anti-quarks.

3 Quantum phase-space distributions

Suppose we have a one-dimensional quantum mechanical system with wave function $\psi(x)$, the Wigner distribution is defined as

$$W(x,p) = \int d\eta e^{ip\eta} \psi^*(x - \eta/2) \psi(x + \eta/2) , \qquad (5)$$

where we have set $\hbar = 1$. When integrating out the coordinate x, one gets the momentum density $|\psi(p)|^2$, which is positive definite. When integrating out p, the positive-definite coordinate space density $|\psi(x)|^2$ follows. For arbitrary p and x, the Wigner distribution is not positive definite and does not have a probability interpretation. Nonetheless, for calculating the physical observables, one can just take averages over the phase-space as if it is a classical distribution

$$\langle \hat{O}(x,p) \rangle = \int dx dp \, W(x,p) O(x,p) \qquad (6)$$

where the operators are ordered according to the Weyl association rule. For a single-particle system, the Wigner distribution contains everything there is in the quantum wave function. For a many-body system, the Wigner distribution can be used to calculate the averages of all one-body operators. Sign changes in the phase-space are a hint that it carries non-trivial quantum phase information.

In QCD, the single-particle wave function must be replaced by (gauge-invariant) quantum fields, and hence it is natural to introduce the *Wigner operator*,

$$\hat{\mathcal{W}}_\Gamma(\mathbf{r},k) = \int d^4\eta e^{ik\cdot\eta} \overline{\Psi}(\mathbf{r} - \eta/2) \Gamma \Psi(\mathbf{r} + \eta/2) , \qquad (7)$$

where $\mathbf{r}$ is the quark phase-space position and k the phase-space four-momentum conjugated to the spacetime separation η. Γ is a Dirac matrix defining the types of quark densities because the quarks are spin-1/2 relativistic particles. Depending on the choice of Γ, we can have vector, axial vector, or tensor density.

For non-relativistic systems for which the center-of-mass is well-defined and fixed, one can define the phase-space distributions by taking the expectation value of the above Wigner operators in the $\mathbf{R} = 0$ state. For the proton for which the recoil effect cannot be neglected, the rest-frame state cannot be uniquely defined. Here we follow Sachs, defining a rest-frame matrix element as that in the Breit frame, averaging over all possible 3-momentum transfers. Therefore, we construct the quantum phase-space quark distribution in the proton as,

$$W_\Gamma(\mathbf{r},k) = \frac{1}{2M_N} \int \frac{d^3\mathbf{q}}{(2\pi)^3} \left\langle \mathbf{q}/2 \left| \hat{\mathcal{W}}_\Gamma(\mathbf{r},k) \right| - \mathbf{q}/2 \right\rangle \qquad (8)$$

$$= \frac{1}{2M_N} \int \frac{d^3\mathbf{q}}{(2\pi)^3} e^{-i\mathbf{q}\cdot\mathbf{r}} \left\langle \mathbf{q}/2 \left| \hat{\mathcal{W}}_\Gamma(0,k) \right| - \mathbf{q}/2 \right\rangle ,$$

where the plane-wave states are normalized relativistically. The most general phase-space distribution depends on *seven* independent variables.

The only way we know how to probe the single-particle distributions is through high-energy processes, in which the light-cone energy $k^- = (k^0 - k^z)/\sqrt{2}$ is difficult to measure, where the z-axis refers to the momentum direction of a probe. Moreover, the leading observables in these processes are associated with the "good" components of the quark (gluon) fields in the sense of light-cone quantization [10], which can be selected by $\Gamma = \gamma^+$, $\gamma^+\gamma_5$, or $\sigma^{+\perp}$ where $\gamma^+ = (\gamma^0 + \gamma^z)/\sqrt{2}$. The direction of the gauge link, n^μ, is then determined by the trajectories of high-energy partons traveling along the light-cone $(1, 0, 0, -1)$ [11,12]. Therefore, from now on, we restrict ourselves to the reduced Wigner distributions by integrating out k^-,

$$W_\Gamma(\mathbf{r},\mathbf{k}) = \int \frac{dk^-}{(2\pi)^2} W_\Gamma(\mathbf{r},k) , \qquad (9)$$

with a light-cone gauge link is now implied. Unfortunately, there is no known experiment at present capable of measuring this 6-dimensional distribution which may be called the *master* or *mother* distribution.

Further phase-space reductions lead to measurable quantities. Integrating out the transverse momentum of partons, we obtain a 4-dimensional quantum distribution

$$\tilde{f}_\Gamma(\mathbf{r},k^+) = \frac{1}{2M_N} \int \frac{d^3\mathbf{q}}{(2\pi)^3} e^{-i\mathbf{q}\cdot\mathbf{r}} \int \frac{d\eta^-}{2\pi} e^{i\eta^- k^+}$$

$$\times \left\langle \mathbf{q}/2 \left| \overline{\Psi}(-\eta^-/2) \Gamma \Psi(\eta^-/2) \right| - \mathbf{q}/2 \right\rangle . \qquad (10)$$

The matrix element under the integrals is what defines the GPDs. More precisely, if one replaces k^+ by Feynman variable xp^+ ($p^+ = E_q/\sqrt{2}$, proton energy $E_q = \sqrt{M^2 + \mathbf{q}^2/4}$) and η^- by λ/p^+, the reduced Wigner distribution becomes the Fourier transformation of the GPD $F_\Gamma(x, \xi, t)$

$$f_\Gamma(\mathbf{r}, x) = \frac{1}{2M_N} \int \frac{d^3\mathbf{q}}{(2\pi)^3} e^{-i\mathbf{q}\cdot\mathbf{r}} F_\Gamma(x, \xi, t) \ . \qquad (11)$$

In the present context, the relation between kinematic variables are $\xi = q^z/(2E_q)$ and $t = -\mathbf{q}^2$. Taking $\Gamma = \sqrt{2}\gamma^+$, F_{γ^+} is the same as that in [13]

$$F_{\gamma^+}(x, \xi, t) \qquad\qquad\qquad (12)$$

$$= \frac{d\lambda}{2\pi} e^{i\lambda x} \left\langle \mathbf{q}/2 \left| \overline{\psi}(-\lambda n/2)\mathcal{L}\sqrt{2}\gamma^+\psi(\lambda n/2) \right| -\mathbf{q}/2 \right\rangle$$

which defines $H(x, \xi, t)$ and $E(x, \xi, t)$.

The phase-space function $f_{\gamma^+}(\mathbf{r}, x)$ can be used to construct 3D images of the quarks for every selected Feynman momentum x in the rest frame of the proton. These images provide the pictures of the proton seen through the Feynman momentum (or "color" or x) filters. They also may be regarded as the result of a quantum phase-space tomography of the proton. We remind the reader again that the Feynman momentum in the rest-frame sense is a special combination of the off-shell energy and momentum along z, namely $E + k^z$. Integrating over the z coordinate, the GPDs are set to $\xi \sim q^z = 0$, and the resulting two-dimensional density $f_{\gamma^+}(\mathbf{r}_\perp, x)$ is just the impact-parameter-space distribution [6]. Further integration over $\mathbf{r}_\perp$ recovers the usual Feynman parton distribution.

The physical content of the above distribution is further revealed by examining its spin structure. Working out the matrix element in (12),

$$\frac{1}{2M_N} F_{\gamma^+}(x, \xi, t) = [H(x, \xi, t) - \tau E(x, \xi, t)]$$

$$+ i[\mathbf{s} \times \mathbf{q}]^z \frac{1}{2M_N}[H(x, \xi, t) + E(x, \xi, t)], \quad (13)$$

where $\tau = \mathbf{q}^2/4M_N^2$. The first term is independent of the proton spin, and is considered as the phase-space charge density

$$\rho_+(\mathbf{r}, x) = \int \frac{d^3\mathbf{q}}{(2\pi)^3} e^{-i\mathbf{q}\cdot\mathbf{r}}[H(x, \xi, t) - \tau E(x, \xi, t)] \ . \quad (14)$$

The second term depends on the proton spin and can be regarded as the third component of the phase-space vector current $j_+^z(\mathbf{r}, x) =$

$$\int \frac{d^3\mathbf{q}}{(2\pi)^3} e^{-i\mathbf{q}\cdot\mathbf{r}} i[\mathbf{s} \times \mathbf{q}]^z \frac{1}{2M_N}[H(x, \xi, t) + E(x, \xi, t)] \ . \quad (15)$$

The E-term generates a convection current due to the orbital angular momentum of massless quarks and vanishes when all quarks are in the s-orbit. The physics in separating f_γ^+ into ρ_+ and j_+^z can be seen from the Dirac matrix γ^+ selected by the high-energy probes, which is a combination of time and space components.

4 Quark charge distribution seen through Feynman-momentum (x) filters

Once the GPDs are extracted from experimental data or lattice QCD calculations, the phase-space charge/current

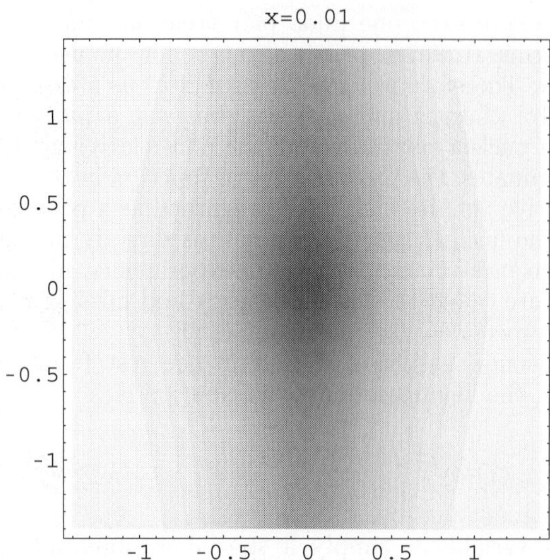

Fig. 1. The up-quark charge density in the proton when Feynman momentum is 0.01

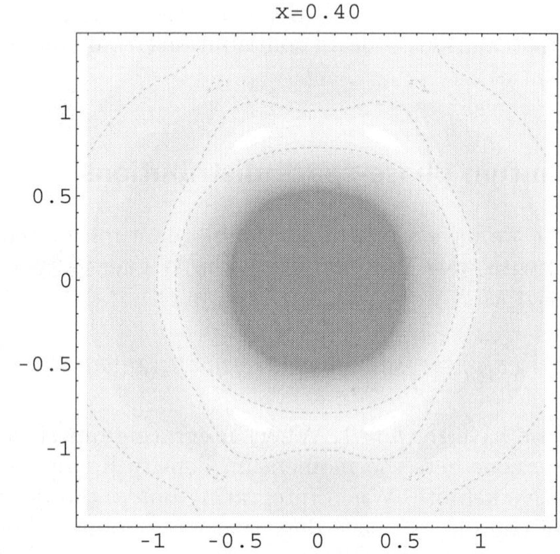

Fig. 2. Same as Fig. 1, the Feynman momentum is 0.4

distributions can be obtained by straightforward Fourier transformations. Without a first-hand knowledge on the GPDs at present, we may be able to learn some general features of the phase-space distributions form GPD models.

The GPDs have been parametrized directly to satisfy various constraints, including 1) the first moments reducing to the measured form factors, 2) the forward limit reproducing the Feynman parton distributions, 3) the x-moments satisfying the polynomiality condition [5], and 4) the positivity conditions [14]. Here, we use a new parametrization without assuming factorized dependence on the t and other variables [3].

In Fig. 1, we show the up-quark charge distributions calculated from $H_u(x, \xi, t)$ for various values of $x = 0.01$. While the intensity of the plots indicates the magnitude of the positive distribution, the lighter areas below the

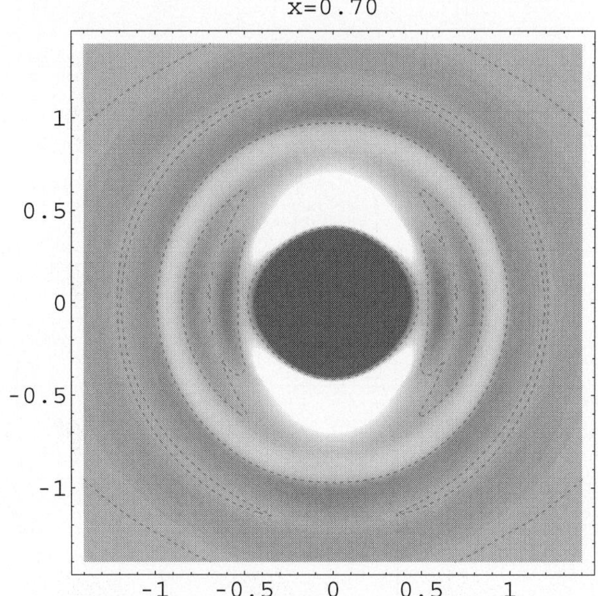

Fig. 3. Same as Fig. 1, the Feynman momentum is 0.7

To summarize, we have introduced the concept of the quantum phase-space distributions for quarks and gluons in the proton. These distributions are measurable through their relations to transverse-momentum dependent parton distributions and generalized parton distributions. They can be used to visualize the phase-space motion of the quarks, and hence allow studying the contribution of the quark orbital angular momentum to the spin of the proton. This work was supported by the U. S. Department of Energy via grant DE-FG02-93ER-40762.

References

1. E.P. Wigner: Phys. Rev. **40**, 749 (1932)
2. M. Hillery, R.F. O'Connell, M.O. Scully, and E.P. Wigner: Phys. Rept. **106**, 121 (1984); H.-W. Lee: Phys. Rept. **259**, 147 (1995)
3. X. Ji: hep-ph/0304037, to appear in Phys. Rev. Lett. (2003); A.V. Belitsky, X. Ji, and F. Yuan: hep-ph/0307383
4. See, e.g., J.C. Collins: Acta Phys. Polon. B **34**, 3103 (2003) and references therein
5. X. Ji: J. Phys. G **24**, 1181 (1998); K. Goeke, M.V. Polyakov, M. Vanderhaeghen: Prog. Part. Nucl. Phys. **47**, 401 (2001); A.V. Radyushkin: hep-ph/0101225; A.V. Belitsky, D. Müller, and A. Kirchner: Nucl. Phys. B **629**, 323 (2002)
6. M. Burkardt: Phys. Rev. D **62**, 071503 (2000); J.P. Ralston and B. Pire: Phys. Rev. D **66**, 111501 (2002); A.V. Belitsky and D. Müller: Nucl. Phys. A **711**, 118 (2002); M. Diehl: Eur. Phys. J. C **25**, 223 (2002)
7. D. Yennie and M. Ravenhall: M. Levy, Rev. Mod. Phys. **29**, 144 (1957)
8. E.J. Ernst, R.G. Sachs, and K.C. Wali: Phys. Rev. **119**, 1105 (1960); R.G. Sachs: Phys. Rev. **126**, 2256 (1962)
9. X. Ji and R. McKeown: Phys. Lett. B **236**, 130 (1990)
10. S.J. Brodsky, H.-C. Pauli, and S.S. Pinsky: Phys. Rept. **301**, 299 (1998); M. Burkardt: Adv. Nucl. Phys. **23**, 1 (1996)
11. J.C. Collins: Phys. Lett. B **536**, 43 (2002)
12. A.V. Belitsky, X. Ji, and F. Yuan: Nucl. Phys. B **656**, 165 (2003)
13. X. Ji: Phys. Rev. Lett. **78**, 610 (1997); Phys. Rev. D **55**, 7114 (1997)
14. P.V. Pobylitsa: Phys. Rev. D **66**, 094002 (2002)

ground-zero contours indicate negative values. The image is rotationally symmetric in the $\mathbf{r}_\perp$-plane (shown as the horizontal axis only). At small x, the distribution extends far beyond the nominal proton size along the vertical z direction. The physical explanation for this is that the position space uncertainty of the quarks is large when x is small, and therefore the quarks are de-localized along the longitudinal direction. This de-localization reflects a very peculiar part of the proton wave function and shows long-range correlations as verified in high-energy scattering. In a nucleus, the parton distributions at small x are strongly modified because of the spatial 1overlap between the protons. Figure 2 shows the charge density at $x = 0.4$ which is roughly round. At larger x, the momentum along z direction is of order proton mass, the quarks are localized to within $1/M_N$. The quantum mechanical nature of the distribution becomes distinct as there are significant changes in the sign at different spatial regions, shown in Fig. 3. There pictures provide a fantastic visualization of the quarks in the proton.

Eur Phys J A (2004) **19**, s01, 29–34

Digital Object Identifier (DOI) 10.1140/epjad/s2004-03-005-1

EPJ A direct

electronic only

Structure of the nucleon from electromagnetic form factors

F. Iachello[1,2]

[1] ECT*, I-38050 Villazzano (Trento), Italy
[2] Center for Theoretical Physics, Sloane Physics Laboratory, Yale University, New Haven, Connecticut 06520-8120, USA

Received: 4 July 2003 / Accepted: 14 Nov 2003 /
Published Online: 6 Feb 2004 – © Società Italiana di Fisica / Springer-Verlag 2004

Abstract. Recent experimental data on the ratio of electric to magnetic elastic form factors are reviewed in light of a model of the nucleon with an intrinsic (quark-like) structure and a meson cloud. The analysis points to the astonishing result that the proton *electric* form factor *vanishes* at $Q^2 \sim 8$ $(\mathrm{GeV/c})^2$ and becomes negative beyond that point. The intrinsic structure is estimated to have a r.m.s. radius of ~ 0.34 fm, much smaller than the proton r.m.s. radius ~ 0.87 fm. The calculations are in perfect agreement with the proton data, but deviate drastically from neutron data at $Q^2 > 1$ $(\mathrm{GeV/c})^2$. Relativistic invariance is a crucial ingredient responsible for the vanishing of G_{E_p}. Symmetry, rather than detailed dynamics, appears to be a determining factor in the structure of the nucleon. Scaling appears to occur at much larger values, $Q^2 \geq 30$ $(\mathrm{GeV/c})^2$, than previously thought.

PACS. 25.30.Bf Elastic electron scattering – 13.40.Gp Electromagnetic form factors – 14.20.Dh Protons and neutrons – 24.85.+p Quarks, gluons, and QCD in nuclei and nuclear processes

1 Introduction

Electromagnetic form factors have played a crucial role in understanding the structure of composite particles. A particularly important composite particle is the nucleon, which forms the basis upon which all matter is built. Studies of the structure of the nucleon with electromagnetic probes begun in the late 50's and early 60's when Hofstadter and collaborators demonstrated that the nucleon was not point-like with a (proton) root-mean square radius $\langle r_p^2 \rangle^{1/2} \sim 0.75$ fm. In the 1970's many experiments were performed, showing that the neutron was a complex particle with a negative r.m.s. radius and $dG_{E_n}/d(Q^2) \sim 0.50$ $(\mathrm{GeV/c})^2$. In 1973, it was suggested that the nucleon has a two component structure with an intrinsic part with form factor $g(Q^2)$ and a meson cloud parametrized in terms of vector mesons, (ρ, ω, φ). In the late 1970's the non-relativistic quark-model was used to describe the properties of hadrons. It was soon realized that this model cannot describe form factors in a consistent way. Also in the late 1970's, QCD emerged as the theory of strong interactions. In a perturbative approach, p-QCD, the asymptotic behavior of the form factors can be derived, yielding the large Q^2 behavior of the nucleon form factors to be $\propto \frac{1}{Q^4}$. Also in the 1980's, experimental groups noted that all form factors, except G_{E_n}, could be described by the empirical dipole form $G_D(Q^2) \propto 1/\left(1 + \frac{Q^2}{0.71}\right)^2$. These observations culminated in the SLAC experiment NE11 on the ratio $\mu_p G_{E_p}/G_{M_p}$ that appeared to be consistent with scaling

up to 10 $(\mathrm{GeV/c})^2$ [1]. However, in 2000-2002 experiments performed at TJNAF [2], [3] using the recoil polarization method have shown the astounding result that the ratio of proton electric to proton magnetic form factor decreases dramatically with Q^2, inconsistent with scaling. In this article, the present situation on electromagnetic form factors of the nucleon will be reviewed.

2 Analysis of form factors

Two basic principles play a crucial role in the analysis of electromagnetic form factors of the nucleon. The first of these is relativistic invariance. This principle fixes the form of the nucleon current to be[4]

$$J^\mu = F_1(Q^2)\gamma^\mu + \frac{\kappa}{2M_N}F_2(Q^2)i\sigma^{\mu\nu}q_\nu \qquad (1)$$

where $F_1(Q^2)$ and $F_2(Q^2)$ are the so-called Dirac and Pauli form factors and κ is the anomalous magnetic moment. This symmetry is expected to be exact. The second is isospin invariance. Although this symmetry is not exact, being of dynamical origin, it is expected to be only slightly broken in a realistic theory of strong interaction. Isospin invariance leads to the introduction of isoscalar, F_1^S and F_2^S, and isovector, F_1^V and F_2^V, form factors, and hence to relations among proton and neutron form factors. The observed Sachs form factors, G_E and G_M can be obtained by the relations

$$G_{M_p} = \left(F_1^S + F_1^V\right) + \left(F_2^S + F_2^V\right)$$
$$G_{E_p} = \left(F_1^S + F_1^V\right) - \tau\left(F_2^S + F_2^V\right)$$
$$G_{M_n} = \left(F_1^S - F_1^V\right) + \left(F_2^S - F_2^V\right)$$
$$G_{E_n} = \left(F_1^S - F_1^V\right) - \tau\left(F_2^S - F_2^V\right) \qquad (2)$$

with $\tau = Q^2/4M_N^2$. These relations also satisfy another constraint, namely the kinematical constraint $G_E(-4M_N^2) = G_M(-4M_N^2)$. This constraint is of crucial importance in the time-like region, while playing a minor role in the space-like region.

Different models of the nucleon correspond to different assumptions for the Dirac and Pauli form factors. In 1973 [5] a model of the nucleon in which the external photon couples to both an intrinsic structure, described by the form factor $g(Q^2)$, and a meson cloud, treated within the framework of vector meson (ρ, ω and φ) dominance, was suggested. In this model the Dirac and Pauli form factors are parametrized as

$$F_1^S(Q^2) = \frac{1}{2}g(Q^2)[(1 - \beta_\omega - \beta_\varphi)$$
$$+ \beta_\omega \frac{m_\omega^2}{m_\omega^2 + Q^2} + \beta_\varphi \frac{m_\varphi^2}{m_\varphi^2 + Q^2}] \qquad (3)$$

$$F_1^V(Q^2) = \frac{1}{2}g(Q^2)[(1 - \beta_\rho) + \beta_\rho \frac{m_\rho^2}{m_\rho^2 + Q^2}]$$

$$F_2^S(Q^2) = \frac{1}{2}g(Q^2)[(-0.120 - \alpha_\varphi)\frac{m_\omega^2}{m_\omega^2 + Q^2}$$
$$+ \alpha_\varphi \frac{m_\varphi^2}{m_\varphi^2 + Q^2}] \qquad (4)$$

$$F_2^V(Q^2) = \frac{1}{2}g(Q^2)[3.706\frac{m_\rho^2}{m_\rho^2 + Q^2}] \qquad (5)$$

In [5] three forms of the intrinsic form factor $g(Q^2)$ were used. The best fit was obtained for $g(Q^2) = (1 + \gamma Q^2)^{-2}$. This form will be used in the remaining part of this talk. Before comparing with the data, an additional modification is needed. In view of the fact that the ρ meson has a non-negligible width, one needs to replace

$$\frac{m_\rho^2}{m_\rho^2 + Q^2} \to \frac{m_\rho^2 + 8\Gamma_\rho m_\pi/\pi}{m_\rho^2 + Q^2 + (4m_\pi^2 + Q^2)\,\Gamma_\rho\alpha(Q^2)/m_\pi} \qquad (6)$$

where

$$\alpha(Q^2) = \frac{2}{\pi}\left[\frac{4m_\pi^2 + Q^2}{Q^2}\right]^{1/2} \ln\left(\frac{\sqrt{4m_\pi^2 + Q^2} + \sqrt{Q^2}}{2m_\pi}\right). \qquad (7)$$

This replacement is important for small Q^2, although, because of the logarithm dependence of the $\pi\pi$ cut expressed by the function $\alpha(Q^2)$, its effect is felt even at moderate and large Q^2.

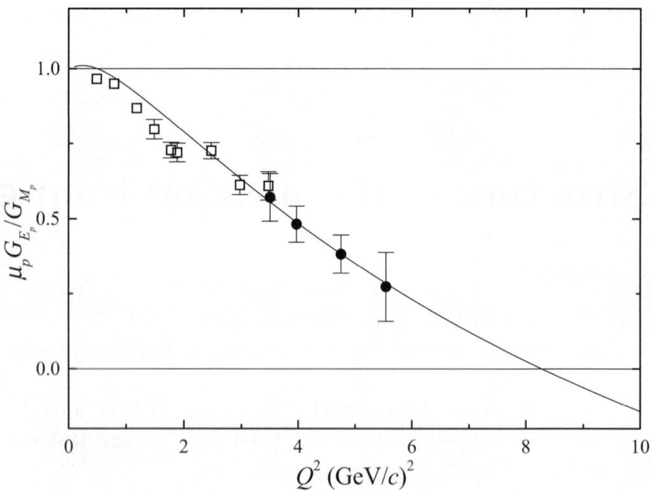

Fig. 1. The measured ratio $\mu_p G_{E_p}/G_{M_p}$ compared with the 1973 prediction [2]: open square [3]: filled circle

2.1 The ratio of electric to magnetic form factors of the proton

By using the coupling constants given in Table 1 of [5] $\beta_\rho = 0.672, \beta_\omega = 1.102, \beta_\varphi = 0.112, \alpha_\varphi = -0.052$, an intrinsic form factor with $\gamma = 0.25$ $(\text{GeV}/c)^{-2}$, standard values of the masses ($m_\rho = 0.765$ GeV, $m_\omega = 0.784$ GeV, $m_\varphi = 1.019$ GeV), and a ρ width $\Gamma_\rho = 0.112$ GeV, one can calculate the ratio $\mu_p G_{E_p}/G_{M_p}$. The result is shown against the new data [2], [3] in Fig. 1. The agreement is astonishing. Figure 1 also shows the remarkable result that the *electric* form factor of the proton crosses zero at $Q^2 \sim 8$ $(\text{GeV}/c)^2$. It would be ot utmost importance to measure the ratio $\mu_p G_{E_p}/G_{M_p}$ at $Q^2 \geq 6$ $(\text{GeV}/c)^2$. A measurement of the zero of the electric form factor, adding to the already measured sharp drop from 1 at $Q^2 = 0$ to ~ 0.27 at $Q^2 = 5.6$ $(\text{GeV}/c)^2$, would unequivocably establish the complex nature of the nucleon. In the model put forward in 1973, the nucleon has both an intrinsic structure (presumably three valence quarks) and additional contributions (presumably $q\bar{q}$ pairs). (The complex nature of the nucleon resulting from electromagnetic form factors is in accord with results obtained by the *EMC* collaboration [6], where the additional, non q^3, components were attributed to gluons.) An estimate of the spatial extent of the intrinsic region (where the fundamental quarks sit) can be obtained from the value of γ in the intrinsic form factor. The r.m.s. of this distribution is ~ 0.34 fm, much smaller that the proton r.m.s. radius ~ 0.87 fm. The zero in the *electric* form factor is a consequence of the two term structure of (2), in particular of the fact that the second term is multiplied by $-Q^2/4M_N^2$. Any model with a two term structure will produce results in qualitative agreement with data. Indeed three of the descriptions considered in [3], a soliton model [7], and two relativistic constituent quark models [8], [9] have this structure and produce results in qualitative agreement with experiment. Also the introduction of relativity in non-relativistic quark models goes in the di-

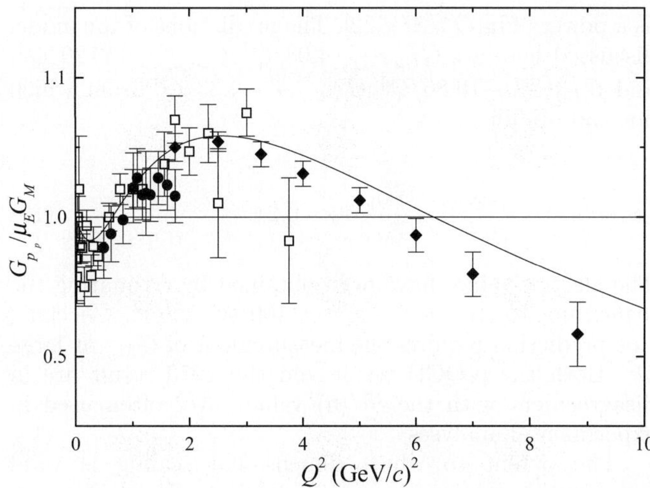

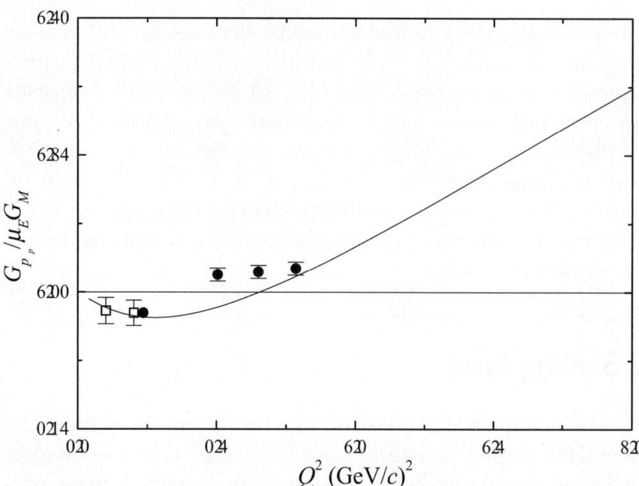

Fig. 2. Experimental values $G_{M_p}/\mu_p G_D$ compared with calculation [11]: open square [12]: filled circle [1]: filled diamond

Fig. 3. Recent experimental values for $G_{M_n}/\mu_n G_D$ compared with calculation [13]: open square [14]: filled circle

rection of reducing the ratio [10]. To discriminate between various models it is necessary to find precisely at which value the zero occurs.

2.2 The magnetic form factor of the proton

The agreement between theory and data for the proton form factors is not limited to the ratio $\mu_p G_{E_p}/G_{M_p}$. Consider the magnetic form factor, G_{M_p}. For convenience of display, normalize it to the so-called dipole form factor, $G_D = (1 + \frac{Q^2}{0.71})^{-2}$. The data [11], [12], [1] in the interval $0 \le Q^2 \le 10$ (GeV/c)^2 are plotted in Fig. 2. They show an ondulation, crossing the value one at $Q^2 \sim 0.6$ (GeV/c)^2 and again at ~ 6 (GeV/c)^2. The calculation is in excellent agreement with the data, with crossing points at precisely the same values ~ 0.6 and 6 (GeV/c)^2. The observed ondulation is proof that vector meson (with masses $\mu^2 \sim 0.5 - 1.0$ (GeV/c)^2) components are important. Without ρ meson component, the form factor should behave smoothly (see Fig. 3 of [5]).

2.3 The magnetic form factor of the neutron

Having established the structure of the proton, I now come to that of the neutron. This is dictated by isospin invariance. Measurements of the neutron form factors are obscured by the knowledge of the wave functions of deuterons or He^3. Older measurements are either in disagreement (for $Q^2 > 1$ (GeV/c)^2) or in marginal agreement ($Q^2 < 1$ (GeV/c)^2) with the 1973 model. However, the situation here appears to be similar to the situation for the proton form factors previous to the experiments of Jones et al [2] and Gayou et al [3]. I consider first the region $Q^2 \le 1$ (GeV/c)^2. An analysis (2001) of recent experiments by J. Golak et al [13] and by H. Anklin et al [14] shows that the new data for G_{M_n}/G_D points to an ondulation with crossing point at ~ 0.6 (GeV/c)^2 as predicted by isospin invariance, and (2). This ondulation was

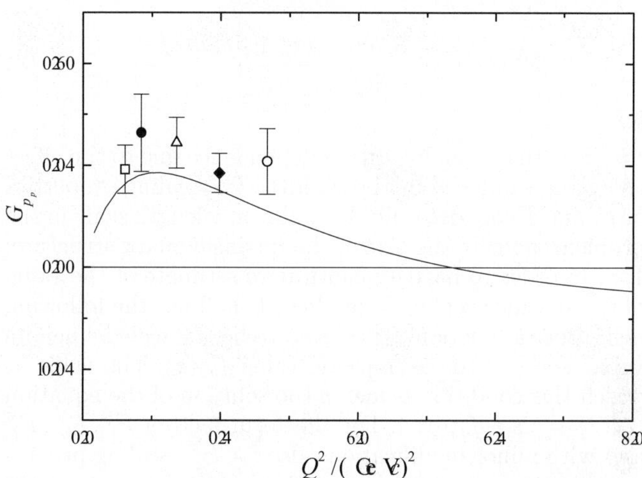

Fig. 4. Recent experimental values for G_{E_n} compared with calculation [15]: open square [16]: filled circle [17]: filled diamond [18]: open up triangle [19]: open circle

absent in the old data. A comparison between the new data and the calculation is shown in Fig. 3. For $Q^2 \ge 1$ (GeV/c)^2 the calculation is in disagreement with the old data. While the data remain close to 1, the calculation keeps increasing. New (unpublished) data at TJNAF appear to indicate that G_{M_n}/G_D does not increase as Q^2 increases. If these data are confirmed, one must conclude that either isospin invariance is broken above 1 (GeV/c)^2 or that there are additional components in the neutron that are not present in the proton.

2.4 The electric form factor of the neutron

A similar situation occurs for new (1999) data for the electric form factor G_{E_n} by Herberg et al [15], Passchier et al [16], Ostrick et al [17], Rohe et al [18], Zhu et al [19]. These are in fair agreement with the calculation as shown in Fig. 4. For $Q^2 \ge 1$ (GeV/c)^2 the calculation is in disagreement with new unpublished data. While the data remain close to 0.05, the calculation keeps decreasing and crosses zero at ~ 1.4 (GeV/c)^2. It would be of the utmost importance to measure G_{M_n} and G_{E_n} at $Q^2 \geq 1 GeV^2$ in a as much as possible model independent way. A measurement of the ratio $\mu_n G_{E_n}/G_{M_n}$ similar to that done for the proton, perhaps using the reaction $d(\vec{e}, e'\vec{n})p$ [20], will be of great value. Similar observations can be made for G_{E_n}. In present analyses this form factor is even more sensitive to models than G_{M_n}.

3 Scaling laws

Another important question is the extent to which the new data support scaling laws [21]. The parametrization of (3) is consistent with scaling laws expected from perturbative QCD, $F_1 \sim 1/Q^4$, $F_2 \sim 1/Q^6$ except for F_2^V whose asymptotic behavior ($Q^2 \to \infty$) is

$$F_2^V(Q^2) \to \frac{3.706}{2\gamma^2 Q^6} \frac{m_\rho^2 + 8\Gamma_\rho m_\pi/\pi}{1 + \frac{\Gamma_\rho}{m_\pi}\frac{2}{\pi}\ln 2\sqrt{\frac{Q^2}{4m_\pi^2}}}, \tag{8}$$

that is with a weak logarithm dependence due to the effective ρ mass induced by the ρ width. The scaling properties of F_1 and F_2 are determined by the only length scale in the problem, namely the size of the intrinsic quark structure, $1/\gamma$. In order to have a quantitative estimate of the value of Q^2 at which scaling is reached, I shall use the following definition: a function $f(z)$ is said to be $x\%$ scaled when its value is $x\%$ of the asymptotic value $f_{as}(z)$. The value at which this condition is met is the solution of the equation $| f(z) | = x | f_{as}(z) |$. For the form factors F_1^S, F_1^V, F_2^S and with minor modifications also for F_2^V, scaling properties are determined by the function $g(Q^2)$. Using the value $\gamma = 0.25$ (GeV/c)$^{-2}$, one obtains an estimate of scaling properties. The function $g(Q^2)$ is 80% scaled at $Q^2 \geq 34$ (GeV/c)^2. This value is much larger than conventionally believed, $Q^2 \sim 4$ (GeV/c)^2. (The dipole form $G_D(Q^2)$ is 80% scaled at $Q^2 \sim 6$ (GeV/c)^2.) The situation for the scaling properties of the form factors G_E and G_M is more complex. The parametrization of (3) is consistent, apart from a weak logarithm dependence, with the scaling laws of perturbative QCD, $G_E \sim G_M \sim 1/Q^4$. However, relativity introduces here another scale, $4M_N^2 = 3.52$ (GeV/c)^2, and, independently from the actual value of the size scale γ, relativistic invariance requires that scaling is not reached unless Q^2 is greater than a few times $4M_N^2$. (This is particularly so for the *electric* form factors). To check scaling properties it would be of utmost importance to measure the ratio $\mu_p G_{E_p}/G_{M_p}$ with the recoil polarization method beyond 10 (GeV/c)^2.

Another prediction from perturbative QCD is that the ratio G_{M_p}/G_{M_n} approaches zero from the negative side for large Q^2,

$$\frac{G_{M_p}}{G_{M_n}} \to 0^- \tag{9}$$

as a power of $\ln(Q^2/\Lambda^2)$ [22]. The predictions of the model discussed here are $G_{Ep} \to -4.08/Q^4$, $G_{M_p} \to 0.9120/Q^4$, and $G_{E_n} \to -10.86/Q^4$, $G_{M_n} \to -4.33/Q^4$ from which one can obtain

$$\frac{G_{M_p}}{G_{M_n}} \to -0.21. \tag{10}$$

The electric values have been obtained by estimating the logarithm dependence at $Q^2 = 100$ (GeV/c)^2. Checking this prediction requires the measurement of G_{M_n} at large Q^2. Both the p-QCD result and the 1973 result are in disagreement with the $SU(6)$ value $-3/2$ often used in experimental analyses.

The extent to which dimensional scaling is valid has been in recent years the subject of many investigations [23]. It has been suggested that the appropriate scaling variable is $QF_{2p}(Q^2)/F_{1p}(Q^2)$ instead of $Q^2 F_{2p}(Q^2)/F_{1p}(Q^2)$. Using (3) one can easily calculate $QF_{2p}(Q^2)/F_{1p}(Q^2)$. From this calculation one can see that the quantity $QF_{2p}(Q^2)/F_{1p}(Q^2)$ remains flat in the interval $2 \leq Q^2 \leq 10$ (GeV/c)^2 and drops from there on, especially after dimensional scaling is reached at $Q^2 \geq 34$ (GeV/c)^2, Fig. 5. The scaling with Q is thus accidental and appropriate only to the intermediate region.

4 Low-Q^2 behavior

The low-Q^2 behavior can also be analyzed in explicit form by using $\lim_{Q^2 \to 0} \alpha(Q^2) = \frac{2}{\pi}(1 + \frac{1}{3}\frac{Q^2}{4m_\pi^2})$. From the slopes of the form factors at $Q^2 \to 0$, one can calculate the mean square radii, defined as $\langle r^2 \rangle = -6\frac{dG}{d(Q^2)}$. The resulting r.m.s. radii are: $\langle r^2 \rangle_{E_p}^{1/2} = 0.817$ fm, $\langle r^2 \rangle_{M_p}^{1/2} = 0.826$ fm and $\langle r^2 \rangle_{M_n}^{1/2} = 0.839$ fm. The neutron electric form factor has a slope at $Q^2 \to 0$ of $\frac{dG_{E_n}}{d(Q^2)} = 0.500$ (GeV/c)^2. These values are in agreement with old experimental data. Accurate proton values are of crucial importance for the interpretation of other experiments, such as muon $g - 2$. The calculated proton electric r.m.s. radius is in disagreement with recent Lamb shift measurements in hydrogen. The origin of this discrepancy must be investigated.

5 Stability against perturbations

In conclusion, the new data clearly point out that the structure of the proton is rather complex and that it contains at least two components. The data appear to be in agreement in the entire measured range with a calculation in which the two component are an intrinsic structure, presumably q^3, and a meson cloud, $q^3 q\bar{q}$, the latter being expressed through vector mesons (ρ, ω, φ). The situation for the neutron is different. The new data are in agreement with the 1973 calculation up to 1 (GeV/c)^2. From there on, they appear to be in disagreement with the new (unpublished) data [24]. One can inquire whether addition of

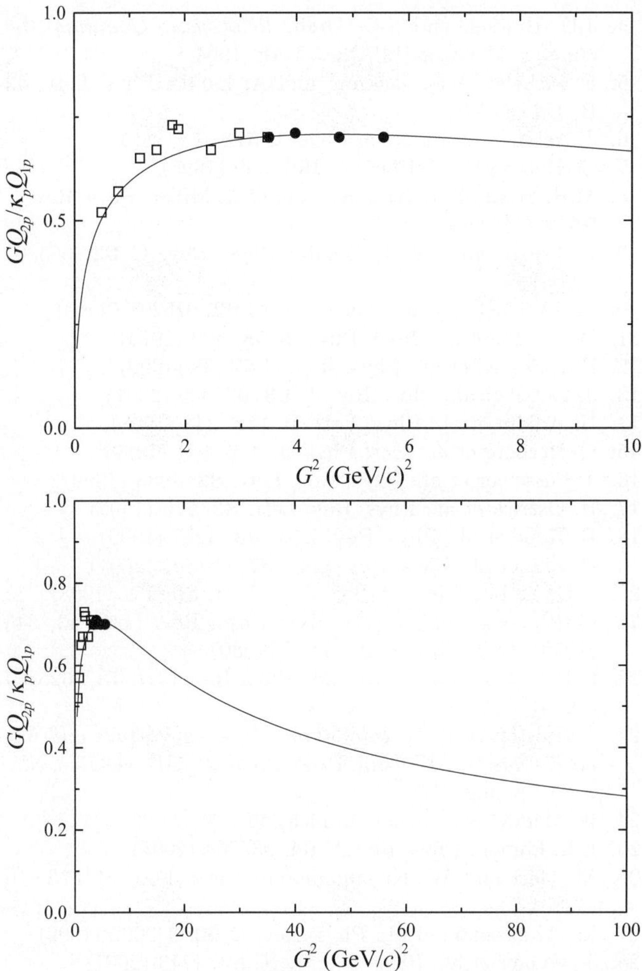

Fig. 5. The experimental ratio QF_{2_p}/F_{1_p} compared with calculation in the range $0 \leq Q^2 \leq 10$ (GeV/c)^2 (top) and $0 \leq Q^2 \leq 100$ (GeV/c)^2 (bottom) [2]: open square [3]: filled circle

other ingredients changes this conclusion. There are three contributions that can be analyzed easily.

(i) The role of additional vector mesons, $\rho(1450), \omega(1390), \varphi(1680)$ [25].

(ii) The addition of an intrinsic piece to the Pauli form factor F_2^V. This can be done by the replacement

$$3.706 \frac{m_\rho^2}{m_\rho^2 + Q^2} \rightarrow (3.706 - \alpha_\rho)\frac{1}{(1 + \gamma Q^2)} + \alpha_\rho \frac{m_\rho^2}{m_\rho^2 + Q^2} \tag{11}$$

The additional piece must be of this type to insure the proper behavior of F_2^V for $Q^2 \rightarrow 0$ and $Q^2 \rightarrow \infty$.

(iii) The role of the widths of ω, φ as well as the effect of changing the width of the ρ meson from the value used in [5].

The *qualitative* features are not affected by these changes, although quantitatively one can make some improvements on the form factor of the neutron. However, because of isospin invariance, an improvement in the neutron form factors produces a deterioration in the descrip-

tion of the proton data. It does not appear that the problem of the neutron form factor at large Q^2 can be solved with these changes. To solve this problem one needs to introduce terms which act only on the neutron, that is terms with $F_S = -F_V$. Work in this direction is in progress.

One can also check whether the logarithm dependence of pertubative QCD

$$Q^2 \rightarrow Q^2 \frac{\ln\left[\left(\Lambda^2 + Q^2\right)/\Lambda_{QCD}^2\right]}{\ln\left[\Lambda^2/\Lambda_{QCD}^2\right]} \tag{12}$$

with $\Lambda = 2.27$ GeV/c and $\Lambda_{CQD} = 0.29$ GeV/c [26] produces major changes in the conclusions. This does not appear to be the case at least up to $Q^2 = 10$ (GeV/c)^2.

6 Consequences of the new experiment

Finally, the experimental results of Jones et al. [2] and Gayou et al. [3], confirming the model calculation of [5], has implications for all hadronic physics.

6.1 Time-like form factors

By an appropriate analytic continuation in the complex plane, the form factor of (3) can be used to analyze form factors in the time-like region. These can be and have been experimentally obtained in the reactions $p\bar{p} \rightarrow e^+e^-$ and $e^+e^- \rightarrow p\bar{p}$. A simple analytic continuation of the intrinsic form factor, $g(Q^2)$, into

$$g(Q^2) = \frac{1}{(1 + \gamma e^{i\theta} Q^2)^2} \tag{13}$$

with $\theta = \pi/4$ appear to indicate that the form factors (3) are in agreement with the data [27].

6.2 Inelastic form factors

The two component structure of the nucleon will reflect itself also in the inelastic form factors. A calculation of the form factor factor $ep \rightarrow e\Delta(1232)$ is in progress.

6.3 Other hadronic form factors

A situation similar to that observed in the nucleon appears to occur also in other hadrons. A calculation of the pion form factor in the two component framework

$$F_\pi(Q^2) = g(Q^2)[(1 - \beta_\rho) + \beta_\rho \frac{m_\rho^2}{m_\rho^2 + Q^2}] \tag{14}$$

with

$$g(Q) = \frac{1}{(1 + \gamma Q^2)} \tag{15}$$

appears to be in excellent agreement with recent experiment [28].

7 Conclusions

The main conclusions that one can draw from the analysis of recent experimental data on electromagnetic form factors are:

(i) the proton appears to have a complex structure with at least two components, an intrinsic component (valence quarks) and a meson cloud ($q\bar{q}$ pairs). The size of the intrinsic structure is r.m.s. ~ 0.34 fm.

(ii) Perturbative QCD is not reached in the proton up to $Q^2 \sim 10$ $(\mathrm{GeV}/c)^2$. Physics up to this scale is dominated by a mixture of hadronic and quark components.

(iii) Symmetry (in particular relativistic invariance), rather than detailed dynamics, appears to be the determining factor in the structure of the proton.

The situation appears to be different for the neutron. Here recent experimental data up to 1 $(\mathrm{GeV}/c)^2$ are consistent with isospin invariance and the structure of the proton, while preliminary data at $Q^2 \geq 1$ $(\mathrm{GeV}/c)^2$ appear to indicate that either isospin invariance is broken or that additional components play a role. It would be of the utmost importance to understand this discrepancy.

Acknowledgements. This work was performed in part under DOE Grant No. DE-FG-02-91ER40608. I wish to thank A.D. Jackson and A. Lande for discussions.

References

1. L. Andivahis et al.: Phys. Rev. D **50**, 5491 (1994)
2. M.K. Jones et al.: Phys. Rev. Lett. **84**, 1398 (2000)
3. O. Gayou et al.: Phys. Rev. Lett. **88**, 092301 (2002)
4. J.D. Bjorken and S.D. Drell: *Relativistic Quantum Mechanics*, McGraw-Hill, New York, 1964
5. F. Iachello, A.D. Jackson, and A. Lande: Phys. Lett. **43** B, 191 (1973)
6. J. Ashman et al.: Nucl. Phys. B **328**, 1 (1989)
7. G. Holzwarth: Z. Phys. A **356**, 339 (1996)
8. M.R. Frank, B.J. Jennings, and G.A. Miller: Phys. Rev. C **54**, 920 (1996)
9. F. Cardarelli and S. Simula: Phys. Rev. C **62**, 065201 (2000)
10. M. DeSanctis et al.: Phys. Rev. C **62**, 025208 (2000)
11. W. Bartel et al.: Nucl. Phys. B **58**, 429 (1973)
12. P.E. Bosted et al.: Phys. Rev. C **42**, 38 (1990)
13. J. Golak et al.: Phys. Rev. C **63**, 034006 (2001)
14. H. Anklin et al.: Phys. Lett. B **428**, 248 (1998)
15. C. Herberg et al.: Eur. Phys. J. A **5**, 131 (1999)
16. J. Passchier et al.: Phys. Rev. Lett. **82**, 4988 (1999)
17. H. Ostrick et al.: Phys. Rev. Lett. **83**, 276 (1999)
18. D. Rohe et al.: Phys. Rev. Lett. **83**, 4257 (1999)
19. H. Zhu et al.: Phys. Rev. Lett. **87**, 081801 (2001)
20. B.D. Milbrath et al.: Phys. Rev. Lett. **80**, 452 (1998)
21. G.P. Lepage and S.J. Brodky: Phys. Rev. Lett. **43**, 545 (1979); Phys. Rev. D **22**, 2157 (1980)
22. G.P. Lepage and S.J. Brodky: Phys. Rev. Lett. **43**, 1625(E) (1979)
23. J. Ralston et al.: in *Interactions between Particle and Nuclear Physics*, AIP Conf. Proc. No. 549 (AIP, Melville, NY, 2000), p. 302
24. R. Madey: private communication
25. E.L. Lomon: Phys. Rev. C **64**, 035204 (2001)
26. M. Gari and W. Krumpelmann: Phys. Lett. B **173**, 10 (1986)
27. M. Ambrogiani et al.: Phys. Rev. D **60**, 032002 (1999)
28. J. Volmer et al.: Phys. Rev. Lett. **86**, 1713 (2001)

Eur Phys J A (2004) **19**, s01, 35–42
Digital Object Identifier (DOI) 10.1140/epjad/s2004-03-006-0

EPJ A direct
electronic only

Structure of the nucleon in chiral perturbation theory

Thomas Fuchs[1] [a], Jambul Gegelia[1,2] [b], and Stefan Scherer[1]

[1] Institut für Kernphysik, Johannes Gutenberg-Universität, D-55099 Mainz, Germany
[2] High Energy Physics Institute, Tbilisi State University, University St. 9, 380086 Tbilisi, Georgia

Received: 23 Sept 20003 / Accepted: 14 Nov 2003 /
Published Online: 6 Feb 2004 – © Società Italiana di Fisica / Springer-Verlag 2004

Abstract. We discuss a renormalization scheme for relativistic baryon chiral perturbation theory which provides a simple and consistent power counting for renormalized diagrams. The method involves finite subtractions of dimensionally regularized diagrams beyond the standard modified minimal subtraction scheme of chiral perturbation theory to remove contributions violating the power counting. This is achieved by a suitable renormalization of the parameters of the most general effective Lagrangian. As applications we discuss the mass of the nucleon, the σ term, and the scalar and electromagnetic form factors.

PACS. 12.39.Fe Chiral Lagrangians – 11.10.Gh Renormalization – 13.40.Gp Electromagnetic form factors

1 Introduction

Starting from Weinberg's pioneering work [1], the application of effective field theory (EFT) to strong interaction processes has become one of the most important theoretical tools in the low-energy regime. The basic idea consists of writing down the most general possible Lagrangian, including *all* terms consistent with assumed symmetry principles, and then calculating matrix elements with this Lagrangian within some perturbative scheme [1]. A successful application of this program thus requires two main ingredients:

(1) a knowledge of the most general effective Lagrangian;
(2) an expansion scheme for observables in terms of a consistent power counting method.

The structure of the most general Lagrangian for both mesonic and baryonic chiral perturbation theory (ChPT) has been investigated for almost two decades. The number of terms in the momentum and quark-mass expansion is given by

$$\underbrace{2}_{\mathcal{O}(q^2)} + \underbrace{10 + 2}_{\mathcal{O}(q^4)} + \underbrace{90 + 4 + 23}_{\mathcal{O}(q^6)} + \cdots$$

for mesonic ChPT [SU(3)×SU(3)] [2,3] and

$$\underbrace{2}_{\mathcal{O}(q)} + \underbrace{7}_{\mathcal{O}(q^2)} + \underbrace{23}_{\mathcal{O}(q^3)} + \underbrace{118}_{\mathcal{O}(q^4)} + \cdots$$

for baryonic ChPT [SU(2)×SU(2)×U(1)] [4,5,6,7,8]. Moreover, the mesonic sector contains at $\mathcal{O}(q^4)$ the Wess-Zumino-Witten action [9,10] taking care of chiral anomalies.

Once the most general effective Lagrangian is known, one needs an expansion scheme in order to perform perturbative calculations of physical observables. In this context one faces the standard difficulties of encountering ultraviolet divergences when calculating loop diagrams. However, since one is working with the most general Lagrangian containing all terms allowed by the symmetries, these infinities can, as part of the renormalization program, be absorbed by a suitable adjustment of the parameters of the Lagrangian [1,11]. Applying dimensional regularization in combination with the modified minimal subtraction scheme of ChPT, in the mesonic sector a straightforward correspondence between the loop expansion and the chiral expansion in terms of momenta and quark masses at a fixed ratio was set up by Gasser and Leutwyler [2]. The situation in the one-nucleon sector turned out to be more complicated [4], since the correspondence between the loop expansion and the chiral expansion seemed to be lost. One of the findings of [4] was that higher-loop diagrams can contribute to terms as low as $\mathcal{O}(q^2)$. A solution to this problem was obtained in the framework of the heavy-baryon formulation of ChPT [12,13] resulting in a power counting analogous to the mesonic sector (for a recent review of ChPT see, e.g., [14]).

Here, we will review some recent efforts to devise a new renormalization scheme leading to a simple and consistent power counting for the renormalized diagrams of a manifestly covariant approach. The basic idea consists in performing additional subtractions of dimensionally regularized diagrams beyond the modified minimal subtraction scheme employed in [4]. As applications we will discuss the mass of the nucleon as well as the scalar and electromagnetic form factors and compare the method with the approach of Becher and Leutwyler [15].

[a] Supported by the Deutsche Forschungsgemeinschaft (SFB 443)
[b] Alexander von Humboldt Research Fellow

Fig. 1. Generic one-loop diagram. The black box denotes some unspecified vertex structure which is irrelevant for the discussion

2 Dimensional regularization

For the regularization of loop diagrams we will make use of dimensional regularization [16], because it preserves algebraic relations between Green functions (Ward identities). We will illustrate the method by considering the following simple example,

$$I(M^2) = \int \frac{d^4k}{(2\pi)^4} \frac{i}{k^2 - M^2 + i0^+}, \quad k^2 = k_0^2 - \mathbf{k}^2, \quad (1)$$

which shows up in the generic diagram of Fig. 1. Naively counting the powers of the momenta, the integral is said to diverge quadratically. In order to regularize (1), we define the integral for n dimensions (n integer) as

$$I_n(M^2, \mu^2) = \mu^{4-n} \int \frac{d^n k}{(2\pi)^n} \frac{i}{k^2 - M^2 + i0^+},$$

where the scale μ ('t Hooft parameter) has been introduced so that the integral has the same dimension for arbitrary n. After a Wick rotation and angular integration, the analytic continuation for complex n reads (see Appendix B of [14] for details)

$$\begin{aligned} I(M^2, \mu^2, n) &= \frac{M^2}{(4\pi)^2} \left(\frac{4\pi\mu^2}{M^2} \right)^{2-\frac{n}{2}} \Gamma\left(1 - \frac{n}{2}\right) \\ &= \frac{M^2}{16\pi^2} \left[R + \ln\left(\frac{M^2}{\mu^2} \right) \right] + O(n-4), \quad (2) \end{aligned}$$

where

$$R = \frac{2}{n-4} - [\ln(4\pi) + \Gamma'(1)] - 1. \quad (3)$$

The idea of renormalization consists of adjusting the parameters of the counterterms of the most general effective Lagrangian so that they cancel the divergences of (multi-) loop diagrams. In doing so, one still has the freedom of choosing a suitable renormalization condition. For example, in the minimal subtraction scheme (MS) one would fix the parameters of the counterterm Lagrangian such that they would precisely absorb the contributions proportional to $2/(n-4)$ in (3), while the modified minimal subtraction scheme ($\overline{\text{MS}}$) would, in addition, cancel the term in square brackets. Finally, in the modified minimal subtraction scheme of ChPT ($\widetilde{\overline{\text{MS}}}$) employed in [2], the seven (bare) coefficients l_i of the $\mathcal{O}(q^4)$ Lagrangian are expressed in terms of renormalized coefficients l_i^r as

$$l_i = l_i^r + \gamma_i \frac{R}{32\pi^2}, \quad (4)$$

where the γ_i are fixed numbers.

3 Mesonic chiral perturbation theory

The starting point of mesonic chiral perturbation theory is a chiral $\text{SU}(2)_L \times \text{SU}(2)_R$ symmetry of the two-flavor QCD Lagrangian in the limit of massless u and d quarks. It is assumed that this symmetry is spontaneously broken down to its isospin subgroup $\text{SU}(2)_V$, i.e., the ground state has a lower symmetry than the Lagrangian. From Goldstone's theorem one expects $6 - 3 = 3$ massless Goldstone bosons which interact "weakly" at low energies, and which are identified with the pions of the "real" world. The explicit chiral symmetry breaking through the quark masses is included as a perturbation. According to the program of EFT the symmetries of QCD are mapped onto the most general effective Lagrangian for the interaction of the Goldstone bosons (pions). The Lagrangian is organized in a derivative and quark-mass expansion [1,2,3]

$$\mathcal{L}_\pi = \mathcal{L}_2 + \mathcal{L}_4 + \mathcal{L}_6 + \cdots, \quad (5)$$

where—in the absence of external fields—the lowest-order Lagrangian is given by [2]

$$\mathcal{L}_2 = \frac{F^2}{4} \text{Tr}\left(\partial_\mu U \partial^\mu U^\dagger \right) + \frac{F^2 M^2}{4} \text{Tr}(U^\dagger + U), \quad (6)$$

with

$$U = \exp\left(i \frac{\tau \cdot \pi}{F} \right)$$

a unimodular unitary (2×2) matrix containing the Goldstone boson fields. In (6), F denotes the pion-decay constant in the chiral limit: $F_\pi = F[1 + \mathcal{O}(\hat{m})] = 92.4$ MeV. Here, we work in the isospin-symmetric limit $m_u = m_d = \hat{m}$, and the lowest-order expression for the squared pion mass is $M^2 = 2B\hat{m}$, where B is related to the quark condensate $\langle \bar{q}q \rangle_0$ in the chiral limit [2].

Using Weinberg's power counting scheme [1] one may analyze the behavior of a given diagram calculated in the framework of (5) under a linear rescaling of all *external* momenta, $p_i \mapsto t p_i$, and a quadratic rescaling of the light quark masses, $m_q \mapsto t^2 m_q$, which, in terms of the Goldstone boson masses, corresponds to $M^2 \mapsto t^2 M^2$. The chiral dimension D of a given diagram with amplitude $\mathcal{M}(p_i, m_q)$ is defined by

$$\mathcal{M}(t p_i, t^2 m_q) = t^D \mathcal{M}(p_i, m_q), \quad (7)$$

where, in n dimensions,

$$D = n N_L - 2 I_\pi + \sum_{k=1}^{\infty} 2k N_{2k}^\pi \quad (8)$$

$$= 2 + (n-2) N_L + \sum_{k=1}^{\infty} 2(k-1) N_{2k}^\pi \quad (9)$$

$$\geq 2 \text{ in 4 dimensions.}$$

Here, N_L is the number of independent loop momenta, I_π the number of internal pion lines, and N_{2k}^π the number of vertices originating from $\mathcal{L}_{2k}$. Clearly, for small enough momenta and masses diagrams with small D, such

Fig. 2. One-loop contribution to the pion self-energy. The number 2 in the interaction blob refers to $\mathcal{L}_2$

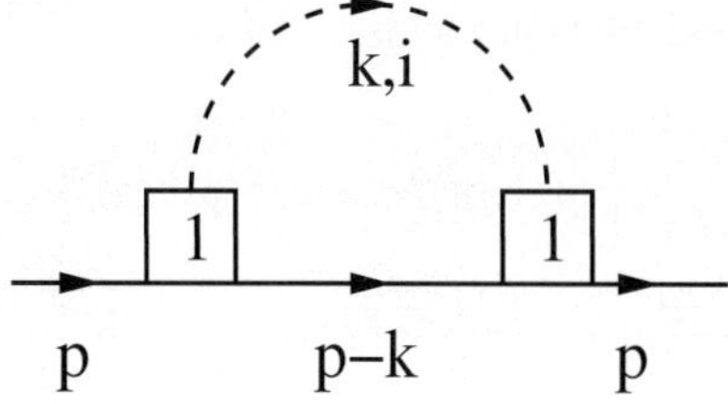

Fig. 3. One-loop contribution to the nucleon self-energy. The number 1 in the interaction blobs refers to $\mathcal{L}_{\pi N}^{(1)}$

as $D = 2$ or $D = 4$, should dominate. Of course, the rescaling of (7) must be viewed as a mathematical tool. While external three-momenta can, to a certain extent, be made arbitrarily small, the rescaling of the quark masses is a theoretical instrument only. Note that, for $n = 4$, loop diagrams are always suppressed due to the term $2N_L$ in (9). In other words, we have a perturbative scheme in terms of external momenta and masses which are small compared to some scale [here $1/(4\pi F)$].

As an example, let us consider the contribution of Fig. 2 to the pion self-energy. According to (8) we expect, in 4 dimensions, the chiral power

$$D = 4 \cdot 1 - 2 \cdot 1 + 2 \cdot 1 = 4.$$

Without going into the details, the explicit result of the one-loop contribution is given by (see, e.g., [14])

$$\Sigma_{\mathrm{loop}}(p^2) = \frac{4p^2 - M^2}{6F^2} I(M^2, \mu^2, n) = \mathcal{O}(q^4),$$

where the integral is given in (2) and is infinite as $n \to 4$. Note that both factors—the fraction and the integral— each count as $\mathcal{O}(q^2)$ resulting in $\mathcal{O}(q^4)$ for the total expression as anticipated.

For a long time it was believed that performing loop calculations using the Lagrangian of (6) would make no sense, because it is not renormalizable (in the traditional sense). However, as emphasized by Weinberg [1, 11], the cancellation of ultraviolet divergences does not really depend on renormalizability; as long as one includes *every one* of the infinite number of interactions allowed by symmetries, the so-called non-renormalizable theories are actually just as renormalizable as renormalizable theories [11]. The conclusion is that a suitable adjustment of the parameters of $\mathcal{L}_4$ [see (4)] leads to a cancellation of the one-loop infinities.

4 Baryonic chiral perturbation theory and renormalization

The extension to processes involving one external nucleon line was developed by Gasser, Sainio, and Švarc [4]. In addition to (5) one needs the most general effective Lagrangian of the interaction of Goldstone bosons with nucleons:

$$\mathcal{L}_{\pi N} = \mathcal{L}_{\pi N}^{(1)} + \mathcal{L}_{\pi N}^{(2)} + \cdots.$$

The lowest-order Lagrangian, expressed in terms of bare fields and parameters denoted by subscripts 0, reads

$$\mathcal{L}_{\pi N}^{(1)} = \bar{\Psi}_0 \left(i\gamma_\mu \partial^\mu - m_0 - \frac{1}{2}\frac{\overset{\circ}{g}_{A0}}{F_0}\gamma_\mu\gamma_5\tau^a\partial^\mu\pi_0^a \right)\Psi_0 + \cdots, \tag{10}$$

where Ψ_0 denotes the (bare) nucleon field with two four-component Dirac fields describing the proton and the neutron, respectively. After renormalization, m and $\overset{\circ}{g}_A$ refer to the chiral limit of the physical nucleon mass and the axial-vector coupling constant, respectively. While the mesonic Lagrangian of (5) contains only even powers, the baryonic Lagrangian involves both even and odd powers due to the additional spin degree of freedom.

Our goal is to propose a renormalization procedure generating a power counting for tree-level and loop diagrams of the (relativistic) EFT which is analogous to that given in [17] (for nonrelativistic nucleons). Choosing a suitable renormalization condition will allow us to apply the following power counting: a loop integration in n dimensions counts as q^n, pion and fermion propagators count as q^{-2} and q^{-1}, respectively, vertices derived from $\mathcal{L}_{2k}$ and $\mathcal{L}_{\pi N}^{(k)}$ count as q^{2k} and q^k, respectively. Here, q generically denotes a small expansion parameter such as, e.g., the pion mass. In total this yields for the power D of a diagram in the one-nucleon sector the standard formula [17, 18]

$$D = nN_L - 2I_\pi - I_N + \sum_{k=1}^{\infty} 2kN_{2k}^\pi + \sum_{k=1}^{\infty} kN_k^N \tag{11}$$

$$= 1 + (n-2)N_L + \sum_{k=1}^{\infty} 2(k-1)N_{2k}^\pi + \sum_{k=1}^{\infty} (k-1)N_k^N \tag{12}$$

≥ 1 in 4 dimensions,

where, in addition to (8), I_N is the number of internal nucleon lines and N_k^N the number of vertices originating from $\mathcal{L}_{\pi N}^{(k)}$. According to (12), one-loop calculations in the single-nucleon sector should start contributing at $\mathcal{O}(q^{n-1})$.

As an example, let us consider the one-loop contribution of Fig. 3 to the nucleon self-energy. According to (11), the renormalized result should be of order

$$D = n \cdot 1 - 2 \cdot 1 - 1 \cdot 1 + 1 \cdot 2 = n - 1. \tag{13}$$

An explicit calculation yields

$$\Sigma_{\text{loop}} = -\frac{3\overset{\circ}{g}_{A0}^{2}}{4F_0^2}\left\{ (\not{p}+m)I_N + M^2(\not{p}+m)I_{N\pi}(-p,0) \right.$$

$$\left. -\frac{(p^2-m^2)\not{p}}{2p^2}\left[(p^2-m^2+M^2)I_{N\pi}(-p,0) + I_N - I_\pi\right]\right\},$$

where the relevant loop integrals are defined as

$$I_\pi = \mu^{4-n}\int\frac{d^nk}{(2\pi)^n}\frac{i}{k^2-M^2+i0^+}, \tag{14}$$

$$I_N = \mu^{4-n}\int\frac{d^nk}{(2\pi)^n}\frac{i}{k^2-m^2+i0^+}, \tag{15}$$

$$I_{N\pi}(-p,0) = \mu^{4-n}\int\frac{d^nk}{(2\pi)^n}\frac{i}{[(k-p)^2-m^2+i0^+]}$$
$$\times\frac{1}{k^2-M^2+i0^+}. \tag{16}$$

Applying the $\widetilde{\text{MS}}$ renormalization scheme—indicated by "r"—one obtains

$$\Sigma_{\text{loop}}^r = -\frac{3g_{Ar}^2}{4F_r^2}\left[-\frac{M^2}{16\pi^2}(\not{p}+m)+\cdots\right] = \mathcal{O}(q^2),$$

i.e., the $\widetilde{\text{MS}}$-renormalized result does not produce the desired low-energy behavior of (13). Gasser, Sainio, and Švarc concluded that loops have a much more complicated low-energy structure if baryons are included. The appearance of another scale, namely, the mass of the nucleon (which does not vanish in the chiral limit), is one of the origins for the complications in the baryonic sector [4]. The apparent "mismatch" between the chiral and the loop expansion has widely been interpreted as the absence of a systematic power counting in the relativistic formulation.

4.1 Heavy-baryon approach

One possibility of overcoming the problem of power counting was provided by the heavy-baryon formulation of ChPT [12,13] resulting in a power counting scheme which follows (11) and (12). The basic idea consists in dividing nucleon momenta into a large piece close to on-shell kinematics and a soft residual contribution: $p = mv + k_p$, $v^2 = 1$, $v^0 \geq 1$ [often $v^\mu = (1,0,0,0)$]. The relativistic nucleon field is expressed in terms of velocity-dependent fields,

$$\Psi(x) = e^{-imv\cdot x}(\mathcal{N}_v + \mathcal{H}_v),$$

with

$$\mathcal{N}_v = e^{+imv\cdot x}\frac{1}{2}(1+\not{v})\Psi, \quad \mathcal{H}_v = e^{+imv\cdot x}\frac{1}{2}(1-\not{v})\Psi.$$

Using the equation of motion for $\mathcal{H}_v$, one can eliminate $\mathcal{H}_v$ and obtain a Lagrangian for $\mathcal{N}_v$ which, to lowest order, reads [13]

$$\widehat{\mathcal{L}}_{\pi N}^{(1)} = \bar{\mathcal{N}}_v(iv\cdot D + g_A S_v \cdot u)\mathcal{N}_v + \mathcal{O}(1/m).$$

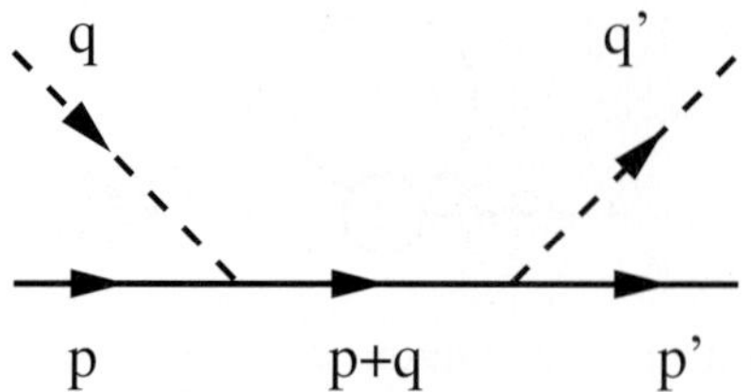

Fig. 4. s-channel pole diagram of πN scattering

The result of the heavy-baryon reduction is a $1/m$ expansion of the Lagrangian similar to a Foldy-Wouthuysen expansion. Now, power counting works along (11) and (12) but the approach has its own shortcomings. In higher orders in the chiral expansion, the expressions due to $1/m$ corrections of the Lagrangian become increasingly complicated.

Moreover—and what is more important—the approach generates problems regarding analyticity. This can easily be illustrated by considering the example of pion-nucleon scattering [19]. The invariant amplitudes describing the scattering amplitude develop poles for $s = m_N^2$ and $u = m_N^2$. For example, the singularity due to the nucleon pole in the s channel (see Fig. 4) is understood in terms of the relativistic propagator

$$\frac{1}{(p+q)^2-m_N^2} = \frac{1}{2p\cdot q + M_\pi^2}, \tag{17}$$

which, of course, has a pole at $2p\cdot q = -M_\pi^2$ or, equivalently, $s = m_N^2$. (Analogously, a second pole results from the u channel at $u = m_N^2$.) Although both poles are not in the physical region of pion-nucleon scattering, analyticity of the invariant amplitudes requires these poles to be present in the amplitudes. Let us compare the situation with a heavy-baryon type of expansion, where, for simplicity, we choose as the four-velocity $p^\mu = m_N v^\mu$,

$$\frac{1}{2p\cdot q + M_\pi^2} = \frac{1}{2m_N}\frac{1}{v\cdot q + \frac{M_\pi^2}{2m_N}}$$
$$= \frac{1}{2m_N}\frac{1}{v\cdot q}\left(1 - \frac{M_\pi^2}{2m_N v\cdot q} + \cdots\right). \tag{18}$$

Clearly, to any finite order the heavy-baryon expansion produces poles at $v\cdot q = 0$ instead of a simple pole at $v\cdot q = -M_\pi^2/(2m_N)$ and will thus not generate the (nucleon) pole structures of the invariant amplitudes. Another example involving loop diagrams will be given in Section 5.

4.2 Infrared regularization

A second solution was offered by Becher and Leutwyler [15] and is referred to as the so-called infrared regularization. The basic idea can be illustrated using the loop integral of (16). To that end, we make use of the Feynman parametrization

$$\frac{1}{ab} = \int_0^1\frac{dz}{[az+b(1-z)]^2}$$

with $a = (k - p)^2 - m^2 + i0^+$ and $b = k^2 - M^2 + i0^+$, interchange the order of integrations, and perform the shift $k \to k + zp$. The resulting integral over the Feynman parameter z is then rewritten as

$$I_{N\pi}(-p, 0) = \int_0^1 dz \cdots = \int_0^\infty dz \cdots - \int_1^\infty dz \cdots,$$

where the first, so-called infrared (singular) integral satisfies the power counting, while the remainder violates power counting but turns out to be regular and can thus be absorbed in counterterms. In the one-nucleon sector, it is straightforward to generalize the method for any one-loop integral consisting of an arbitrary number of nucleon and pion propagators [15] (see also [20]).

4.3 Extended on-mass-shell scheme

In the following, we will concentrate on yet another solution which has been motivated in [21] and has been worked out in detail in [22] (for other approaches, see [23]). The central idea consists of performing additional subtractions beyond the $\widetilde{\text{MS}}$ scheme such that renormalized diagrams satisfy the power counting. Terms violating the power counting are analytic in small quantities and can thus be absorbed in a renormalization of counterterms. In order to illustrate the approach, let us consider as an example the integral

$$H(p^2, m^2; n) = \int \frac{d^n k}{(2\pi)^n} \frac{i}{[(k-p)^2 - m^2 + i0^+][k^2 + i0^+]},$$

where

$$\Delta = \frac{p^2 - m^2}{m^2} = \mathcal{O}(q)$$

is a small quantity. We want the (renormalized) integral to be of order

$$D = n - 1 - 2 = n - 3.$$

The result of the integration is of the form (see [22] for details)

$$H \sim F(n, \Delta) + \Delta^{n-3} G(n, \Delta),$$

where F and G are hypergeometric functions and are analytic in Δ for any n. Hence, the part containing G for noninteger n is proportional to a noninteger power of Δ and satisfies the power counting. The part proportional to F can be obtained by first expanding the integrand in small quantities and *then* performing the integration for each term [24]. It is this part which violates the power counting, but, since it is analytic in Δ, the power-counting violating pieces can be absorbed in the counterterms. This observation suggests the following procedure: expand the integrand in small quantities and subtract those (integrated) terms whose order is smaller than suggested by the power counting. In the present case, the subtraction term reads

$$H^{\text{subtr}} = \int \frac{d^n k}{(2\pi)^n} \frac{i}{[k^2 - 2p \cdot k + i0^+][k^2 + i0^+]}\Bigg|_{p^2 = m^2}$$

and the renormalized integral is written as

$$H^R = H - H^{\text{subtr}} = \mathcal{O}(q^{n-1}).$$

Using our EOMS scheme it is also possible to include (axial) vector mesons explicitly [25]. Moreover, the infrared regularization of Becher and Leutwyler can be reformulated in a form analogous to the EOMS renormalization scheme and can thus be applied straightforwardly to multi-loop diagrams with an arbitrary number of particles with arbitrary masses [26] (see also [20]).

5 Applications

As the first application, we discuss the result for the mass of the nucleon at $\mathcal{O}(q^3)$. Within the $\widetilde{\text{MS}}$ scheme of [4] the result is given by

$$m_N = m - 4c_1^r M^2 + \frac{3g_{Ar}^2 M^2}{32\pi^2 F_r^2} m \left(1 + 8c_1^r m\right) - \frac{3g_{Ar}^2 M^3}{32\pi F_r^2}, \tag{19}$$

where r indicates $\widetilde{\text{MS}}$-renormalized quantities, and where we have used the renormalization scale $\mu = m$ with m the $\text{SU}(2) \times \text{SU}(2)$ chiral limit of the nucleon mass (at fixed $m_s \neq 0$). The third term on the r. h. s. of (19) violates the power counting of (11), because it is proportional to M^2, i.e., $\mathcal{O}(q^2)$, while it is obtained from the diagram of Fig. 3 which should generate contributions of $\mathcal{O}(q^3)$. On the other hand, the result in the EOMS scheme is given by [22]

$$m_N = m - 4c_1 M^2 - \frac{3g_A^2 M^3}{32\pi F^2} + \mathcal{O}(M^4), \tag{20}$$

where all parameters are understood to be taken in the EOMS scheme. Clearly, this expression satisfies the power counting, because the renormalized loop contribution of Fig. 3 is of $\mathcal{O}(M^3)$. The relation between the $\widetilde{\text{MS}}$-renormalized and the EOMS-renormalized coefficients is given by

$$c_1^r = c_1 + \frac{3mg_A^2}{128\pi^2 F^2}[1 + 8mc_1] + \cdots.$$

A full calculation of the nucleon mass at $\mathcal{O}(q^4)$ yields [22]

$$m_N = m + k_1 M^2 + k_2 M^3 + k_3 M^4 \ln\left(\frac{M}{m}\right) + k_4 M^4 + \mathcal{O}(M^5), \tag{21}$$

where the coefficients k_i are given by

$$k_1 = -4c_1, \quad k_2 = -\frac{3\overset{\circ}{g}_A^2}{32\pi F^2},$$

$$k_3 = \frac{3}{32\pi^2 F^2}\left(8c_1 - c_2 - 4c_3 - \frac{\overset{\circ}{g}_A^2}{m}\right),$$

$$k_4 = \frac{3\overset{\circ}{g}_A^2}{32\pi^2 F^2 m}(1 + 4c_1 m) + \frac{3}{128\pi^2 F^2}c_2 + \frac{1}{2}\alpha. \tag{22}$$

Here, $\alpha = -4(8e_{38} + e_{115} + e_{116})$ is a linear combination of $\mathcal{O}(q^4)$ coefficients [8]. In order to obtain an estimate for the various contributions of (21) to the nucleon mass, we make use of the set of parameters c_i of [27],

$$c_1 = -0.9\, m_N^{-1}, \quad c_2 = 2.5\, m_N^{-1},$$
$$c_3 = -4.2\, m_N^{-1}, \quad c_4 = 2.3\, m_N^{-1}. \qquad (23)$$

These numbers were obtained from a (tree-level) fit to the πN scattering threshold parameters of [28]. Using the numerical values

$$g_A = 1.267, \quad F_\pi = 92.4\,\mathrm{MeV}, \quad m_N = m_p = 938.3\,\mathrm{MeV},$$
$$M_\pi = M_{\pi^+} = 139.6\,\mathrm{MeV}, \qquad (24)$$

we obtain for the mass of nucleon in the chiral limit (at fixed $m_s \neq 0$):

$$m = m_N - \Delta m$$
$$= [938.3 - 74.8 + 15.3 + 4.7 + 1.6 - 2.3]\,\mathrm{MeV}$$
$$= 882.8\,\mathrm{MeV}$$

with $\Delta m = 55.5\,\mathrm{MeV}$. Here, we have made use of an estimate for α obtained from the σ term (see the following discussion).

Similarly, an analysis of the σ term yields

$$\sigma = \sigma_1 M^2 + \sigma_2 M^3 + \sigma_3 M^4 \ln\left(\frac{M}{m}\right) + \sigma_4 M^4 + \mathcal{O}(M^5), \qquad (25)$$

with

$$\sigma_1 = -4c_1, \quad \sigma_2 = -\frac{9\overset{\circ}{g}_A^{\,2}}{64\pi F^2},$$

$$\sigma_3 = \frac{3}{16\pi^2 F^2}\left(8c_1 - c_2 - 4c_3 - \frac{\overset{\circ}{g}_A^{\,2}}{m}\right),$$

$$\sigma_4 = \frac{3}{8\pi^2 F^2}\left[\frac{3\overset{\circ}{g}_A^{\,2}}{8m} + c_1(1 + 2\overset{\circ}{g}_A^{\,2}) - \frac{c_3}{2}\right] + \alpha. \qquad (26)$$

We obtain [with $\alpha = 0$ in (26)]

$$\sigma = (74.8 - 22.9 - 9.4 - 2.0)\,\mathrm{MeV} = 40.5\,\mathrm{MeV}. \qquad (27)$$

The result of (27) has to be compared with the dispersive analysis $\sigma = (45 \pm 8)$ MeV of [29] which would imply, neglecting higher-order terms, $\alpha M^4 \approx 4.5$ MeV. As has been discussed, e.g., in [15], a fully consistent description would also require to determine the low-energy coupling constant c_1 from a complete $\mathcal{O}(q^4)$ calculation of, say, πN scattering.

The results of (22) and (26) satisfy the constraints as implied by the application of the Hellmann-Feynman theorem to the nucleon mass [2,4]

$$\sigma = M^2\, \frac{\partial m_N}{\partial M^2}. \qquad (28)$$

A chiral low-energy theorem [30,31] relates the scalar form factor at $t = 2M_\pi^2$ to the πN scattering amplitude at

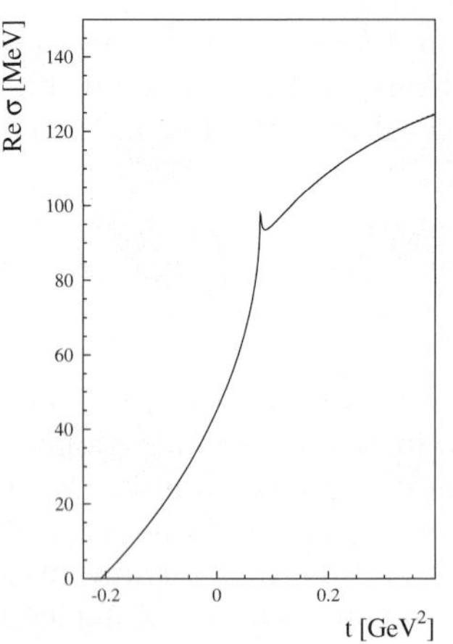
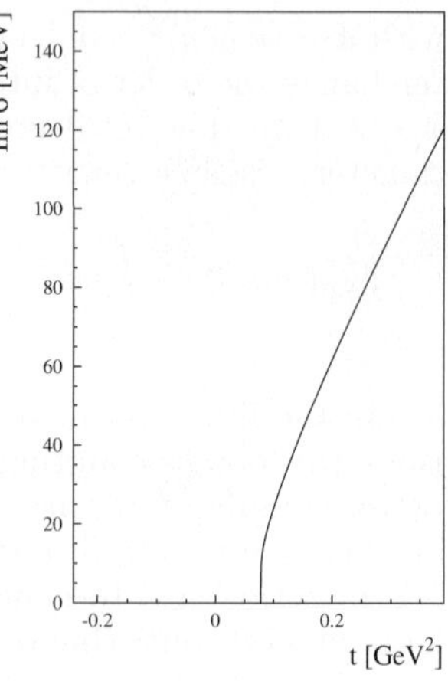

Fig. 5. Scalar form factor $\sigma(t)$ as a function of t at $\mathcal{O}(q^4)$

the unphysical point $\nu = 0$, $t = 2M_\pi^2$ (for a recent discussion of the corrections, see [27]). Defining the difference $\Delta_\sigma = \sigma(2M_\pi^2) - \sigma(0)$, one obtains a similar expansion for Δ_σ as for the nucleon mass and the σ term [15]

$$\Delta_\sigma = \Delta_1 M^3 + \Delta_2 M^4 \ln\left(\frac{M}{m}\right) + \Delta_3 M^4 + \mathcal{O}(M^5), \qquad (29)$$

where

$$\Delta_1 = \frac{3\overset{\circ}{g}_A^{\,2}}{64\pi F^2}, \quad \Delta_2 = \frac{1}{16\pi^2 F^2}\left(\frac{3\overset{\circ}{g}_A^{\,2}}{m} + c_2 + 6c_3\right),$$

$$\Delta_3 = 8e_{22} - \frac{c_1\overset{\circ}{g}_A^{\,2}}{4\pi^2 F^2} + \frac{3(\pi - 2)\overset{\circ}{g}_A^{\,2}}{128\pi^2 F^2 m} + \frac{3c_1(\pi - 4)}{16\pi^2 F^2}$$
$$+ \frac{c_2(14 - 3\pi)}{192\pi^2 F^2} + \frac{3c_3}{16\pi^2 F^2}, \qquad (30)$$

where e_{22} is an $\mathcal{O}(q^4)$ coefficient [8]. Using the parameters and numerical values of (23) and (24), respectively, we obtain [with $e_{22} = 0$ in (30)]

$$\Delta_\sigma = (7.6 + 10.2 - 0.9)\,\mathrm{MeV} = 16.9\,\mathrm{MeV}, \qquad (31)$$

which has to be compared with the dispersive analysis $\Delta_\sigma = (15.2 \pm 0.4)$ MeV of [29] resulting in the estimate $8e_{22}M^4 \approx -1.7$ MeV.

Next we discuss the scalar form factor which is defined as

$$\langle N(p')|\hat{m}[\bar{u}(0)u(0) + \bar{d}(0)d(0)]|N(p)\rangle = \bar{u}(p')u(p)\sigma(t).$$

The numerical results for the real and imaginary parts of the scalar form factor at $\mathcal{O}(q^4)$ are shown in Fig. 5 for the extended on-mass-shell scheme (solid lines) and the infrared regularization scheme (dashed lines). While the imaginary parts are identical in both schemes, the differences in the real parts are practically indistinguishable. Note that for both calculations $\sigma(0)$ and Δ_σ have been fitted to the dispersion results of [29]. Figure 6 contains an enlargement near $t \approx 4M_\pi^2$ for the results at $\mathcal{O}(p^3)$ which clearly displays how the heavy-baryon calculation

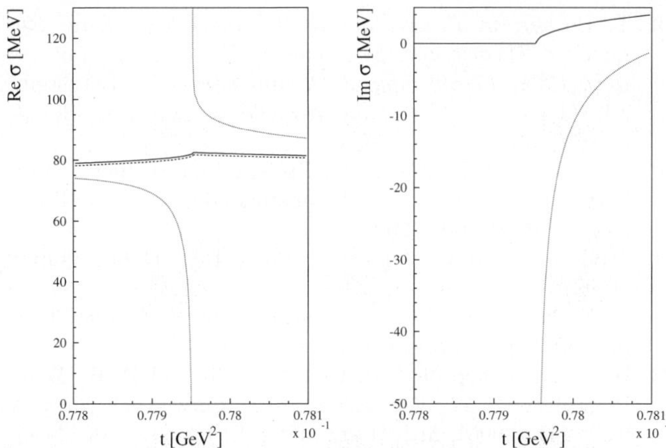

Fig. 6. Real and imaginary parts of the scalar form factor as a function of t at $\mathcal{O}(q^3)$ in the vicinity of $t = 4M_\pi^2$. Solid lines: EOMS scheme; dashed lines: infrared regularization (IR) of [15]; dotted lines: HBChPT calculation of [13]. On this scale the (unphysical) divergence of both real and imaginary parts of the heavy-baryon result becomes visible

fails to produce the correct analytical behavior. Both real and imaginary parts diverge as $t \to 4M_\pi^2$.

As the final example, we consider the electromagnetic form factors of the nucleon which are defined via the matrix element of the electromagnetic current operator as

$$\langle N(p_f) | J^\mu(0) | N(p_i) \rangle =$$

$$\bar{u}(p_f) \left[\gamma^\mu F_1^N(Q^2) + \frac{i\sigma^{\mu\nu}q_\nu}{2m_N} F_2^N(Q^2) \right] u(p_i), \quad N = p, n,$$

where $q = p_f - p_i$ is the momentum transfer and $Q^2 \equiv -q^2 = -t \geq 0$. Instead of the Dirac and Pauli form factors F_1 and F_2 one commonly uses the electric and magnetic Sachs form factors G_E and G_M defined by

$$G_E^N(Q^2) = F_1^N(Q^2) - \frac{Q^2}{4m_N^2} F_2^N(Q^2),$$

$$G_M^N(Q^2) = F_1^N(Q^2) + F_2^N(Q^2).$$

At $Q^2 = 0$, these form factors are given by the electric charges and the magnetic moments in units of the charge and the nuclear magneton, respectively:

$$G_E^p(0) = 1, \quad G_E^n(0) = 0, \quad G_M^p(0) = 1 + \kappa_p = 2.793,$$

$$G_M^n(0) = \kappa_n = -1.913.$$

Figure 7 shows the results for the Sachs form factors at $\mathcal{O}(q^4)$ in the EOMS scheme (solid lines) [32] and the infrared regularization (dashed lines) [33]. The description of G_E^p, G_M^p, and G_M^n turns out to be only marginally better than that of the $\mathcal{O}(q^3)$ calculation [32]. For the very-small Q^2 region the improvement is due to additional free parameters which have been adjusted to the magnetic radii. As can be seen from Fig. 7, the $\mathcal{O}(q^4)$ results only provide a decent description up to $Q^2 = 0.1\,\text{GeV}^2$ and do not generate sufficient curvature for larger values of Q^2. Moreover, the situation for G_E^n seems to be even worse, where

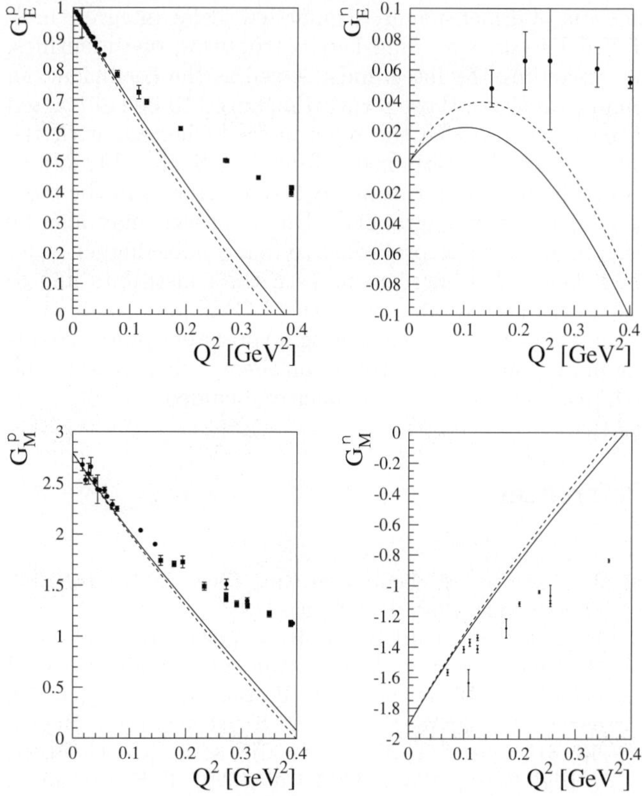

Fig. 7. The Sachs form factors of the nucleon at $\mathcal{O}(q^4)$. The solid and dashed lines refer to the results in the EOMS scheme [32] and the infrared regularization [33], respectively. The experimental data for G_E^p, G_E^n, G_M^p, and G_M^n are taken from [34], [35], [36], and [37], respectively

we found better agreement with the experimental data for the $\mathcal{O}(q^3)$ results [32]. We conclude that the perturbation series converges, at best, slowly and that higher-order contributions must play an important role.

6 Summary

We have discussed renormalization in the framework of mesonic and baryonic chiral perturbation theory. While the combination of dimensional regularization and the modified minimal subtraction scheme (of ChPT) leads to a straightforward power counting in terms of momenta and quark masses at a fixed ratio in the mesonic sector, the situation in the baryonic sector proves to be more complicated. At first sight, the correspondence between the loop expansion and the chiral expansion seems to be lost. Solutions to this problem have been given in terms of the heavy-baryon formulation and, more recently, the infrared regularization approach.

Here, we have discussed the so-called extended on-mass-shell renormalization scheme which allows for a simple and consistent power counting in the single-nucleon sector of manifestly Lorentz-invariant chiral perturbation theory. In this scheme a given diagram is assigned a chiral order D according to (11). After reducing the diagram to the sum of dimensionally regularized scalar integrals multiplied by corresponding Dirac structures, one identifies, by expanding the integrands as well as the coefficients in small quantities, those terms which need to be subtracted in order to produce the renormalized diagram with the chiral order D determined beforehand. Such subtractions can be realized in terms of local counterterms in the most general effective Lagrangian. Our approach may also be used in an iterative procedure to renormalize higher-order loop diagrams in agreement with the constraints due to chiral symmetry. Moreover, the EOMS renormalization scheme allows for implementing a consistent power counting in baryon chiral perturbation theory when vector (and axial-vector) mesons are explicitly included.

References

1. S. Weinberg: Physica A **96**, 327 (1979)
2. J. Gasser and H. Leutwyler: Ann. Phys. (N.Y.) **158**, 142 (1984); Nucl. Phys. B **250**, 465 (1985)
3. D. Issler: SLAC-PUB-4943-REV (1990) (unpublished); R. Akhoury and A. Alfakih: Ann. Phys. (N.Y.) **210**, 81 (1991); S. Scherer and H.W. Fearing: Phys. Rev. D **52**, 6445 (1995); H.W. Fearing and S. Scherer: Phys. Rev. D **53**, 315 (1996); J. Bijnens, G. Colangelo, and G. Ecker: J. High Energy Phys. **9902**, 020 (1999); T. Ebertshäuser, H.W. Fearing, and S. Scherer: Phys. Rev. D **65** (2002) 054033; J. Bijnens, L. Girlanda, and P. Talavera: Eur. Phys. J. C **23**, 539 (2002)
4. J. Gasser, M.E. Sainio, and A. Švarc: Nucl. Phys. B **307**, 779 (1988)
5. A. Krause: Helv. Phys. Acta **63**, 3 (1990)
6. G. Ecker and M. Mojžiš: Phys. Lett. B **365**, 312 (1996)
7. U.-G. Meißner, G. Müller, and S. Steininger: Ann. Phys. (N.Y.) **279**, 1 (2000)
8. N. Fettes, U.-G. Meißner, M. Mojžiš, and S. Steininger: Ann. Phys. (N.Y.) **283** (2001) 273 (2001); *ibid.* **288**, 249 (2001)
9. J. Wess and B. Zumino: Phys. Lett. B **37**, 95 (1971)
10. E. Witten: Nucl. Phys. B **223**, 422 (1983)
11. S. Weinberg: *The Quantum Theory Of Fields. Vol. 1: Foundations* (Cambridge University Press, Cambridge 1995)
12. E. Jenkins and A.V. Manohar: Phys. Lett. B **255**, 558 (1991); ibid. **259**, 353 (1991)
13. V. Bernard, N. Kaiser, J. Kambor, and U.-G. Meißner: Nucl. Phys. B **388**, 315 (1992)
14. S. Scherer: in *Advances in Nuclear Physics, Vol. 27*, edited by J.W. Negele and E.W. Vogt (Kluwer Academic/Plenum Publishers, New York 2003)
15. T. Becher and H. Leutwyler: Eur. Phys. J. C **9**, 643 (1999)
16. G. 't Hooft and M.J. Veltman: Nucl. Phys. B **44**, 189 (1972); M.J. Veltman: *Diagrammatica. The Path to Feynman Rules* (Cambridge University Press, Cambridge 1994)
17. S. Weinberg: Nucl. Phys. B **363**, 3 (1991)
18. G. Ecker: Prog. Part. Nucl. Phys. **35**, 1 (1995)
19. T. Becher: in *Proceedings of the Workshop Chiral Dynamics: Theory and Experiment III, Jefferson Laboratory, USA, 17–20 July, 2000*, edited by A.M. Bernstein, J.L. Goity, and U.-G. Meißner (World Scientific, Singapore, 2002)
20. J.L. Goity, D. Lehmann, G. Prezeau, and J. Saez: Phys. Lett. B **504**, 21 (2001); D. Lehmann and G. Prezeau: Phys. Rev. D **65**, 016001 (2002)
21. J. Gegelia and G. Japaridze: Phys. Rev. D **60**, 114038 (1999)
22. T. Fuchs, J. Gegelia, G. Japaridze, and S. Scherer: hep-ph/0302117; to appear in Phys. Rev. D.
23. H.-B. Tang: hep-ph/9607436; P.J. Ellis and H.-B. Tang: Phys. Rev. C **57**, 3356 (1998); M.F. Lutz: Nucl. Phys. A **677**, 241 (2000); M.F. Lutz and E.E. Kolomeitsev: Nucl. Phys. A **700**, 193 (2002)
24. J. Gegelia, G.S. Japaridze, and K.S. Turashvili: Theor. Math. Phys. **101**, 1313 (1994)
25. T. Fuchs, M.R. Schindler, J. Gegelia, and S. Scherer: hep-ph/0308006
26. M.R. Schindler, J. Gegelia, and S. Scherer: hep-ph/0309005
27. T. Becher and H. Leutwyler: J. High Energy Phys. **0106**, 017 (2001)
28. R. Koch: Nucl. Phys. A **448**, 707 (1986)
29. J. Gasser, H. Leutwyler, and M.E. Sainio: Phys. Lett. B **253**, 252 (1991)
30. T.P. Cheng and R.F. Dashen: Phys. Rev. Lett. **26**, 594 (1971)
31. L.S. Brown, W.J. Pardee, and R.D. Peccei: Phys. Rev. D **4**, 2801 (1971)
32. T. Fuchs, J. Gegelia, and S. Scherer: nucl-th/0305070
33. B. Kubis and U.-G. Meißner: Nucl. Phys. A **679**, 698 (2001)
34. L.E. Price, J.R. Dunning, M. Goitein, K. Hanson, T. Kirk, and R. Wilson: Phys. Rev. D **4**, 45 (1971); C. Berger, V. Burkert, G. Knop, B. Langenbeck, and K. Rith: Phys. Lett. B **35**, 87 (1971); K.M. Hanson, J.R. Dunning, M. Goitein, T. Kirk, L.E. Price, and R. Wilson: Phys. Rev. D **8**, 753 (1973); G.G. Simon, C. Schmitt, F. Borkowski, and V.H. Walther: Nucl. Phys. A **333**, 381 (1980)
35. T. Eden et al.: Phys. Rev. C **50**, 1749 (1994); I. Passchier et al.: Phys. Rev. Lett. **82**, 4988 (1999); M. Ostrick et al.: Phys. Rev. Lett. **83**, 276 (1999); C. Herberg et al.: Eur. Phys. J. A **5**, 131 (1999); J. Becker et al.: Eur. Phys. J. A **6**, 329 (1999)
36. T. Janssens, R. Hofstadter, E.B. Hughes, and M.R. Yearian: Phys. Rev. **142**, 922 (1966); C. Berger, V. Burkert, G. Knop, B. Langenbeck, and K. Rith: Phys. Lett. B **35**, 87 (1971); K.M. Hanson, J.R. Dunning, M. Goitein, T. Kirk, L.E. Price, and R. Wilson: Phys. Rev. D **8**, 753 (1973); G. Höhler, E. Pietarinen, I. Sabba Stefanescu, F. Borkowski, G.G. Simon, V.H. Walther, and R.D. Wendling: Nucl. Phys. B **114**, 505 (1976)
37. P. Markowitz et al.: Phys. Rev. C **48**, 5 (1993); H. Anklin et al.: Phys. Lett. B **336**, 313 (1994); E.E. Bruins et al.: Phys. Rev. Lett. **75**, 21 (1995); H. Anklin et al.: Phys. Lett. B **428**, 248 (1998); W. Xu et al.: Phys. Rev. Lett. **85**, 2900 (2000); G. Kubon et al.: Phys. Lett. B **524**, 26 (2002)

Eur Phys J A (2004) **19**, s01, 43–45
Digital Object Identifier (DOI) 10.1140/epjad/s2004-03-007-y

EPJ A direct

electronic only

Double heavy baryons and dimesons

M. Rosina[1,2] and D. Janc[2]

[1] Faculty of Mathematics and Physics, University of Ljubljana,
  P.O. Box 2964, SI-1000 Ljubljana, Slovenia
[2] Jožef Stefan Institute,
  P.O. Box 3000, SI-1000 Ljubljana, Slovenia

Received: 30 Oct 2003 / Accepted: 14 Nov 2003 /
Published Online: 6 Feb 2004 – © Società Italiana di Fisica / Springer-Verlag 2004

Abstract. We critically examine the question whether the $cc\bar{q}\bar{q}$ dimeson is bound or not.

PACS. 12.39.Pn Potential models – 12.39.Jh Nonrelativistic quark model – 12.40.Yx Hadron mass models and calculations

1 Introduction

There is a revived interest to study double heavy baryons and dimesons, both due to the theoretical urge of understanding better the quark-quark effective interaction, as well as due to new experimental opportunities in Fermilab and LHC.

The effective interaction between heavy quarks (and antiquarks) is expected to be cleaner than between light quarks. For heavy particles the nonrelativistic constituent quark model is more acceptable, the perturbative QCD contributions (such as one-gluon-exchange) are more adequate and chiral fields are less important. The effective interaction between a heavy quark and a heavy antiquark has been reasonably well studied and fitted by the charmonium and bottomium spectra. There is, however, no free diquark to study the effective interaction between two heavy quarks; one has to dress the diquark in order to obtain a colour singlet object QQq or $QQ\bar{q}\bar{q}$ (Q=c or b, q=u,d, or s).

It is straightforward to extrapolate the one-gluon-exchange (OGE) interaction from $Q\bar{Q}$ to QQ (Q= any quark). The charge conjugation changes the $\bar{Q}$ antitriplet to Q triplet. Then the colour factor $\lambda \cdot \lambda/4 = -4/3$ for the $Q\bar{Q}$ singlet changes to $-2/3$ for the QQ antitriplet (the "$V_{QQ} = \frac{1}{2}V_{Q\bar{Q}}$ rule"). On the other hand, it is questionable whether the (linear) confining potential should also possess such a colour factor and obey the $V_{QQ} = \frac{1}{2}V_{Q\bar{Q}}$ rule. The fact that the ground state energies and some excited states of light and heavy baryons are reasonably well reproduced with such a "universal" OGE + confining effective interaction is encouraging [1] but not conclusive. There may be other mechanisms for the $V_{QQ} = \frac{1}{2}V_{Q\bar{Q}}$ rule. For example, the flux tubes in a Y configuration can be mimicked by twice weaker two-body flux lines since the length of the arms of the Y is approximately half the length of the circumference of the triangle. The colour singlet 3-quark system is insensitive to the features of the *colour · colour* operator since it is just a constant in the 3-body singlet representation. To explore the colour structure of the effective interaction one has to go beyond mesons and baryons to dimesons and other exotics.

The study of double-heavy baryons and of double-heavy dimesons are complementary. The double-heavy baryons help to study the QQ interaction, while the dimesons also test the pion exchange between light quarks [2] and are more sensitive to three-body forces.

Our constituent quark model calculation [3] has shown the bb-dimeson to be bound by more than 100 MeV and the cc-dimeson to be unbound, which is consistent with some other calculations, for example [4]. We have proposed to look for the bb-dimesons at LHC assuming a mechanism of double $b\bar{b}$ production by double gluon gluon fusion $(g+g)+(g+g) \rightarrow (b+\bar{b})+(b+\bar{b})$ which has been described at this Conference by Danielle Treleani [5]. The two b-quarks then join into a diquark which gets dressed with a light quark or two light antiquarks to become a double heavy baryon or a dimeson. However, the production rate bb-dimesons has been estimated to be rather low [6,7], about 5 events/hour, and there seem to be no characteristic decays.

Therefore it is of utmost importance to look also for the cc-dimesons since their production rate might be as much as 10^4 events/hour if the same mechanism applies. They would also be easier to detect, for example by $cc\bar{u}\bar{d} \rightarrow D^+ + K^- + \pi^+$. There is, of course, a grat risk that they do not exist. If they, however, do exist they would be very exciting – we would have to revise our ideas about the effective quark-quark interaction, and/or introduce many-quark forces.

2 Can the $cc\bar{u}\bar{d}$ dimeson be bound?

We have obtained a phenomenological estimate for the binding energy of the cc-dimeson ($ISP = 01+$) with respect to the DD* by assuming a compact structure like in the $\bar{\Lambda}_c$ or $\bar{\Lambda}_b$ baryon with the cc-diquark playing the role of the heavy antiquark. For the cc binding in the diquark we assumed the $V_{QQ} = \frac{1}{2}V_{Q\bar{Q}}$ rule and no 3-body forces. We compared the following hadrons [3]

$$m_{cc\bar{u}\bar{d}} = 2m_c + m_u + m_d + E_{cc} + E_{\bar{u}\bar{d}[cc]}$$
$$m_{J/\psi} = 2m_c + E_{c\bar{c}}$$
$$m_{\bar{\Lambda}_c} = m_c + m_u + m_d + E_{\bar{u}\bar{d}\bar{c}}$$

where $E_{\bar{u}\bar{d}[cc]} \approx E_{\bar{u}\bar{d}\bar{c}}$ is the potential plus kinetic energy contribution of the two light antiquarks in the field of a heavy diquark or antiquark, respectively, and it cancels in the difference in the limit where the mass of the b quark goes to infinity and the heavy diquark is point-like so that we can neglect the size of the heavy diquark in the dimeson.

We estimated the diquark binding energy by using the theorem [3] $V_{cc} = \frac{1}{2}V_{c\bar{c}} \Rightarrow E_{cc}(m_{red}) = \frac{1}{2}E_{c\bar{c}}(\frac{1}{2}m_{red})$. Since meson binding energies lie on a smooth curve as a function of their reduced masses, it is easy to interpolate for the "fictitious meson" with $m_{red}/2$ and we get [3] $E_{cc} - \frac{1}{2}E_{c\bar{c}} = 134 \pm 20\,\mathrm{MeV}$ yielding

$$\Delta E_{cc\bar{u}\bar{d}} = m_{\Lambda_c} + m_{J/\psi}/2 + E_{cc} - E_{c\bar{c}}/2 - m_D - m_{D^*}$$
$$= (-42 + 134)\,\mathrm{MeV} = +92\,\mathrm{MeV}.$$

This means that such a compact structure is not bound with respect to the DD* threshold. Also detailed four-body calculations with OGE+linear potential with Bhaduri or Grenoble parameters [4] did not yield a bound state.

An alternative estimate lies considerably lower but is still unbound:

$$\Delta E_{cc\bar{u}\bar{d}} = m_{\Lambda_b} - m_b + m_c + m_{J/\psi}/2 + E_{cc} - E_{c\bar{c}}/2$$
$$- m_D - m_{D^*} = (-94 + 134)\,\mathrm{MeV} = +40\,\mathrm{MeV}.$$

The actual cc-diquark mass lies midway between the masses of the c and b quark (appearing in the center of Λ_c and Λ_b, respectively), therefore the answer is inbetween the two estimates which still means no binding.

The question arises whether the parameters in the OGE+linear confinement model could be stretched so as to bind cc-dimeson without spoiling the fit to mesons and baryons. If the $V_{QQ} = \frac{1}{2}V_{Q\bar{Q}}$ rule applies smaller quark masses could do the job. For Bhaduri masses, half of reduced mass od the cc diquark ($m_c/4 = 467\,\mathrm{MeV}$) coincides with the reduced mass of D_s, $m_c m_s/(m_c + m_s) = 454\,\mathrm{MeV}$ so that $E_{cc} = \frac{1}{2}E_{c\bar{s}}$. If we decrease all quark masses by 200 MeV, the reduced mass of D_s, would decrease by 132 MeV and $m_c/4$ only by 50 MeV. Higher reduced mass of cc compared to D_s means better binding of cc (by about 40 MeV). This is still not quite enough but might work in cooperation with additional effects.

A three-body interaction of the type

$$V_{ijk} = -\frac{U_0}{8}d^{abc}\lambda_i^a\lambda_j^b\lambda_k^c \exp(-(r_i^2 + r_j^2 + r_k^2)/a^2) \quad (1)$$

with at most $U_0 = 20$ MeV and $a = 2.3$ fm would bind. The choice of $a < 1$ fm gives small effect, and above 2.3 fm the effect saturates. Due to the combinatorics, a three-body interaction is more effective for tetraquarks than for baryons and the proposed one spoils baryons only by few MeV.

The pion exchange between D and D* leads to a coulomb-like long-range force because the exchanged pion is almost on the mass shell [8]: $(D^* \to D+\pi)$, $(D+\pi \to D^*)$. (Note that $m_{D^{*+}} - m_{D^+} - m_{\pi^0} = 5.6\,\mathrm{MeV}$, $m_{D^{*0}} - m_{\pi^0} = 7.1\,\mathrm{MeV}$, $m_{D^{*+}} - m_{D^0} - m_{\pi^+} = 5.8\,\mathrm{MeV}$.) This should in principle give a (weak) binding. We are studying the conflicting effects of short-range QQ interaction and this long-range DD* interaction.

3 A speculation using the ccu and ccd signals

Recent SELEX experiments and analysises [9] gave some more and some less convincing signals about the ccu(3460 and 3541) and ccd(3443 and 3520) baryons. If confirmed, they would have a dramatic effect on our estimates about the binding of the $cc\bar{u}\bar{d}$ dimeson. If refuted, the present section remains a piece of science fiction.

Our expectations about the ccq baryon are consistent with the ~ 3530 MeV isodoublet but would need a lot of stretching to accommodate the ~ 3450 isodoublet (if this one is confirmed as the spin=1/2 ground state). A phenomenological estimate similar as in the previous section gives for s=1/2 (assuming an S=1 cc-diquark) the value inbetween

$$m_{ccq} = \frac{1}{2}m_{J/\psi} + E_{cc} - \frac{1}{2}E_{c\bar{c}} + \frac{3}{4}m_D + \frac{1}{4}m_{D^*} = 3584\,\mathrm{MeV}$$

and

$$m_{ccq} = \frac{1}{2}m_{J/\psi} + E_{cc} - \frac{1}{2}E_{c\bar{c}} + m_c - m_b$$
$$+ \frac{1}{4}m_B + \frac{3}{4}m_{B^*} - \frac{1}{2}(m_{D^*} - m_D) = 3535\,\mathrm{MeV}$$

The predicted spin 3/2 state lies higher by $\frac{3}{4}(m_{D^*} - m_D) = 106$ MeV Such spin-spin splitting is noticeably larger than the difference 80 MeV between the 3530 and 3450 MeV SELEX levels and it will be some surprise if the 3450 level is confirmed as a ground state and the 3530 level gets an 3/2 assignment.

Then follows a phenomenological estimate for the cc-dimeson

$$\Delta E_{cc\bar{u}\bar{d}} = m_{ccu} - (\frac{3}{4}m_D + \frac{1}{4}m_{D^*})$$
$$+ m_{\Lambda_c} - m_D - m_{D^*}$$
$$= -42 \quad \text{or} \quad +38\,\mathrm{MeV}$$

assuming the 3450 or 3530 MeV level, respectively, to be the ccu ground state

The alternative estimate is very similar.

$$\Delta E_{c c \bar{u} \bar{d}} = m_{ccu} - \left(\frac{1}{4} m_{\mathrm{B}} + \frac{3}{4} m_{\mathrm{B}^*} \right)$$
$$+ \frac{1}{2}(m_{\mathrm{D}^*} - m_{\mathrm{D}}) + m_{\varLambda_{\mathrm{b}}} - m_{\mathrm{D}} - m_{\mathrm{D}^*}$$
$$= -45 \quad \text{or} \quad +35 \, \mathrm{MeV}$$

4 Conclusion

There are several subtle effects each of which separately is not likely to bind the $c c \bar{u} \bar{d}$ dimeson with respect to the DD* threshold. However, their cooperative effect might just bind it or just fail to bind it. Therefore we join and support those researchers who propose the detection of the $c c \bar{u} \bar{d}$ dimeson as a crucial experiment.

References

1. B. Silvestre-Brac: Few-Body Systems **20**, 1 (1996)
2. L. Glozman, W. Plessas, K. Varga, and R.F. Wagenbrunn: Phys. Rev. D **58**, 094030–1 (1998)
3. D. Janc and M. Rosina: Few-Body Systems **31**, 1 (2001)
4. B. Silvestre-Brac and C. Semay: Z. Phys. C **57**, 273 (1993)
5. D. Treleani and A. Del Fabbro: these Proceedings
6. D. Janc, M. Rosina, D. Treleani, and A. Del Fabbro: Few-Body Systems Suppl. **14**, 25 (2003)
7. M. Rosina, D. Janc, D. Treleani, and A. Del Fabbro: in *Hadron Physics*, A.H. Blin, B. Hiller, A.A. Osipov, M.C. Ruivo, E. van Beveren (eds.) (American Institute of Physics, Melville, New York 2003) p. 377
8. J.-M. Richard: *Double Charm Physics*, hep-ph/0212224 (2002)
9. M. Mattson et al. (SELEX Collaboration): Phys. Rev. Lett. **89**, 112001-1-5 (2002); also many internal reports

Eur Phys J A (2004) **19**, s01, 47–53
Digital Object Identifier (DOI) 10.1140/epjad/s2004-03-008-x

EPJ A direct

electronic only

Feasibility study of deeply virtual compton scattering using COMPASS at CERN

N. d'Hose, E. Burtin, P.A.M. Guichon, and J. Marroncle

CEA-Saclay, SPhN-DAPNIA-DSM, F91191 Gif-sur-Yvette Cedex, France

Received: 5 Aug 2003 / Accepted: 14 Nov 2003 /
Published Online: 6 Feb 2004 – © Società Italiana di Fisica / Springer-Verlag 2004

Abstract. This paper presents the reactions which can be performed at COMPASS to study the Generalized Parton Distributions (GPDs). The high energy muon beam at CERN allows to measure Hard Exclusive Meson Production or Deeply Virtual Compton Scattering (DVCS) in the Bjorken regime in a large range of Q^2 and x_{Bj} ($1.5 \leq Q^2 \leq 7.5$ GeV^2 and $0.03 \leq x_{Bj} \leq 0.25$). Exploratory measurements dedicated to ρ^0 or π^0 production can be investigated with the present setup. DVCS measurement require an upgrade of the COMPASS setup.

PACS. 13.60.Fz – 13.60.Hb – 13.60.Le – 14.20.Dh – 24.85.+p

1 Goal of an experiment with the high energy muon beam

In the quest for understanding the structure of the nucleon, Generalized Parton Distributions (GPDs) [1,2] have emerged as a very promising tool. They provide a unified description of the nucleon by interpolating between the parton distributions and the hadronic form factors. Moreover GPDs complete the nucleon puzzle as they give a measurement of total angular momentum contribution of the quarks to the nucleon spin. Experimentally the GPDs can be accessed in exclusive measurements such as Hard Exclusive Meson ($\rho, \pi...$) Production (HEMP) and Deeply Virtual Compton Scattering (DVCS). The latter reaction is the simplest from the theoretical point of view but also the most difficult experimentally because one has to select perfectly the final state (one lepton, one proton and one photon) among all the possible reactions. In pratice Meson Production can be investigated with the present COMPASS setup at CERN while a DVCS measurement at COMPASS would require an upgrade of the existing apparatus.

Experiments have already been undertaken at very high energy with the HERA collider [3,4] to study mainly the gluon GPDs at very small x_{Bj} ($\leq 10^{-2}$). Larger values of x_{Bj} have been investigated in fixed target experiments at JLab [5] (at 6 GeV, with plans for an upgrade at 11 GeV) and HERMES [6] (at 27 GeV). The experimental program using COMPASS at CERN (at 100 and/or 190 GeV) would enlarge the kinematical domain to a large range of Q^2 and x_{Bj} ($1.5 \leq Q^2 \leq 7$ GeV^2 and $0.03 \leq x_{Bj} \leq 0.25$) (see Fig. 1). A large range in Q^2 is re-

quired to control the factorisation in a hard, pertubatively calculable amplitude and a soft amplitude which is parametrized by the generalized parton distributions $H, E, \tilde{H}, \tilde{E}$. The GPDs depend on three kinematical variables: x and ξ parameterize the longitudinal momentum fractions of the partons, while t relates to the transverse momentum transfer.

Since the theoretical proof of factorization assumes that the transfer t is finite (that is $t/Q^2 \to 0$) [7], we consider in the following $|t|$ smaller than 1 GeV^2. Another condition of factorization concerns the helicity of the virtual photon. In case of Hard Exclusive Meson Production it is mandatory to impose that the virtual photon be longitudinal in order to select the perturbative gluon exchange. Experimentally we should consider Rosenbluth separation for π^0 production, while for ρ^0 production we can select longitudinal ρ^0s through the angular distribution of the decay products and assume the s-channel helecity conservation. Hard Exclusive Meson Productions seem more complex to analyze as they contain non perturbative information on both the target and the produced meson. Nevertheless they offer the possibility to disentangle different GPDs (vector meson production depends on H and E only; pseudo-scalar production depends on $\tilde{H}$ and $\tilde{E}$ only) and to separate contributions from different flavors. Forward differential longitudinal ρ_L^0 electroproduction cross section measurements which provide the largest counting rates, have already been undertaken and are presented in Fig. 2 as a function of c.m. energy W for three values of Q^2 (5.6, 9 and 27 GeV^2). The theoretical curve is an incoherent sum of the quark and gluon contributions [8]. No measurement have been done at x_{Bj} larger than 0.05. A

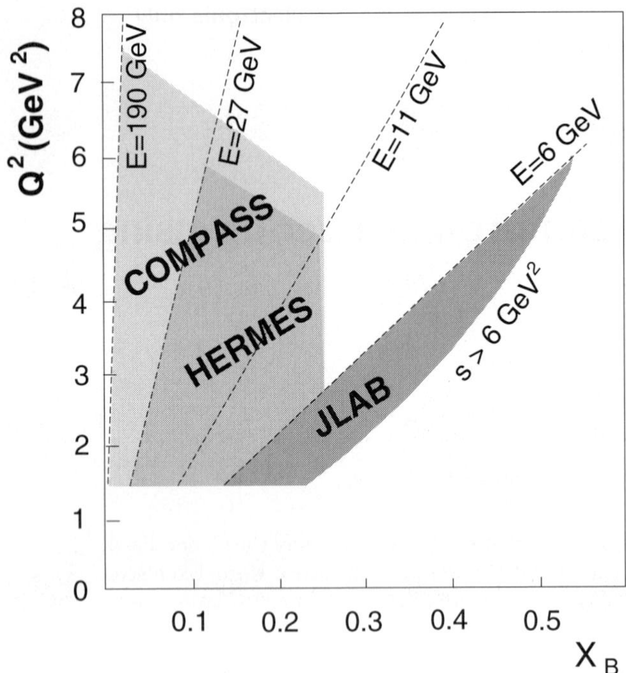

Fig. 1. Kinematical coverage for various planned or proposed experiments. The limit $s \geq 6$ GeV^2 assures to be above the resonance domain, and $Q^2 > 1.5$ GeV^2 allows to reach the Deep Inelastic regime

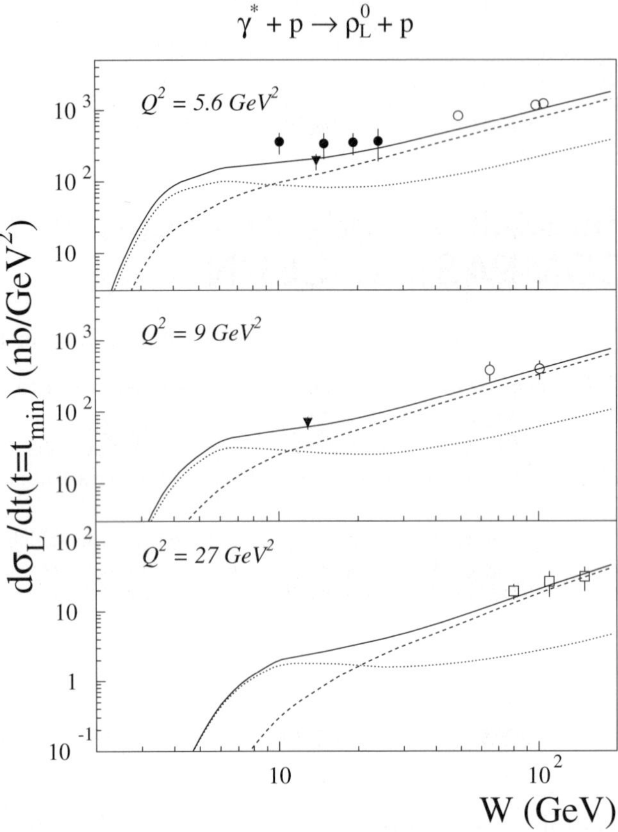

Fig. 2. Longitudinal forward differential cross section for ρ_L^0 production (Fig. from [8]). Predictions reproduce quark contributions (dotted lines), gluon contributions (dashed lines) and the sum of both (full lines). The data are from NMC (triangles) [9], E665 (solid circles) [10], ZEUS 93 (open circles) [11] and ZEUS 95 (open squares) [12]

larger domain in x_{Bj}, Q^2 and t could be explored with the muon beam available at CERN.

Deeply virtual Compton scattering is accessed by photon lepto-production: $lp \rightarrow l'p'\gamma$. In this reaction, the final photon can be emitted either by the leptons (Bethe-Heitler process) or by the proton (genuine DVCS process). If the lepton energy is large enough (see Fig. 3 with E_μ = 190 GeV, Q^2 = 4 GeV^2, x_{Bj}=0.1), the DVCS contribution dominates over the BH contribution so that the cross section is essentially the square of the DVCS amplitude which, at leading order, has the form:

$$\mathcal{A}(\gamma_T^*) \quad \sim \quad \int_{-1}^{+1} \frac{H(x,\xi,t)}{x - \xi + i\epsilon} dx...$$

$$\sim \quad \mathcal{P} \int_{-1}^{+1} \frac{H(x,\xi,t)}{x - \xi} dx... - i\pi H(\xi,\xi,t)....$$

(where $\xi \sim x_{Bj}/2$ and t are fixed by the experiment). At smaller lepton energy (see Fig. 3 with $E_\mu = 100$ GeV and same values of Q^2 and x_{Bj} as above), the interference between BH and DVCS becomes large and offers a unique opportunity to study Compton scattering amplitude including its phase. A careful analysis of the dependence of the cross section on the azimuthal angle ϕ between the leptonic and hadronic planes and on Q^2 allows one to disentangle higher-twist effects and to select the real or imaginary parts of the DVCS amplitude [13,14]. Considering an unpolarized target, the dependence of the cross section on the angle φ, and on the charge e_ℓ and longitudinal polarization P_ℓ of the beam, can be written as follows:

$$\frac{d\sigma(\ell p \rightarrow \ell p \gamma)}{d\varphi}$$

$$= A_{BH}(cos(\varphi), cos(2\varphi), cos(3\varphi), cos(4\varphi))$$

$$+ A_{INT}(cos(\varphi), cos(2\varphi))$$

$$[\, e_\ell\, [c_1 cos(\varphi)\Re e\mathcal{A}(\gamma_T^*) + c_2 cos(2\varphi)\Re e\mathcal{A}(\gamma_L^*) + ...]$$

$$+ e_\ell P_\ell\, [s_1 sin(\varphi)\Im m\mathcal{A}(\gamma_T^*) + s_2 sin(2\varphi)\Im m\mathcal{A}(\gamma_L^*)]]$$

$$+ A_{VCS}(cos(\varphi), cos(2\varphi), P_\ell\, sin(\varphi))$$

where A_{BH}, A_{INT}, c_i, s_i are known expressions and $\mathcal{A}$ represents $\gamma^* p \rightarrow \gamma p$ amplitudes for different γ^* polarization. The scaling predictions give leading twist-2 and twist-3 contributions for $\mathcal{A}(\gamma_T^*)$ and $\mathcal{A}(\gamma_L^*)$ respectively. The dotted points in brackets stand for a φ-independent term and a term with $cos(3\varphi)$ which contains twist-4 and twist-2 contributions (the latter is related to the gluon double helicity-flip distributions). Both are predicted to be small in the kinematics under study but can readily be

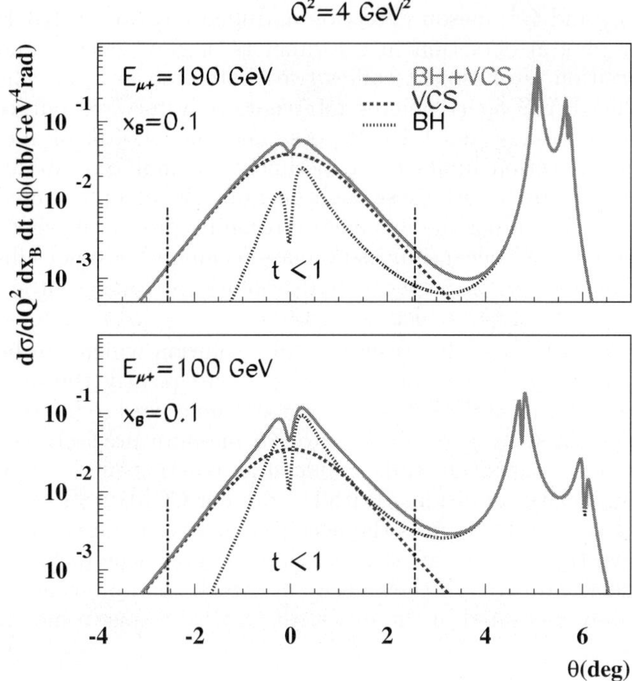

Fig. 3. Cross sections for the photon leptoproduction $\mu p \to \mu p \gamma$ as a function of the outgoing real photon angle (relative to the virtual photon direction). Comparison between BH (dotted lines), DVCS (dashed lines) and the total cross sections (full lines) for 2 energies of the muon beam available at CERN: 190 and 100 GeV. The interesting domain is limited by a transfer $|t|$ smaller than 1 GeV^2 i.e. θ investigating a small region around 0 degree

includied in a full analysis. If a longitudinally polarized lepton beam and an unpolarized target are used, the angular analysis and the Q^2 dependence of the cross section difference $\sigma(e^\uparrow) - \sigma(e^\downarrow)$ allow one to select the imaginary part of the DVCS amplitude and thus the GPDs at the specific values $x = \xi$. This study is being investigated at HERMES [6] and JLab [5]. If two muon beams of opposite charge and polarization are used, the angular analysis and the Q^2 dependence of the sum of cross sections $\sigma(\mu^{+\downarrow}) + \sigma(\mu^{-\uparrow})$ allow also one to select the imaginary part of the DVCS amplitude. Moreover the same method applied to the difference of cross sections $\sigma(\mu^{+\downarrow}) - \sigma(\mu^{-\uparrow})$ allows one to select the real part of the DVCS amplitude which, for a given ξ, is sensitive to the complete dependence on x of the GPDs. The deconvolution (over x) of this formula to extract the GPDs is not yet clearly solved, but comparison to model predictions can easily be made. It is clear that the muon beam of high energy at CERN can offer many possibilities in order to investigate the many-faceted problem of the GPDs knowledge.

Figure 4 shows the azimuthal distribution of the charge asymmetry which could be measured at COMPASS and the strong sensitivity to two different models [15]. The first one is based on a simple parametrization of the GPDs:

$$H^f(x, \xi, t) = H^f(x, \xi, 0) F_1^f(t)/2$$

where $F_1^f(t)$ represents the elastic Dirac form factor for the quark flavor f in the nucleon. The second one [16,17, 18,19] relies on the fact that the GPDs measure the contribution of quarks with longitudinal momentum fraction x to the corresponding form factor as is suggested by the sum rule:

$$\int_{-1}^{+1} H^f(x, \xi, t) dx = F_1^f(t).$$

As one can associate the Fourier transform of form factors with charge distributions in position space, one can expect that the GPDs contain information about the distribution of partons in transverse position space. In fact it has been demonstrated that, when t is purely transverse which amounts to $\xi = 0$, then $H(x, 0, t)$ is the Fourier transform of the probability density to find a quark with momentum fraction x at a given distance from the center of momentum in the transverse plane. Qualitatively one expects that quarks with a large x come essentially from the small valence "core" of the nucleon, while the small x region should receive contributions from the much wider meson "cloud". Therefore one expects a gradual increase of the t-dependence of $H(x, 0, t)$ as one goes from larger to smaller values of x. This suggests the parametrization: $H(x, 0, t) = q(x) e^{t < b_\perp^2 >} = q(x)/x^{\alpha t}$ where $< b_\perp^2 > = \alpha \cdot ln 1/x$ represents the increase of the nucleon transverse size with energy. The domain of small x_{Bj} reacheable at COMPASS is related to the observation of sea quarks or meson "cloud" or also gluons and it provides a large sensitivity to this three-dimensional picture of partons inside a hadron.

2 General requirements for COMPASS

The highest luminosity reachable at COMPASS is required to investigate these exclusive measurements. The experiment will use 100-190 GeV/c muons from the M2 beam line. Limits on radio-protection in the experimental hall imply that the maximum flux of muon to be expected is of $2 \cdot 10^8$ muons per SPS spill (5.2s spill duration, repetition each 16.8s). Under these circumstances, we can reach a luminosity of $\mathcal{L} = 5 \cdot 10^{32}$ cm^{-2}s^{-1} with the present polarized ^{6}LiD or NH$_3$ target of 1.2 meter long, and only $\mathcal{L} = 1.3 \cdot 10^{32}$ cm^{-2}s^{-1} with a new liquid hydrogen target of 2.5 meter long.

In order to get useful cross sections with positive and negative muon beams, it is necessary to perform a precise absolute luminosity measurement. This has already been achieved by the NMC Collaboration within a 1% accuracy [20]. The integrated muon flux was measured continuously by two methods: either by sampling the beam with a random trigger (provided by the α emitter Am^{241}) or by sampling the counts recorded in 2 scintillators hodoscope planes used to determine incident beam tracks. The beam tracks were recorded off-line, in the same way as the scattered muon tracks to determine exactly the integrated usable muon flux.

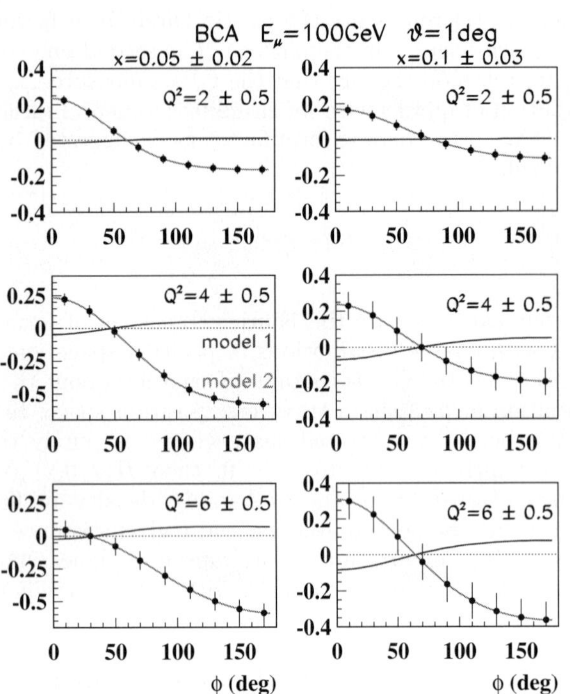

Fig. 4. Projected error bars for a measurement of the azimuthal angular distribution of the beam charge asymmetry measurable at COMPASS at $E_\mu = 100$ GeV and $|t| \leq 0.6$ GeV^2 for 2 domains of x_{Bj} ($x_{Bj} = 0.05 \pm 0.02$ and $x_{Bj} = 0.10 \pm 0.03$) and 3 domains of Q^2 ($Q^2 = 2 \pm 0.5$ GeV^2, $Q^2 = 4 \pm 0.5$ GeV^2 and $Q^2 = 6 \pm 0.5$ GeV^2) obtained in 6 months of data taking with a global efficiency of 25% and with $2 \cdot 10^8$ μ per SPS spill ($P_{\mu+} = -0.8$ and $P_{\mu-} = +0.8$) and a 2.5m long liquid hydrogen target

Moreover μ^+ and μ^- beams of 100 GeV energy, with the same and as large as possible intensity as well as exactly opposite polarization (to a few %) are required. The muons are provided by pion and kaon decay and are naturally polarized. The pions and kaons come from the collision of the SPS 400 GeV proton beam on a Be primary target. A solution is under study [21]. It consists in:
1) selecting 110 GeV pion beams from the collision and 100 GeV muon beams after the decay section in order to maximize the muon flux;
2) keeping constant the collimator settings which define the pion and muon momentum spreads (both the collimator settings in the hadron decay section and the scrapper settings in the muon cleaning section) in order to fix the μ^+ and μ^- polarizations at exactly the opposite value ($P_{\mu+} = -0.8$ and $P_{\mu-} = +0.8$);
3) fixing $N_{\mu-}$ to $2 \cdot 10^8$ μ per SPS spill with the longest 500mm Be primary target; 4) using a shorter target to find $N_{\mu+}$ close to $2 \cdot 10^8$ μ per SPS spill.

This paragraph presents the experimental procedure to select the exclusive HEMP or DVCS channel and the difference equipments that are required. They are mostly part of the existing high resolution COMPASS spectrometer: muon detection which insures a good resolution in x_{Bj} and Q^2, meson detection and identification in RICH or photon detection in calorimeters of good energy and position resolutions to allow two photons separation. The COMPASS spectrometer intercepts only forward outgoing particles (until 10 degrees) and the photon or meson detection limits the experiment to small x_{Bj} values ($x_{Bj} \leq 0.15$). At these high energies the complete final state, including the low energy recoiling proton, needs to be detected because missing mass techniques are not efficient due to the experimental resolutions (the resolution in missing mass which is required is $(m_p + m_\pi)^2 - m_p^2 = 0.25$ GeV^2 and the experimental resolution which can be achieved is larger than 1 GeV^2). Consequently the high resolution COMPASS spectrometer would need to be complemented by a recoil detector to measure precisely the proton momentum and exclude other reactions under high luminosity conditions. With the present COMPASS setup we can try to by-pass the necessity of a recoil detector to investigate the cleanest channel: $\mu p \rightarrow \mu p \rho^0$ where ρ^0s are identified through their decay in two charged pions accurately measured in the forward COMPASS spectrometer.

3 An ideal solution with a completed setup

It is clear that only a recoil detector which allows the low energy recoiling proton detection will help to select exclusive channels as HEMP or DVCS. The latter reaction is surely the most delicate because one has to select a final state with one muon, one photon and one low energy proton among many competing reactions listed below:
1) Hard Exclusive π^0 Production $\mu p \rightarrow \mu p \pi^0$ where π^0 decays in two photons, for which the photon with higher energy imitates a DVCS photon, and the photon with smaller energy is emitted at large angle outside of the acceptance or its energy is below the photon detection threshold.
2) Diffractive dissociation of the proton $\mu p \rightarrow \mu \gamma N^*$ with the subsequent decay of the excited state N^* in $N + k\pi$. (The low energy pions are emitted rather isotropically).
3) Inclusive Deep Inelastic Scattering with, in addition to the reconstructed photon, other particles produced outside the acceptance or for which tracks are not reconstructed due to inefficiency.
Moreover one has to take into account a background which includes beam halo tracks with hadronic contamination, beam pile-up, particles from the secondary interactions and external Bremsstrahlung.

A simulation has been realized in order to define the proper geometry of the detector complementing the present COMPASS setup and to analyze the operational conditions. The goal was to maximize the ratio of DVCS events over DIS events for a sample of events with one muon and one photon in the COMPASS spectrometer acceptance plus only one proton of momentum smaller than 750 MeV/c and angle larger than 40 degrees (it is the typical kinematics of a DVCS event at small t). The simulation relies on the event generator program PYTHIA 6.1 [22] which includes most of the known processes [23] such as Deeply Inelastic Scattering and Deeply Meson Production.

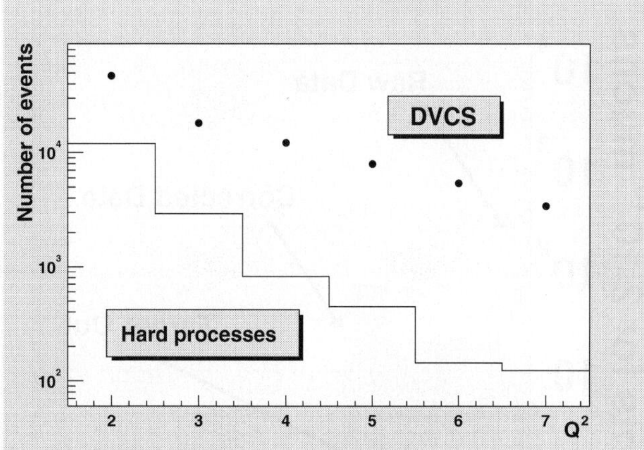

Fig. 5. Number of events for DVCS (dots) and DIS (histogram) processes as a function of Q^2 for selection of events with only one muon, one photon and one recoiling proton and condition for charged particle detection up to 40 degrees and for photon detection up to an angle of 24 degrees and above a threshold of 50 MeV

The experimental parameters such as maximum angle and energy threshold for photon detection and maximum angle for charged particle detection could then be tuned. With photon detection extended up to 24 degrees and above an energy threshold of 50 MeV and with charged particle detection up to 40 degrees, one observes that the number of DVCS events as estimated with models is more than an order of magnitude larger than the number of DIS events over the whole useful Q^2 range (see Fig. 5).

The COMPASS setup will be instrumented with two electromagnetic calorimeters ECAL1 and ECAL2 [24,25]. They are mainly constituted of lead-glass blocks called GAMS. They are cells of $38.4 \times 38.4 \times 450$ mm^3. Typical characteristics of such calorimeter are:
- energy resolution: $\sigma P_\gamma / P_\gamma = 0.055/\sqrt{P_\gamma} + 0.015$
- position resolution: $\sigma_x = 6.0/\sqrt{P_\gamma} + 0.5$ in mm
- high rate capability: 90% of signal within 50ns gate with no dead time
- effective light yield: about 1 photoelectron per MeV; hence low energy photons of down 20 MeV can be reconstructed.
The separation of the overlapping electromagnetic showers in the cellular GAMS calorimeter is carefully studied in the [26]. The result of the study shows that at 10 GeV one can reach a 100% level of the separation efficiency for a minimum distance between 2 photon tracks at the entrance of the calorimeter of D = 4 cm. The last value is slightly shifted to D = 5 cm at 40 GeV.
This excellent performance of the calorimeters will provide a key role in the perfect separation between DVCS events and Hard π^0 events.

One possible solution to complement the present COMPASS setup is presented in Fig. 6. It consists of one recoil detector described below, an extended calorimetry from 10 to 24 degrees, and a veto for charged forward particles until 40 degrees. This calorimeter has to work in a

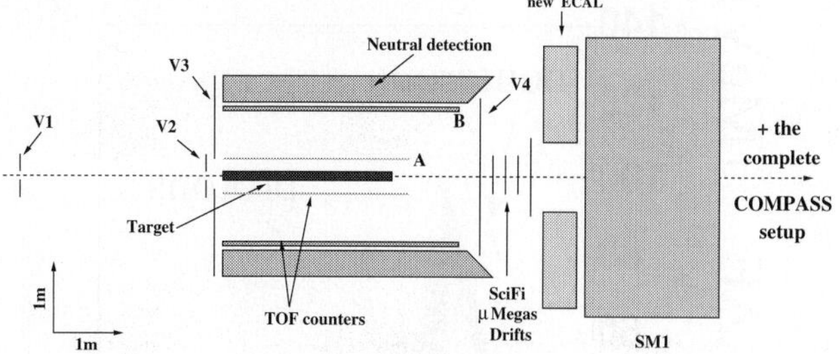

Fig. 6. Proposition for a detector complementing the COMPASS setup. A recoil detector, an extended calorimetry from 10 to 24 degrees, and a veto (V4) for charged forward particles until 40 degrees have been added

crowdy environment and in a magnetic fringe field of SM1 and therefore it has to be studied further.

At the present time our studies have focused on the possibility to design and successfully operate a dedicated recoil detector. One goal is to identify and measure the protons momenta between a minimum value and 750 MeV/c. A solution consists in a large time of flight setup between a thin segmented cylindrical layer of scintillator counters surrounding the 2.5 meter long target, and a thick layer at about 1m distance from the first layer. The thickness of the first layer has to be as small as possible in order to detect protons of minimum momentum. With an hydrogen target of 3cm diameter, target wall thickness of about 3mm of equivalent scintillator and a first layer of 4mm, a minimum momentum of 270 MeV/c is reached. All the counters are read at both sides by photomultiplier counters to determine time and position with very accurate resolutions better than 200ps and 1.5cm respectively. The consequent resolution in momentum varies from 2 to 5%. The resolution in t is twice this value, thus it is very desirable to further study all the parameters which can be improved. Moreover the exclusion of extra particles have to be studied with kinematical fits depending on the experimental resolution and/or with low energy π^0 detection. This detector has to work in a high rate environment. It has to be as large and hermetic as possible within a reasonable cost. The actual realisation of such performances is under active investigation. An efficiency study of such a recoil detector is being performed.

We have tested the concept of this detector using the already existing muon beam and a simplified setup (one sector of scintillators with reduced length). The muon beam was scattered off a 10 cm long polyethylene target, mostly equivalent in radiation length to the foreseen long liquid hydrogen target. We used three scintillators read-out at both sides, a 4mm thick close to the target (A), a 5 cm thick 80 cm away from the target (B) and an extra scintillator (C) to know if particles go through B or are stopped in B. The rates observed in the scintillator close to the target, using the nominal intensity of $2 \cdot 10^8$ muons per spill, is of the order of 1 MHz (mainly due to Möller electrons). It demonstrates that the background environment is acceptable for the time of flight system.

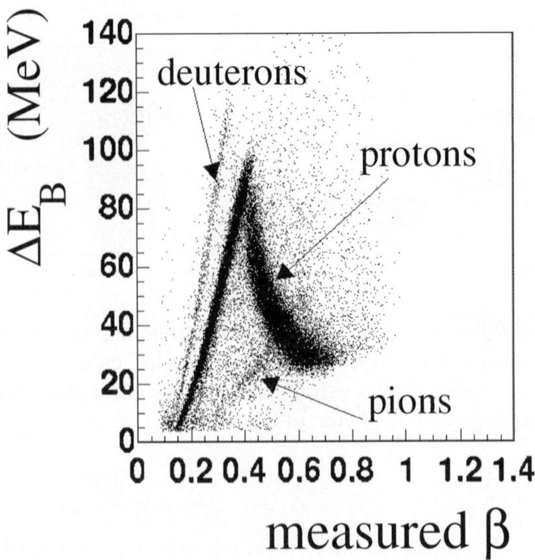

Fig. 7. Energy lost in the thick scintillator (B) as a function of the measured β

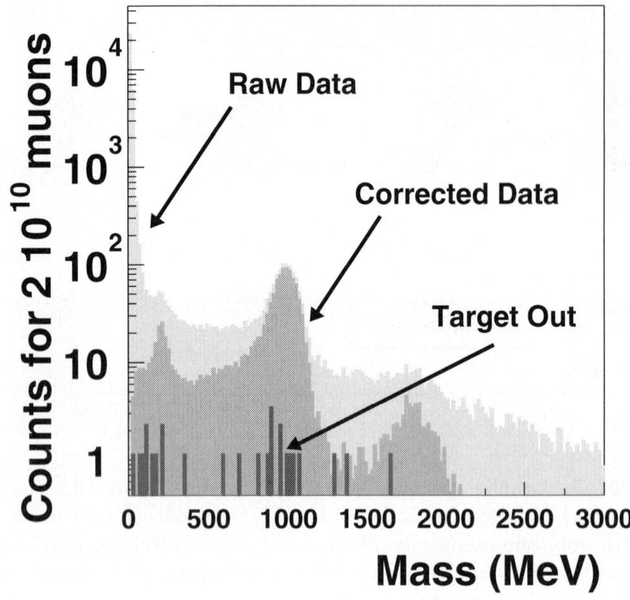

Fig. 8. Mass distribution of particles stopped in B. The three peak are pions, protons and deuterons respectively

The result of the time of flight operation (see Fig. 7) shows a clear proton signal. With the knowledge of the β velocity and the energy lost in the thick scintillator (B) for stopped particles one can reconstruct their masses. It is done in Fig. 8 where one can see pions, protons and deuterons for raw data, corrected data and the target out contribution which is about two order of magnitude smaller. The position resolution obtained on A and B and the time of flight resolution are better than 1.8 cm and 300 ps respectively. The performance of this ToF system is limited by the number of photoelectrons that are collected and dispersion due to counter lenght. Extension to long (about 4 meter) and thin scintillators have to be studied carefully and technology has to be improved to achieve still better resolution. The construction of a fully functional prototype segment of the appropriate length is foreseen [27].

4 Conclusion

This study points out the advantages of the COMPASS spectrometer and the high energy of the muon beam available at CERN which would give access to a large Q^2 and x_{Bj} range and encourages us for the following roadmap. Hard Exclusive Meson Production have to be undertaken as soon as possible with the present setup. The ρ^0 channel which decays in $\pi^+\pi^-$ is the easiest channel to isolate, the π^0 channel is more difficult but very important to test the calorimetry performances. A complete experiment with both Hard Exclusive Meson Production with a large set of mesons and Deeply Virtual Compton Scattering has to be envisaged in a next step with a completed COMPASS setup. For this purpose one needs a "long" hydrogen target, a recoil detector and an extension of the calorimetry at larger angles.

COMPASS is the unique place which provides μ^+ and μ^- of 100 GeV in order to study carefully two scales of observation $x_{Bj} = 0.05 \pm 0.02$ and $x_{Bj} = 0.10 \pm 0.03$ on a large domain of Q^2 from 2 to 7 GeV^2 and to measure the azimuthal distribution of the Beam Charge Asymmetry which seems very promising to test the geometrical interpretation of GPDs.

References

1. X. Ji: Phys. Rev. Lett. **78**, 610 (1997); Phys. Rev. D **55**, 7114 (1997)
2. A.V. Radyushkin: Phys. Rev. D **56**, 5524 (1997)
3. (H1 Collaboration), C. Adloff et al.: Phys. Lett. B **517**, 47 (2001)
4. (ZEUS Collaboration), S. Chekanov et al.: DESY-03-059, submitted to Phys. Lett. **B**
5. (CLAS Collaboration), S. Stepanyan et al.: Phys. Rev. Lett. **87**, 182002 (2001)
6. (HERMES Collaboration), A. Airapetian et al.: Phys. Rev. Lett. **87**, 182001 (2001)
7. J.C. Collins, L. Frankfurt, and M. Strikman: Phys. Rev. D **56**, 2982 (1997)
8. M. Vanderhaeghen, P.A.M. Guichon, and M. Guidal: Phys. Rev. D **60**, 094017 (1999)
9. (NMC Collaboration), M. Arneodo et al.: Nucl. Phys. B **429**, 503 (1994)
10. (E665 Collaboration), M.R. Adams et al.: Z. Phys. C **74**, 237 (1997)
11. (ZEUS Collaboration), M. Derrick et al.: Phys. Lett. B **356**, 601 (1995)
12. (ZEUS Collaboration), J. Breitweg et al.: Eur. Phys. J. C **6**, 603 (1999)
13. M. Diehl, T. Gousset, B. Pire, and J. Ralston: Phys. Lett. B **411**, 193 (1997)

14. A.V. Belitsky, D. Müller, and A. Kirchner: Nucl. Phys. B **629**, 323 (2002)
15. Model following K. Goeke, M.V. Polyakov, and M. Vanderhaeghen: Prog. Part. in Nucl. Phys. **47**, 401 (2001). Implementation by L. Mossé
16. M. Burkardt: Phys. Rev. D **62**, 07503 (2000); hep-ph/0207047
17. J.P. Ralston and B. Pire: hep-hp/0110075
18. M. Diehl: Eur. Phys. J. C **25**, (2002)
19. A.V. Belitsky and D. Müller: Nucl. Phys. A **711**, 118 (2002); hep-hp/0206306
20. (NMC Collaboration), P. Amaudruz et al.: Phys. Lett. B **295**, 159 (1992); R.P. Mount: Nucl. Instrum. Methods **187**, 401 (1981)
21. L. Gatignon: private communication
22. PYTHIA 6.1, User's manual, T. Sjöstrand et al.: "High Energy Physics Event Generation with PYTHIA 6.1", Comput. Phys. Commun. **135**, 238 (2001); hep-ph/0010017
23. C. Friberg and T. Sjöstrand: hep-ph/0007314
24. A Proposal for a Common Muon and Proton apparatus for Structure and Spectroscopy, CERN/SPSLC 96–14 and wwwcompass.cern.ch
25. V. Poliakov: Presentation of ECAL1 and ECAL2, January 25, 2001
26. A.A. Lednev: Separation of the overlapping electromagnetic showers in the cellular GAMs-type calorimeters, Preprint IHEP 93–153 (1993), Protvino, Russia
27. Joint Research Activity on Generalized Parton Distributions in the framework of the European Integrated Infrastructure Initiative on Hadronic Physics (I3HP)

Hadronic resonances

Eur Phys J A (2004) **19**, s01, 55–60
Digital Object Identifier (DOI) 10.1140/epjad/s2004-03-009-9

EPJ A direct
electronic only

Electroproduction of nucleon resonances

L. Tiator[1], D. Drechsel[1], S. Kamalov[1,2], M.M. Giannini[3], E. Santopinto[3], and A. Vassallo[3]

[1] Institut für Kernphysik, Johannes Gutenberg-Universität, D-55099 Mainz, Germany
[2] JINR Dubna, 141980 Moscow Region, Russia
[3] Dipartimento di Fisica dell'Universita di Genova and I.N.F.N., Sezione di Genova, I-16164 Genova, Italy

Received: 30 Oct 2003 / Accepted: 14 Nov 2003 /
Published Online: 6 Feb 2004 – © Società Italiana di Fisica / Springer-Verlag 2004

Abstract. The unitary isobar model MAID has been extended and used for a partial wave analysis of pion photo- and electroproduction in the resonance region $W < 2$ GeV. Older data from the world data base and more recent experimental results from Mainz, Bates, Bonn and JLab for Q^2 up to 4.0 $(\text{GeV}/c)^2$ have been analyzed and the Q^2 dependence of the helicity amplitudes have been extracted for a series of four star resonances. We compare single-Q^2 analyses with a superglobal fit in a new parametrization of Maid2003 together with predictions of the hypercentral constituent quark model. As a result we find that the helicity amplitudes and transition form factors of constituent quark models should be compared with the analysis of bare resonances, where the pion cloud contributions have been subtracted.

PACS. 13.40.Gp – 13.60.Le – 14.20.Gk – 25.20.Lj – 25.30.Rw

1 Introduction

Our knowledge of nucleon resonances is mostly given by elastic pion nucleon scattering [1]. All resonances that are given in the Particle Data Tables [2] have been identified in partial wave analyses of πN scattering both with Breit-Wigner analyses and with speed-plot techniques. From these analyses we know very well the masses, widths and the branching ratios into the πN and $\pi\pi N$ channels. These are reliable parameters for all resonances in the 3 and 4 star categories. There remain some doubts for the two prominent resonances, the Roper $P_{11}(1440)$ which appears unusually broad and the $S_{11}(1535)$ that cannot uniquely be determined in the speed-plot due to its position close to the ηN threshold. Both resonances, however, have recently been found on the lattice in a very precise calculation with a pion mass as low as 180 MeV and converge very close to the empirical masses [3]. This has been achieved in a quenched calculation giving rise to the conclusion that both resonance states are simply qqq states.

Starting from these firm grounds, using pion photo- and electroproduction we can determine the electromagnetic γNN^* couplings. They can be given in terms of electric, magnetic and charge transition form factors $G_E^*(Q^2)$, $G_M^*(Q^2)$ and $G_C^*(Q^2)$ or by linear combinations as helicity amplitudes $A_{1/2}(Q^2)$, $A_{3/2}(Q^2)$ and $S_{1/2}(Q^2)$. So far, we have some reasonable knowledge of the $A_{1/2}$ and $A_{3/2}$ amplitudes at $Q^2 = 0$, which are tabulated in the Particle Data Tables. For finite Q^2 the information found in the literature is very scarce and practically does not exist at all for the longitudinal amplitudes $S_{1/2}$. But even for the transverse amplitudes only few results are

firm, these are the G_M^* form factor of the $\Delta(1232)$ up to $Q^2 \approx 10$ GeV^2, the $A_{1/2}(Q^2)$ of the $S_{11}(1535)$ resonance up to $Q^2 \approx 5$ GeV^2 and the asymmetry $A(Q^2) = (A_{1/2}^2 - A_{3/2}^2)/(A_{1/2}^2 + A_{3/2}^2)$ for the $D_{13}(1520)$ and $F_{15}(1680)$ resonance excitation up to $Q^2 \approx 3$ GeV^2 which change rapidly between -1 and $+1$ at small $Q^2 \approx 0.5$ GeV^2 [4]. Frequently also data points for other resonance amplitudes, e.g. for the Roper are shown together with quark model calculations but they are not very reliable. Their statistical errors are often quite large but in most cases the model dependence is as large as the absolute value of the data points. In this context it is worth noting that also the word 'data point' is somewhat misleading because these photon couplings and amplitudes cannot be measured directly but can only be derived in a partial wave analysis. Only in the case of the $\Delta(1232)$ resonance this can and has been done directly in the experiment by Beck et al. at Mainz [5]. For the Delta it becomes possible due to two important theoretical facts, the Watson theorem and the well confirmed validity of the $s+p$ – wave truncation. Within this assumption a complete experiment was done with polarized photons and with the measurement of both π^0 and π^+ in the final state, allowing also for an isospin separation. For other resonances neither the theoretical constraints are still valid nor are we any close to a complete experiment. The old data base was rather limited with large error bars and no data with either target or recoil polarization was available. Even now we do not have many data points with double polarization, however, the situation for unpolarized $e + p \rightarrow e' + p + \pi^0$ has considerably improved, mainly by the new JLab experiments in all three halls A,B and C. These data cover a large energy range from the Delta up to the third resonance region with a wide angular range in θ_π. Due to the 2π coverage in the ϕ angle a separation of the unpolarized cross section

$$\frac{d\sigma_v}{d\Omega} = \frac{d\sigma_T}{d\Omega} + \varepsilon\frac{d\sigma_L}{d\Omega} + \sqrt{2\varepsilon(1+\varepsilon)}\frac{d\sigma_{LT}}{d\Omega}\cos\phi + \varepsilon\frac{d\sigma_{TT}}{d\Omega}\cos 2\phi \tag{1}$$

in three parts becomes possible and is very helpful for the partial wave analysis. Even without a Rosenbluth separation of $d\sigma_T$ and $d\sigma_L$ we have an enhanced sensitivity of the longitudinal amplitudes due to the $d\sigma_{LT}$ interference term. Such data are the basis of our new partial wave analysis with an improved version of the Mainz unitary isobar model MAID.

2 Photo- and electroproduction

For our analysis of pion electroproduction we will use the dynamical model DMT [6] and the unitary isobar model MAID [7]. In the dynamical approach to pion photo- and electroproduction [8], the t-matrix is expressed as

$$t_{\gamma\pi}(E) = v_{\gamma\pi} + v_{\gamma\pi}\,g_0(E)\,t_{\pi N}(E)\,, \tag{2}$$

where $v_{\gamma\pi}$ is the transition potential operator for $\gamma^* N \to \pi N$, and $t_{\pi N}$ and g_0 denote the πN t-matrix and free propagator, respectively, with $E \equiv W$ the total energy in the CM frame. A multipole decomposition of (2) gives the physical amplitude in channel α [8],

$$t_{\gamma\pi}^{(\alpha)}(q_E, k; E + i\epsilon) = \exp\left(i\delta^{(\alpha)}\right)\cos\delta^{(\alpha)} \times [v_{\gamma\pi}^{(\alpha)}(q_E, k)$$
$$+ P\int_0^\infty dq'\frac{q'^2 R_{\pi N}^{(\alpha)}(q_E, q'; E)\, v_{\gamma\pi}^{(\alpha)}(q', k)}{E - E_{\pi N}(q')}], \tag{3}$$

where $\delta^{(\alpha)}$ and $R_{\pi N}^{(\alpha)}$ are the πN scattering phase shift and reaction matrix in channel α, respectively; q_E is the pion on-shell momentum and $k = |\mathbf{k}|$ is the photon momentum. The multipole amplitude in (3) manifestly satisfies the Watson theorem and shows that the $\gamma\pi$ multipoles depend on the half-off-shell behavior of the πN interaction.

In a resonant channel the transition potential $v_{\gamma\pi}$ consists of two terms

$$v_{\gamma\pi}(E) = v_{\gamma\pi}^B + v_{\gamma\pi}^R(E), \tag{4}$$

where $v_{\gamma\pi}^B$ is the background transition potential and $v_{\gamma\pi}^R(E)$ corresponds to the contribution of the bare resonance excitation. The resulting t-matrix can be decomposed into two terms [9]

$$t_{\gamma\pi}(E) = t_{\gamma\pi}^B(E) + t_{\gamma\pi}^R(E), \tag{5}$$

where

$$t_{\gamma\pi}^B(E) = v_{\gamma\pi}^B + v_{\gamma\pi}^B\,g_0(E)\,t_{\pi N}(E), \tag{6}$$
$$t_{\gamma\pi}^R(E) = v_{\gamma\pi}^R + v_{\gamma\pi}^R\,g_0(E)\,t_{\pi N}(E). \tag{7}$$

Here $t_{\gamma\pi}^B$ includes the contributions from the nonresonant background and renormalization of the vertex $\gamma^* N R$. The advantage of such a decomposition is that all the processes which start with the excitation of a bare resonance are summed up in $t_{\gamma\pi}^R$. Note that the multipole decomposition of both $t_{\gamma\pi}^B$ and $t_{\gamma\pi}^R$ would take the same form as (3).

As in MAID [7], the background potential $v_{\gamma\pi}^{B,\alpha}(W, Q^2)$ was described by Born terms obtained with an energy dependent mixing of pseudovector-pseudoscalar πNN coupling and t-channel vector meson exchanges. The mixing parameters and coupling constants were determined from an analysis of nonresonant multipoles in the appropriate energy regions. In the new version of MAID, the S, P, D and F waves of the background contributions are unitarized in accordance with the K-matrix approximation,

$$t_{\gamma\pi}^{B,\alpha}(\text{MAID}) = \exp\left(i\delta^{(\alpha)}\right)\cos\delta^{(\alpha)} v_{\gamma\pi}^{B,\alpha}(W, Q^2). \tag{8}$$

From (3) and (8), one finds that the difference between the background terms of MAID and of the dynamical model is that off-shell rescattering contributions (principal value integral) are not included in MAID. To take account of the inelastic effects at the higher energies, we replace $\exp\left(i\delta^{(\alpha)}\right)\cos\delta^{(\alpha)} = \frac{1}{2}(\exp\left(2i\delta^{(\alpha)}\right)+1)$ in (3) and (8) by $\frac{1}{2}(\eta_\alpha\exp\left(2i\delta^{(\alpha)}\right)+1)$, where η_α is the inelasticity. In our actual calculations, both the πN phase shifts $\delta^{(\alpha)}$ and inelasticity parameters η_α are taken from the analysis of the GWU group [10].

Following [7], we assume a Breit-Wigner form for the resonance contribution $\mathcal{A}_\alpha^R(W, Q^2)$ to the total multipole amplitude,

$$\mathcal{A}_\alpha^R(W, Q^2) = \bar{\mathcal{A}}_\alpha^R(Q^2)\frac{f_{\gamma R}(W)\Gamma_R\,M_R\,f_{\pi R}(W)}{M_R^2 - W^2 - iM_R\Gamma_R}\,e^{i\phi}, \tag{9}$$

where $f_{\pi R}$ is the usual Breit-Wigner factor describing the decay of a resonance R with total width $\Gamma_R(W)$ and physical mass M_R. The expressions for $f_{\gamma R}$, $f_{\pi R}$ and Γ_R are given in [7]. The phase $\phi(W)$ in (9) is introduced to adjust the phase of the total multipole to equal the corresponding πN phase shift $\delta^{(\alpha)}$. While in the original version of MAID [7] only the 7 most important nucleon resonances were included with mostly only transverse e.m. couplings, in our new version all four star resonances below $W = 2$ GeV are included. These are $P_{33}(1232)$, $P_{11}(1440)$, $D_{13}(1520)$, $S_{11}(1535)$, $P_{33}(1232)$, $S_{31}(1620)$, $S_{11}(1650)$, $D_{15}(1675)$, $F_{15}(1680)$, $D_{33}(1700)$, $P_{13}(1720)$, $F_{35}(1905)$, $P_{31}(1910)$ and $F_{37}(1950)$.

The resonance couplings $\bar{\mathcal{A}}_\alpha^R(Q^2)$ are independent of the total energy and depend only on Q^2. They can be taken as constants in a single-Q^2 analysis, e.g. in photoproduction, where $Q^2 = 0$ but also at any fixed Q^2, where enough data with W and θ variation is available. Alternatively they can also be parametrized as functions of Q^2 in an ansatz like

$$\bar{A}_\alpha(Q^2) = \bar{A}_\alpha(0)\frac{1 + c_1^\alpha Q^2}{(1 + c_2^\alpha Q^2)^n} \tag{10}$$

with $n \geq 2$. Also other parameterizations with an asymptotic fall-off for large Q^2, e.g. of Gaussian form work equally well and can effectively be rewritten in either form.

Table 1. Recent experimental data of π^0 electroproduction on the proton. The Mainz experiment was done with beam and recoil polarization, all others are unpolarized measurements. From JLab Hall B data sets at fixed Q^2 of 0.4, 0.525, 0.65, 0.75 0.90, 1.15 and 1.45 GeV^2 have been used

laboratory	$Q^2(GeV^2)$	$W_{cm}(MeV)$	$\theta_\pi^{cm}(deg)$
Mainz [20]	0.121	1232	$180°$
Bates [21]	0.126	1152 - 1322	$0 - 38°$
Bonn [22]	0.630	1153 - 1312	$5 - 175°$
JLab, Hall A[23]	1.0	1110 - 1950	$146 - 167°$
JLab, Hall B[24]	$0.4 - 1.8$	1100 - 1680	$26 - 154°$
JLab, Hall C[25]	2.8, 4.0	1115 - 1385	$25 - 155°$

With such an ansatz it is possible to determine the parameters $\bar{A}_\alpha(0)$ from a fit to the world database of photoproduction, while the parameters c_1^α and c_2^α can be obtained from a combined fitting of all electroproduction data at different Q^2. The latter procedure we call the 'superglobal fit'. In MAID the photon couplings $\bar{A}_\alpha$ are direct input parameters.

Equation (9) can also be used for a general definition. At the resonance position, $W = M_R$ we obtain

$$\mathcal{A}_\alpha^R(M_R, Q^2) = i\,\bar{A}_\alpha(Q^2)f_{\gamma R}(M_R)f_{\pi R}(M_R)c_{\pi N}e^{i\phi(M_R)}$$
$$= \mathcal{A}_\alpha^{res}(M_R, Q^2)\,e^{i\phi(M_R)} \qquad (11)$$

with $f_{\gamma R}(M_R) = 1$ and

$$f_{\pi R}(M_R) = \left[\frac{1}{(2j+1)\pi}\frac{k_W}{|q|}\frac{m_N}{M_R}\frac{\Gamma_{\pi N}}{\Gamma_{tot}^2}\right]^{1/2}. \qquad (12)$$

The factor $c_{\pi N}$ is $\sqrt{3/2}$ and $-1/\sqrt{3}$ for the isospin 3/2 and isospin 1/2 multipoles, respectively. This leads to the definition

$$\bar{A}_\alpha(Q^2) = \frac{1}{c_{\pi N}f_{\pi R}(M_R)}\mathrm{Im}\mathcal{A}_\alpha^{res}(M_R, Q^2). \qquad (13)$$

It is important to note that by this definition the phase factor $e^{i\phi}$ in (9 and 11) is not considered as part of the resonant amplitude but rather as an artifact of the unitarization procedure. In the case of the $\Delta(1232)$ resonance this phase vanishes at the resonance position due to Watson's theorem, however, for all other resonances it is finite and in some extreme cases it can reach values of about 60^0. $\bar{A}_\alpha$ is a short-hand notation for the electric, magnetic and longitudinal multipole photon couplings of a given partial wave α. As an example, for the P_{33} partial wave the specific couplings are denoted by $\bar{E}_{1+}$, $\bar{M}_{1+}$ and $\bar{S}_{1+}$. By linear combinations they are connected with the more commonly used helicity photon couplings $A_{1/2}$, $A_{3/2}$ and $S_{1/2}$.

For resonances with total spin $j = \ell + 1/2$ we get

$$A_{1/2}^{\ell+} = -\frac{1}{2}[(\ell+2)\bar{E}_{\ell+} + \ell\bar{M}_{\ell+}],$$
$$A_{3/2}^{\ell+} = \frac{1}{2}\sqrt{\ell(\ell+2)}(\bar{E}_{\ell+} - \bar{M}_{\ell+}),$$

$$S_{1/2}^{\ell+} = -\frac{\ell+1}{\sqrt{2}}\bar{S}_{\ell+} \qquad (14)$$

and for $j = (\ell + 1) - 1/2$

$$A_{1/2}^{(\ell+1)-} = \frac{1}{2}[(\ell+2)\bar{M}_{(\ell+1)-} - \ell\bar{E}_{(\ell+1)-}],$$
$$A_{3/2}^{(\ell+1)-} = -\frac{1}{2}\sqrt{\ell(\ell+2)}(\bar{E}_{(\ell+1)-} + \bar{M}_{(\ell+1)-}),$$
$$S_{1/2}^{(\ell+1)-} = -\frac{\ell+1}{\sqrt{2}}\bar{S}_{(\ell+1)-}. \qquad (15)$$

With some care, as will be discussed below, the photon couplings can be directly compared to matrix elements of the electromagnetic current calculated in quark models between the nucleon and the excited resonance states,

$$A_{1/2} = -\sqrt{\frac{2\pi\alpha_{fs}}{k_W}} < R, \frac{1}{2}\,|\,J_+\,|\,N, -\frac{1}{2} > \zeta\,,$$
$$A_{3/2} = -\sqrt{\frac{2\pi\alpha_{fs}}{k_W}} < R, \frac{3}{2}\,|\,J_+\,|\,N, \frac{1}{2} > \zeta\,,$$
$$S_{1/2} = -\sqrt{\frac{2\pi\alpha_{fs}}{k_W}} < R, \frac{1}{2}\,|\,\rho\,|\,N, \frac{1}{2} > \zeta\,, \qquad (16)$$

where $J_+ = -\frac{1}{\sqrt{2}}(J_x + iJ_y)$. However, these couplings are only defined up to a phase ζ. Since the sign of the pionic decay of the resonance has been ignored in the empirical definition of these amplitudes, (13), it must be taken into account in a model calculation in order to make comparison with the empirical data. Therefore the phases ζ have to be individually calculated in each model. In the calculations found in the literature this has often been ignored and causes some confusion in comparing these numbers, especially in critical cases as for the Roper resonance $P_{11}(1440)$, where the correct sign cannot simply be guessed.

3 The hypercentral constituent quark model

As an application for evaluating the photon couplings we have used the hypercentral Constituent Quark Model (hCQM) [11]. It consists of a hypercentral quark interaction containing a linear plus coulomb-like term, as suggested by lattice QCD calculations [12]

$$V(x) = -\frac{\tau}{x} + \alpha x \;, \quad \text{with} \quad x = \sqrt{\rho^2 + \lambda^2}\,, \qquad (17)$$

where x is the hyperradius defined in terms of the standard Jacobi coordinates ρ and λ. We can think of this potential both as a two-body potential in the hypercentral approximation or as a true three-body potential. A hyperfine term of the standard form [13] is added and treated as a perturbation. The parameters α, τ and the strength of the hyperfine interaction are fitted to the spectrum ($\alpha = 1.61\ fm^{-2}$, $\tau = 4.59$ and the strength of the hyperfine interaction is determined by the Δ - Nucleon mass difference). Recently, isospin dependent terms have been

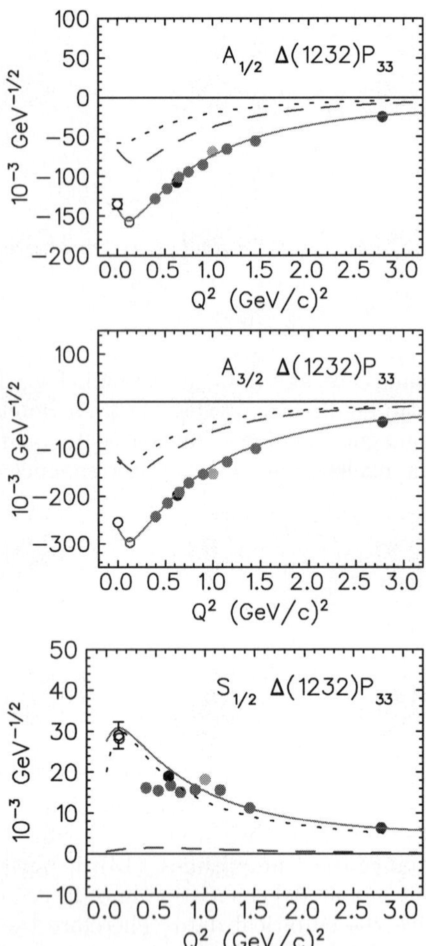

Fig. 1. The Q^2 dependence of the $N \to \Delta$ helicity amplitudes. The solid and dashed curves are the results of the superglobal fit with MAID and the predictions of the hyperspherical constituent quark model. The dotted lines show the pion cloud contributions calculated with DMT. The data points at finite Q^2 are the results of our single-Q^2 fits, see Table 1 for references. At $Q^2 = 0$ for $A_{1/2}$ and $A_{3/2}$ the photon couplings from PDG are shown [2]

introduced [14] in the hCQM hamiltonian. The complete interaction used is given by

$$H_{int} = V(x) + H_{\mathrm{S}} + H_{\mathrm{I}} + H_{\mathrm{SI}} \; . \tag{18}$$

Having fixed the values of all parameters, the resulting wave functions have been used for the calculation of the photocouplings [15], the transition form factors for the negative parity resonances [16], the elastic form factors [17,18] and now also for the longitudinal and tranverse transition form factors for all the 3- and 4-star and the missing resonances.

4 Data analysis

The unitary isobar model MAID was used to analyze the world data base of pion photoproduction and recent

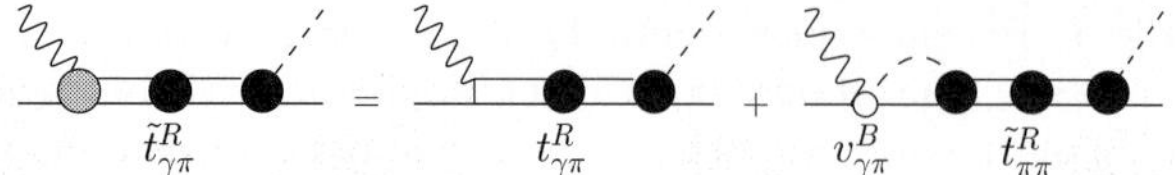

Fig. 2. Resonances with dressed and bare electromagnetic vertices

differential cross section data on $p(e, e'p)\pi^0$ from Mainz, Bates, Bonn and JLab. These data cover a Q^2 range from $0.1 \cdots 4.0$ $(\mathrm{GeV}/c)^2$ and an energy range $1.1 < W < 2.0$ GeV, see Table 1. In a first attempt we have fitted each data set at a constant Q^2 value separately. This is similar to a partial wave analysis of pion photoproduction and only requires additional longitudinal couplings for all the resonances. The Q^2 evolution of the background, Born terms and vector meson exchange, is described with a standard dipole form factor. In a second attempt we have introduced a Q^2 evolution of the transition form factors of the nucleon to N^* and Δ resonances and have parameterized each of the transverse ($A_{1/2}$ and $A_{3/2}$) and longitudinal ($S_{1/2}$) helicity amplitudes. In a combined fit with all electroproduction data from the world data base of GWU/SAID [19] and the data of our single-Q^2 fit we obtained a Q^2 dependent solution (superglobal fit). In Fig. 1 we show our results for the $\Delta(1232)$ excitation. Our superglobal fit agrees very well with our single-Q^2 fits, except for the 2 lowest points of $S_{1/2}$ from our analysis of the Hall B data. Whether this is an indication for a different Q^2 dependence has still to be investigated. Generally, all our single-Q^2 points are shown with statistical errors from χ^2 minimization only. A much bigger error has to be considered for model dependence.

We also compare our empirical analyses with the predictions of the hypercentral constituent quark. It turns out that the transverse amplitudes of the quark model are about half of the magnitudes and for the longitudinal amplitude $S_{1/2}(Q^2)$ the quark model gives essentially zero. This is due to the fact that in the empirical analysis the resonance contribution is fully dressed by the pion cloud which is not the case in a constituent quark model. As depicted in Fig. 2 a fully dressed resonance contribution is renormalized on each vertex and in the propagator. The baryonic states of the hCQM including hyperfine interaction can be considered as resonances dressed by hadronic interaction giving rise to the empirical masses. However, the electromagnetic vertex correction (third part of Fig. 2) is not included and has to be calculated separately. We have already started to do this and in a first attempt we have used the dynamical model DMT and extracted the pion loop contributions for the s- and p-waves. This estimate of the pion cloud vertex correction is shown as dotted lines in the figures. For the longitudinal Delta excitation the entire amplitude is practically given by the pion cloud contribution and only a negligible part arises from a bare Delta. The same is true for the electric amplitude which is given by the combination $A_{1/2} - A_{3/2}/\sqrt{3}$. This calculation also explains why none of the constituent quark models was ever able to give the empirical strength of the M1 excitation (or the transition moment $\mu_{N\Delta}$) of the $\Delta(1232)$. While in a simple $SU(2)$ calculation the transition mo-

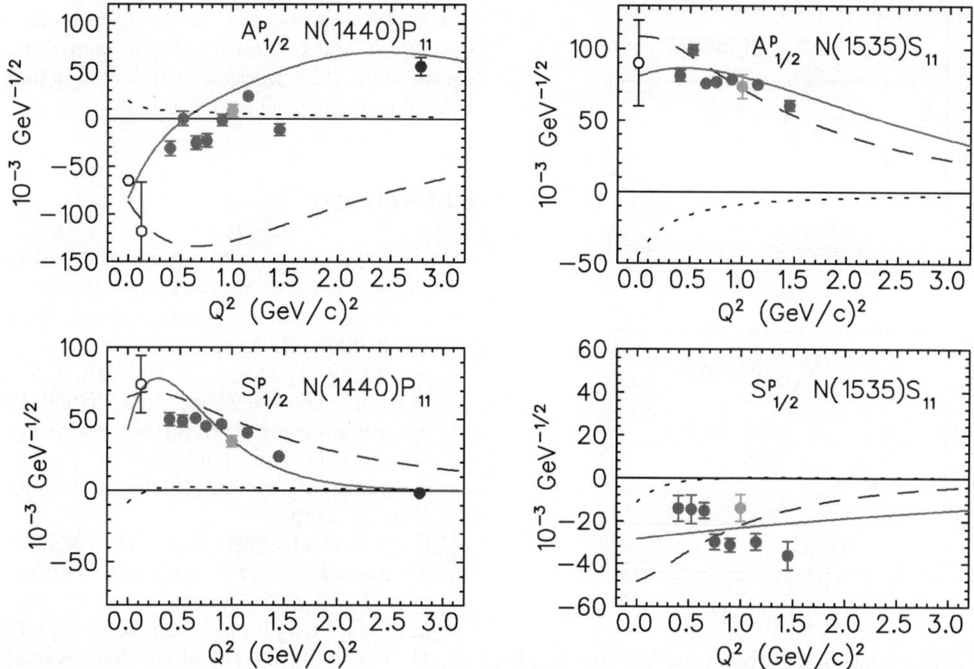

Fig. 3. The Q^2 dependence of the transverse and longitudinal helicity amplitudes for the $P_{11}(1440)$ and the $S_{11}(1535)$ resonance excitation. The notation of the curves and the data is the same as in Fig. 1

ment is lower by about 30% in more refined calculations it can be as low as only half of the empirical value. This is also the case here in our hCQM, even if it has more realistic wave functions.

In Fig. 3 we show our results for the helicity amplitudes of the Roper resonance $P_{11}(1440)$ and the $S_{11}(1535)$. For these resonance amplitudes the pion cloud contributions are most important near the photon point and become already negligible around $Q^2 = 0.5$ GeV^2. The comparison between the hCQM and the empirical amplitudes is reasonably good, except for the $A_{1/2}$ amplitude of the Roper. This finding has to be further investigated both in the framework of the quark model and also in the empirical analysis. Certainly, for the Roper resonance the existing data is not very sensitive to this partial wave. Further experiments with double polarization could be very helpful to solve this problem.

Finally, in Figs. 4 and 5 we show our results for the $D_{13}(1520)$ and the $F_{15}(1680)$ resonances. Here the largest discrepancies between our quark model calculations and the empirical analysis appear in the helicity 3/2 amplitudes at small Q^2. So far, the dynamical model calculations have only been done for s- and p- waves, therefore we cannot give a pion cloud calculations for these partial waves. However, our findings encourages very strongly such extensions of the dynamical model. Furthermore we also have some empirical results for the partial waves that are not shown here, but most of them come out of the fit with rather large errors bars in the single-Q^2 analysis. This gives us less confidence also for our superglobal fit. The reason for it is mainly that we have fewer data points to analyze at higher energies.

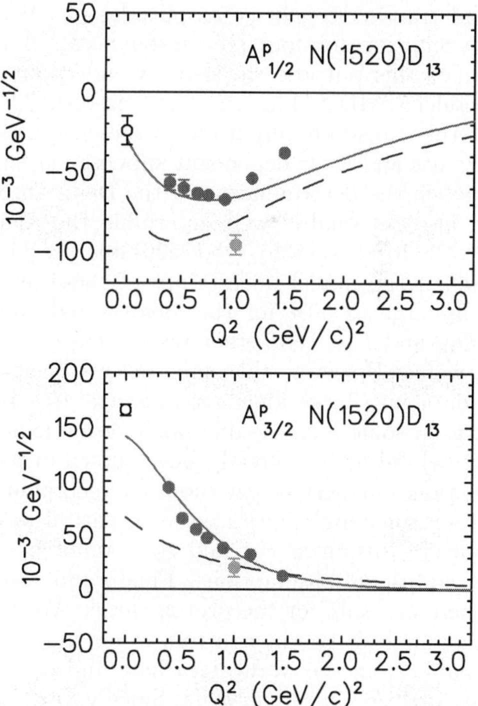

Fig. 4. The Q^2 dependence of the transverse helicity amplitudes for the $D_{13}(1520)$ resonance excitation. The notation of the curves and the data is the same as in Fig. 1

5 Conclusions

Using the world data base of pion photo- and electroproduction and recent data from Mainz, Bonn, Bates and JLab we have made a first attempt to extract all longitu-

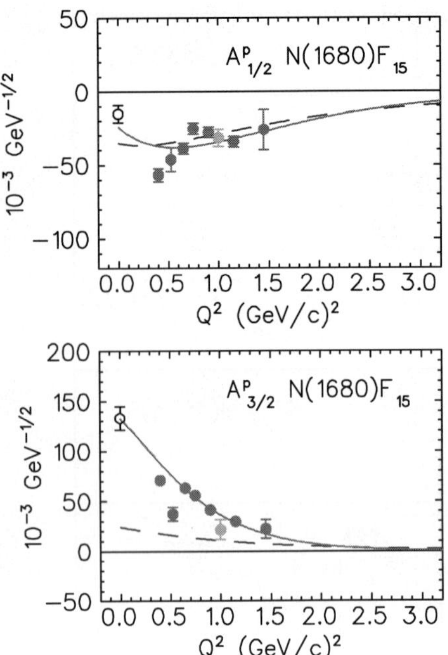

Fig. 5. The Q^2 dependence of the transverse helicity amplitudes for the $F_{15}(1680)$ resonance excitation. The notation of the curves and the data is the same as in Fig. 1

dinal and transverse helicity amplitudes of nucleon resonance excitation for four star resonances below $W = 2$ GeV. For this purpose we have extended our unitary isobar model MAID and have parametrized the Q^2 dependence of the transition amplitudes. Comparisons between single-Q^2 fits and a Q^2 dependent superglobal fit give us confidence in the determination of the Delta amplitudes. We can also reasonably well determine the amplitudes of the $P_{11}(1440)$, $S_{11}(1535)$, $D_{13}(1520)$ and the $F_{15}(1680)$, even though the model uncertainty of these amplitudes can be as large as 50% for the longitudinal amplitudes of the D_{13} and F_{15}. For other resonances the situation is even worse. However, this only reflects the fact that precise data in a large kinematical range are absolutely necessary. In some cases double polarization experiments are very helpful as has already been shown in pion photoproduction. Furthermore, without charged pion electroproduction, some ambiguities between partial waves that differ only in isospin as S_{11} and S_{31} cannot be resolved without additional assumptions. Finally, all results discussed here are only for the proton target. We have also started an analysis for the neutron, where much less data are available from the world data base and no new data has been analyzed in recent years. Since we can very well rely on isospin symmetry, only the electromagnetic couplings of the neutron resonances with isospin $1/2$ have to be determined. We have found such a solution for the neutron and will implement it in the next version of MAID (MAID2003). It will be a challenge for the experiment to investigate also the neutron resonances in the near future.

Acknowledgements. We wish to thank T. Bantes, G. Laveissiere and C. Smith for their contribution on the experimental data. This work was supported in part by the Deutsche Forschungsgemeinschaft (SFB443).

References

1. G. Höhler et al.: *Handbook of Pion-Nucleon Scattering, Physics Data 12–1* (Karlsruhe, 1979)
2. K. Hagiwara et al. (Particle Data Group): Phys. Rev. D **66**, 010001 (2002)
3. S.J. Dong, T. Draper, I. Horvath, F.X. Lee, K.F. Liu, N. Mathur, and J.B. Zhang: hep-ph/0306199
4. For an overview and further references see S. Boffi, C. Giusti, F.D. Pacatiand, and M. Radici: Electromagnetic Response of Atomic Nuclei, Clarendon Press, Oxford, 1996, p. 114ff
5. R. Beck et al.: Phys. Rev. C **61**, 035204 (2000)
6. S. Kamalov, S.N. Yang, D. Drechsel, O. Hanstein, and L. Tiator: Phys. Rev. C **64**, 032201 (2001); http://www.kph.uni-mainz.de/MAID/DMT/
7. D. Drechsel, O. Hanstein, S.S. Kamalov, and L. Tiator: Nucl. Phys. A **645**, 145 (1999); http://www.kph.uni-mainz.de/MAID/
8. S.N. Yang: J. Phys. G **11**, L205 (1985)
9. S.S. Kamalov and S.N. Yang: Phys. Rev. Lett. **83**, 4494 (1999)
10. R.A. Arndt, I.I. Strakovsky and R.L. Workman: Phys. Rev. C **56**, 577 (1997)
11. M. Ferraris, M.M. Giannini, M. Pizzo, E. Santopinto, and L. Tiator: Phys. Lett. B **364**, 231 (1995)
12. Gunnar S. Bali: Phys. Rep. **343**, 1 (2001)
13. N. Isgur and G. Karl: Phys. Rev. D **18**, 4187 (1978); D **19**, 2653 (1979); D **20**, 1191 (1979); S. Godfrey and N. Isgur: Phys. Rev. D **32**, 189 (1985)
14. M.M. Giannini, E. Santopinto, and A. Vassallo: Eur. Phys. J. A **12**, 447 (2001); Nucl. Phys. A **699**, 308 (2002)
15. M. Aiello, M. Ferraris, M.M. Giannini, M. Pizzo, and E. Santopinto: Phys. Lett. B **387**, 215 (1996)
16. M. Aiello, M. M. Giannini, and E. Santopinto: J. Phys. G: Nucl. Part. Phys. **24**, 753 (1998)
17. M. De Sanctis, E. Santopinto, and M.M. Giannini: Eur. Phys. J. **A1**, 187 (1998).
18. M. De Sanctis, M.M. Giannini, L. Repetto, and E. Santopinto: Phys. Rev. **C62**,025208 (2000).
19. R.A. Arndt, W.J. Briscoe, I.I. Strakovsky, and R.L. Workman: http://gwdac.phys.gwu.edu/
20. Th. Pospischil et al.: Phys. Rev. Lett. **86**, 2959 (2001)
21. C. Mertz et al.: Phys. Rev. Lett. **86**, 2963 (2001)
22. T. Bantes and R. Gothe: private communication
23. G. Laveissiere et al.: Proc. of NSTAR2001, Mainz, World Scientific 2001, p 271 and private communications
24. K. Joo et al.: Phys. Rev. Lett. **88**, 122001-1 (2002)
25. V.V. Frolov et al.: Phys. Rev. Lett. **82**, 45 (1999)

Eur Phys J A (2004) **19**, s01, 61–65

Digital Object Identifier (DOI) 10.1140/epjad/s2004-03-010-4

EPJ A direct

electronic only

Impact of recent data on N* structure

Steven A. Dytman

Department of Physics and Astronomy, University of Pittsburgh, Pittsburgh, PA 15260 USA

Received: 29 July 2003 / Accepted: 14 Nov 2003 /
Published Online: 6 Feb 2004 – © Società Italiana di Fisica / Springer-Verlag 2004

Abstract. Many modern experiments are posed with the issue of physics interpretation of their data when the theory is complicated. Certainly, experiments studying N* resonances are in this category. This short paper presents examples of interpretation made by inspection of the data, Breit-Wigner analyses, and coupled channels analysis. There are significant advantages to all three, but only a coupled channels analysis can provide the checks needed for a complete analysis. Examples from the S_{11} and P_{13} partial waves are discussed.

PACS. 13.30.-a Baryon decays – 13.60.Le Meson production in electromagnetic reactions – 13.20.Gk Baryon resonances with S=0

1 Introduction

The study of N* resonances has gotten a large boost from results from CEBAF, GRAAL, Mainz, Bonn and Spring8 in the last few years. These labs were all represented in talks at this conference. Experimenters are blessed with many reactions that feed into N^* states, but this also makes analysis more of a challenge. It is possible to learn about a given N^* state from a single reaction as long as there is a large enough coupling to the initial and final channels that a bump in the total cross section is produced. However, what if 2 reactions give contradictory results? (Is there a way to choose a correct interpretation?) What if a state is not seen in another reaction? (Does that mean the state doesn't really exist?) Unitary coupled channel analyses provide the formalism to get the necessary understanding. They simultaneously account for all possible decay channels in all reactions with various theoretical constraints.

Data alone can also provide insights into the underlying states. Is there a peak in the total cross section? This most likely has a resonance associated with it. However, peaks can come from threshold effects and low statistics and/or broad peaks have a way of disappearing when more accurate experiments are done. Simple Breit Wigner analyses are the natural way to first analyze experimental results. This gives a common language to present results. However important these studies may be, assumptions must be made in absence of the bigger picture because there is no well-established link between the Breit-Wigner models and coupled channel models. This talk provides a few examples where new data gets a clearer interpetation when viewed through a coupled channel picture [1]. Although the new data can seldom be directly incorporated into a coupled channel fit, comparisons with full fits can

provide important tests of tentative interpretations of new data. Unfortunately, the simple Breit-Wigner interpretation is shown to be incomplete in some cases.

2 S_{11} partial wave

The S_{11} resonances are important to study. The lowest state ($S_{11}(1535)$) is so close to ηN threshold (1.486 GeV for a proton target) that many papers assumed the strong features in the data come from strong final state interactions. If so, the long-standing interpretation of the data as the quark model state would be in doubt. It also has a transition form factor with a very slow falloff with Q^2 that is very hard for quark models to reproduce. Finally, this partial wave is an excellent place to search for N^* states beyond the quark model because it has no states in the mass range 1.7-2.0 GeV. Thus, experiments can provide many ways to test models of these states.

Unfortunately, the lowest state has a very unusual energy dependence in πN elastic scattering and photoproduction because of the strong threshold effect when the ηN channel opens up. A Breit-Wigner shape is wrong for these interactions unless the threshold cusp is properly included. A speed plot is misleading because the most rapid change comes at ηN threshold (also where the total cross section peaks). Thus, one must be very careful in the choice of analyses to be used for this state. The PDG summary [2] shows significant doubt in its properties showing a strong need for new, better data and for more consistent analyses. On the other hand, the 2nd state ($S_{11}(1650)$) is prominent in πN elastic scattering. A third state ($S_{11}(2090)$) was seen in older πN data.

This confusing picture at first became more uncertain with recent data. The Mainz eta photoproduction data at

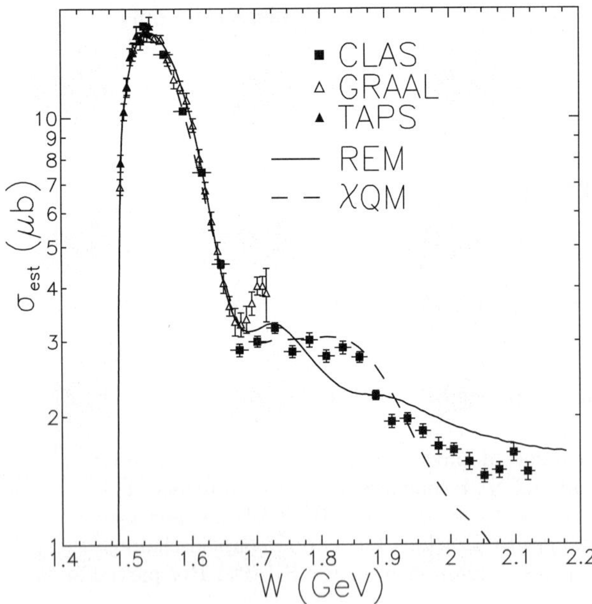

Fig. 1. η photoproduction total cross section data from Mainz, CLAS, and GRAAL

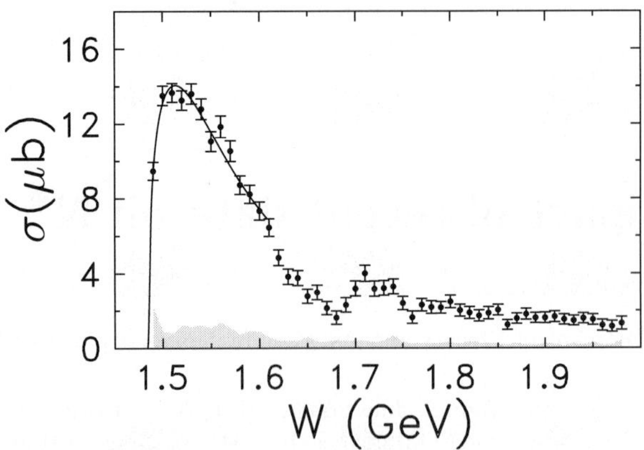

Fig. 2. Published η electroproduction total cross section data from CLAS for $Q^2=0.625$ $(\mathrm{GeV/c})^2$

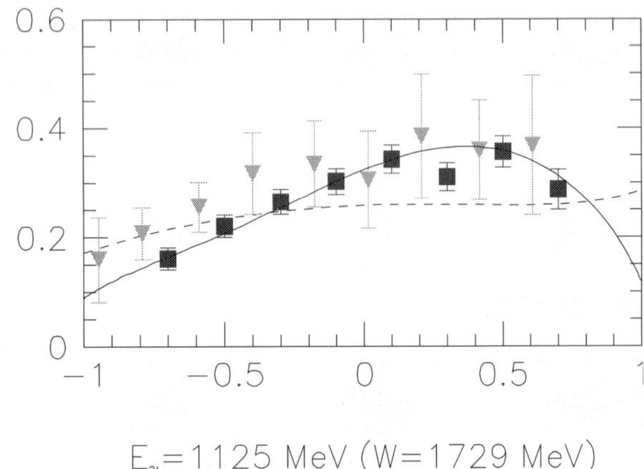

Fig. 3. η differential cross sections at W=1730 MeV for CLAS (squares) and GRAAL (diamonds), the energy where their total cross sections disagree

threshold (see Fig. 1) was one of the first reactions studied in the new generation of experiments. Quality was much higher than for the previous data. Although the energy region studied just reached the peak of $S_{11}(1535)$, these new data in conjunction with older data showed a prominent peak in the total cross section just above threshold that had the energy dependence of a single Breit-Wigner resonance. Breit-Wigner interpretations of these data gave a very large total width and proton photocoupling ($A_{1/2}$). These values were twice as large as the results of most previous πN analyses and caused an *apparent* controversy. The correct picture comes from a coupled channel analysis. The Breit-Wigner energy dependence is only valid near the peak and the coupled channel effects are large because the πN channel couples strongly to the ηN channel. Since the first two S_{11} states overlap, the quantum mechanical interference must be taken into account, also difficult to do with Breit-Wigner amplitudes. Thus, the πN and ηN asymptotic states and the 1535 and 1650 MeV intermediate states must all be included. Only a coupled channels model can do all of this with appropriate theoretical constraints. Therefore, only a coupled channel analysis can provide a consistent description of all the data. Our most recent result gives a full width of 122±20 MeV and $A_{1/2}^p = 91\pm6$ GeV$^{-1/2}$ [1].

The 3rd PDG state (at high energy) is not seen in newer πN elastic partial wave analyses [4]. Instead, tentative evidence for a bump was seen at $\sim$1.8 GeV in both GRAAL eta photoproduction [5] (see Fig. 1) and CLAS electroproduction [8] (see Fig. 2) cross section data. (At the same time, both the full width and photocoupling of $S_{11}(1535)$ got smaller in Breit-Wigner analyses as the higher energy data became available.) This would be a particularly interesting state as the quark model has no place for it. The picture for the 3rd state became quickly confusing when photoproduction data from CLAS [7] (Fig. 1) further extended the energy range covered. No peak in the *total cross section* was seen even though the *differential cross sections* from GRAAL and CLAS agreed at these energies. (see Fig. 3). The apparent contradiction is explained by the fact that neither of the experiments had complete angular coverage. Thus, a model was required to extrapolate to the most forward angles in order to determine the total cross section. GRAAL used a polynomial and CLAS used a model [6]. This issue is settled by looking at the more recent CLAS electroproduction data [9]. A very mild peak is seen in the total cross section (see Fig. 4) and the GRAAL extrapolation is shown to be more correct in the angular distributions (see Fig. 5. No strong conclusion can be made at this time.

A new third state should have repercussions in the interpretation of other data, especially the older, high quality πN elastic data. This author studied this issue in a series of fits with and without a third state. These fits included $\pi N \to \pi N$, $\pi N \to \eta N$, and $\pi N \to \pi\pi N$, $\gamma N \to \pi N$, and $\gamma p \to \eta p$ data including the recent GRAAL data. It gives very weak evidence for any third S_{11} state. Very inconclusive evidence for a state at $\sim$1.85 GeV is found. The only feature in the old data is a slight peak in the

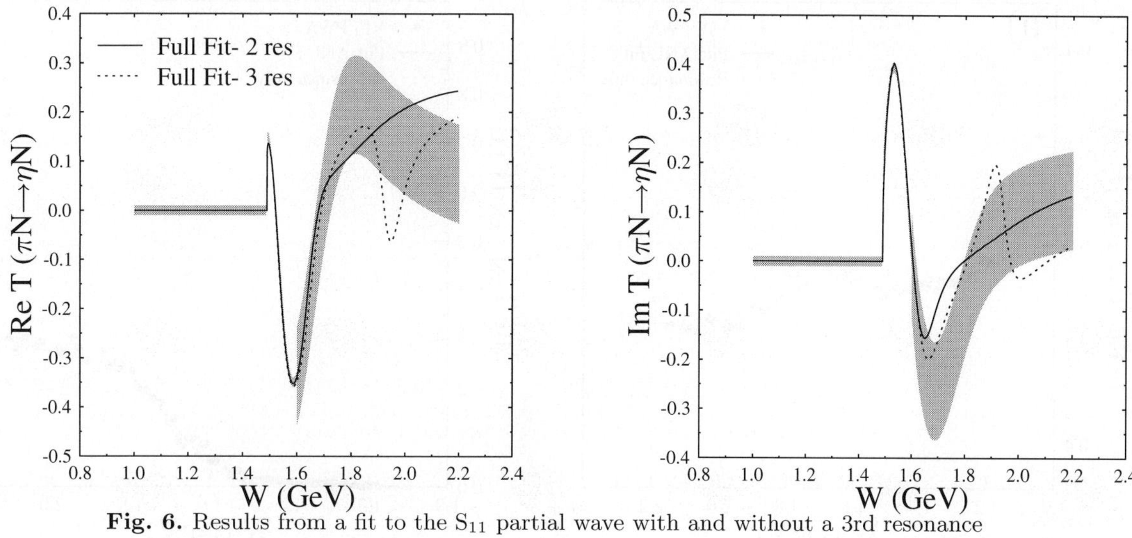

Fig. 6. Results from a fit to the S_{11} partial wave with and without a 3rd resonance

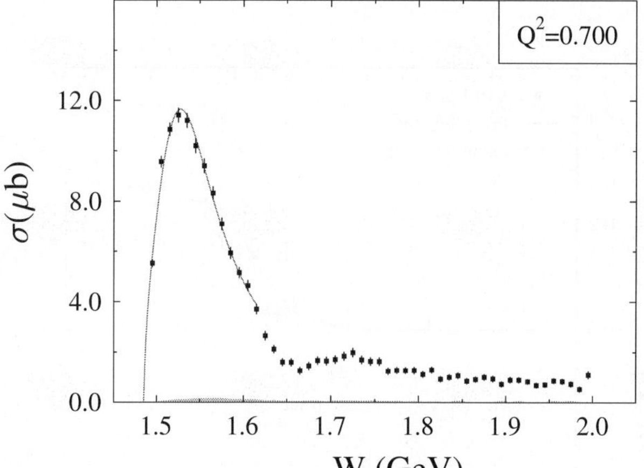

Fig. 4. New η electroproduction total cross section data from CLAS for Q^2=0.7 $(\mathrm{GeV/c})^2$

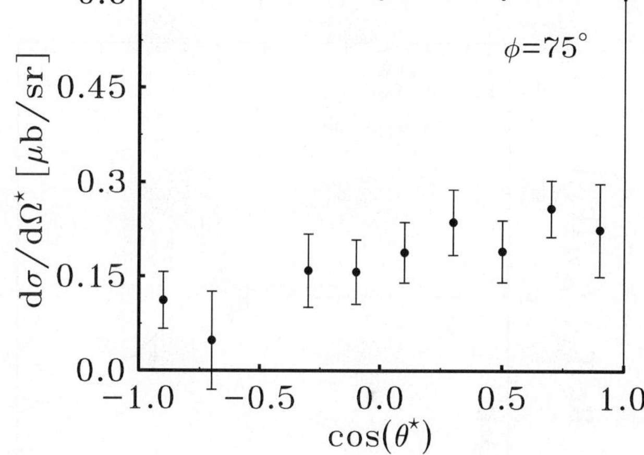

Fig. 5. New η electroproduction differential cross section data from CLAS for Q^2=0.8 $(\mathrm{GeV/c})^2$, W=1.72 GeV, and ϕ^*=75 degrees

$\pi N \to \eta N$ amplitude at about 1.8 GeV that cannot be fit except with a new state (see Fig. 6). However, the overall impact on χ^2 is very small and the mass and width of this new state are very uncertain in this study. As the reader can see, the fit has more structure than the data. At present, the status of this tentative 3rd S_{11} state is still unclear. Full understanding will await a partial wave analysis of the eta photoproduction data.

3 P$_{13}$ partial wave

States in this partial wave have been highlighted as a result of recent $\pi^+\pi^-$ electroproduction data from CLAS [10]. The data is far more accurate and has a much larger kinematical range than previous data.

They made a careful Breit-Wigner prediction for these data. This isn't trivial because the interference of the ρN, $\pi\Delta$ (both as decays from N^* resonances and as part of nonresonant amplitudes) and $\pi\pi N$ final states must be

accounted for. Fits to the $\pi^+\pi^- N$ photoproduction data gave a good description of the nonresonant rho production in the t-channel. However, the modeling of the electromagnetic and hadronic couplings to the various N^* states had to be taken from previous results. (This probably introduces significant model dependence since the previous studies used different models.)

The prediction was a good match to the data except for a significant shortfall in a prominent peak in the total cross section at $W \sim 1.75$ GeV. They noted the large ρN coupling to P$_{13}$ in PDG was probably inconsistent with the new data. Either the existing P$_{13}$ state is greatly modified or there is a new P$_{x3}$ (isospin is not determined) state in close proximity.

Potential uncertainties in the Breit-Wigner model cannot be ignored, but the uncertainties in the states at $W \sim$ 1.8 GeV are also large, leading us to prefer the more conservative interpretation. About 5 N^* states of different quantum numbers will all contribute to most observables,

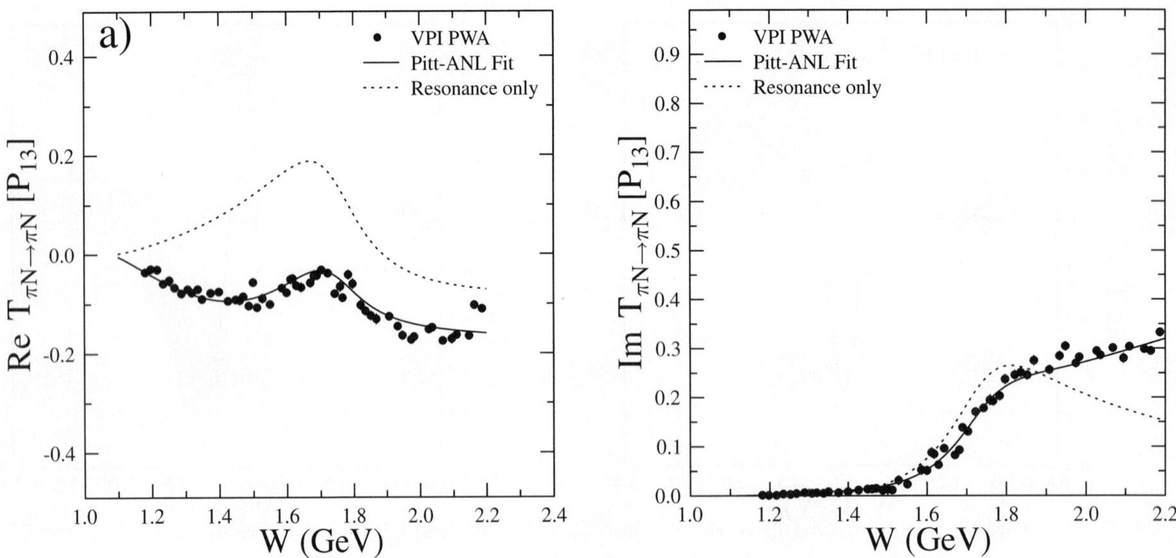

Fig. 7. P_{13} partial wave amplitude for πN elastic scattering. The full fit and the amplitude including only resonance couplings are shown

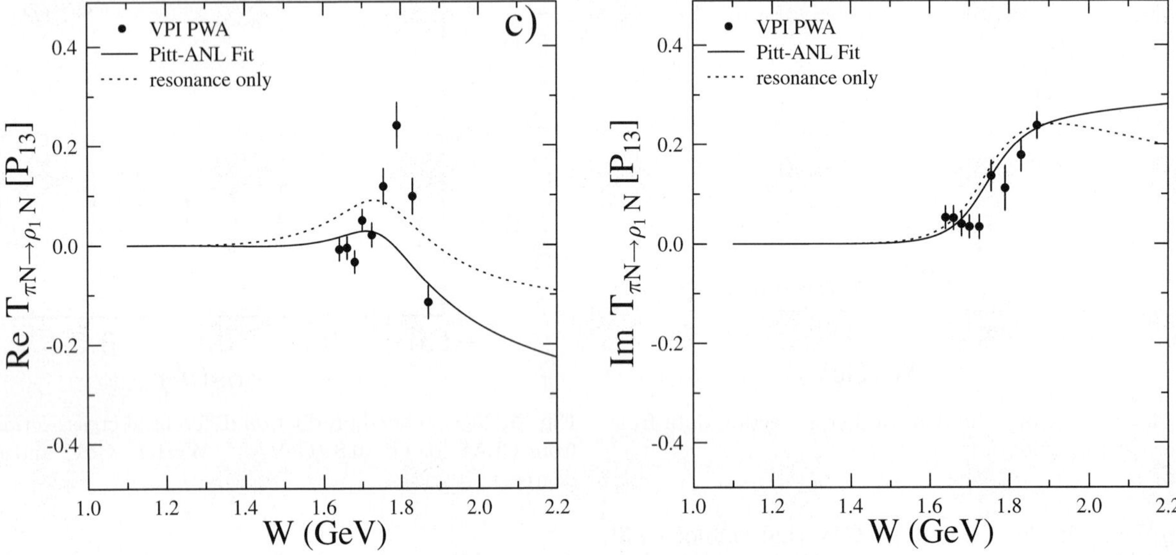

Fig. 8. P_{13} partial wave amplitude for $\pi N \to \rho N$ with $S_{\rho N}=1/2$

making unambiguous identification very difficult in low quality older data. The most recent partial wave analysis of $\pi N \to \pi\pi N$ data [3] surprisingly found no contribution of $\pi\Delta$ to P_{13}. In addition, the $\pi N \to \rho N$ amplitudes are very uncertain (see Fig. 8); the energy range covered is very small and the structure in the real part is surprisingly sharp. (We've increased the error bars by a factor of 2 because any model has great difficulty fitting such sharp features.) Therefore, a likely result of any resonance analysis is an uncertain branching fraction for $P_{13} \to \rho N$. The PDG result is a very large BF for this decay with a surprisingly small uncertainty. It is also surprising that P_{13} is the only state in this region to have a large ρN decay while most of the others have a dominant $\pi\Delta$ decay. Fits to this partial wave are very difficult for these reasons despite doubling the estimated error bars.

Difficulties with this partial wave led us to try new fits to the full set of partial wave amplitudes allowing a $\pi\Delta$ channel to be open even though there is no PWA result available. This fit gave better results than fits without the $\pi\Delta$ channel because the interferences were less complicated. In addition, the branching fractions for P_{13} decay to the two $\pi\pi N$ channels was much closer to the new values found in the Breit-Wigner analysis in the Ripani, et al. paper. Thus, we feel there is enough uncertainty in the old data to create significant doubt that a new state is seen in the new CLAS data. Nevertheless, the new data provide convincing evidence that the properties of $P_{13}(1720)$ require significant changes.

Another feature of the coupled channel fits is the importance of πN elastic scattering amplitudes (see Fig. 7), giving further concern for the Breit-Wigner fits. Although there is a prominent 'feature' at $W \sim 1.8$ GeV, the real amplitude has the peak while the normal Breit-Wigner shape has a peak in the imaginary amplitude. That and the large off-resonance amplitude are strong signals that the nonresonant amplitudes are also important. The πN nonresonant amplitude will therefore couple strongly into all inelastic reactions. While coupled channel models have this valuable constraint. the Breit-Wigner analysis of Ripani, et al. [10] has no ties to πN elastic scattering amplitudes.

4 Lessons learned

As new data come out, there will be many new opportunities for advancing knowledge of N^* states. There will also be potential mistakes to be made. Although Breit-Wigner analyses provide the obvious first way to extract physics from new data, the pitfalls of this approach should be recognized and appropriate caveats provided. The full problem is unfortunately complicated and much more sophisticated models are *sometimes* needed to learn the right physics interpretation. Previous coupled channel analyses provide an excellent guide to what truncations in channel space are appropriate. For example, the $S_{11}(1535)$ state can be studied with a minimum of πN and ηN channels because the $\pi \pi N$ channels aren't prominent. There are of course well established resonances such as the Delta and $D_{13}(1520)$ which can be studied using simple models with no theoretical difficulties as Breit-Wigner and coupled channels models give the same properties for these states. However, more complete studies are often required to know which simplifications are possible.

In the 2 examples given, the first Breit-Wigner analyses of exciting new data obtained results that are quite different than those obtained in published coupled channel work. This is due to the important constraints on any analysis that are provided by the new data, but also to the lack of theoretical constraints in any Breit-Wigner model. The coupled channel picture of $S_{11}(1535)$ is surprisingly different than the Breit-Wigner picture. The more complicated picture of S_{11} comes as the natural consequence of the strong coupling of $S_{11}(1535)$ to both πN and ηN. Further data and analyses will provide further clarification. In the case of P_{13}, even a careful Breit-Wigner analysis has to use both electromagnetic and hadronic couplings.

This would provide a problem in any case. Here, the lack of good data for $\pi N \to \pi \pi N$ makes a Breit-Wigner interpretation more difficult. The first priority should be to understand the real characteristics of $P_{13}(1720)$; with this uncertainty it is difficult to suggest a new state. In addition, there is enough flexibility in the old data to allow a coupled channel interpretation which is consistent with the new data.

5 Conclusions

Although it is an interesting time for N^* structure studies with all the data coming out, the interpretations will not always be simple. In the long run coupled channel analyses have significantly more theoretical and data constraints and will be more viable. Since this can take a year or more to come about, a reasonable balance between single channel Breit-Wigner and coupled channel analyses will be most fruitful for the first results. The Breit-Wigner analyses can be improved by checking interpretations against what is known from previous coupled channel work.

Acknowledgements. This work was supported by NSF grant PHY-0140116.

References

1. T.P. Vrana, S.A.Dytman, and T.-S.H. Lee: Phys. Rept. **328**, 181 (2000). Photoproduction results to be published
2. Particle Data Group: Eur. Phys. J. C **15**, 1 (2000)
3. D.M. Manley, R.A. Arndt, Y. Goradia, and V.L. Teplitz: Phys. Rev. D **30**, 904 (1984)
4. R.A. Arndt, I.I. Strakovsky, and R.L. Workman: Phys. Rev. C **53**, 430 (1996)
5. F. Renard et al.: Phys. Lett. B **528**, 215 (2002)
6. D. Drechsel, O. Hanstein, S.S. Kamalov, and L. Tiator: Nucl. Phys. **645**, 145 (1999)
7. M. Dugger et al.: Phys. Rev. Lett. **89**, 222002 (2002)
8. R. Thompson et al.: Phys. Rev. Lett. **86**, 1702 (2001)
9. J. Mueller: contribution to the Proceedings of PANIC03, to be published
10. M. Ripani et al.: accepted for publication, Phys. Rev. Lett. (2003). See also contribution of M. Ripani to these proceedings

Eur Phys J A (2004) **19**, s01, 67–69
Digital Object Identifier (DOI) 10.1140/epjad/s2004-03-011-3

EPJ A direct
electronic only

Introduction to BRAG: The Baryon Resonance Analysis Group

D.M. Manley

Department of Physics, Kent State University, Kent, OH 44242 USA, e-mail: `manley@kent.edu`

Received: 28 July 2003 / Accepted: 14 Nov 2003 /
Published Online: 6 Feb 2004 – © Società Italiana di Fisica / Springer-Verlag 2004

Abstract. The Baryon Resonance Analysis Group (BRAG) was established in 1999 as a network of researchers dedicated to the extraction of resonance information from electromagnetic and hadronic facilities worldwide. This article presents an overview of the group and its activities.

PACS. 14.20.Gk Baryon resonances with S=0 – 14.20.Jn Hyperons

1 Introduction

A few years ago, a group of experimentalists, theorists, and phenomenologists met together and established the Baryon Resonance Analysis Group (BRAG). This group was formed in part because it was felt that a more concerted effort was needed to analyze and extract useful physics from the new N^* data then beginning to become available from JLab, Mainz, Bonn, Bates, BES, BNL-AGS, and other facilities where research on baryon resonances is performed. The operation of BRAG is the responsibility of its elected Steering Committee. The Steering Committee is responsible for scheduling, announcing, and presiding over meetings, and for inviting outside speakers. The Steering Committee also conducts the nomination and balloting process, and serves to maintain the roster and e-mail lists.

According to the group charter, the Steering Committee is responsible for maintaining contact with the individual Working Groups (see Sect. 2) and for soliciting regular updates on Working Group activities and progress every 4-6 months. The Steering Committee also oversees regular dissemination and publication of BRAG meetings minutes and BRAG results, maintenance of the BRAG web site, and contact with experimental collaborations. Finally, the Steering Committee is responsible for the search for external funding for BRAG. The current BRAG Steering Committee consists of Mark Manley (Chair), Bill Briscoe (Secretary), Cornelius Bennhold (Working Group Liaison), and Mauro Giannini (Publisher).

BRAG has four main goals: (1) To carry out a regular review of progress and remaining problems in the field of N^* physics; (2) To carry out a coordinated dissemination of data, existing results, and conventions to assist the field to grow and reduce duplication of efforts; (3) To facilitate the discussion of controversial results, thus allowing important new ideas to be more quickly recognized, and errors to be more quickly corrected; and (4) To initiate studies of problems not feasible for individual researchers. A regular review of progress and remaining problems in the field is needed given the proliferation of new measurements and analyses mentioned in the opening paragraph.

Membership in BRAG is defined as being listed on the membership roster. Those wishing to participate should supply their address and interest lists to the Secretary. No fees are charged and members may withdraw at any time. Voting members must either attend one meeting or respond to one e-mail survey every two years. The most recent BRAG meeting was held on October 8, 2002 at the University of Pittsburgh as a pre-meeting associated with the NSTAR 2002 Conference.

BRAG currently has about 117 members from 52 institutions in 18 different countries. The institutions having the largest numbers of members in BRAG are Jefferson Laboratory and the George Washington University in the U.S., and Universität Bonn and Universität Mainz in Germany. Additional information is available on the group website, http://cnr2.kent.edu/ manley/BRAG.html.

2 Working groups

Working groups are formed between BRAG members with similar interests and common goals in a particular aspect of N^* physics. BRAG members may belong to several working groups with overlapping interests. As the field evolves, the interest and the focus of a working group may adjust. A working group exists when at least three BRAG members have decided to collaborate on a particular issue and have notified the Secretary of their intention. Working group membership is open to all BRAG members and exists after it has been reported to the Secretary. One member acts as the main contact or chair.

There are three standard BRAG working groups: (1) Extraction and Interpretation of Resonance Parameters; (2) Partial-Wave Analysis; and (3) Database Issues.

2.1 Working Group on Extraction and Interpretation of Resonance Parameters

The Working Group on Extraction and Interpretation of Resonance Parameters was established to study the extraction and interpretation of resonance parameters within different approaches. One goal is to ascertain the minimum theoretical requirements that different approaches should include. Another question addressed by this group is how should background be separated from resonance contributions. This group also is concerned with the proper way to compare extracted resonance parameters with quark-model results. Its current chair is Cornelius Bennhold of the George Washington University.

One of the projects that this working group carried out was an investigation of nucleon resonance properties in different multichannel analyses [1]. A goal of this work was to study how the extraction of resonance parameters is affected by different methods of separating resonance and background contributions. Three independent university groups (GW-Giessen, KSU, Pitt-ANL) carried out independent unitary multichannel fits and investigated background contributions to selected waves (*e.g.*, S11, P13, P33, D13) by turning off the direct coupling of the resonances to γN channels. One of the results of this study was the identification of a "first tier" of states in which all three analyses agreed with each other at 20-30% level. For pion photoproduction multipole amplitudes, the three different groups found fairly good agreement in some cases (*e.g.*, E_{1+} and M_{2-} proton multipoles for P33 and D13 waves, respectively), but rather large differences for other cases (*e.g.*, M_{1+} proton multipoles for P13 wave).

A natural extension of the work described above and a possible future project for this working group would be for independent university groups to investigate the model dependency of resonance parameters by performing unitary multichannel fits of a specified set of partial-wave amplitudes (PWAs) fitted over a pre-specified energy region. These fits should be limited to the same number of channels and resonances in each partial wave. This working group has also discussed [2] another possible project in which the goal would be to determine a set of *objective* criteria for ranking resonances. This BRAG rating system would replace or augment the scheme used by the Particle Data Group in the *Review of Particle Physics* [3].

2.2 Working Group on Partial-Wave Analysis

The Working Group on Partial-Wave Analysis was established to study the model-dependence of partial-wave analyses or multipole analyses of data associated with the N* program. A goal of this group is to find a set of PWAs that are as model independent as possible. In addition, it seeks to determine the optimum set of experiments necessary to achieve this goal. Its current chair is Ron Workman of the George Washington University.

An important project carried out by this working group was the investigation of the inherent model dependence of pion photoproduction multipole analyses in the low-energy region 180-450 MeV, which is dominated by the $\Delta(1232)$ resonance [4]. Several different groups (GWU, Mainz, RPI, Yerevan, and Kharkov) carried out independent multipole analyses by using a selected benchmark database. From the combined set of benchmark fits, the $E2/M1$ ratio was found to be $-2.38 \pm 0.27\%$. The most important quantity here is the uncertainty. This project also identified a particular data set for neutral pion photoproduction as problematic for all groups. (These data have since been remeasured.)

A possible future project for this working group would be to extend the investigation summarized above to electroproduction. The goal would be to try to understand the model independence of the E_{1+}/M_{1+} and S_{1+}/M_{1+} ratios as a function of Q^2. Another possible project for the future would be for different university groups to carry out unitary, multichannel partial-wave analyses of selected πN elastic and inelastic reactions ($\pi N \to K\Lambda$, $\pi N \to \eta N$, $\gamma N \to \pi N$, *etc.*). This project should directly fit physical observables (*e.g.*, cross sections and spin observables) as opposed to fitting partial-wave amplitudes obtained from single-reaction partial-wave analyses.

2.3 Working Group on Database Issues

The Working Group on Database Issues was established to explore and advance database issues related to N* physics. The ideal BRAG database would contain all experimental data below c.m. energy $W = 3$ GeV and momentum transfer squared $Q^2 = 5$ (GeV/c)^2 using electromagnetic (photon and electron) and hadron (pion, kaon, and proton) probes with hydrogen and deuterium (neutron) targets. Its current chair is Igor Strakovsky of the George Washington University.

A. Bellachia of the George Washington University discussed one of the key projects of this group at the most recent BRAG meeting [5]. The project, now well underway, involves replacing the well-known SAID experimental database (developed at Virginia Tech by R. Arndt and L. D. Roper) with a web-based system (iSAID). While SAID uses "flat files," the new iSAID system uses a Relational Database Management System (RDMS). The iSAID system enhances the older SAID with tools to analyze, search for, and integrate new data. Data representation in iSAID is done with XML: eXtensible Markup Language. This is a flexible, platform independent language that separates content and presentation (somewhat like HTML). Free tools are available for processing and web support.

Two different types of users are expected to benefit from and utilize iSAID: a general user and a data provider. A general user will be able to compare existing data with models encoded into the iSAID system. Built-in tools will allow a comprehensive view of the current status of theory and experiment. The other type of user, a data provider, will typically be an experimentalist with preliminary data who will be able to add data to the iSAID database to allow comparisons between models and existing data. Data from this source will be private and not accessible to general users. This will also allow exploratory fits before publication.

The present SAID database connects models and data for 10 different reactions (30 reaction subtypes). As of January 2003, data have been transferred from flat files for the 10 reaction types to a single unified RDMS in iSAID. A prototype iSAID is available for testing from the site, http://128.164.158.188/isaid/index.jsp.

3 Summary

BRAG exists as a useful community of experts within the N^* field. All interested physicists (especially junior faculty, postdocs, and students) are encouraged to join and take active roles. Those wishing to subscribe to the BRAG listserver need only go to the site

https://mailer.csit.fsu.edu/mailman/listinfo/brag

and follow the instructions. The BRAG website includes many useful links and webpages, including a frequently updated page with conference links. Members are encouraged to join working groups or start new ones. Ideas for projects are needed and several leadership opportunities are available.

Acknowledgements. The author thanks the organizers of the Fourth International Conference on Perspectives in Hadronic Physics for their invitation to make a presentation at the conference and for their hospitality. This work was supported in part by the U.S. Department of Energy under grant number DE-FG02-01ER41194.

References

1. C. Bennhold et al.: "Nucleon Resonance Properties in Multichannel Approaches," in NSTAR 2001, Proceedings of the Workshop on The Physics of Excited Nucleons, (World Scientific, 2001), edited by D. Dreschel and L. Tiator, p. 109
2. D.M. Manley et al.: "Star Ratings of Baryons," in NSTAR 2002, Proceedings of the Workshop on The Physics of Excited Nucleons (in press)
3. (Particle Data Group), K. Hagiwara et al.: Phys. Rev. D **66**, 010001–1 (2002)
4. R.A. Arndt et al.: "Multipole Analysis of a Benchmark Data Set for Pion Photoproduction," in NSTAR 2001, Proceedings of the Workshop on The Physics of Excited Nucleons, (World Scientific, 2001), edited by D. Dreschel and L. Tiator, p. 467
5. A. Bellachia et al.: "ISAID: A Web-Based Implementation of the SAID System," in NSTAR 2002, Proceedings of the Workshop on The Physics of Excited Nucleons (in press)

Eur Phys J A (2004) **19**, s01, 71–76
Digital Object Identifier (DOI) 10.1140/epjad/s2004-03-012-2

EPJ A direct

electronic only

Baryon resonance analysis from two pion electroproduction at Jefferson Laboratory

Marco Ripani

INFN, Sezione di Genova, Via Dodecaneso 33, I-16146, Genova, Italy

Received: 31 July 2003 / Accepted: 14 Nov 2003 /
Published Online: 6 Feb 2004 – © Società Italiana di Fisica / Springer-Verlag 2004

Abstract. The cross section for the reaction $ep \to e'p\pi^+\pi^-$ was measured in the resonance region for $1.4 < W < 2.1$ GeV and $0.5 < Q^2 < 1.5$ GeV^2/c^2 using the CLAS detector at Jefferson Laboratory. The data show resonant structures not visible in previous experiments. The comparison of our data to a phenomenological prediction using available information on N^* and Δ states shows an evident discrepancy. A better description of the data is obtained either by a sizeable change of the properties of the $P_{13}(1720)$ resonance or by introducing a new baryon state, not reported in published analyses.

PACS. 13.60.Le Meson production – 13.40.Gp Electromagnetic form factors – 14.20.Gk Baryon resonances with S=0

1 Introduction

All existing information on baryon resonances has been obtained in experiments where a pion is present either in the incoming or in the outgoing channel ($\pi N \to \pi N$, $\gamma N \to \pi N$, etc.). It is therefore clear how these experiments are not suited to look for states with a weak pion coupling. Actually, many of the nucleon excited states in the mass region around and above 1.7 GeV tend to decouple from the single-pion and eta channels, while decaying predominantly in multipion channels, such as $\Delta\pi$ or $N\rho$[1]. A similar situation is expected for the "missing states"[2,3,4,5], predicted by symmetric quark models but lacking experimental evidence, whose search is carried on at several labs. Other models, with different symmetry properties and a reduced number of degrees of freedom, as e.g. in [6], predict fewer states. By using an electromagnetic probe to produce multipion final state, we have both the possibility of enhancing poorly known states in the mass region above 1.5 GeV, and of discovering new states. At the same time, the electromagnetic probe allows to study the transition form factors, that are predicted in quark models and whose knowledge is essential in understanding the degrees of freedom and the symmetries involved in the baryon wavefunctions[7,8]. It is therefore very important to extend our knowledge about the excited nucleon states by using the new powerful electron and photon beams in conjunction with multiparticle detection in the final state, to select different decay channels and explore their resonance content in much more detail than in the past. These capabilites are available at Jefferson Laboratory, where the high intensity and high quality continuous electron beam has been used in a variety of experiments aimed at a vast improvement of our understanding of the light quark baryon resonance properties[9], the so-called N* program. In this contribution, I will focus on a particular topic of this program , the measurement of baryon resonance properties in the double pion electroproduction channel, classified as JLab experiment E-93-006. In this experiment, the goal was to use electroproduction of pion pairs as a tool to investigate resonances of higher mass and achieve a better understanding of their properties, in particular the electromagnetic transition form factors, completely unknown for some states, and at the same time attempt to discover new states, predicted by quark models but still lacking experimental evidence. The cross sections measured with CLAS show clear resonant structures, not visible in the previous limited data, and calculations based on the known resonance properties or on quark model predictions are not able to fully account for the features observed, leading to a possible evidence for a contribution from a new state, not reported in the existing Particle Data Group (PDG) listing [1].

2 Generalities

Studying high-lying resonances involves new features and open issues. The presence of many broad, overlapping states makes it necessary to use appropriate filters to enhance particular states or a group of states. This is naturally accomplished using different decay channels to study selected states that manifest through a correspondingly large branching fraction. In particular, an important part of the CLAS experimental program at Jefferson Laboratory is devoted to the study of multipion channels, like

$\Delta\pi$ and ρN, with the goal of extracting information on the electromagnetic excitation of high-lying states weakly visible in single pion production, like the $D_{33}(1700)$, and of establishing the existence of the "missing" states predicted by the quark models.

In this talk I report on a measurement of the $ep \to e'p\pi^+\pi^-$ reaction studied with the CEBAF Large Acceptance Spectrometer (CLAS) at Jefferson Lab (JLab experiment E-93-006). More details on the experimental and physical analysis can be found in [10]. Beam currents of a few nA were delivered to Hall B on a liquid-hydrogen target, corresponding to luminosities up to 4×10^{33} cm^{-2}s^{-1}. Data were taken in 1999 for about two months at beam energies of 2.6 and 4.2 GeV. Important features of the CLAS [11] are its large kinematic coverage for multi-charged-particle final states and its good momentum resolution ($\Delta p/p \sim 1\%$). Using an inclusive electron trigger based on a coincidence between the forward electromagnetic shower calorimeter and the gas Cerenkov detector, many exclusive hadronic final states were measured simultaneously. Scattered electrons were identified through cuts on the calorimeter energy loss and the Cerenkov photo-electron distribution. Different channels were separated through particle identification using time-of-flight information and other kinematic cuts. We used the missing-mass technique, requiring detection in CLAS of at least $ep\pi^+$. The good resolution allowed selection of the exclusive final state, $ep\pi^+\pi^-$. After applying all cuts, our data sample included about 2×10^5 two-pion events.

The range of invariant hadronic center-of-mass (CM) energy W (in 25 MeV bins) was 1.4-1.9 GeV for the first two bins in the invariant momentum transfer Q^2, 0.5-0.8 (GeV/c)^2 and 0.8-1.1 (GeV/c)^2, and 1.4-2.1 GeV for the highest Q^2 bin, 1.1-1.5 (GeV/c)^2. Data were corrected for acceptance, reconstruction efficiency, radiative effects, and empty target counts [10]. In particular, a specifically developed Monte Carlo code was used to calculate the acceptance and efficiency. To this purpose, event distributions were generated in a realistic way and then processed through the GEANT-based code simulating the detector response. The same Monte Carlo event generator was used to perform extrapolations to kinematic regions where the acceptance vanishes. This type of corrections was typically only a few percent of the total cross section measured.

A particularly convenient set of kinematic variables for the analysis of resonance decay into $\Delta^{++}\pi^-$ (the dominant decay channel in the energy region considered here) is made up from the invariant mass of the $p\pi^+$ and the $\pi^+\pi^-$ pair, then the polar angles θ and ϕ of the π^- and finally the residual rotation freedom ψ of the $p\pi^+$ pair[13]. We derived experimental cross sections by binning and correcting the data in the full kinematic space defined by these variables, instead of applying a maximum likelihood procedure. The cross sections derived in our method can then be compared to any model calculation, knowledge of specific experimental features of the detector being unnecessary. We chose this approach as it is not sure *a priori* that models employed in the physics analysis (see below) are correctly describing the measured data distributions. On the other hand, a general partial-wave expansion is certainly very complicated and affected by strong ambiguities. We believe that our approach, being simple and direct, can provide a first interpretation of the data, revealing the most important features as the presence of isobars in the final state and the most prominent resonance excitations. In particular, missing state contributions with strong excitation in this channel can certainly be detected. Our analysis can provide a basis for future, more sophisticated searches based on likelihood functions and/or coupled channel calculations.

3 Physical analysis

Since existing theoretical models [14] are limited to W <1.6 GeV, we have employed a phenomenological calculation [15,16] for a first interpretation of the data. This model describes the reaction $\gamma_v p \to p\pi^+\pi^-$ in the kinematic range of interest as a sum of amplitudes for $\gamma_v p \to \Delta\pi \to p\pi^+\pi^-$ and $\gamma_v p \to \rho^0 p \to p\pi^+\pi^-$, according to the structures observed in the final state invariant mass distributions, while all other possible mechanisms are parameterized as phase space. A detailed treatment was developed for the non-resonant contributions to $\Delta\pi$, while for ρp production they were described through a diffractive ansatz. For the resonant part, a total of 12 states, classified with 3 and 4 stars [1], with sizeable $\Delta\pi$ and/or ρp decays, were included based on a Breit-Wigner ansatz. A few model parameters in non-resonant production were fitted to CLAS data at high W, where the non-resonant part creates a forward peaking in the angular distributions, and kept fixed in the subsequent analysis. The phase between resonant and non-resonant $\Delta\pi$ mechanisms was fitted to the CLAS data as well.

Resonance electromagnetic excitation was initially described through a Single Quark Transition Model (SQTM) fit[17], partial decay branches were taken from a previous analysis of hadronic data[18], and total widths were taken from PDG[1]. Even though our approach was that of a single channel analysis, we emphasize that, by using hadronic couplings from existing coupled channel analyses, we ensured that our resonance behavior was as compatible as possible with the information coming from other channels. This means that our resonance content was not completely derived from the CLAS data, with all inevitable ambiguities, but was constrained to the existing knowledge derived from the more sophisticated hadronic coupled channel techniques. Another important point regarding our analysis is that when calculating non-resonant terms for the $\Delta\pi$ channel, we applied an effective treatment of unitarity[15, 16], thereby taking into account the coupling to all other competing inelastic channels. Using this model approach, we first analysed the single-differential cross sections most sensitive to the dynamical content of our measurement[15, 16], i.e. $\frac{d\sigma}{dM_{p\pi^+}}$, $\frac{d\sigma}{dM_{\pi^+\pi^-}}$, and $\frac{d\sigma}{d\cos\theta_{\pi^-}}$, obtained by integrating over the other hadronic variables. These three 1-D distributions were then analysed or fitted simultaneously.

Starting from the above mentioned ingredients, we first produced a reference curve to be compared with the data (step (A) in text and figures). By reference curve, we mean

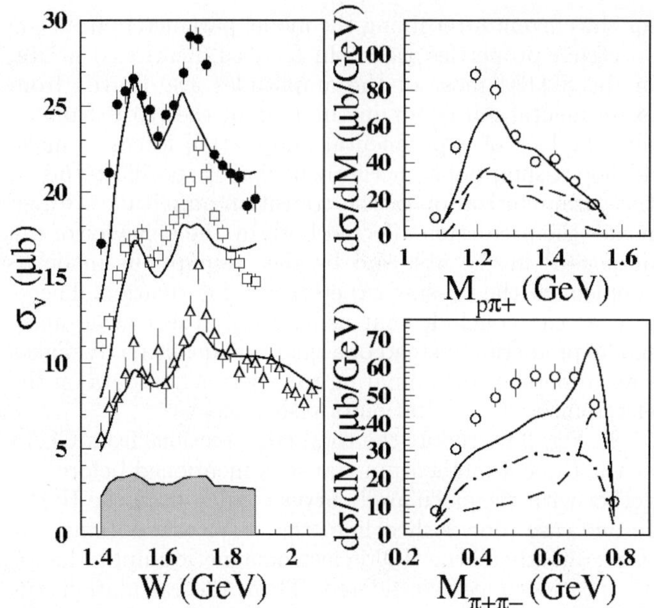

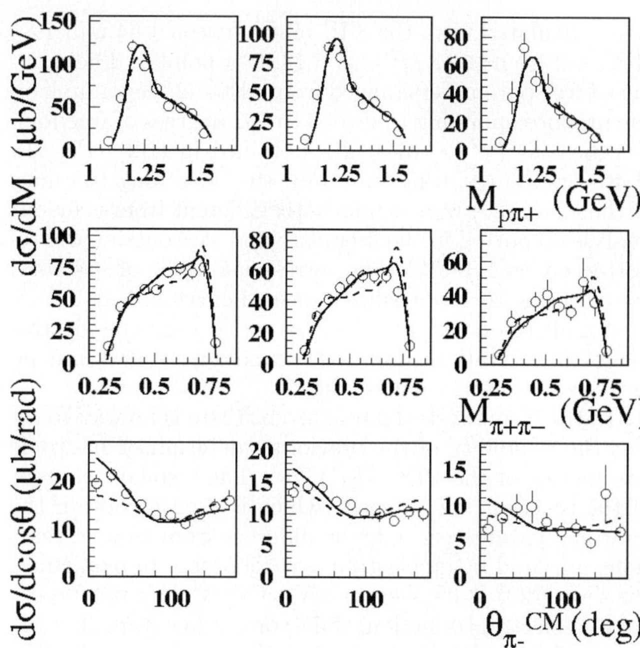

Fig. 1. *Left:* Total cross section for $\gamma_v p \to p\pi^+\pi^-$ as a function of W. Data from CLAS are shown at Q^2=0.5-0.8 $(\mathrm{GeV/c})^2$ (full points), Q^2=0.8-1.1 $(\mathrm{GeV/c})^2$ (open squares), and Q^2=1.1-1.5 $(\mathrm{GeV/c})^2$ (open triangles). Error bars are statistical only, while the bottom band shows the sytematic uncertainties for the lowest Q^2 bin. *Right:* $\frac{d\sigma_v}{dM_{p\pi+}}$ (top) and $\frac{d\sigma_v}{dM_{\pi+\pi-}}$ (bottom) from CLAS at Q^2=0.8-1.1 $(\mathrm{GeV/c})^2$ and W=1.7-1.725 GeV (statistical error bars only). The curves represent our step (A) reference calculation: the dashed line includes all resonances, the dot-dashed line includes only the non-resonant part, and the solid line is the full calculation

Fig. 2. $\frac{d\sigma_v}{dM_{p\pi+}}$, $\frac{d\sigma_v}{dM_{\pi+\pi-}}$, and $\frac{d\sigma_v}{dcos\theta_{\pi-}}$ from CLAS (from top to bottom) at W=1.7-1.725 GeV and for the three mentioned Q^2 intervals (left to right). The error bars include statistical errors only. Curves (see text) correspond to the fits (B2) (solid), and (B4) (dashed)

a set of total and differential cross sections obtained from our model without any parameter modification or fit. This reference calculation on one hand indicates how well we do in modeling non-resonant processes and how well the existing knowledge on all baryon resonances that contribute significantly to double pion electroproduction is reflected in the data. Any deviation of the data from this reference curve may therefore indicate either that our approach to non-resonant processes has basic problems, or that some resonance properties are not correctly accounted for in the existing database, or that some anomalous behavior is manifesting in the data, which in turn may signal the presence of new states or an unexpected microscopic structure of a known resonance. Results for step (A) are reported in Fig. 1. The total cross section strength for $W < 1.65$ GeV (except for the region close to threshold), and for $W > 1.8$ GeV is well reproduced. Instead, a strong discrepancy is evident at W around 1.7 GeV. Moreover, at this energy the reference curve exhibits a lack of $\Delta\pi$ strength in the $p\pi^+$ invariant mass (Fig. 1, right top), and a strong peak in the $\pi^+\pi^-$ invariant mass (Fig. 1, right bottom), connected to sizeable ρ meson production. The latter was traced back to the 70-91% branching ratio of the $P_{13}(1720)$ into this channel[1,18,19].

Considering the resonance properties given by the PDG, and our limited knowledge on the Q^2 dependence of the photocouplings, the best ordinary candidates in providing the missing strength seen in the bump at 1.71 GeV are the $D_{13}(1700)$, the $P_{13}(1720)$, and the $P_{11}(1710)$ (the latter was not included in step (A)). Indeed, following the PDG, it seems that the bump at about W=1.7 GeV cannot be due to the $D_{15}(1675)$, $F_{15}(1680)$, or $D_{33}(1700)$ states; the first because its well known position cannot match the peak; the second because of its well known position and photocouplings [20]; the third due to its large width ($\sim$300 MeV). Instead, there are no data available on the Q^2 dependence of $A_{1/2}$ or $A_{3/2}$ for the $D_{13}(1700)$, $P_{13}(1720)$, $P_{11}(1710)$ [20], and the hadronic couplings of the $D_{13}(1700)$ and the total width of the $P_{11}(1710)$ are poorly known. It is interesting to notice that, if no configuration mixing occurs, the $D_{13}(1700)$ cannot be excited in the SQTM from proton targets, while the SQTM prediction for the $P_{13}(1720)$ relies on ad hoc assumptions, since input data from states in the same multiplet are not sufficient [17].

Therefore we first performed three separate fits, (B1), (B2), and (B3), where the photo- and hadronic couplings of only one of those resonances at a time were widely varied, specifically the $D_{13}(1700)$ for (B1), the $P_{13}(1720)$ for (B2), and the $P_{11}(1710)$ for (B3). Our investigation at this stage was including the possibility of accounting for the 1.7 GeV structure via interference effects, although the peaking of such an interference pattern at the same W for all Q^2 bins would be rather surprising. Before proceeding with such fits, we performed slight variations of the initial curves from step (A), as allowed by the uncertainties in the knowledge of a number of states. All fit χ^2/ν values were calculated from the 8 W bins between 1.64 and 1.81 GeV and from the 3 Q^2 bins (624 data points). The number of free parameters ranged from 11 to 32, depending on the fit, corresponding to ν=613 to 592 degrees of freedom. The best fit (χ^2/ν =3.4) was obtained in (B2) (Fig. 2). However, the resulting values for the branching fractions of the $P_{13}(1720)$ were significantly different from previous analyses reported in the literature and well outside the reported errors[1,18,19]. The corresponding fit of the total cross sections is reported as a dashed curve in Fig. 3. In a final multiresonance fit (B4), we varied only the electromagnetic excitation of all three candidate states, but no better solution was found (Fig. 2).

In the framework of our analysis, there is no way to assess the reliability of the previously determined hadronic parameters of the PDG $P_{13}(1720)$. The resonant content of the reaction $\pi N \to \pi\pi N$, which is used to obtain the hadronic parameters, may be different from that of reactions initiated by an electromagnetic probe. In particular, the $P_{13}(1720)$ state seen in $\pi N \to \pi\pi N$ may not be excited in electroproduction, while some other state that decouples from πN may be excited electromagnetically. We verified indeed that, if we introduced a new state as responsible for the observed bump, a good fit was obtained by a P_{I3} state (isospin could not be determined) with mass of about 1720 MeV, a width of about 90 MeV, about 40 % decay to $\Delta\pi$ and 17 % to ρp. The visual quality of the fit was very similar to the (B2) curves in Fig. 2. It is important to remark that such a good fit was achieved by using the published hadronic parameters for the ordinary $P_{13}(1720)$, while the electromagnetic amplitudes had to be suppressed with respect to the SQTM by about a factor 2 to avoid the mentioned unobserved ρ peak in the $\pi\pi$ invariant mass distribution. The quality of this fit was indistinguishable from the curves (B2) in Fig. 2. The quality of the fit of the total cross sections was also comparable to (B2), reported as a dashed curve in Fig. 3.

4 Using a Quark Model

To broaden the scope of our analysis and make cross section predictions using electromagnetic matrix elements from a fully theoretical picture, we introduced in our calculation the electromagnetic transition amplitudes from the Genova Hypercentral Quark Model (HQM) [21,8,22], including also the recently calculated longitudinal couplings, normally neglected in the resonance studies. The HQM incorporates the basic features of the SU(6) symmetry broken by an spin-isospin specific term. The potential used is based on the Coulomb attraction that dominates the short distances, together with a long distance linear confinement term. The form factors are derived from a one-body quark current operator and have been shown to describe quite well the existing data on the first excited resonances, specifically the $D_{13}(1520)$ and the $S_{11}(1535)$.

Using a quark model prediction instead of a fit, we do not change the basic assumptions about the degrees of freedom or the symmetries involved, but the electromagnetic amplitudes are predicted using a theoretical form

for the current after fixing the model parameters from the spectrum properties alone. In fact, as mentioned above, in the SQTM most of the amplitudes are derived from experimental data, but in the case of the [70,1$^-$] multiplet the lack of experimental information forces to make ad hoc assumptions on the behavior of specific terms in the parameterization of the transition amplitudes, which makes the prediction affected both by the behavior of the data used in the fits, and by the assumptions made to supplement the missing experimental information. Therefore we can conclude that using a fully theoretical quark model prediction instead of a quark model fit really makes some difference in comparing to experimental data on the electromagnetic excitation of resonances.

In Fig. 3 we report the total cross sections from CLAS at the three 4-momentum transfers mentioned before, together with the calculated curves coming from the HQM. In one case (dot-dashed line) the curve was obtained by using directly all the HQM electromagnetic amplitudes for all the resonances considered. Then, the calculation was modified increasing the strength of the Roper resonance as indicated also by our first analysis based on the SQTM (solid line), and as also indicated by the lack of strength at low W in the HQM calculation. It is important to remark that we also included the prediction from the HQM about the longitudinal resonance couplings, often neglected in many analyses. Our resonance fit based on the SQTM is also shown for comparison (dashed line).

As clearly shown in Fig. 3, we still observe a considerable lack of strength in the region around 1.7 GeV, where the strong bump in the data is barely visible in the calculation. It seems therefore that two different analyses, one based on a general model-independent form for the transition matrix elements with parameters derived from existing data, the other based on a specific theoretical form for the quark transition current with no constraints from the data besides those necessary to reproduce the spectrum, are not able to explain the structure seen in the new CLAS data. To test once again the hypothesis that the observed peak may come from a new state, we took the parameters for the new resonance obtained in our previous fits and included the new state on top of the others calculated with the HQM amplitudes, with slight parameter adjustments. In Fig. 4, the result, plotted as the solid line, is compared to the HQM calculation containing only ordinary states and the modified Roper, plotted as the dashed line. It seems that the inclusion of the new state is compatible with the HQM resonance amplitudes, too.

At this point, it is natural to ask whether the differential cross sections are reproduced. In Fig. 5, I report such comparison between the CLAS data on top of the bump at 1.71 GeV and the HQM calculation with the addition of the new state: once again, the inclusion of the new state on top of the theoretical N* prediction provides a rather good match to the CLAS data.

From the two analyses performed, there emerges an indication that the electromagnetic excitation strength shown by the CLAS data is underestimated by single quark transition models, both phenomenological as well as fully theoretical. This may indicate different things: 1)

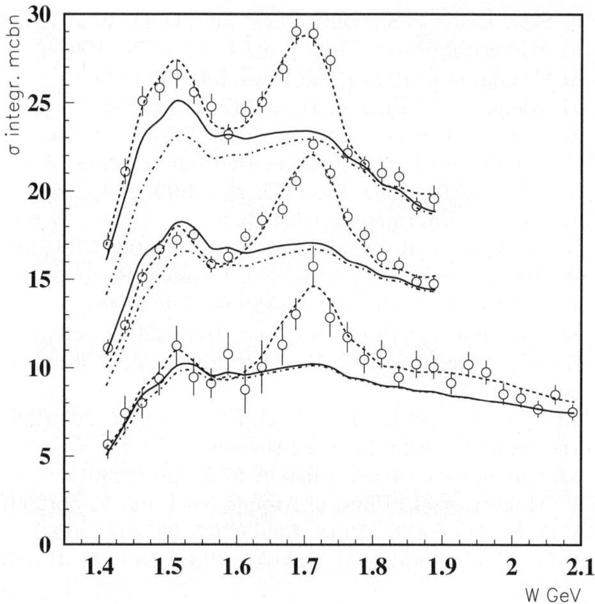

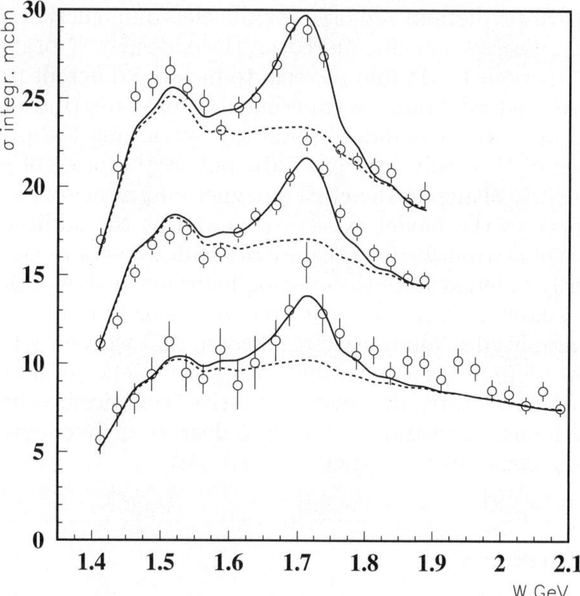

Fig. 3. Comparison of the CLAS data to a calculation based on resonant electromagnetic amplitudes from the HQM. The data are from CLAS at the three mentioned momentum transfers. The dot-dashed line is the calculation within the Genova-Moscow model using all resonance electromagnetic amplitudes (including longitudinal ones) from the HQM. The solid line is the calculation within the Genova-Moscow model using all resonance electromagnetic amplitudes (including longitudinal ones) from the HQM except for the Roper, which is taken from our previous fits. For comparison, we report as the dashed line our final fit starting from the SQTM amplitudes

Fig. 4. Comparison of the CLAS data to a calculation based on resonant electromagnetic amplitudes from the HQM. The data are from CLAS at the three mentioned momentum transfers. The dashed line is the calculation within the Genova-Moscow model using all resonance electromagnetic amplitudes (including longitudinal ones) from the HQM except for the Roper, which is taken from our previous fits. The solid line is obtained by adding to the ordinary N^* states calculated by the HQM the new state found in our previous fits

the single quark transition picture may be wrong, and we may be observing more complicated excitation mechanisms, with a redistribution of the strength; 2) additional degrees of freedom, like $q\bar{q}$ pairs, may modify the electromagnetic excitation, although their effect should disappear at higher Q^2, at variance with our observations; an admixture of exotic components cannot be excluded, either [23]; 3) the missing strength in the model analysis may be really due to the presence of one or more additional resonances, not observed before.

5 Conclusions

In conclusion, two pion production is one of the main subjects of investigation in Hall B at Jefferson Lab, as a key channel for extending our knowledge of light-quark baryons. The new high quality data from CLAS have shown the clear presence of resonance structures, not visible in the previous, technically limited experiments. In analysing the new data from CLAS, we adopted a phenomenological approach, based on the database of measured electromagnetic and hadronic resonance properties, and with effective unitarity constraints on the most relevant non-resonant mechanisms. Our analysis of the data has revealed that the prominent structure at 1.7 GeV seems difficult to explain when quark model fits or the-

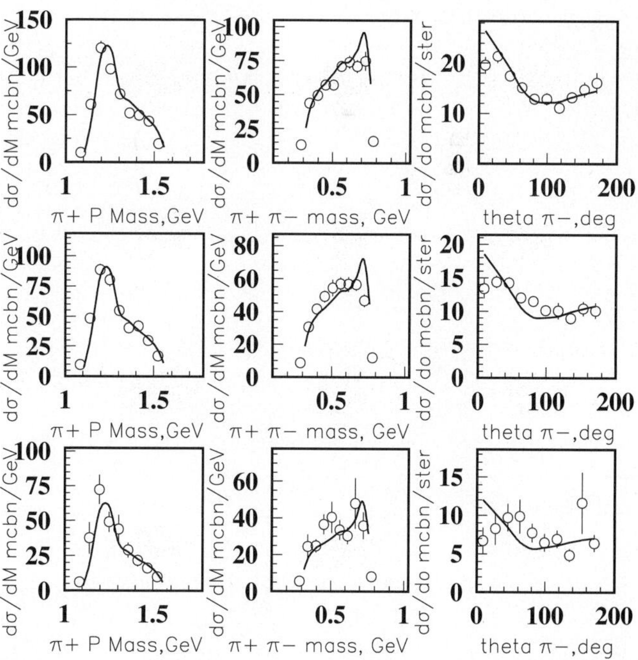

Fig. 5. $\frac{d\sigma_v}{dM_{p\pi^+}}$, $\frac{d\sigma_v}{dM_{\pi^+\pi^-}}$, and $\frac{d\sigma_v}{dcos\theta_{\pi^-}}$ from CLAS (left to right) at W=1.7-1.725 GeV and for the three mentioned Q^2 intervals (from top to bottom). The error bars include statistical errors only. The solid line is obtained by adding to the ordinary N^* states calculated by the HQM the new state found in our previous fits

oretical predictions are used for the electromagnetic transition matrix elements. Moreover, the expected N* branching fractions to $\Delta\pi$ and ρp seem to be quite different from those inferred from the experimental mass distributions. A good data fit is indeed obtained by varying the properties of the ordinary $P_{13}(1720)$, but at the price of significantly changing the electromagnetic form factors with respect to the model expectations and at the additional price of attributing to this state branching fractions significantly different from the existing hadronic analyses. Both our analyses, based on quark model fits or fully theoretical amplitudes, alternatively indicated that the new CLAS cross sections are compatible with the presence of a new, unobserved state, although alternative scenarios involving multiquark transition, additional degrees of freedom, or exotic components cannot be ruled out.

References

1. K. Hagiwara et al.: Phys. Rev. D **66**, 010001 (2002)
2. R. Koniuk and N. Isgur: Phys. Rev. Lett. **44**, 845 (1980); Phys. Rev. D **21**, 1868 (1980)
3. R. Koniuk: Nucl. Phys., B **195**, 452 (1982)
4. S. Capstick and W. Roberts: Phys. Rev. D **49**, 4570 (1994)
5. F. Stancu and P Stassart: Phys. Rev. D **47**, 2140 (1993)
6. M. Kirchbach: Mod. Phys. Lett. A **12**, 3177 (1997)
7. M.M. Giannini: Rep. Prog. Phys. **54**, 453 (1991)
8. M. Aiello et al.: Phys. Lett. B **387**, 215 (1996)
9. M. Ripani: in Proceedings of the ICTP 3rd International Conference on Perspectives in Hadronic Physics, Trieste, Italy, 7–11 May 2001; Nucl. Phys. A **699**, 270 (2002)
10. M. Ripani et al.: hep-ex/0304034
11. B. Mecking et al.: Nucl. Inst. Meth. A **503**, 513 (2003); W. Brooks: Proc. of PANIC '99, Uppsala, Sweden, 10–16 June 1999, Nucl. Phys. A **663&664**, 29c (2000)
12. M. Ripani et al.: Phys. Rev. Lett. **91**, 022002 (2003)
13. Byckling and Kajantie: Particle kinematics, Wiley and Sons (1973)
14. J.C. Nacher and E. Oset: Nucl. Phys. A **674**, 205 (2000), and related work referenced therein
15. M. Ripani et al.: Nucl. Phys. A **672**, 220 (2000)
16. V. Mokeev et al.: Phys. of Atomic Nucl. **64**, 1292 (2001)
17. V.D. Burkert et al.: Phys. Rev. C **67**, 035204 (2003)
18. D.M. Manley and E.M. Salesky: Phys. Rev. D **45**, 4002 (1992)
19. T.P. Vrana et al.: Phys. Rept. **328**, 181 (2000)
20. V. Burkert: Nucl. Phys. A **684**, 16c (2001)
21. E. Santopinto: in these Proceedings
22. M. Aiello et al.: J. Phys. **G** 24, 753 (1998)
23. H. Walliser and V.B. Kopeliovich: hep-ph/0304058

Eur Phys J A (2004) **19**, s01, 77–79
Digital Object Identifier (DOI) 10.1140/epjad/s2004-03-013-2

EPJ A direct
electronic only

Generalized sum rules of the nucleon in the constituent quark model

M. Gorchtein[1], D. Drechsel[2], M. Giannini[1], and E. Santopinto[1]

[1] Universit a di Genova, Sezione INFN di Genova, via Dodecaneso 33, 16142 Genova (Italy)
[2] Institut für Kernphysik, Universität Mainz, Staudingerweg 45, 55099 Mainz (Germany)

Received: 16 Oct 2003 / Accepted: 14 Nov 2003 /
Published Online: 6 Feb 2004 – © Società Italiana di Fisica / Springer-Verlag 2004

Abstract. The sum rules serve a powerful tool to study the nucleon structure by providing a bridge between the statical properties of the nucleon (such as electrical charge, and magnetic moment) and the dynamical properties (e.g. the transition amplitudes to excited states) in a wide range of energy and momentum transfer Q^2. We study the generalized sum rules of the nucleon in the framework of the constituent quark model. We use two different CQM, the one with the hypercentral potential [1,2,3], and with the harmonic oscillator potential [4], both with only few parameters fixed to the baryonic spectrum. We confront our results to the model independent sum rules and to the predictions of the phenomenological MAID [5] model and find that in all the cases considered, in the intermediate Q^2 range (0.2-1.5 GeV^2), both CQM models provide a good description of the sum rules on the neutron.

PACS. 12.39.Jh Nonrelativistic quark model – 14.20.Gk Baryon resonances and helicity amplitudes

1 Introduction

In the recent years, precise measurements of single and double polarization observables for the photo- and electorabsorption have become possible. The inclusive cross section for the process $(ep \to eX)$ can be written in terms of the four partial cross sections,

$$\frac{d\sigma}{d\Omega dE'} = \Gamma_V \left[\sigma_T + \sigma_L - hP_x\sqrt{2\epsilon(1-\epsilon)}\sigma_{LT} \right.$$
$$\left. - hP_z\sqrt{1-\epsilon^2}\sigma_{TT} \right], \qquad (1)$$

with $\sigma_T = \frac{\sigma_{1/2}+\sigma_{3/2}}{2}$, $\sigma_{TT} = \frac{\sigma_{1/2}-\sigma_{3/2}}{2}$, the virtual photon flux factor $\Gamma_V = \frac{\alpha_{em}}{2\pi^2}\frac{E'}{E}\frac{K}{Q^2}\frac{1}{1-\epsilon}$, and the photon polarization $\epsilon = \frac{1}{1+2(1+\nu^2/Q^2)\tan^2(\Theta/2)}$, where $E(E')$ denote the initial (final) electron energy, $\nu = E - E'$ the energy transfer to the target, Θ the electron c.m. scattering angle, and $Q^2 = 4EE'\sin^2(\Theta/2)$ the four momentum transfer. The virtual photon spectrum normalization factor is chosen to be $K = \frac{s-M^2}{2M}$, $h = \pm 1$ refers to the electron helicity, while P_z and P_x are the components of the target polarization. For a general and complete consideration of the nucleon sum rules we readress the reader to the review [6].

2 Constituent quark model

We study the generalized sum rules for the nucleon within a hyper central constituent quark model (HCCQM) pre-

viously reported in [1,2,3]. The model based on the lattice QCD inspired potential of the form $V(x) = -\frac{\alpha}{x} + \beta x + V_{hyp}$ and allows for a consistent description of the baryonic spectrum with a minimal number of parameters. Furthermore, to display the model dependence of a CQM calculation, we list also the results within the CQM with a harmonic oscillator (HO) potential [4]. Within this model, we for the first time report a calculation of the longitudinal amplitudes and their contribution to the sum rules, details of which will be reported in an upcoming article.

The electromagnetic transition helicity amplitudes are defined as

$$A_{1/2} = -\frac{e}{\sqrt{2\omega}} < R, \frac{1}{2}|J_+|N, -\frac{1}{2} >,$$

$$A_{3/2} = -\frac{e}{\sqrt{2\omega}} < R, \frac{3}{2}|J_+|N, \frac{1}{2} >,$$

$$S_{1/2} = \frac{e}{\sqrt{2\omega}} < R, \frac{1}{2}|\rho|N, \frac{1}{2} >, \qquad (2)$$

where $\frac{1}{2}$, $\frac{3}{2}$ stands for the spin projection of the initial (nucleon) and final (resonance) hadronic state, and the definition was used, $J_+ \equiv \varepsilon^+ \cdot \mathbf{J} = -\frac{J_x+iJ_y}{\sqrt{2}}$.

We will present the results for the sum rules within the $zero - width$ approximation where one has for the contribution of a single resonance R to the partial cross sections

$$\sigma_L^R = 2\pi\delta(\nu - \nu_R)\frac{Q^2}{q_R^2}|S_{R,1/2}|^2$$

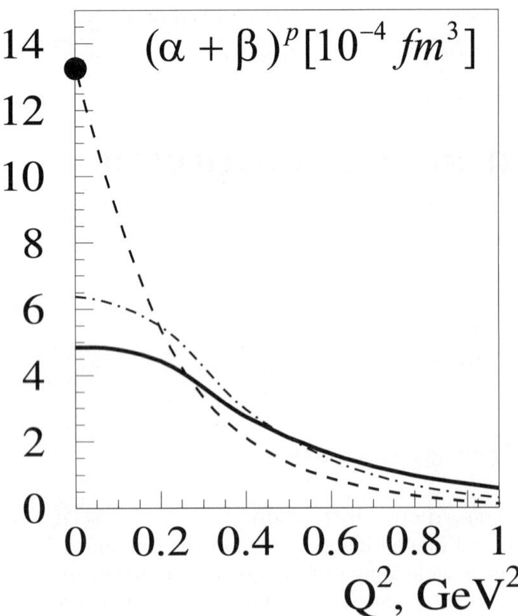

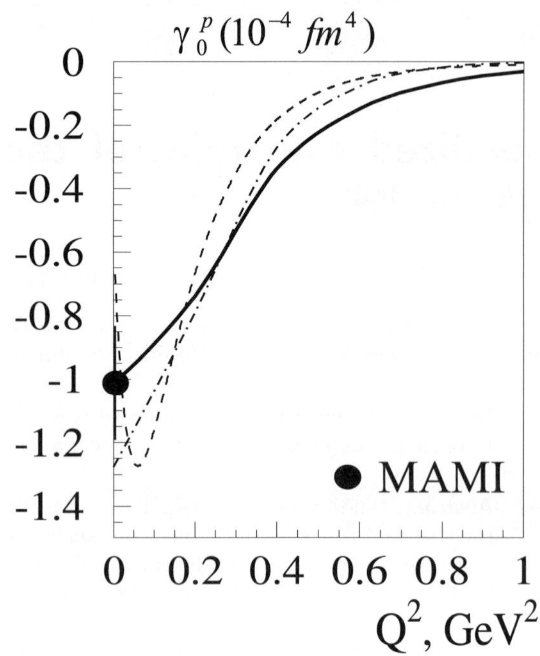

Fig. 1. Results for the sum of the proton polarizabilities $\alpha + \beta$ calculated in HCCQM (solid line), HO model (dashed-dotted line) in comparison with MAID (dashed line). The solid circle corresponds to the Baldin sum rule value at $Q^2 = 0$ [7]

Fig. 2. Results for γ_0 on proton. Notation as in Fig. 1. The data point at $Q^2 = 0$ is from [7]

$$\sigma_{LT}^R = \sqrt{2}\pi\delta(\nu - \nu_R)\frac{Q}{q_R}\left(S_{R,1/2} \cdot A_{R,1/2}\right)$$

$$\sigma_{1/2,3/2}^R = 2\pi\delta(\nu - \nu_R)|A_{R,1/2,3/2}|^2\,, \tag{3}$$

with $\nu_R = \frac{M_R^2 - M^2 + Q^2}{2M}$, $q_R = \sqrt{\nu_R^2 + Q^2}$, and M_R the resonance mass.

3 Sumrules for the forward polarizabilities of the proton

We start with the Baldin sum rule which relates the sum of the electromagnetic polarizabilities to the integral over the total photoabsorption cross section,

$$\alpha(Q^2) + \beta(Q^2) = \frac{1}{2\pi^2}\int_{\nu_0}^{\infty}\frac{K}{\nu}\frac{\sigma_T(\nu, Q^2)}{\nu^2}d\nu\,, \tag{4}$$

where $\nu_0 = m_\pi + \frac{m_\pi^2 + Q^2}{2M}$ is the pion production threshold.

As it can be seen from Fig. 1, both constituent quark models fall short at $Q^2 = 0$ by a factor of 3, which is a consequence of lacking the large contribution of pion production. However, starting from $Q^2 = 0.2$ GeV^2 all three models give similar results.

We next turn to the sum rules with the helicity flip cross section σ_{TT}. In Fig. 2, we show the results for the forward spin polarizability,

$$\gamma_0(Q^2) = \frac{1}{2\pi^2}\int_{\nu_0}^{\infty}\frac{K}{\nu}\frac{\sigma_{TT}(\nu, Q^2)}{\nu^3}d\nu \tag{5}$$

In the case of γ_0, the small value phenomenologically comes about due to a strong cancellation of a large negative contribution of the $\Delta(1232)$ resonance, and a large positive contribution of near threshold pion production. Though the latter is not present in neither of the two presented quark model calculations, both do surprisingly well for this sum rule, as can be seen in Fig. 2, since almost all the transition helicity amplitudes in a CQM lack strenght, as compared to the phenomenological analysis. Therefore, the fact that the quark model results are consistent with the results of MAID in the shown range of Q^2 should be seen rather as a coincidence. It is interesting to note that, due to the characteristical for the HO potential gaussian form factors, the HO model closely reproduces the slope of the MAID curve.

4 Generalized GDH sum rule

The GDH sum rule relates the anomalous magnetic moment of the nucleon to the integral over its excitation spectrum,

$$-\frac{\kappa^2}{4} = \frac{M^2}{2\pi e^2}\int_{\nu_0}^{\infty}d\nu\frac{\sigma_{1/2} - \sigma_{3/2}}{\nu}\,, \tag{6}$$

thus providing a test of a quark model, since both left and right hand sides of this sum rule can be calculated. One of the possible generalizations of this integral to the case of finite Q^2 is

$$I_A(Q^2) = \frac{M^2}{\pi e^2}\int_{\nu_0}^{\infty}d\nu\frac{K}{\nu}\frac{\sigma_{TT}(\nu, Q^2)}{\nu}\,, \tag{7}$$

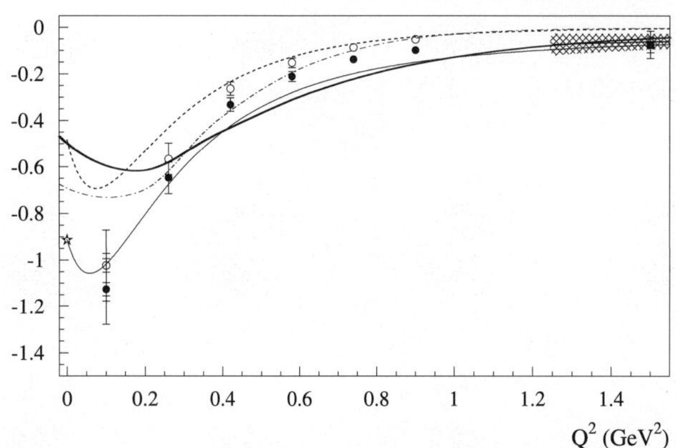

Fig. 3. Results for the generalized GDH integral I_A^n. Notation as in Fig. 1. The thin solid line corresponds to the experimental fit, as described in text. The solid line starting from 1.25 GeV^2 corresponds to the evaluation of the GDH integral using the data on the DIS structure functions (for the details, see [6]). The solid star represents the sum rule value at $Q^2 = 0$. The data points are from [8] (solid squares) and [9] (solid (full result) and open (resonant part with $W \leq 2$ GeV) circles)

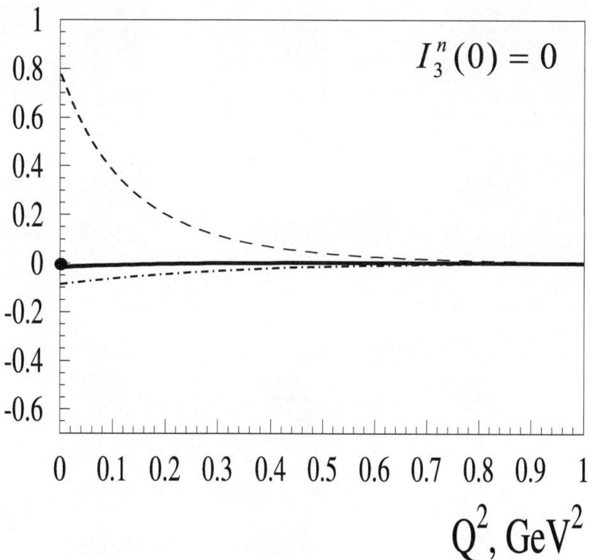

Fig. 4. Results for the I_3 integral for the neutron. Notation as in Fig. 1. Solid cicrcle corresponds to the sum rule value

In Fig. 3, we show the results for the GDH integral I_A on the neutron. As one can see, the sum rule at the real photon point is not obeyed in either quark model.

Starting from $Q^2 = 0.3$ GeV^2, the HCCQM practically follows the experimental fit, which assumes the following form (for details, see [6]):

$$I_A(Q^2) = I_A^{res}(Q^2) + 2M^2\Gamma_A^{as}\left[\frac{1}{Q^2 + \mu^2} - \frac{c\mu^2}{(Q^2 + \mu^2)^2}\right],$$

$$c = 1 + \frac{\mu^2}{2M^2\Gamma_A^{as}}\left[\frac{\kappa^2}{4} + I_A^{res}(0)\right], \tag{8}$$

with $\mu = m_\rho$ and the resonance part as calculated with MAID. Due to the characteristical HO gaussian form factors having a more steep Q^2 dependence, the HO model is able to reproduce the data up to 1 GeV2, but falls short beyond this region.

5 Sum rule for the integral $I_3(Q^2)$

The only sum rule containing a prediction for the electric charge of the nucleon is

$$I_3(0) = \int_{\nu_0}^{\infty} d\nu \frac{K}{\nu}\frac{\sigma_{LT}}{Q} \underset{Q^2 \to 0}{\to} \frac{e_N\kappa_N}{4}. \tag{9}$$

In Fig. 4, we show the results for the I_3 integral on the neutron. While the MAID prediction is in complete disagreement with the sum rule value, one can see that the HCCQM result differs only slightly from zero, as required by the sum rule, and HO model gives a small negative value. Apart from the different potential of the two CQM models, the presented HO calculation does not take account of the hyperfine mixing of the wave functions which is responsible for the cancellation within this integral.

References

1. M. Ferraris, M.M. Giannini, M. Pizzo, E. Santopinto, and L. Tiator: PL B **364**, 231–238 (1995)
2. M. Aiello, M. Ferraris, M.M. Giannini, M. Pizzo, and E. Santopinto: PL B **387**, 215-221 (1996)
3. M. Aiello, M.M. Giannini, and E. Santopinto: J. Phys. G: Nucl. Part. Phys. **24**, 753–762 (1998)
4. N. Isgur and G. Karl: Phys. Rev. D **18**, 4187
5. D. Drechsel, O. Hanstein, S. Kamalov, and L. Tiator: Nucl. Phys. A **645**, 145 (1999)
6. D. Drechsel, B. Pasquini, and M. Vanderhaeghen: Phys. Rept. **378**, 99–205 (2003), and references therein
7. J. Ahrens et al. (GDH and A2 Collaboration): Phys. Rev. Lett. **84**, 5950 (2000)
8. A. Airapetian et al. (HERMES Collaboration): hep-ex/ 0210047
9. M. Amarian et al. (JLab E94010 collaboration): Phys. Rev. Lett. (2002) in press

Eur Phys J A (2004) **19**, s01, 81–85
Digital Object Identifier (DOI) 10.1140/epjad/s2004-03-014-0

Electromagnetic form factors in the hypercentral CQM

M. De Sanctis[1], M.M. Giannini[2], E. Santopinto[1], and A. Vassallo[1]

[1] Universita di Genova e INFN, Sezione di Genova, via Dodecaneso 33, 16142 Genova, Italy
[2] INFN,Sezione di Roma1, Piazzale Aldo Moro, Roma, Italy

Received: 14 Oct 2003 / Accepted: 14 Nov 2003 /
Published Online: 6 Feb 2004 – © Società Italiana di Fisica / Springer-Verlag 2004

Abstract. We report on the recent results of the hypercentral Constituent Quark Model (hCQM). The model contains a spin independent three-quark interaction which is inspired by Lattice QCD calculations and reproduces the average energy values of the $SU(6)$ multiplets. The splittings are obtained with a $SU(6)$-breaking interaction, which can include also an isospin dependent term. Concerning Constituent Quark models, we have shown for the first time that the decreasing of the ratio of the elastic form factors of the proton is due to relativistic effects using relativistic corrections to the e.m. current and boosts. Now the elastic nucleon form factors have been recalculated, using a relativistic version of the hCQM and a relativistic quark current showing a very detailed reproduction of all the four form factor existing data over the complete range of $0 - 4 \ GeV^2$. Futhermore, the model has been used for predictions concerning the electromagnetic transverse and longitudinal transition form factors giving a good description of the medium Q^2 behaviour. We show that the discrepancies in the reproduction of the helicity amplitudes at low Q^2 are due to pion loops. We have calculated the helicity amplitudes for all the 3 and 4 star resonances opening the possibility of application to the evaluation of cross sections.

PACS. 13.40.Gp Electromagnetic form factors – 12.39.Jh Nonrelativistic quark model – 14.20.Gk Baryon resonances and helicity amplitudes – 12.39.Ki Relativistic quark model – 12.39.Pn Potential Models – 13.40.Gp Electromagnetic Form Factors – 14.30.Gk Baryon Resonances with S=0 – 25.20.Lj Photoproduction reactions – 25.30.Rw Electroproduction reactions

1 Introduction

In recent years much attention has been devoted to the description of the internal nucleon structure in terms of constituent quark degrees of freedom. Besides the now classical Isgur-Karl model [1], the Constituent Quark Model has been proposed in quite different approaches: the Capstick and Isgur model [2], the hypercentral formulation [3] and the chiral model [4,5]. In the following the hypercentral Constituent Quark Model (hCQM), which has been used for a systematic calculation of various baryon properties, will be briefly reviewed. The four electromagnetic elastic form factors of the nucleon have been recently calculated using both a relativistic version of the hCQM and a relativistic current and some results are presented compared with the new data. We have calculated in a systematic way the transverse and longitudinal electromagnetic form factors for all the 3 and 4 star resonances. This effort opens the possibility to many applications for calculations of cross sections (see Ripani contribution for an application to the Jlab two pions data [6]). Finally we will also show how it is possible to reproduce in great detail also the behaviour at low Q^2 of the helicity amplitudes for the nucleon resonances considering chiral corrections

to this model (see L. Tiator contribution for more details on pion loop corrections to the hCQM [7]).

2 The hypercentral model

The experimental 4 and 3 star non strange resonances can be arranged in $SU(6)-$multiplets. This means that the quark dynamics has a dominant $SU(6)-$ invariant part, which accounts for the average multiplet energies. In the hCQM it is assumed to be [3]

$$V(x) = -\frac{\tau}{x} + \alpha x, \tag{1}$$

where x is the hyperradius

$$x = \sqrt{\rho^2 + \lambda^2} \ , \tag{2}$$

where ρ and λ are the Jacobi coordinates describing the internal quark motion. The dependence of the potential on the hyperangle $\xi = arctg(\frac{\rho}{\lambda})$ has been neglected. Interactions of the type linear plus Coulomb-like have been used since long time for the meson sector, e.g. the

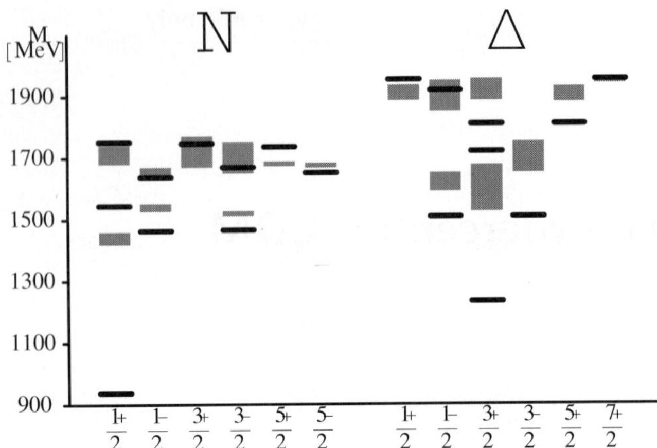

Fig. 1. The spectrum obtained with the hypercentral model (3) and the parameters (4) (*full lines*)), compared with the experimental data of PDG [13] (*grey boxes*)

Cornell potential. This form has been supported by recent Lattice QCD calculations [8].

In the case of baryons a so called hypercentral approximation has been introduced [9,10], this approximation amounts to average any two-body potential for the three quark system over the hyperangle ξ and works quite well, specially for the lower part of the spectrum [11]. In this respect, the hypercentral potential 1 can be considered as the hypercentral approximation of the Lattice QCD potential. On the other hand, the hyperradius x is a collective coordinate and therefore the hypercentral potential contains also three-body effects.

The hypercoulomb term $1/x$ has important features [3, 12]: it can be solved analytically and the resulting form factors have a power-law behaviour, at variance with the widely used harmonic oscillator; moreover, the negative parity states are exactly degenerate with the first positive parity excitation, providing a good starting point for the description of the spectrum.

The splittings within the multiplets are produced by a perturbative term breaking the $SU(6)$ symmetry, which, as a first approximation, can be assumed to be the standard hyperfine interaction H_{hyp} [1]. The three quark hamiltonian for the hCQM is then:

$$H = \frac{p_\lambda^2}{2m} + \frac{p_\rho^2}{2m} - \frac{\tau}{x} + \alpha x + H_{hyp}, \qquad (3)$$

where m is the quark mass (taken equal to $1/3$ of the nucleon mass). The strength of the hyperfine interaction is determined in order to reproduce the $\Delta - N$ mass difference, the remaining two free parameters are fitted to the spectrum, reported in Fig. 1, leading to the following values:

$$\alpha = 1.61\ fm^{-2}, \qquad \tau = 4.59 . \qquad (4)$$

Keeping these parameters fixed, the model has been applied to calculate various physical quantities of interest: the photocouplings [14], the electromagnetic transition amplitudes [15], the elastic nucleon form factors [16]

and the ratio between the electric and magnetic proton form factors [17]. Some results of such parameter free calculations are presented in the next section.

3 The results

The electromagnetic transition amplitudes are defined as the matrix elements of the electromagnetic interaction, between the nucleon, N, and the resonance, B, states:

$$A_{1/2} = \langle B, J', J'_z = \tfrac{1}{2}\ |H^t_{em}|N, J\ =\ \tfrac{1}{2}, J_z = -\tfrac{1}{2}\ \rangle\zeta$$

$$A_{3/2} = \langle B, J', J'_z = \tfrac{3}{2}\ |H^t_{em}|N, J\ =\ \tfrac{1}{2}, J_z = \tfrac{1}{2}\ \rangle\ \zeta$$

$$S_{1/2} = \langle B, J', J'_z = \tfrac{1}{2}\ |H^l_{em}|N, J\ =\ \tfrac{1}{2}, J_z = \tfrac{1}{2}\ \rangle\ \zeta$$

$$(5)$$

where ζ is the sign of the $N\pi$ amplitude.

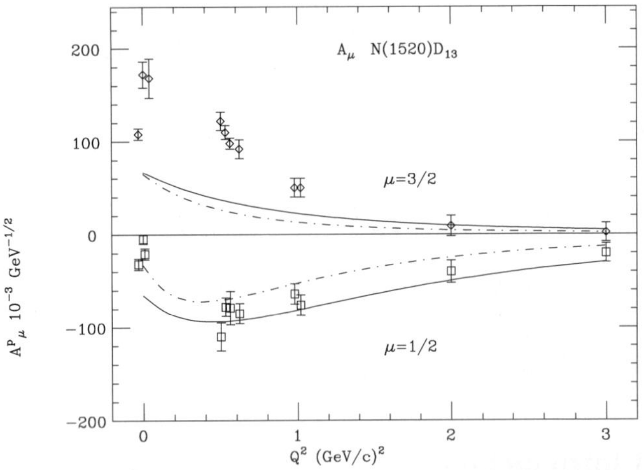

Fig. 2. The helicity amplitudes for the $D_{13}(1520)$ resonance, calculated with the hCQM of (3) and (4) (*full curve*, [15]). The dashed curve is obtained with the analytical version of the hCQM ([12]), where the behaviour of the quark wave function is determined mainly by the hypercoulomb potential. The data are from the compilation of [21]

The proton photocouplings of the hCQM [14] ((5) calculated at the photon point), in comparison with other calculations [19,24], have the same overall behaviour, having the same SU(6) structure in common, but in many cases they all show a lack of strength.

Taking into account the Q^2 behaviour of the transition matrix elements of (5), one can calculate the hCQM helicity amplitudes in the Breit frame [15]. The hCQM results for the $D_{13}(1520)$ and the $S_{11}(1535)$ resonances [15] are given in Fig. 2 and 3, respectively. The agreement in the case of the S_{11} is remarkable, the more so since the hCQM curve has been published three years in advance with respect to the recent TJNAF data [22]. We have completed our program in order to calculate in a systematic way the helicity amplitudes, transverse and longitudinal ones, for

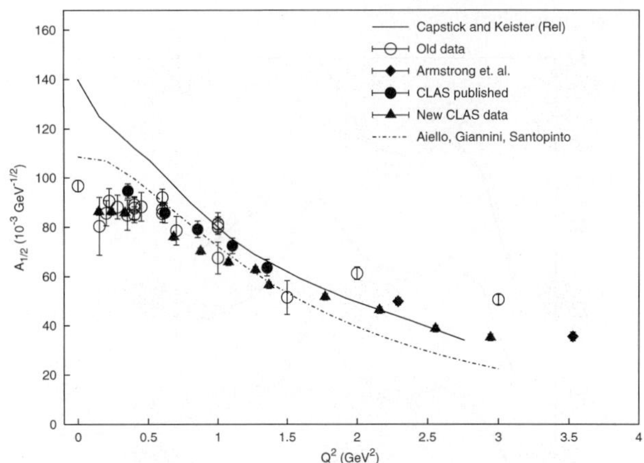

Fig. 3. The helicity amplitudes for the $S_{11}(1535)$ resonance, calculated with the hCQM of (3) and (4) (*dotted curve*, [15]) and the model of [24] (*full curve*). The data are taken from the compilation of [25]

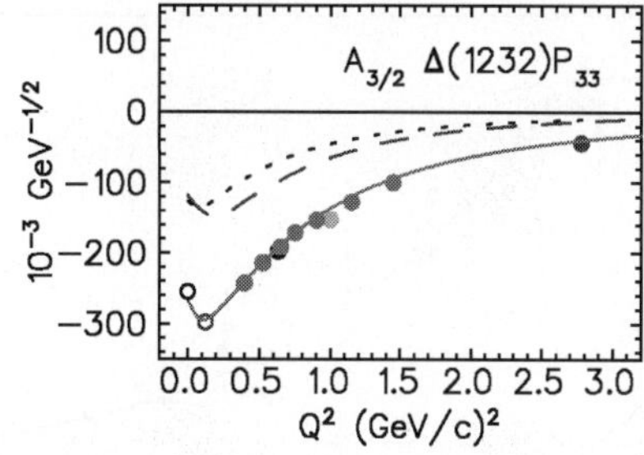

Fig. 4. The transverse $A_{3/2}$ helicity amplitude for the $\Delta(1232)$ resonance. The *dotted line* corresponds to the hCQM results, the *dashed line* to the pion loop contributions and the full line to a fit of the existing data (see L.Tiator contribution for a complete explanation)

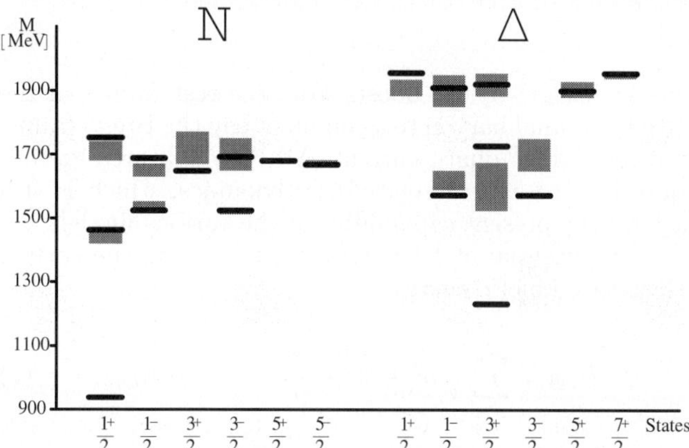

Fig. 5. The spectrum obtained with the hypercentral model containing isospin dependent terms (7) [27] (*full lines*)), compared with the experimental data of PDG [13] (*grey boxes*)

all the 3 and 4 star resonances (the results are at disposal under request) [23]. In general the Q^2 behaviour is reproduced, except for discrepancies at small Q^2, especially in the $A^p_{3/2}$ amplitude of the transition to the $D_{13}(1520)$ state. These discrepancies, as the ones observed in the photocouplings, can be ascribed either to the non-relativistic character of the model or to the lack of explicit quark-antiquark configurations, which may be important at low Q^2 . The kinematical relativistic corrections at the level of boosting the nucleon and the resonance states to a common frame are not responsible for these discrepancies, as we have demonstrated in [26]. Similar results are obtained for the other negative parity resonances [15].

It should be mentioned that the r.m.s. radius of the proton corresponding to the parameters of (4) is 0.48 fm, which is the same value obtained in [18] in order to reproduce the D_{13} photocoupling. Therefore the missing strength at low Q^2 can be ascribed to the lack of quark-antiquark effects, probably more important in the outer region of the nucleon.

For example, for the Delta resonance the contribution of the pion cloud is very important [7]. For the transverse amplitudes $A_{1/2}$ and $A_{3/2}$ it is about 50% at low Q^2 and for the longitudinal amplitude as well as for the electric amplitude the pion cloud is absolutely dominant. In Fig. 4 we show for the $A_{3/2}$ amplitude that only the sum of both contribution will get close to the empirical results.

4 The isospin dependence

In the chiral Constituent Quark Model [4,5], the non confining part of the potential is provided by the interaction with the Goldstone bosons, giving rise to a spin- and flavour-dependent part, which is crucial in this approach for the description of the lower part of the spectrum. More generally, one can expect that the quark-antiquark pair

production can lead to an effective residual quark interaction containing an isospin (flavour) dependent term.

Therefore, we have introduced isospin dependent terms in the hCQM hamiltonian. The complete interaction used is given by [27]

$$H_{int} = V(x) + H_{\mathrm{S}} + H_{\mathrm{I}} + H_{\mathrm{SI}} , \qquad (6)$$

where $V(x)$ is the linear plus hypercoulomb SU(6)-invariant potential of 1, while the remaining terms are the residual SU(6)-breaking interaction, responsible for the splittings within the multiplets. H_{S} is a smeared standard hyperfine term, H_{I} is isospin dependent and H_{SI} spin-isospin dependent. The resulting spectrum for the 3 and 4 star resonances is shown in Fig. 5 [27]. The contribution of the hyperfine interaction to the $N - \Delta$ mass difference is in this case only about 35%, while the remaining splitting comes from the spin-isospin term, (50%), and from the isospin one, (15%). It should be noted that the position of the Roper and the negative parity states is well reproduced.

5 Relativity

The relativistic effects that one can introduce starting from a non relativistic quark model are: a) the relativistic

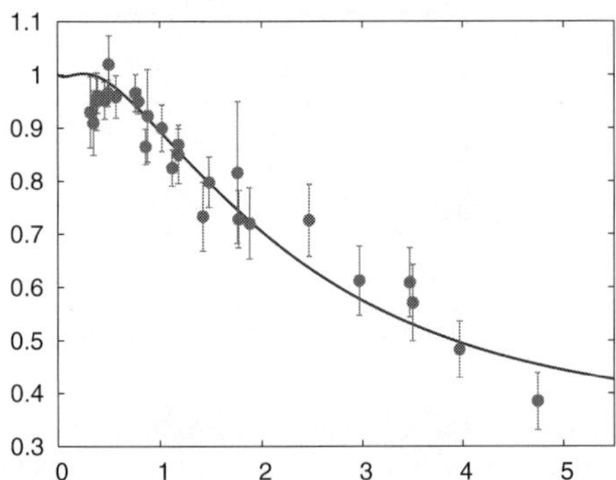

Fig. 6. The ratio between the electric and magnetic proton form factors, calculated with the relativistic hCQM of (8), a relativistic current and a small constituent quark form factor [29], compared with the TJNAF data [28,30]

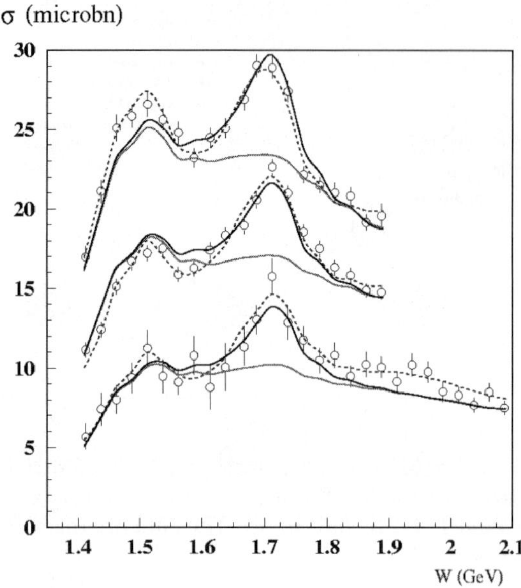

Fig. 7. Comparison of the new CLAS data for the two pion total cross section with the calculation based on the electromagnetic helicity amplitudes from the hCQM, transverse and longitudinal ones, the *solid line* is the result obtained including also the new state found by their previous analysis. The explanation of this picture and of the work can be found on Ripani's contribution

kinetic energy; b) the boosts from the rest frames of the initial and final baryon to a common (say the Breit) frame; c) a relativistic quark current. All these features are not equivalent to a fully relativistic dynamics, which is still beyond the present capabilities of the various models.

The potential of 1 has been refitted using the correct relativistic kinetic energy

$$H_{rel} = \sum_{i=1}^{3} \sqrt{p_i^2 + m^2} - \frac{\tau}{x} + \alpha x + H_{hyp}. \qquad (7)$$

The resulting spectrum is not much different from the non relativistic one and the parameters of the potential are only slightly modified.

The boosts and a relativistic quark current expanded up to lowest order in the quark momenta has been used both for the elastic form factors of the nucleon [16] and the helicity amplitudes [26]. In the latter case, as already mentioned, the relativistic effects are quite small and do not alter the agreement with data discussed previously. For the elastic form factors, the relativistic effects are quite strong and bring the theoretical curves much closer to the data; in any case they are responsible for the decrease of the ratio between the electric and magnetic proton form factors, as it has been shown for the first time in [17], in qualitative agreement with the recent Jlab data [28].

A relativistic quark current, with no expansion in the quark momenta, and the boosts to the Breit frame have been applied to the calculation of the elastic form factors in the relativistic version of the hCQM (7) [29]. The resulting theoretical form factors of the proton, calculated, it should be stressed, without free parameters and assuming pointlike quarks, are good with some discrepancies at low Q^2, which, as discussed previously, can be attributed to the lacking of the quark-antiquark pair effects. Concerning the ratio between the electric and magnetic proton form factors the deviation from unity reaches almost the 50% level, not far from the new TJNAF data [30].

Nevertheless to obtain a full description of the existing data on the elastic form factors one still has to take into account the $q\bar{q}$ degrees of freedom. A way to describe effectively the effect of these extra degrees of freedom is

to introduce constituent quark form factors. Adding the effect of small contituent quark form factor to the hCQM results, the curve shown in Fig. 6 is obtained.

6 Conclusions

The hCQM is a generalization to the baryon sector of the widely used quark-antiquark potential containing a coulomb plus a linear confining term. The three free parameters have been adjusted to fit the spectrum [3] and then the model has been used for a systematic calculation of various physical quantities: the photocouplings [14], the helicity amplitudes for the electromagnetic excitation of negative parity baryon resonances [15,26,23], the elastic form factors of the nucleon [16,29] and the ratio between the electric and magnetic proton form factors [17,29]. The agreement with data is quite good, specially for the helicity amplitudes, which are reproduced in the medium-high Q^2 behaviour, leaving some discrepancies at low (or zero) Q^2, where the lacking quark-antiquark contributions are expected to be effective. It should be noted that the hypercoulomb term in the potential is the main responsible of such an agreement [12], while for the spectrum a further fundamental aspect is provided by the isospin dependent interactions [27]. We have completed our program calculating the transverse and longitudinal helicity amplitudes for all the resonances [23], opening in this way the possibility of applications to the calculation of cross sections, as for example in the two pion case [6] (see Fig. 7). Finally, we have calculated the chiral corrections to our helicity amplitudes showing an impressive reproduction of all the existing data [7].

References

1. N. Isgur and G. Karl: Phys. Rev. D **18**, 4187 (1978); D **19**, 2653 (1979); D **20**, 1191 (1979); S. Godfrey and N. Isgur: Phys. Rev. D **32**, 189 (1985)
2. S. Capstick and N. Isgur: Phys. Rev. D **34**, 2809 (1986)
3. M. Ferraris, M.M. Giannini, M. Pizzo, E. Santopinto, and L. Tiator: Phys. Lett. B **364**, 231 (1995); E. Santopinto: PhD Thesis, Universitá di Genova (1995)
4. L. Ya. Glozman and D.O. Riska: Phys. Rep. C **268**, 263 (1996)
5. L. Ya. Glozman, Z. Papp, W. Plessas, K. Varga, and R.F. Wagenbrunn: Phys. Rev. C **57**, 3406 (1998); L. Ya. Glozman, W. Plessas, K. Varga, and R. F. Wagenbrunn, Phys. Rev. D **58**, 094030 (1998)
6. M. Ripani: to be published on Eur. Phys. J. (2004)
7. L.Tiator, D. Drechsel, S. Kamalov, E. Santopinto, M.M. Giannini, and A. Vassallo: to be published on Eur. Phys. J. (2004)
8. G. Bali et al.: Phys. Rev. D **51**, 5165 (1995); G. Bali: Phys. Rept. **343**, 1 (2001)
9. P. Hasenfratz, R.R. Horgan, J. Kuti, and J.M. Richard: Phys. Lett. B **94**, 401 (1980)
10. J.-M. Richard: Phys. Rep. C **212**, 1 (1992)
11. M. Fabre de la Ripelle and J. Navarro: Ann. Phys. (N.Y.) **123**, 185 (1979)
12. E. Santopinto, F. Iachello, and M.M. Giannini: Nucl. Phys. A **623**, 100c (1997); Eur. Phys. J. A **1**, 307 (1998)
13. Particle Data Group: Eur. Phys. J. C **15**, 1 (2000)
14. M. Aiello, M. Ferraris, M.M. Giannini, M. Pizzo, and E. Santopinto: Phys. Lett. B **387**, 215 (1996)
15. M. Aiello, M.M. Giannini, and E. Santopinto: J. Phys. G: Nucl. Part. Phys. **24**, 753 (1998)
16. M. De Sanctis, E. Santopinto, and M.M. Giannini: Eur. Phys. J. A **1**, 187 (1998)
17. M. De Sanctis, M.M. Giannini, L. Repetto, and E. Santopinto: Phys. Rev. C **62**, 025208 (2000)
18. L.A. Copley, G. Karl, and E. Obryk: Phys. Lett. **29**, 117 (1969)
19. R. Koniuk and N. Isgur: Phys. Rev. D **21**, 1868 (1980)
20. F.E. Close and Z. Li: Phys. Rev. D **42**, 2194 (1990); Z. Li and F.E. Close: Phys. Rev. D bf 42, 2207 (1990)
21. V.D. Burkert: private communication
22. R.A. Thompson et al.: Phys. Rev. Lett. 86, 1702 (2001)
23. M.M. Giannini, E. Santopinto, and A. Vassallo: to be published
24. S. Capstick and B.D. Keister: Phys. Rev. D **51**, 3598 (1995)
25. V.D. Burkert: arXiv:hep-ph/0207149
26. M. De Sanctis, E. Santopinto, and M.M. Giannini: Eur. Phys. J. A **2**, 403 (1998)
27. M.M. Giannini, E. Santopinto, and A. Vassallo: Eur. Phys. J. A **12**, 447 (2001)
28. M.K. Jones et al.: Phys. Rev. Lett. B **84**, 1398 (2000)
29. M. De Sanctis, M.M. Giannini, E. Santopinto, and A. Vassallo: to be published
30. O. Gayon et al.: Phys. Rev. Lett. **88**, 092301 (2002)
31. M.M. Giannini, E. Santopinto, A. Vassallo, and M. De Sanctis: to be published on Eur. Phys. J. (2004)

Eur Phys J A (2004) **19**, s01, 87–92
Digital Object Identifier (DOI) 10.1140/epjad/s2004-03-015-y

EPJ A direct
electronic only

Transverse asymmetry of $^{3}\overrightarrow{\mathrm{He}}$ and the magnetic form factor of the neutron

A. Kievsky[1], E. Pace[2], and G. Salmè[3]

[1] Istituto Nazionale di Fisica Nucleare, Sezione di Pisa, Via Buonarroti 2, I-56100 Pisa, Italy
[2] Dipartimento di Fisica, Università di Roma "Tor Vergata" and Istituto Nazionale di Fisica Nucleare, Sezione Tor Vergata, Via della Ricerca Scientifica 1, I-00133 Roma, Italy
[3] Istituto Nazionale di Fisica Nucleare, Sezione Roma I, P. le A. Moro 2, I-00185 Roma, Italy

Received: 3 Oct 2003 / Accepted: 14 Nov 2003 /
Published Online: 6 Feb 2004 – © Società Italiana di Fisica / Springer-Verlag 2004

Abstract. Preliminary calculations of the inclusive electromagnetic responses by a polarized ^{3}He target, including final state interaction effects, relativistic kinematics and a relativistic electron-nucleon cross section, are presented. The features of our approach will be briefly illustrated, in order to yield an insight on the model dependence affecting the extraction of the neutron magnetic form factor, G_M^n, from the inclusive scattering of polarized electrons by a polarized ^{3}He target, recently explored at TJLAB for $0.1\ (GeV/c)^2 \leq Q^2 \leq 0.6\ (GeV/c)^2$.

PACS. 13.40Gp, – 13.60.Hb – 21.45+v – 24.70.+s

1 Introduction

In the last few years, an impressive amount of experimental work (see, for a recent review, K. de Jager [1]) has been devoted to an accurate investigation of nucleon electromagnetic (em) form factors, obtaining unexpected results like the puzzling ratio G_E/G_M for the proton (see, e.g., [2]). In order to have a complete knowledge of the nucleon form factors, one has to face with the particularly difficult problem represented by the extraction of the neutron em form factors, since free neutron targets do not exist in nature. To overcome such a difficulty, light nuclei, such as deuteron or polarized ^{3}He, have been considered as an effective neutron target. As a matter of fact, within a naive model for a polarized ^{3}He target with only a symmetric S-wave component in the bound-state, the two protons have opposite spins and therefore only the neutron spin contributes to the em polarized responses of the target. Within a realistic description of ^{3}He, it is not a trivial task to disentangle the neutron information from the nuclear-structure effects, given the many effetcs playing relevant roles, like: i) the "small" components of the bound-state wave function; ii) the Δ excitation; iii) the inclusion of the final state interaction (FSI), between the knocked-out nucleon and the interacting spectator pair; iv) the meson exchange currents (MEC); v) the relativistic corrections. As a first step, the analysis of inclusive responses of polarized ^{3}He was carried out within the *plane wave impulse approximation* (PWIA) in [3,4,5,6], where realistic nucleon-nucleon interactions were adopted and in the final state only the interaction between the spectator pair and the knocked-out nucleon was disregarded. Moreover a relativistic electron-nucleon cross section [7] was adopted.

Recently, a step forward has been performed through calculations of the em responses which include FSI, but within a non relativistic framework [8,9].

In this contribution, a review of recent applications of our PWIA description of the polarized responses will be given, along with a presentation of preliminary calculations which include FSI in the two-body break-up channel.

In Sect. 2 the polarized cross section in PWIA will be briefly reviewed and relevant applications will be presented; in Sect. 3 the fully-interacting wave function for a three-nucleon system in the continuous spectrum will be introduced; in Sect. 4, preliminary results with FSI for the transverse response of a polarized ^{3}He target will be shown; Sect. 5 contains summary and perspectives.

2 The polarized cross section in PWIA

The inclusive scattering of polarized electrons (with helicity h) by a polarized ^{3}He target ($\overrightarrow{e}\ +\ ^{3}\overrightarrow{\mathrm{He}}\ \to e' + X$) is given by

$$\frac{d^2\sigma(h)}{d\Omega_2 d\omega} = \Sigma + h\,\Delta \qquad (1)$$

with

$$\Sigma = \sigma_{Mott} \left[\left(\frac{Q^2}{|\mathbf{q}|^2} \right)^2 A_L(Q^2,\omega) + \right.$$
$$\left. + \left(\frac{Q^2}{2|\mathbf{q}|^2} + tan^2\frac{\theta_e}{2} \right) A_T(Q^2,\omega) \right] \quad (2)$$

$$\Delta = -\sigma_{Mott}\, tan\frac{\theta_e}{2} \times$$
$$\left\{ cos\theta^*\, A_{T'}(Q^2,\omega)\, \frac{(\epsilon_i + \epsilon_f)}{|\mathbf{q}|} tan\frac{\theta_e}{2} + \right.$$
$$\left. - \frac{Q^2}{|\mathbf{q}|^2\,\sqrt{2}}\, sin\theta^* cos\phi^*\, A_{TL'}(Q^2,\omega) \right\} \quad (3)$$

where θ_e is the scattering angle, $\epsilon_{i(f)}$ the energy of the initial (final) electron, θ^* and ϕ^* are the azimuthal and polar angles of the target polarization vector, with respect to the direction of the three-momentum transfer $\mathbf{q}$; $Q^2 = |\mathbf{q}|^2 - \omega^2$, with ω the energy transfer. The unpolarized (A_L, A_T) and polarized ($A_{T'}$, $A_{TL'}$) inclusive responses contain the nuclear-structure effects. The key point in our PWIA calculation of the inclusive responses is the following approximation of the three-nucleon wave function describing the final state, viz.

$$|j, j_z, T, T_z, \pi, \epsilon_{int}, \alpha; \mathbf{P}\rangle \rightarrow \frac{1}{\sqrt{3}} \times$$
$$|\mathbf{p}_f, \sigma_f \tau_f\rangle |j_{23}m_{23}, T_{23}\tau_{23}, \pi_{23}, \lambda_{23}, \epsilon_{23}; \mathbf{P}_{23}\rangle \quad (4)$$

with $j(j_z)$ the total angular momentum (third component), $T(T_z)$ the total isospin (third component), π the total parity, ϵ_{int} the intrinsic energy of the three-nucleon system, $\alpha \equiv \{j_{23}T_{23}\lambda_{23}, \pi_{23}, \epsilon_{23}\}$, $\mathbf{P} = \mathbf{p}_f + \mathbf{P}_{23}$ the three-momentum of the three-nucleon centre of mass (CM), $|\mathbf{p}_f, \sigma_f \tau_f\rangle$ the plane wave describing the knocked-out nucleon. In (4), the wave function of the fully-interacting spectator pair is given by $|j_{23}m_{23}, T_{23}\tau_{23}, \pi_{23}, \epsilon_{23}; \mathbf{P}_{23}\rangle$. The other terms, that properly antisymmetrize the approximated three-nucleon wave function, are dropped out, since only the direct interaction between the virtual photon and the struck nucleon is taken into account.

Following [4], the nuclear response functions can be expressed through a 2x2 matrix, $\hat{\mathbf{P}}_{\mathcal{M}}^N(\mathbf{p}, E)$, representing the spin dependent spectral function of a nucleon, N, inside a nucleus with component of the total angular momentum along the polarization $\mathbf{S}_A$ equal to $\mathcal{M}$. The elements of the matrix $\hat{\mathbf{P}}_{\mathcal{M}}^N(\mathbf{p}, E)$ are given by

$$\dot{P}_{\sigma,\sigma',\mathcal{M}}^N(\mathbf{p}, E) = \sum_{f(A-1)} {}_N\langle\mathbf{p},\sigma; \psi_{f(A-1)}|\psi_{j\mathcal{M}}\rangle$$
$$\langle\psi_{j\mathcal{M}}|\psi_{f(A-1)}; \mathbf{p}, \sigma'\rangle_N\, \delta(E - E_{f(A-1)} + E_A) \quad (5)$$

where E and $\mathbf{p}$ are the missing energy and the three-momentum of the nucleon in the bound state, respectively, $|\psi_{j\mathcal{M}}\rangle$ is the ground state of the target nucleus with polarization $\mathbf{S}_A$, and $|\psi_{f(A-1)}; \mathbf{p}, \sigma'\rangle_N$ a shorthand

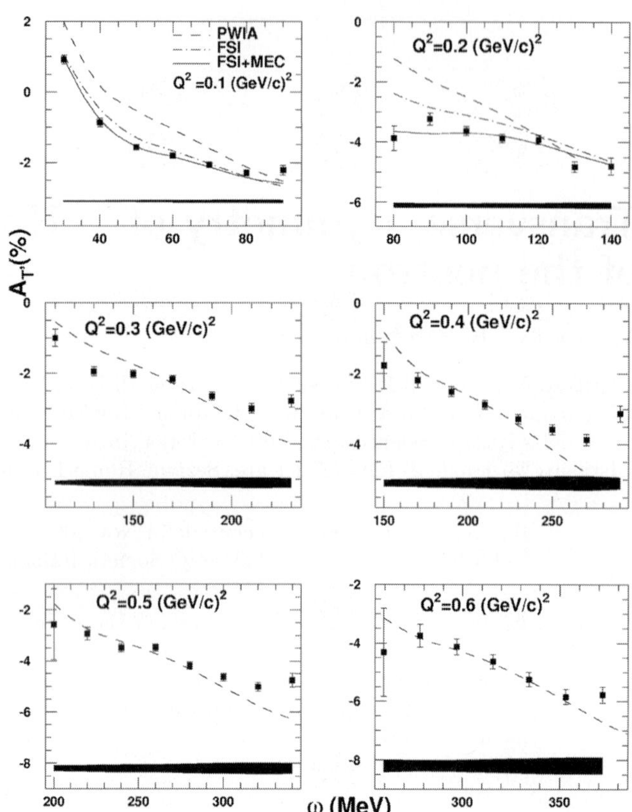

Fig. 1. The transverse asymmetry $A_{T'}$ vs the energy transfer, ω, for different values of Q^2. *Dashed lines:* PWIA calculations within our approach [5]; *dash-dotted lines* and *solid lines:* Faddeev calculations with FSI only and with FSI + MEC, respectively [8,9]. (After W. Xu et al. [10])

notation for the *plane wave impulse* approximation of a three-nucleon system in the continuum, given by (4). For $j = 1/2$, $\hat{\mathbf{P}}_{\mathcal{M}}^N(\mathbf{p}, E)$ can be written in a more compact form as follows [4]

$$\hat{\mathbf{P}}_{\mathcal{M}}^N(\mathbf{p}, E) = \frac{1}{2} \left\{ B_{0,\mathcal{M}}^N(|\mathbf{p}|, E) + \right.$$
$$\left. \sigma \cdot \left[\mathbf{S}_A\, B_{1,\mathcal{M}}^N(|\mathbf{p}|, E) + \hat{p}\, (\hat{p} \cdot \mathbf{S}_A)\, B_{2,\mathcal{M}}^N(|\mathbf{p}|, E) \right] \right\} (6)$$

where $B_{0,\mathcal{M}}^N(|\mathbf{p}|, E)$ is the trace of $\hat{\mathbf{P}}_{\mathcal{M}}^N(\mathbf{p}, E)$ and is the usual unpolarized spectral function, while the other two functions, $B_{1,\mathcal{M}}^N(|\mathbf{p}|, E)$ and $B_{2,\mathcal{M}}^N(|\mathbf{p}|, E)$, describe the spin structure of the probability distribution of finding a nucleon in the nucleus with a given momentum, missing energy and polarization. Then, the calculation of the inclusive responses in (2) and (3) proceeds through the evaluation of the functions $B_{i,\mathcal{M}}^N$ [3,4]. It should be pointed out that, in our PWIA, we take into account relativistic effects, like the relativistic treatment of the final state kinematics and the relativistic electron-nucleon cross section CC1 proposed in [7]. In Figs. 1 and 2 calculations of the transverse (polarized) response $A_{T'}$, obtained with the AV18 nucleon-nucleon potential [11], are compared with the data recently measured at TJLAB, both in the region of the quasielastic peak (Fig. 1) and in the low

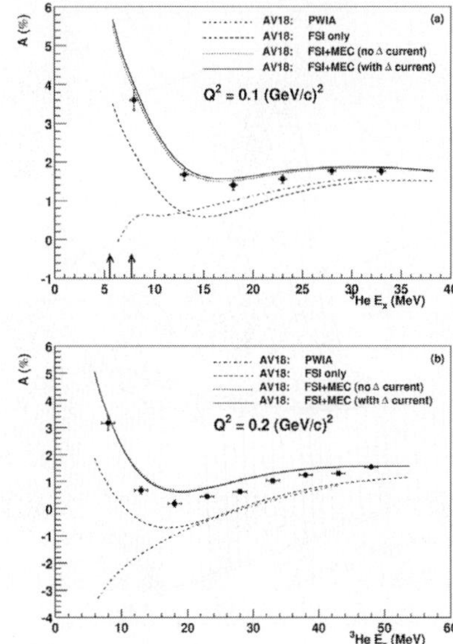

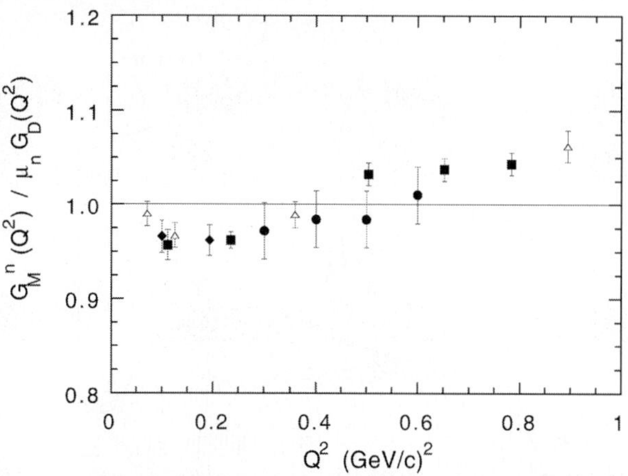

Fig. 2. The transverse asymmetry $A_{T'}$ vs $E_X = \sqrt{M_{He}^2 + 2M_{He}\omega - Q^2} - M_{He}$ (in the non relativistic limit $E_X = \epsilon_{int} + B_{He}$). *Dash-dotted lines:* our PWIA calculations [5]; *dashed and solid lines:* FSI and FSI + MEC calculations [8,9], respectively. Experimental data belong to a kinematical region where the energy transfer is lower than the value corresponding to the quasielastic peak (after F. Xiong et al. [14])

energy-transfer region (Fig. 2). At low values of Q^2, calculations [8,9] including FSI and MEC, but within a non relativistic approach, are also shown. On one side, the disagreement, at low Q^2, between our PWIA calculations [5] and both experimental data and calculations including FSI illustrates the relevance of nuclear-structure effects beyond PWIA. On the other side, at higher values of Q^2, in the region close to the quasielastic peak, the quite reasonable description of the data achieved by our PWIA (that contains relativistic effects) confirms the physical expectation of a minor role played by FSI, when the nucleon rapidly gets out. Experimental data in the region of the quasielastic peak and theoretical results (necessary for an estimate of the model dependence of the procedure) have been exploited for extracting the neutron form factor, $G_M^n(Q^2)$ in the range $0.1 \leq Q^2 \leq 0.6$ $(GeV/c)^2$, see Fig. 3. For $Q^2 \leq 0.2$ $(GeV/c)^2$ (diamonds) the calculations by the Bochum group [8,9] were used, while for $Q^2 > 0.2$ $(GeV/c)^2$ our PWIA calculations have been employed in the procedure of extraction [13] (full dots). At the quasielastic peak, the uncertainties due to FSI and MEC, estimated from the calculations of the Bochum group, amount to few percent for $Q^2 > 0.2$ $(GeV/c)^2$, see W. Xu et al. [13].

3 Three-nucleon scattering states

The fully-interacting, intrinsic wave function, for a three-nucleon system in the continuous spectrum, can be de-

Fig. 3. The most recent experimental determinations of $G_M^n/\mu_n G_D$ vs Q^2 ($G_D = (1 + Q^2/0.71)^{-2}$). *Triangles:* G. Kubon et al. [15]; *diamonds:* W. Xu et al. [10]; *black squares:* H. Anklin et al. [12]; *full dots:* W. Xu et al. [13]. The full dots have been extracted [13] using our PWIA calculations [5]

composed [16] as follows

$$\Phi^{jj_zTT_z\pi} = \Psi_A^{jj_zTT_z\pi} + \Psi_C^{jj_zTT_z\pi} =$$
$$= \sum_{i=3} \left[\psi_A^{jj_zTT_z\pi}(i) + \psi_C^{jj_zTT_z\pi}(i) \right] \quad (7)$$

where the intrinsic energy is understood; $\Psi_A^{jj_zTT_z\pi}$ is the solution of the Schrödinger equation in the asymptotic region, with two well separated clusters, $\Psi_C^{jj_zTT_z\pi}$ describes the system when the three nucleons are close each other. The functions $\psi_A^{jj_zTT_z\pi}(i)$ and $\psi_C^{jj_zTT_z\pi}(i)$ are Faddeev-like amplitudes, corresponding to the three permutations of the intrinsic coordinates ($\equiv \{\mathbf{r_1}, \mathbf{r_2}, \mathbf{r_3}\}$).

The asymptotic component, $\Psi_A^{jj_zTT_z\pi}$, can be recast in a different way, in order to emphasize its physical content. If one considers the case of a N-d scattering state, for the sake of concreteness, one has

$$\Psi_A^{LXjj_zTT_z\pi} = \Omega_{LXj}^R(\mathbf{x}_1, \mathbf{y}_1) +$$
$$+ \sum_{L'X'} {}^j\mathcal{L}_{LL'}^{XX'} \left[i\, \Omega_{L'X'j}^R(\mathbf{x}_1, \mathbf{y}_1) + \Omega_{L'X'j}^I(\mathbf{x}_1, \mathbf{y}_1) \right] +$$
$$+ \left[\psi_A^{LXjj_zTT_z}(2) + \psi_A^{LXjj_zTT_z}(3) \right] \quad (8)$$

where L is the relative orbital angular momentum of N with respect to the deuteron, X is the intermediate coupling of the spin of the nucleon with the total angular momentum of the deuteron, and the intrinsic coordinates $\{\mathbf{x}_1, \mathbf{y}_1\}$ are defined as follows

$$\mathbf{x}_1 = \mathbf{r}_2 - \mathbf{r}_3 \qquad \mathbf{y}_1 = \frac{1}{\sqrt{3}}\left[\mathbf{r}_2 + \mathbf{r}_3 - 2\mathbf{r}_1\right]. \quad (9)$$

In (8), $\Omega_{LXJ}^{R(I)}$ represents the regular ("irregular", but properly regularized at small distances [16,17]) solution describing the free scattering of a nucleon by an interacting pair (in this case a deuteron); the matrix $\mathcal{L}$ is given by

$$\mathcal{L} = \frac{\mathcal{S} - 1}{2i} = -\pi\mathcal{T} \quad (10)$$

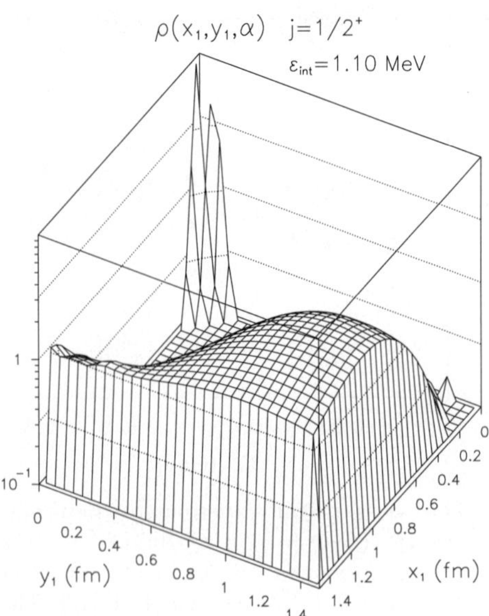

Fig. 4. The density function defined in (13) vs the two Jacobi coordinates, $\{|\mathbf{x}_1|,|\mathbf{y}_1|\}$ for $j = 1/2$, $T = 1/2$, $\pi = +1$ and $\epsilon_{int} = 1.1$ MeV

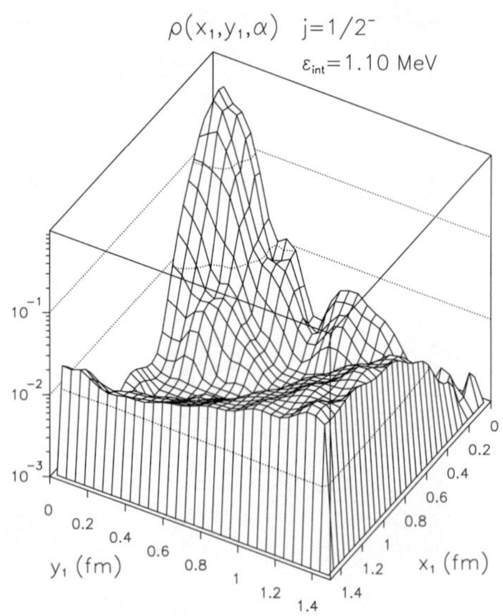

Fig. 5. The density function defined in (13) vs the two Jacobi coordinates, $\{|\mathbf{x}_1|,|\mathbf{y}_1|\}$ for $j = 1/2$, $T = 1/2$, $\pi = -1$ and $\epsilon_{int} = 1.1$ MeV

with $\mathcal{S}$ and $\mathcal{T}$ the S-matrix and the T-matrix, respectively. Similar expressions hold for $\psi_A^{LXjj_zTT_z}(2)$ and $\psi_A^{LXjj_zTT_z}(3)$. Three terms can be recognized in (8):

- the first one produces the PWIA, i.e. contains an interacting pair and a free particle;
- the second term describes the rescattering between the interacting pair and the asymptotically free particle;
- the third term, $\left[\psi_A^{LXjj_zTT_z}(2) + \psi_A^{LXjj_zTT_z}(3)\right]$, takes care of the correct antisymmetrization of $\Psi_A^{jj_zTT_z\pi}$.

The core component, $\Psi_C^{jj_zTT_z\pi}$, goes to zero for large interparticle distances and energies below the deuteron breakup threshold, while for higher energies, must reproduce a three outgoing particle state. In the approach developed in [16], $\Psi_C^{jj_zTT_z\pi}$ is explicitly written as an expansion on a basis of Hyperspherical Harmonics Polynomials, with the inclusion of pair-correlation functions, to be determined along with the elements of the S-matrix (see (10)), through a variational procedure (complex Kohn variational principle).

In the coordinate space the state (7) can be written

$$\langle \mathbf{y}_1, \mathbf{x}_1 | j, j_z; TT_z, \pi, \epsilon_{int}; \beta \rangle = \frac{1}{\sqrt{3}} \sum_{L'X'} \sum_{j'_{23}S'_{23}} \sum_\ell$$

$$\mathcal{Y}_{L'X'jj_zTT_z}^{j'_{23}S'_{23}\ell}(\hat{x}_1,\hat{y}_1)\, \mathcal{J}_{L'X'jT}^{j'_{23}S'_{23}\ell}(|\mathbf{y}_1|,|\mathbf{x}_1|,\epsilon_{int};\beta) \quad (11)$$

where $\beta \equiv \{\alpha; LX\}$ is the set of quantum numbers of the incoming wave. The "trivial" dependence upon angular, spin and isospin variables is described by the functions

$$\mathcal{Y}_{L'X'jj_zTT_z}^{j'_{23}S'_{23}\ell}(\hat{x}_1,\hat{y}_1) = \sum_{M\sigma} \sum_{M_Xm_{23}} \langle j'_{23}m_{23}\tfrac{1}{2}\sigma|X'M_{X'}\rangle\, \chi_{\frac{1}{2}}^{\sigma}$$

$$\langle X'M_{X'}L'M|jj_z\rangle\, Y_{L'M'}(\Omega_{\hat{y}_1}) \sum_m \sum_{m_{S'_{23}}} \chi_{S'_{23}}^{m_{S'_{23}}}\, Y_{\ell m}(\Omega_{\hat{x}_1})$$

$$\langle \ell m S'_{23}m_{S'_{23}}|j'_{23}m'_{23}\rangle \sum_{\tau'\tau'_{23}} \langle T'_{23}\tau'_{23}\tfrac{1}{2}\tau'|TT_z\rangle\, \mathcal{T}_{\frac{1}{2}}^{\tau'}\, \mathcal{T}_{T'_{23}}^{\tau'_{23}} \quad (12)$$

that form a complete basis for a two-fermion system (T'_{23} is determined through the parity of S'_{23} and ℓ), while the "non trivial" structure of the final state wave function is described by $\mathcal{J}_{L'X'jT}^{j'_{23}S'_{23}\ell}(|\mathbf{y}_1|,|\mathbf{x}_1|,\epsilon_{int};\beta)$. In order to give an insight on the three-nucleon wave function, it is useful to introduce density functions defined as follows

$$\rho(|\mathbf{y}_1|,|\mathbf{x}_1|,\alpha) =$$

$$= \sum_{LX} \sum_{L'X'} \sum_{j'_{23}S'_{23}} \sum_\ell \left\{ \left| \Re e\left[\mathcal{J}_{L'X'jT}^{j'_{23}S'_{23}\ell}(|\mathbf{y}_1|,|\mathbf{x}_1|,\epsilon_{int};\beta) \right] \right|^2 + \right.$$

$$\left. + \left| \Im m\left[\mathcal{J}_{L'X'jT}^{j'_{23}S'_{23}\ell}(|\mathbf{y}_1|,|\mathbf{x}_1|,\epsilon_{int};\beta) \right] \right|^2 \right\} \quad (13)$$

Such density functions yield information on the probability distribution of a nucleon in the final state, taking into account the caveat that the wave functions in the continuum are not square integrable. Examples of the density function for a state with an asymptotic $p + d$ cluster is presented in Figs. 4 and 5 for $j = 1/2^+$ and $j = 1/2^-$, respectively, with $\epsilon_{int} = 1.1$ MeV.

4 Beyond PWIA

In order to include the full interaction between the three nucleons in the final state, one has to abandon the elegant and compact language of the spectral functions, reverting to a lenghty calculation of matrix elements of the current operator of the three-body system. We approximate the current operator as a sum of one-body operators, i.e. we do not consider for the moment two-body contributions, viz

$$\langle j', j'_z; T', T'_z, \pi', \epsilon'_{int}; \beta'; \mathbf{q} | J^\mu_{IA}(0) | \tfrac{1}{2}, j_z; \tfrac{1}{2}, T_z, \pi_b, \epsilon_b; \mathbf{0} \rangle =$$

$$= 3 \sum_{\sigma_1} \sum_{\sigma'_1} \sum_{\sigma_2} \sum_{\sigma_3} \int d\mathbf{k}_1 \, d\mathbf{k}_2 \times$$

$$\langle j', j'_z; T', T'_z, \pi', \epsilon'_{int}; \beta | \mathbf{k}'_1, \mathbf{k}'_2, \sigma'_1, \sigma_2, \sigma_3 \rangle \times$$
$$\langle \mathbf{q} + \mathbf{k}_1, \sigma'_1 | J^\mu_{1,free}(0) | \mathbf{k}_1, \sigma_1 \rangle \times$$

$$\langle \mathbf{k}_1, \mathbf{k}_2, \sigma_1, \sigma_2, \sigma_3 | \tfrac{1}{2}, j_z; \tfrac{1}{2}, T_z, \pi_b, \epsilon_b \rangle \qquad (14)$$

where $\langle \mathbf{k}_1, \mathbf{k}_2, \sigma_1, \sigma_2, \sigma_3 | j, j_z; T, T_z, \pi, \epsilon \rangle$ indicates the intrinsic three-body wave function, related to the state containing the centre of mass motion by the following simplified expression

$$\langle \mathbf{p}_1, \mathbf{p}_2, \mathbf{p}_3, \sigma_1, \sigma_2, \sigma_3 | j, j_z; T, T_z, \pi, \epsilon; \mathbf{P} \rangle =$$
$$\delta(\mathbf{p}_1 + \mathbf{p}_2 + \mathbf{p}_3 - \mathbf{P}) \times$$
$$\langle \mathbf{k}_1, \mathbf{k}_2, \sigma_1, \sigma_2, \sigma_3 | j, j_z; T, T_z, \pi, \epsilon \rangle \qquad (15)$$

where $\mathbf{k}_i$ is the spatial part of the four-momentum $k^\mu_i \equiv \{ \sqrt{m^2 + |\mathbf{k}_i|^2}, \mathbf{k}_i \} = B^\mu_\nu(\mathbf{P}/M) p^\mu_i$, with B^μ_ν the appropriate boost for the chosen Hamiltonian dynamics (see, e.g., [19]). The expression in (15) is an approximate one, since the Wigner functions corresponding to the boost transformations, as well as other kinematical factors, are dropped out in this introductory presentation of our approach. Our preliminary calculations based on the matrix elements of the em current of ^{3}He, (14), both for the unpolarized (A_L and A_T) and the polarized ($A_{T'}$) responses are presented in Figs. 6,7,8 and 9. FSI is taken into account in the two-body break-up channel (where a fully interacting three-nucleon state behaves asymptotically like a pd system) for $j' \leq 5/2$, while PWIA is used in all other channels. The AV18 nucleon-nucleon interaction [11] has been adopted, without Coulomb effects nor three-body forces (included in the future). Comparisons with the experimental data and the PWIA calculations are also shown. The strong effect of FSI is fully confermed and qualitatively we obtain the same results by the Bochum group [8,9].

5 Summary and perspectives

Recently the inclusive scattering of polarized electrons by a polarized ^{3}He target have been measured at TJLAB, both in the region of the quasi elastic peak and in the low-ω wing. The direct comparison of calculations corresponding to different theoretical approaches with these experimental results allows one to better understand the model dependence in the extraction of the neutron magnetic form factor G^n_M. Indeed for a satisfactory interpretation of the data one needs accurate theoretical calculations that include effects beyond the PWIA, such as i) FSI, ii) MEC, iii) relativistic effects and iv) contributions from the explicit presence of the Δ excitation in the ground state of ^{3}He. In our calculations we have adopted both relativistic kinematics and a relativistic electron-nucleon

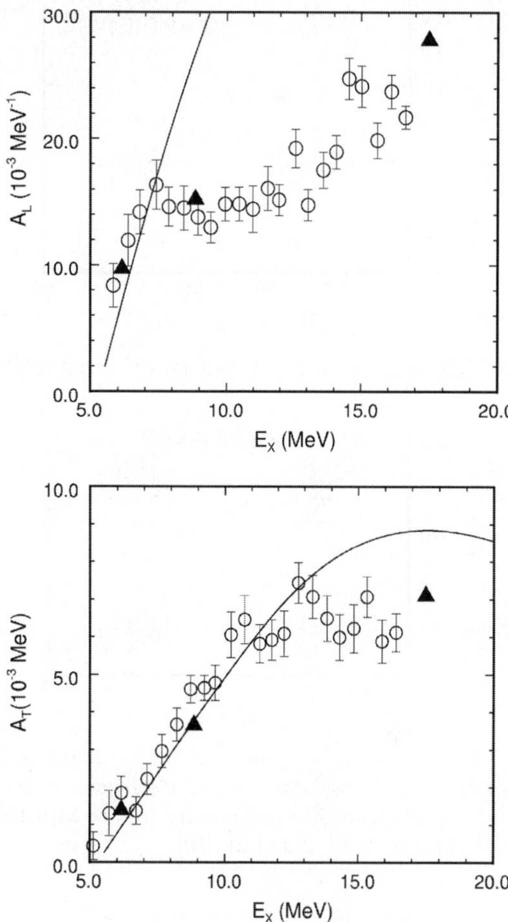

Fig. 6. Unpolarized response functions, A_L and A_T, vs the the missing energy $E_X = \sqrt{(\omega + M_{He})^2 - |\mathbf{q}|^2} - M_{He} \simeq B_3 + \epsilon'_{int}$ for $|\mathbf{q}| \sim 175 \; MeV$. *Solid curves:* PWIA calculations, with unpolarized spectral function obtained [5] from the AV18 nucleon-nucleon interaction [11]. *Triangles:* preliminary calculations with FSI. Experimental data from [18]

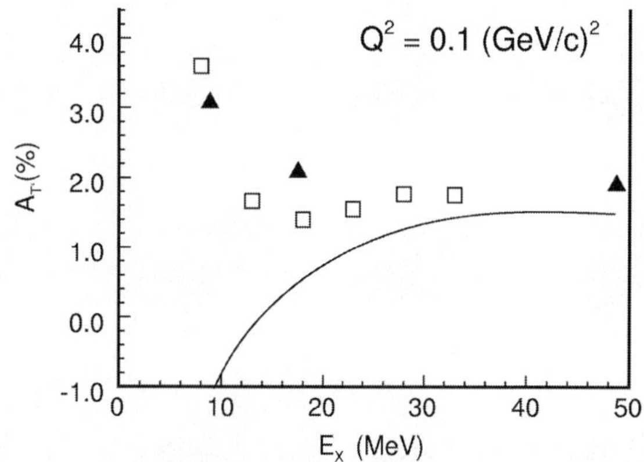

Fig. 7. The transverse response $A_{T'}$ for $Q^2 = 0.1 \; (GeV/c)^2$ vs the excitation energy, E_X, at low energy transfer, as in Fig. 2. *Solid line:* PWIA; *triangles:* preliminary results with FSI. Experimental data from F. Xiong et al. [14]

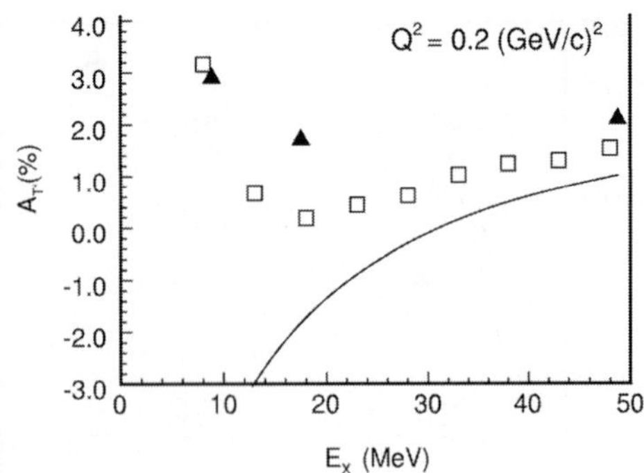

Fig. 8. The same as in Fig. 7, but for $Q^2 = 0.2\ (GeV/c)^2$

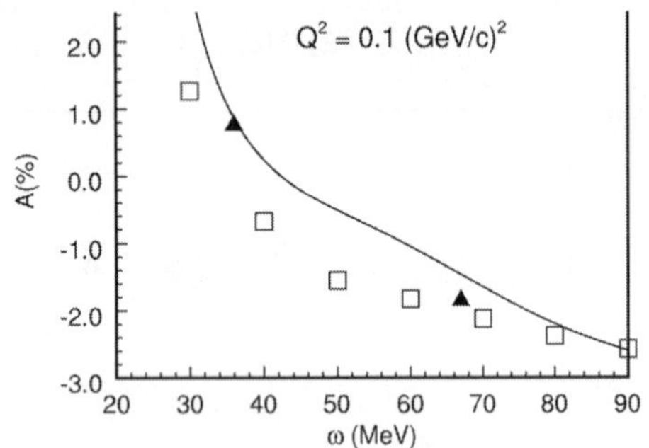

Fig. 9. The transverse response $A_{T'}$ for $Q^2 = 0.1\ (GeV/c)^2$ vs the energy transfer, ω, in the region of the quasielastic peak. *Solid line:* PWIA; triangles: preliminary results with FSI. Experimental data from W. Xu et al. [10]

cross section [7], and we have taken into account exactly the FSI in the two-body break-up channel, by using the three-nucleon wave functions obtained by the Pisa group [16,17] within a variational approach for both the bound and the excited states.

The development of our approach will follow two distinct paths: i) a better treatment of the relativistic effects within the so called Relativistic Hamiltonian Dynamics (see, e.g., B.D. Keister and W. Polyzou [19]), that allows a Poincaré covariant description of an interacting system with a fixed number of particles; ii) the inclusion of FSI in the three-body break-up channels, Coulomb effects and three-body forces.

References

1. K. de Jager: e.g. this Conference
2. O. Gayou et al.: Phys. Rev. Lett. **88**, 092301 (2003) and references therein quoted
3. C. Ciofi degli Atti, E. Pace, and G. Salmè: Phys. Rev. C **46**, R1591 (1992)
4. C. Ciofi degli Atti, E. Pace, and G. Salmè, Phys. Rev. C **51**, 1108 (1995)
5. A. Kievsky, E. Pace, G. Salmè, and M. Viviani: Phys. Rev. C **56**, 64 (1997)
6. R.W. Schültze and P.U. Sauer: Phys. Rev. C **48**, 38 (1993)
7. T. De Forest, Jr.: Nucl. Phys. A **392**, 232 (1983)
8. S. Ishikawa, J. Golak, H. Witala, H. Kamada, and W. Glöckle: Phys. Rev. C **57**, 39 (1998)
9. J. Golak, G. Ziemer, H. Kamada, H. Witala, and W. Glöckle: Phys. Rev. C **63**, 034006 (2001)
10. W. Xu et al.: Phys. Rev. Lett. **85**, 2900 (2000)
11. R.B. Wiringa, R.A. Smith, and T.A. Ainsworth: Phys. Rev. C **29**, 1207 (1984)
12. H. Anklin et al.: Phys. Lett. B **428**, 248 (1998)
13. W. Xu et al.: C **67**, 012201 (2003)
14. F. Xiong et al.: Phys. Rev. Lett. **87**, 242501 (2001)
15. G. Kubon et al.: Phys. Lett. B **524**, 26 (2002)
16. A. Kievsky, M. Viviani, and S. Rosati: Phys. Rev. C **64**, 024002 (2001)
17. A. Kievsky, M. Viviani, and S. Rosati: Nucl. Phys. A **577**, 511 (1994)
18. G.A. Retzlaff et al.: Phys. Rev. C **49**, 1263 (1994)
19. B.D. Keister and W.N. Polyzou: Adv. Nucl. Phys. **20**, 225 (1991)

Eur Phys J A (2004) **19**, s01, 93–97
Digital Object Identifier (DOI) 10.1140/epjad/s2004-03-016-x

EPJ A direct
electronic only

Electromagnetic splittings for hadrons with dressed constituent quarks

B. Silvestre-Brac[1], C. Semay[2], and F. Brau[2]

LPSC,53, Av. des Martyrs, F-38026 Grenoble-Cedex, France
Université de Mons-Hainaut, Place du Parc, 20, B-7000 Mons, Belgium

Received: 28 July 2003 / Accepted: 14 Nov 2003 /
Published Online: 6 Feb 2004 – © Società Italiana di Fisica / Springer-Verlag 2004

Abstract. Electromagnetic splittings for hadrons are calculated in a formalism where the constituent quarks are considered as dressed quasiparticles. The electromagnetic interaction, which contains coulomb, contact and hyperfine terms, is folded with the quark electrical density. The strong potential is a modification of the well known funnel potential. Our model contains only one free parameter and the agreement with experimental data is reasonable although it seems very difficult to obtain a perfect description in any case.

PACS. 12.39Pn Potential models – 13.40Dk Electromagnetic process and properties

1 Introduction

Quantum Chromodynamics (QCD) is believed to be the good theory of strong interaction, but mesons and baryons belong to the non perturbative regime, and, in this case, the theory is very complicated. This explains why a number of alternative simpler models were invented.

Among them, the non relativistic quark model (NRQM) is very appealing: it is simple concerning both the formulation and the numerical calculation, it allows a good treatment of the center of mass motion and it has met with a lot of successes in many domains (see [1]).

In those models, the degrees of freedom are called constituent quarks; they are complicated objects and they must be considered as quasi-particles with some spatial extension.In this picture, the constituent quarks must have some spatial density and the strong interaction should be folded with some gluonic density and the electromagnetic potential with some electromagnetic quark density ([2],[3],[6],[7]). There is no reason that the strong density and the electromagnetic density should be the same.

One knows that it is very difficult to obtain a good description for the spectra of mesons and baryons in a unified treatment([8], [9]). In general, a model good for mesons fails for the description of baryons and the other way round. Here we need to have a correct description of both and we use a strong potential that is a good compromise for that.

But spectra are not enough to test completely a model, and one must rely on more sensitive observables ([10]). The electromagnetic splittings of isospin multiplets are very well suited to make such a study. The origin of the splitting is a mass difference between the up and down

quark (probably already present in the original QCD lagrangian) and also an electromagnetic potential containing a coulomb term and relativistic corrections to it.

In this paper, we want to deal with all the known splittings both in mesons and in baryons in a consistent approach and to push the NRQM study further in several domains. First we want to perform a precise and complete treatment, avoiding perturbative expressions. Second we introduce the contact term, that is usually neglected, in the electromagnetic potential. Lastly, our most important improvement is the use of a dressed electromagnetic interaction among the quarks.

The first section is devoted to the description of strong and electromagnetic potentials that will be employed. The second section presents the results for the splittings of mesons and baryons and the last section deals with conclusions.

2 Strong and electromagnetic potentials

2.1 AL1 potential

Since we are interested by meson and baryon splittings, we have the necessity to use a strong potential that allows a good description of spectra for both types of hadrons. The AL1 potential ([11],[12]) is a slight modification of the well known funnel potential. It looks like :

$$V_{ij}(r) = -\frac{3}{16}\lambda_i \cdot \lambda_j \left[V_C^{(ij)}(r) + V_H^{(ij)}(r) \right]. \qquad (1)$$

Basically it contains two terms with the same colour dependence, that comes in fact from one gluon exchange.

The central part

$$V_C^{(ij)}(r) = -\frac{\kappa}{r} + ar + C. \tag{2}$$

is coulomb+linear and it has also a constant contribution, necessary to reproduce the absolute masses. There is no reason, except simplicity, that the colour dependence is kept unchanged for the confining and constant parts.

The hyperfine term

$$V_H^{(ij)}(r) = \frac{8\pi}{3m_i m_j}\kappa'\frac{\exp(-r^2/\sigma_{ij}^2)}{\pi^{3/2}\,\sigma_{ij}^3}\mathbf{s}_i\cdot\mathbf{s}_j. \tag{3}$$

is flavour dependent through the interacting masses; beside the usual $1/m_i m_j$ factor, the range σ_{ij} is also mass dependent: σ_{ij} does depend on the flavor

$$\sigma_{ij} = A\left(\frac{2m_i m_j}{m_i + m_j}\right)^{-B}. \tag{4}$$

The parameters are completely phenomenological and have been determined essentially on meson spectra. In this sense, they already contain the dressing by the gluonic density. This is manifest from the expression of the hyperfine term, in which the usual $\delta(\mathbf{r})$ term is regularized by a gaussian function.

This potential, used in a Schroedinger equation, gives an overall good description of hadronic spectra, both in the mesonic and baryonic sectors.

There exist many algorithms to compute radial wave functions. Due to the extreme sensitivity of the splitting on the numerical treatment, it is very important to adopt a method which is very precise (and if possible a fast one). In the mesonic sector, we used a method based on Lagrange mesh, which is very simple, very precise and very fast. Technical details of this method can be found in [4]. The number of significant digits is around 10. In the baryonic sector, we used a variational method based on harmonic oscillator basis with different sizes for the various Jacobi coordinates. This method was described in [5] and was proved to be competitive with the stochastic method, if pushed to a number of quanta in the basis equal to 20. The number of significant digits is estimated to 5.

2.2 Electromagnetic quark density

In NRQM, the constituent quarks are extended objects. This means that for a quark at an average position r, there exists a certain probability to be at position r': this probability is more or less a density $\rho(r-r',\gamma)$, where the parameter γ represents the size of the object. This density must be a peaked function reducing to a delta function at the limit of a vanishing size. Another natural property is that the density is isotropic. Lastly, we also require that its integral over the whole space is unity. The most popular densities are of lorentzian, gaussian or Yukawa type. Here we adopt a Yukawa form. There is a precise reason for that: the density is the leading ingredient of the meson

charge form factor. It is an experimental fact that the data accomodate rather nicely a Yukawa density, which has the good asymptotic behaviour. The chosen form is such as :

$$\rho_i(\mathbf{u}) = \frac{1}{4\pi\gamma_i^2}\frac{e^{-u/\gamma_i}}{u}, \tag{5}$$

Within this framework, the two body potentials are obtained from the bare ones with help of a double convolution, one convolution for each of the interacting particle. In fact, just by a simple change of variables, such a double convolution can be reduced to a single convolution:

$$U_{ij}(\mathbf{r}) = \int d\mathbf{r}'\,U_{ij}^{(b)}(\mathbf{r}')\,\rho_{ij}(\mathbf{r}-\mathbf{r}'), \tag{6}$$

In the pecular case of our Yukawa density, the expression for the interacting density is :

$$\rho_{ij}(\mathbf{u}) = \frac{1}{4\pi(\gamma_i^2-\gamma_j^2)}\left(\frac{e^{-u/\gamma_i}}{u} - \frac{e^{-u/\gamma_j}}{u}\right) \tag{7}$$

This expression is valid for interacting particles with different sizes. For particles with identical sizes, the true expression is just the limit of the previous one for $\gamma_j \to \gamma_i$.

2.3 Bare potential

To describe the splittings, one needs first a weak $SU(2)$ breaking allowing a different mass for the u and d quark, but also the presence of the electromagnetic potential. Its traditionnal form originates from relativistic corrections to the Coulomb potential.

$$U_{ij}^{(b)}(\mathbf{r}) = (U_{\text{coul}})_{ij}^{(b)}(\mathbf{r}) + (U_{\text{cont}})_{ij}^{(b)}(\mathbf{r}) + (U_{\text{hyp}})_{ij}^{(b)}(\mathbf{r}), \tag{8}$$

The important terms are the Coulomb, hyperfine and contact terms whose expressions are :

$$(U_{\text{coul}})_{ij}^{(b)}(\mathbf{r}) = \quad Q_i Q_j \frac{\alpha}{r}, \tag{9}$$

$$(U_{\text{cont}})_{ij}^{(b)}(\mathbf{r}) = -\frac{\pi}{2}Q_i Q_j\left(\frac{1}{m_i^2} + \frac{1}{m_j^2}\right)\alpha\,\delta(\mathbf{r}), \tag{10}$$

$$(U_{\text{hyp}})_{ij}^{(b)}(\mathbf{r}) = -\frac{8\pi Q_i Q_j}{3m_i m_j}\alpha\,\delta(\mathbf{r})\mathbf{s}_i\cdot\mathbf{s}_j, \tag{11}$$

where α is the fine structure constant.

In principle it contains also Darwin, spin-orbit and tensor contributions, but their effects are presumably weak and they are neglected in the following. Usually people also neglect the contact term, keeping only the Coulomb and dipole-dipole interactions. Here we include the effect of the contact term in order to grasp quantitatively its effect and to justify a posteriori the validity (or not) of neglecting it.

2.4 Dressed potential

From the bare potential of the previous section 8, and the interacting density 7, the convolution 6 gives the expression of the dressed electromagnetic potential. For each contribution, the various terms look like this :

$$(U_{\mathrm{coul}})_{ij}(r) =$$

$$\alpha Q_i Q_j \left(\frac{1}{r} - \frac{\gamma_i^2}{\gamma_i^2 - \gamma_j^2} \frac{e^{-r/\gamma_i}}{r} + \frac{\gamma_j^2}{\gamma_i^2 - \gamma_j^2} \frac{e^{-r/\gamma_j}}{r} \right), \quad (12)$$

$$(U_{\mathrm{cont}})_{ij}(r) =$$

$$-\frac{\alpha Q_i Q_j}{8(\gamma_i^2 - \gamma_j^2)} \left(\frac{1}{m_i^2} + \frac{1}{m_j^2} \right) \frac{e^{-r/\gamma_i} - e^{-r/\gamma_j}}{r}, \quad (13)$$

$$(U_{\mathrm{hyp}})_{ij}(r) =$$

$$-\frac{2\alpha Q_i Q_j}{3 m_i m_j (\gamma_i^2 - \gamma_j^2)} \frac{e^{-r/\gamma_i} - e^{-r/\gamma_j}}{r} \mathbf{s}_i \cdot \mathbf{s}_j. \quad (14)$$

and the corresponding limits in the case of particles with identical sizes.

Very often such a potential is treated perturbatively. Here we have the ability to perform an exact treatment. A comparison with a perturbative calculation is very instructive.

3 Results

3.1 Determination of the parameters

The first thing to do is to determine the parameters. We do not want to introduce a lot of new parameters; here we restrict the number of free parameters to the minimum unavoidable. In particular, the parameters of the strong potential are maintained without modification. The first parameters to be introduced are the electromagnetic sizes of the quarks γ_i. Indeed the meson form form factors, or alternatively the charge mean square radii, are very dependent from those parameters. The dressed radii are just the sum of the bare radii plus a contribution entirely due to the dressing, which can be expressed, with the Yukawa density, as:

$$\langle r^2 \rangle = \langle r^2 \rangle^{(b)} + 6 \sum_{i=1}^{2} e_i \gamma_i^2. \quad (15)$$

Thus the pion radius is well suited to get the size for the u and d quarks; I suppose that these two sizes are identical. The kaon radius depends both on the size of the ordinary quark and the size of the strange quark. Since the size of the ordinary quark is already known from the pion radius, the kaon radius provides us with the size of the strange quark. Typically we have $\gamma_u = 1.225 GeV^{-1}$ and $\gamma_s = 0.200 GeV^{-1}$. We don't have any experimental data concerning charmed and bottom quarks; we choose values that are smoothly drecreasing with size, namely $\gamma_c = 0.04 GeV^{-1}$ and $\gamma_b = 0.013 GeV^{-1}$.

Concerning the parameters appearing in the strong potential, we decided to let them unchanged, except the masses of the u and d quarks that are now split. In order to minimize the number of free parameters, we also maintain their average value; thus we have at our disposal only one free parameter, the mass difference $\Delta = m_d - m_u$, to try to explain the totality of all known splittings. This parameter is determined on a precise and sensitive value, namely the mass difference for the sigma multiplet.

3.2 Experimental sample

For experimental sample, we consider 10 splittings in the mesonic sector, and 16 splittings in the baryonic sector. They belong to the light and heavy quark sectors. They are typically of order of few MeV. The hierarchy of most of the multiplets can be explained naively just by supposing $m_d > m_u$. Nevertheless, it remains some puzzling questions that cannot be interpreted in this naive scheme. I list some of them here.

- $n - p$ is a positive value (this is normal), but $\pi^+ - \pi^0$ is a larger quantity while naively it should be smaller;
- $\pi^+ - \pi^0$ is positive, while $\rho^+ - \rho^0$ is negative;
- One multiplet in the D sector, namely the D_2, is much smaller than all others in the same sector;
- Σ_c^{++} is experimentally the highest member of the multiplet while a naive argument would have expected it the smallest one;
- in the Ξ_c sector, there is a multiplet which does not fit with the pattern of the others.

Indeed, the electromagnetic splitting is a small quantity (around $1\ MeV$ or less) that is obtained by a difference of large quantities of order $1\ GeV$. This means that it is a very sensitive quantity which represents a subbtle balance between various ingredients.

In particular, it is of first importance to perform a very precise numerical treatment both in the mesonic and the baryonic sector. An error of 10^{-3} GeV on the absolute masses, which can be considered as good for such an observable, would lead to very incorrect values concerning the splittings. We feel that an accuracy of order 10^{-5} on the absolute masses should be reached in order to insure reliable conclusions. This is why we take a special care in our numerical treatment.

It is also important to take care of all the terms of the electromagnetic potential, because all of them are equally important. In the same spirit, it is crucial to have a good wave function. Changing the strong potential may affect seriously the wave function, without spoiling too much the spectra, and this may have a dramatic repercussion on the theoretical splittings.

3.3 Some approximations

Usually the dressing of the electromagnetic potential is not considered, or only partly, and often the bare one is taken into account in a perturbative way. Of course,

Table 1. Electromagnetic splittings (in MeV) for mesons obtained with an exact treatment based on wave functions resulting from AL1 potential. The total electromagnetic hamiltonian is considered. For information, the experimental data, extracted from [13], are given in column "Exp"

Splitting	Exp	Theo
$\pi^+ - \pi^0$	4.594 ± 0.001	1.69
$\rho^+ - \rho^0$	-0.5 ± 0.7	0.71
$K^0 - K^+$	3.995 ± 0.034	8.11
$K^{0*} - K^{+*}$	6.7 ± 1.2	1.44
$K_2^0 - K_2^+$	6.8 ± 2.8	-0.76
$D^+ - D^0$	4.78 ± 0.1	2.31
$D^{+*} - D^{0*}$	2.6 ± 1.8	1.04
$D_1^+ - D_1^0$	6.8 ± 5	-2.16
$D_2^+ - D_2^0$	0.1 ± 4	-2.23
$B^0 - B^+$	0.33 ± 0.28	-1.71

Table 2. Same as Table 1 for baryons. The theoretical uncertainty may affect only the last digit

Splitting	Exp	Theo
$n - p$	1.293	1.15
$\Delta^0 - \Delta^{++}$	2.25 ± 0.68	3.72
$\Delta^+ - \Delta^{++}$	1.2 ± 0.6	1.35
$\Sigma^- - \Sigma^0$	4.81 ± 0.04	4.76
$\Sigma^- - \Sigma^+$	8.08 ± 0.08	8.55
$\Sigma^{-*} - \Sigma^{0*}$	2.0 ± 2.4	2.94
$\Sigma^{-*} - \Sigma^{+*}$	0 ± 4	1.96
$\Xi^- - \Xi^0$	6.48 ± 0.24	7.38
$\Xi^{-*} - \Xi^{0*}$	3.2 ± 0.6	2.66
$\Sigma_c^{++} - \Sigma_c^0$	0.35 ± 0.18	0.37
$\Sigma_c^0 - \Sigma_c^+$	0.9 ± 0.4	0.33
$\Sigma_c^{++*} - \Sigma_c^{0*}$	1.9 ± 1.7	0.19
$\Xi_c^0 - \Xi_c^+$	5.5 ± 1.8	3.42
$\Xi_c^{0\prime} - \Xi_c^{+\prime}$	$\simeq 4.2 \pm 3.5$	0.20
$\Xi_c^{+*} - \Xi_c^{0*}$	$\simeq 2.9 \pm 2.0$	-0.25
$\Xi_c^{0**} - \Xi_c^{+**}$	$\simeq 4.1 \pm 2.5$	3.51

changing the free parameters can give correct results, but we think that using a dressed expression is more satisfactory. It allows to have a complete consistency between the treatments for spectra, for charge radii, and for the electromagnetic splittings.

The use of the dressed potential, instead of the bare one is crucial. Indeed, the bare interaction cannot be used in an exact calculation, because it leads to a collapse, due to the presence of the Dirac term. But even in a perturbative treatment, we checked that the differences between bare and dressed potentials can raise 50 MeV, a very important value. We also think that, since the strong potential is already dressed with a gluonic density, it is much more consistent and satisfactory to dress also the electromagnetic potential.

Using a perturbative procedure can give catastrophic results for certain states. In fact the fault is not due to the electromagnetic potential, but mainly to the strong hyperfine term that depends strongly on the mass difference Δ.

Also, all terms of the electromagnetic potential are important, including the contact term that is usually neglected, but it is impossible to say which term is the leading one. Nevertheless, we have remarked that the contact term has a tendancy to spoil the result. We have no explanation for this bad feature.

All these remarks are the consequence that the splittings are a small quantity obtained from large contributions.

3.4 Meson splittings

The results for the splittings in the meson sector are presented in Table 1.

All the terms of the electromagnetic potential are taken into account and an exact numerical treatment is performed. The results are not very good. This is in part due to the fact that the only free parameter has been fitted in the baryonic sector (on the sigma multiplet). We could have obtained much better results, by fitting for example the pion multiplet, but in that case the baryonic sector would have been spoilt a lot. Let us also stress that the results very much depend on the strong potential. Using a different potential gives results appreciably different. In this case, none of the mentioned puzzles can find a solution, except may be the charmed resonances. Once again, we are faced to the problem of having a consistent description of mesons and baryons.

3.5 Baryon splittings

The results for the splittings in the baryonic sector are presented in Table 2, with the same conventions. The results look much more in agreement with data. The order of magnitude is the correct one, and when differences are significant, this always correspond to experimental values affected of a large uncertainty. The order in the multiplet is practically always respected, in particular for the Σ_c multiplet, solving thus one puzzle. Owing to the fact that we have only one free parameter at our disposal, these results can be considered as very encouraging.

Let us mention that the absolute masses of the baryons are also in nice agreement with the data. A small discrepancy is seen for light baryons, but it can be attributed to three-body forces that are not considered here.

4 Conclusions

The electromagnetic splittings are very sensitive observables depending upon a lot of ingredients. In consequence, it is of first importance

- to dress the potentials
- to take into account all influent terms in the potentials
- to perform an exact treatment and not only a perturbative one
- to use numerical algorithm that allow precise calculations for both the two-body and the three-body case

The results also depend significantly on the wave function, and it is important to employ a strong potential that gives a good description of both mesaon and baryon sectors.

We were very cautious in all these respects, and are very confident in our conclusions.

Some puzzles have found a solution, but others are still opened questions, and this means that this subject merits further investigations.

References

1. W. Lucha, F.F. Schöberl, and D. Gromes: Phys. Rep. **200**, 127 (1991)
2. C. Semay and B. Silvestre-Brac: Nucl. Phys A **618**, 455 (1997)
3. C. Semay and B. Silvestre-Brac: Nucl. Phys A **647**, 72 (1999)
4. C. Semay, D. Baye, M. Hesse, and B. Silvestre-Brac: Phys. Rev. E **64**, 016703 (2001)
5. B. Silvestre-Brac, R. Bonnaz, C. Semay, and F. Brau: ISN Grenoble, **ISN-00-66**, 2000 (unpublished)
6. F. Brau and C. Semay: Phys. Rev. D **58**, 034015 (1998)
7. C. Semay, F. Brau, and B. Silvestre-Brac: Phys. Rev. C **64**, 055202 (2001)
8. W.H. Blask, U. Bohn, M.G. Huber, B.C. Metsch, and H.R. Petry: Z. Phys. A **337**, 327 (1990)
9. F. Brau, C. Semay, and B. Silvestre-Brac: Phys. Rev. C **66**, 055202 (2002)
10. M. Genovese, J.M. Richard, B. Silvestre-Brac, and K. Varga: Phys. Rev. D **59**, 014012 (1998); and references there in.
11. B. Silvestre-Brac and C. Semay: ISN Grenoble, **ISN-93-69**, 1993 (unpublished)
12. C. Semay and B. Silvestre-Brac: Z. Phys. C **61**, 271 (1994)
13. K. Hagiwara et al. (Particle Data Group): Phys. Rev. D **66**, 01001 (2002)

Eur Phys J A (2004) **19**, s01, 99–104
Digital Object Identifier (DOI) 10.1140/epjad/s2004-03-017-9

EPJ A direct
electronic only

A microscopic $NN \to NN^*(1440)$ potential

B. Juliá-Díaz[1,2], A. Valcarce[2], P. González[3], and F. Fernández[2]

[1] Department of Physical Sciences, University of Helsinki and Helsinki Institute of Physics, P.O. Box 64, 00014 Helsinki, Finland
[2] Grupo de Física Nuclear, Universidad de Salamanca, E-37008 Salamanca, Spain
[3] Dpto. de Física Teórica and IFIC, Universidad de Valencia - CSIC, E-46100 Burjassot, Valencia, Spain

Received: 1 July 2003 / Accepted: 14 Nov 2003 /
Published Online: 6 Feb 2004 – © Società Italiana di Fisica / Springer-Verlag 2004

Abstract. By means of a $NN \to NN^*(1440)$ transition potential derived in a parameter-free way from a quark-model based NN potential, we determine simultaneously the $\pi NN^*(1440)$ and $\sigma NN^*(1440)$ coupling constants. We also present a study of the target Roper excitation diagram contributing to the $p(d, d')$ reaction.

PACS. 12.39.Jh Nonrelativistic quark model – 13.75.Cs Nucleon-nucleon interactions

1 Introduction

The $N^*(1440)$ (Roper) is a broad resonance which couples strongly (60–70%) to the πN channel and significantly (5–10%) to the σN channel [1]. These features suggest that the Roper resonance should play an important role in nuclear dynamics as an intermediate state. Graphs involving the excitation of $N^*(1440)$ appear in different systems, as for example the three-nucleon interaction mediated by π and σ exchange contributing to the triton binding energy [2]. The excitation of the Roper resonance has also been used to explain the missing energy spectra in the $p(\alpha, \alpha')$ reaction [3] or the $np \to d(\pi\pi)^0$ reaction [4]. The coupling of the $N^*(1440)$ to πN and σN channels could also be important in heavy ion collisions at relativistic energies [5,6]. Finally, pion electro- and photoproduction may take place through the $N^*(1440)$ excitation [7]. However the use of a $NN \to NN^*(1440)$ transition potential as a straightforward generalization of some pieces of the $NN \to NN$ potential plus the incorporation of resonance width effects may have serious shortcomings specially concerning the short-range part of the interaction [8].

In this talk we present some applications of a recently derived $NN \to NN^*(1440)$ interaction [9], obtained by means of the same quark-model approach previously used to study the NN system and transition potentials involving the Δ. A main feature of the quark treatment is its universality in the sense that all the baryon-baryon interactions are treated on an equal footing. Moreover, once the model parameters are fixed from NN data there are no free parameters for any other case. This allows a microscopic understanding and connection of the different baryon-baryon interactions that is beyond the scope of any analysis based only on effective hadronic degrees of freedom. These studies are instructive inasmuch as they

are expected to lead to a deeper understanding of the nuclear potential and entail a rethinking of basic nuclear concepts from the point of view of the fundamental quark substructure. We center our attention in the derivation of the $\pi NN^*(1440)$ and $\sigma NN^*(1440)$ coupling constants and in the study of a reaction mediated by the excitation of the Roper resonance, the $p(d, d')$ reaction.

2 $\pi NN^*(1440)$ and $\sigma NN^*(1440)$ coupling constants

The usual way to determine meson$-NN$ coupling constants is trough the fitting of NN scattering data with phenomenological meson exchange models. Therefore, a consistent way to obtain meson$-NN^*$ coupling constants is from a transition $NN \to NN^*$ potential, in particular when ratios over meson$-NN$ coupling constants are to be considered. In order to derive the transition potential we shall follow the same quark model approach previously used for NN scattering [10]. Explicitly, the $NN \to NN^*(1440)$ potential at interbaryon distance R is obtained by sandwiching the qq potential, V_{qq}, between NN and $NN^*(1440)$ states, written in terms of quarks, for all the pairs formed by two quarks belonging to different baryons. The qq potential contains a confining term taken to be linear (r_{ij}), the usual perturbative one-gluon-exchange (OGE) interaction containing Coulomb ($1/r_{ij}$), spin-spin ($\sigma_i \cdot \sigma_j$) and tensor (S_{ij}) terms, and pion and sigma exchanges as a consequence of the breaking of chiral symmetry. The wave function of the Roper, $N^*(1440)$, and nucleon, N, states can be written as $|N^*(1440)\rangle = \left\{ \sqrt{\frac{2}{3}} |[3](0s)^2(1s)\rangle - \sqrt{\frac{1}{3}} |[3](0s)(0p)^2\rangle \right\} \otimes [1^3]_c$ and $|N\rangle = |[3](0s)^3\rangle \otimes [1^3]_c$ where $[1^3]_c$ is the completely antisymmetric color state, [3] is the completely symmetric spin-isospin state and $0s$, $1s$, and $0p$, stand for harmonic oscillator orbitals.

The transition potential obtained can be written at all distances in terms of baryonic degrees of freedom [11]. One should realize that a qq spin and isospin independent potential as for instance the scalar one-sigma exchange (OSE), gives rise at the baryon level, apart from a spin-isospin independent potential, to a spin-spin, an isospin-isospin and a spin-isospin dependent interactions [9]. Nonetheless for distances $R \geq 4$ fm, where quark antisymmetrization interbaryon effects vanish, we are only left with the direct part, i.e. with a scalar OSE at the baryon level. The same kind of arguments can be applied to the one-pion exchange (OPE) potential. Thus asymptotically ($R \geq 4$ fm) OSE and OPE have at the baryon level the same spin-isospin structure than at the quark level. Hence we can parametrize the asymptotic central interactions as

$$V^{OPE}_{NN \to NN^*(1440)}(R) = \frac{1}{3} \frac{g_{\pi NN}}{\sqrt{4\pi}} \frac{g_{\pi NN^*(1440)}}{\sqrt{4\pi}} \frac{m_\pi}{2M_N}$$

$$\frac{m_\pi}{2(2M_r)} \frac{\Lambda^2}{\Lambda^2 - m_\pi^2} \left[(\sigma_N.\sigma_N)(\tau_N.\tau_N) \right] \frac{e^{-m_\pi R}}{R}, \quad (1)$$

and

$$V^{OSE}_{NN \to NN^*(1440)}(R) = - \frac{g_{\sigma NN}}{\sqrt{4\pi}} \frac{g_{\sigma NN^*(1440)}}{\sqrt{4\pi}}$$

$$\frac{\Lambda^2}{\Lambda^2 - m_\sigma^2} \frac{e^{-m_\sigma R}}{R}, \quad (2)$$

where g_i stands for the coupling constants at the baryon level and M_r is the $NN^*(1440)$ reduced mass.

By comparing the baryonic potentials with the asymptotic behavior of the ones previously obtained from the quark-model calculation we can extract the $\pi NN^*(1440)$ and $\sigma NN^*(1440)$ coupling constants. As the parameters at the quark level are fixed once for all from the NN interaction our results allow a prediction of these constants in terms of the elementary πqq coupling constant and the one-baryon model dependent structure. The sign obtained for the meson-$NN^*(1440)$ coupling constants and for their ratios to the meson-NN coupling constants is ambiguous since it comes determined by the arbitrarily chosen relative sign between the N and $N^*(1440)$ wave functions. Only the ratios between the $\pi NN^*(1440)$ and $\sigma NN^*(1440)$ would be free of this uncertainty. This is why we will quote absolute values except for these cases where the sign is a clear prediction of the model. To get such a prediction we can use any partial wave. We shall use for simplicity the $^1S_0$ wave, this is why we only wrote the central interaction in (1).

The $[\Lambda^2/(\Lambda^2 - m_i^2)]$ factor comes from the vertex form factor chosen at momentum space as a square root of monopole $[\Lambda^2/(\Lambda^2 + \mathbf{q}^2)]^{1/2}$, the same choice taken at the quark level, where chiral symmetry requires the same form for pion and sigma. A different choice for the form factor at the baryon level, regarding its functional form as well as the value of Λ, would give rise to a different vertex factor and eventually to a different functional form

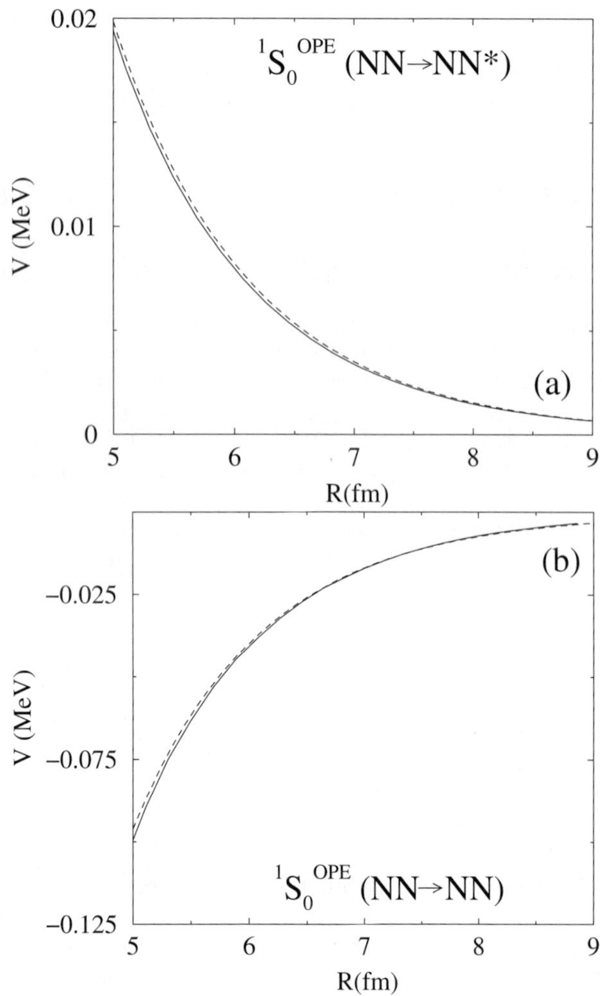

Fig. 1. a Asymptotic behavior of the one-pion exchange $^1S_0$ $NN \to NN^*(1440)$ potential (*solid line*). The *dashed line* denotes the fitted curve according to (1). **b** Same as **a** but for the one-pion exchange $^1S_0$ $NN \to NN$ potential

for the asymptotic behavior. For instance, for a modified monopole form, $[(\Lambda^2 - m^2)/(\Lambda^2 - \mathbf{q}^2)]^{1/2}$, where m is the meson mass (m_π or m_σ), the vertex factor would be 1, i.e. $[(\Lambda^2 - m^2)/(\Lambda^2 - m^2)]$, keeping the potential the same exponentially decreasing asymptotic form. Then it is clear that the extraction from any model of the meson-baryon-baryon coupling constants depends on this choice. We shall say they depend on the coupling scheme.

For the one-pion exchange and for our value of $\Lambda = 4.2$ fm^{-1}, $[\Lambda^2/(\Lambda^2 - m_\pi^2)] = 1.03$, pretty close to 1. As a consequence, in this case the use of our form factor or the modified monopole form at baryonic level makes little difference in the determination of the coupling constant. This fact is used when fixing $g^2_{\pi qq}/4\pi$ from the experimental value of $g^2_{\pi NN}/4\pi$ extracted from NN data.

To get $g_{\pi NN^*(1440)}/\sqrt{4\pi}$ we turn to our results for the $^1S_0$ OPE potential, Fig. 1, and fit its asymptotic behavior

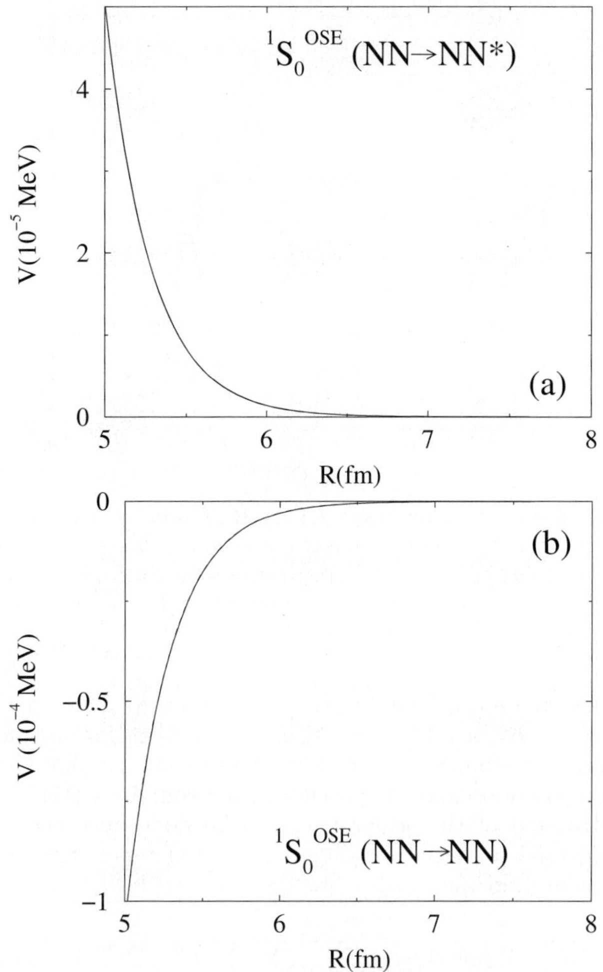

Fig. 2. a Asymptotic behavior of the one-sigma exchange $^1S_0$ $NN \rightarrow NN^*(1440)$ potential (*solid line*). The *dashed line* denotes the fitted curve according to (2). **b** Same as **a** but for the one-sigma exchange $^1S_0$ $NN \rightarrow NN$ potential

(in the range $R : 5 \rightarrow 9$ fm) to (1). We obtain

$$\frac{g_{\pi NN}}{\sqrt{4\pi}} \frac{g_{\pi NN^*(1440)}}{\sqrt{4\pi}} \frac{\Lambda^2}{\Lambda^2 - m_\pi^2} = -3.73\,, \qquad (3)$$

i.e. $g_{\pi NN^*(1440)}/\sqrt{4\pi} = -0.94$. As explained above only the absolute value of this coupling constant is well defined. Let us note that in [12] a different sign with respect to our coupling constant is obtained what is a direct consequence of the different global sign chosen for the $N^*(1440)$ wave function. The coupling scheme dependence can be explicitly eliminated if we compare $g_{\pi NN^*(1440)}$ with $g_{\pi NN}$ extracted from the $NN \rightarrow NN$ potential within the same quark model approximation, Fig. 1. Thus we get

$$\left| \frac{g_{\pi NN^*(1440)}}{g_{\pi NN}} \right| = 0.25\,. \qquad (4)$$

By proceeding in the same way for the OSE potential, i.e. by fitting the potential given in Fig. 2(a) to (2), and following an analogous procedure for the NN case,

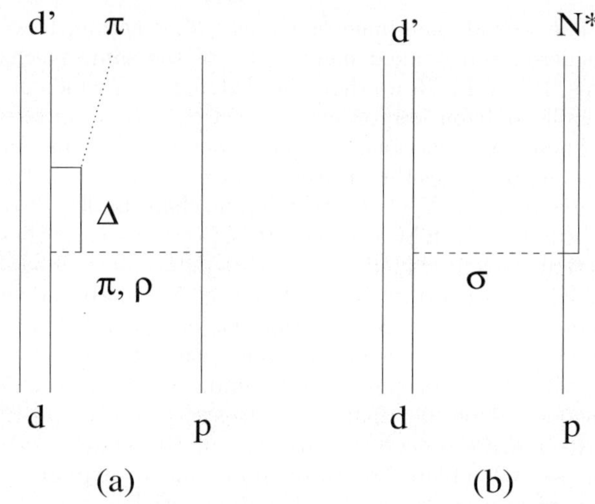

Fig. 3. Dominant mechanisms contributing to the $p(d, d')$ reaction [16]

Fig. 2(b), we can write

$$\left| \frac{g_{\sigma NN^*(1440)}}{g_{\sigma NN}} \right| = 0.47\,. \qquad (5)$$

The ratio given in (4) is similar to that obtained in [12] and a factor 1.5 smaller than the one obtained from the analysis of the partial decay width. Nonetheless one can find in the literature values for $f_{\pi NN^*(1440)}$ ranging between $0.27-0.47$ coming from different experimental analyses with uncertainties associated to the fitting of parameters [4,6,7].

Regarding the ratio obtained in (5), our result agrees quite well with the only experimental available result, obtained in [13] from the fit of the cross section of the isoscalar Roper excitation in $p(\alpha, \alpha')$ in the $10-15$ GeV region, where a value of 0.48 is given. Furthermore, we can give a very definitive prediction of the magnitude and sign of the ratio of the two ratios,

$$\frac{g_{\pi NN^*(1440)}}{g_{\pi NN}} = 0.53 \frac{g_{\sigma NN^*(1440)}}{g_{\sigma NN}}\,, \qquad (6)$$

which is an exportable prediction of our model.

3 Roper excitation in pd scattering

There are two experiments where the contribution from the $N^*(1440)$ resonance has been isolated by means of model-dependent theoretical methods. The first one is the $p(\alpha, \alpha')$ reaction carried out in Saclay [14] already ten years ago. The data showed two peaks in the cross section that were not understood for some years. The most prominent one was attributed to a Δ excitation in the projectile (DEP) [15]. The second peak was explained when a Roper excitation in the target (RET) was considered [3] giving a plausible explanation to the measured differential cross section.

The second experiment is the $p(d, d')$ reaction. It was considered and studied making use of the same mechanisms [16]. In Fig. 3 we show the two diagrams which give the bulk contribution to the cross section of the processes.

These two reactions are particularly interesting because in both cases the projectile (d or α) has $T = 0$. This ensures that the $N^*(1440)$ reaction mechanism, Fig. 3(b), can only be driven by a scalar interaction. Therefore these reactions have provided a method to determine the baryonic coupling constant between the N, $N^*(1440)$ and the σ meson once the Δ contribution has been fixed. The results for the coupling constants obtained in this way from $p(\alpha, \alpha')$ have been quoted and compared to ours in the previous section. Our purpose in this section is the study of the $p(d, d')$ process making use of the quark model $NN \to NN^*(1440)$ transition potential, to explore the mechanism proposed in [16]. The differential cross section for the process is given by:

$$\frac{d^2\sigma}{dE_{d'} d\Omega_{d'}^L} = \frac{p_{d'}}{(2\pi)^5} \frac{M_d^2 M^2}{\lambda^{1/2}(s, M^2, M_d^2)} \times$$

$$\int \frac{d^3 p_\pi}{E_{N'} \omega_\pi} \bar{\Sigma}\Sigma|T|^2 \delta(E_d + E_N - E_{d'} - E_{N'} - \omega_\pi), \quad (7)$$

where $M(M_d)$ is the nucleon (deuteron) mass, s is the invariant mass of the $p-d$ system, $\lambda(x, y, z) = x^2 + y^2 + z^2 - 2xy - 2yz - 2xz$ and $\bar{\Sigma}\Sigma|T|^2$ is the amplitude for the elementary process of $N^*(1440)$ production. This amplitude can be written in terms of the scalar transition potential $(V_0)_{NN \to NN^*}$ [16]:

$$\bar{\Sigma}\Sigma|T|^2 = 12 F_d^2 \left(\frac{f'}{m_\pi}\right)^2 |G^*|^2 |(V_0)_{NN \to NN^*}(q_{cm})|^2 q_{cm}^2 .$$
$$(8)$$

The function $F_d(\mathbf{k})$ is the deuteron form factor defined as

$$F_d(\mathbf{k}) = \int d\mathbf{r} \, \phi^*(\mathbf{r}) \, e^{i\frac{\mathbf{k} \cdot \mathbf{r}}{2}} \, \phi(\mathbf{r}) \quad (9)$$

where $\phi(\mathbf{r})$ is the deuteron S-wave function, and the momentum $\mathbf{k} = \mathbf{p}_d - \mathbf{p}_{d'}$ is taken in the initial deuteron rest frame. q_{cm} is the momentum transfer between the nucleons in the center of mass system and $f' \equiv f_{\pi NN^*}$. G^* is the $N^*(1440)$ propagator as given in [16].

We evaluate the cross section in the center of mass system and then relate the result to the one which is shown by the experimentalists making use of:

$$\frac{d^2\sigma}{dE_{d'} d\Omega_{d'}^L} = \frac{d^2\sigma}{dE_{d'} d\Omega_{d'}^{cm}} \frac{d\Omega_{d'}^{cm}}{d\Omega_{d'}^L} . \quad (10)$$

For the kinematics considered it can be shown that

$$\frac{d\Omega_{d'}^{cm}}{d\Omega_{d'}^L} = \frac{p_d^L p_{d'}^L}{p_d^{cm} p_{d'}^{cm}} \left(1 - \frac{E_d^{cm}}{\sqrt{s}}\right) + \frac{\cos(\theta^{cm})}{p_{d'}^{cm\,2}} \frac{E_{d'}^{cm}}{\sqrt{s}} p_d^L p_{d'}^L .$$
$$(11)$$

In order to perform the calculation using our quark model, we need to extract the genuine scalar potential at all distances from our $NN \to NN^*(1440)$ transition potential. At short distances, $R < 2$ fm, the quark model

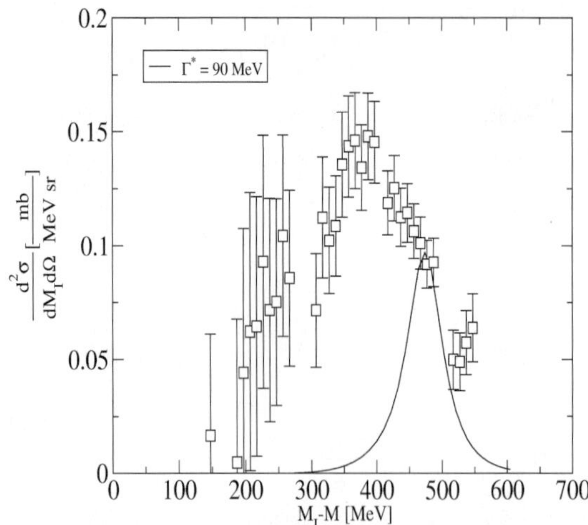

Fig. 4. Quark model result for the RET process contributing to the $p(d, d')$ reaction. M_I is the invariant mass of the target system. Experimental data correspond to $T_d = 2.3$ GeV and $\theta^L = 1.1$ deg. They were obtained in [16] by means of a theoretical subtraction of the Δ contribution

based potential has a non-trivial structure. Due to the presence of the antisymmetrizer we have that, for instance, a scalar coupling at quark level gives rise to a scalar, spin-spin, pseudoscalar and pseudovector couplings [17]. The extraction of the scalar part can be done once the unprojected potential for different (ST) channels has been evaluated in the form:

$$V_{NN \to NN^*}^{(S,T)}(\mathbf{q}) = \frac{4\pi}{\mathcal{N}^{ST}} \sum_{JM_J}^{J_{MAX}} \sum_{LM_L} \sum_{M_S} C_{JM_J}^{LSM_L M_S} Y_{LM_L}(\hat{q})$$

$$\int_0^\infty r^2 \, dr \, j_L(qr) \, \hat{V}_{NN \to NN^*}^{L,S;JT}(r), \quad (12)$$

where we are adding up the partial waves up to a certain J_{MAX}. $\mathcal{N}^{ST}$ is the unprojected norm of the $NN^*(1440)$ system and $\hat{V}^{L,S;JT}$ is the projected potential multiplied by the norm of the corresponding partial wave.

Then we write down the most general form for the interaction:

$$V_{NN \to NN^*}^{(S,T)} = V_0 + V_1 \, \boldsymbol{\sigma_1} \cdot \boldsymbol{\sigma_2} + V_2 \, \boldsymbol{\tau_1} \cdot \boldsymbol{\tau_2} +$$
$$V_3 \, \boldsymbol{\sigma_1} \cdot \boldsymbol{\sigma_2} \, \boldsymbol{\tau_1} \cdot \boldsymbol{\tau_2} . \quad (13)$$

where V_i are functions of the interbaryon momentum, σ_i and τ_i are spin and isospin matrices of the baryons. V_0 is the scalar part of the total potential which is the only part that can be included in our process of $N^*(1440)$ excitation in $p(d, d')$ reactions.

Finally, if we consider different (ST) channels, we obtain the following system of equations,

$$V_{NN \to NN^*}^{(0,0)} = V_0 - 3V_1 - 3V_2 + 9V_3$$
$$V_{NN \to NN^*}^{(1,0)} = V_0 + V_1 - 3V_2 - 3V_3$$
$$V_{NN \to NN^*}^{(0,1)} = V_0 - 3V_1 + V_2 - 3V_3$$
$$V_{NN \to NN^*}^{(1,1)} = V_0 + V_1 + V_2 + V_3 , \quad (14)$$

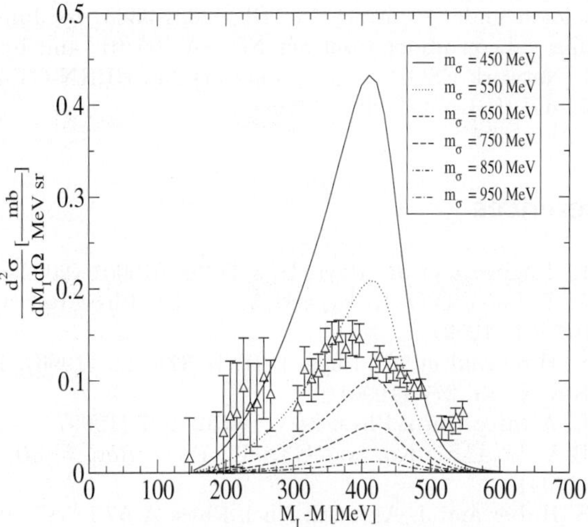

Fig. 5. Dependence of the $p(d, d')$ cross section calculated in [16] on the mass of the σ meson. M_I is the invariant mass of the target system. Experimental data correspond to $T_d = 2.3$ GeV and $\theta^L = 1.1$ deg. They were obtained in [16] by means of a theoretical subtraction of the Δ contribution

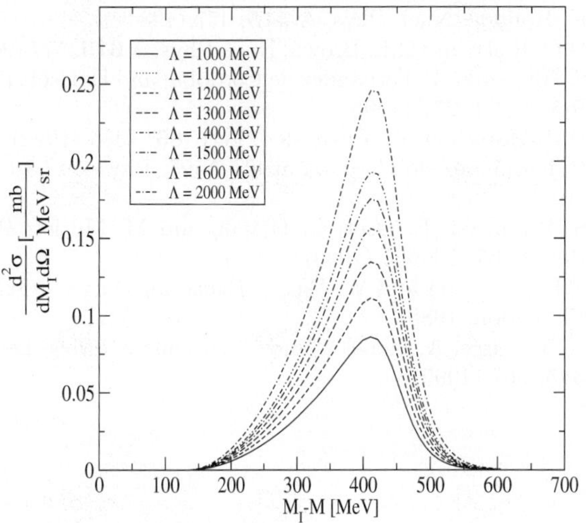

Fig. 6. Dependence of the $p(d, d')$ cross section calculated in [16] on the baryonic cut-off mass. M_I is the invariant mass of the target system

and solving for the scalar part,

$$V_0 = \frac{1}{16} \left[V_{NN \to NN^*}^{(0,0)} + 3 V_{NN \to NN^*}^{(0,1)} + 3 V_{NN \to NN^*}^{(1,0)} + 9 V_{NN \to NN^*}^{(1,1)} \right] . \tag{15}$$

We focus our attention on the target Roper excitation process. To compare to data it is necessary to subtract the Δ contribution and the interference term from the experimental points. The parameters of the Δ excitation on the projectile used in the phenomenological model were settled in the (α, α') reaction. We assume this process to be

correctly described. Therefore we consider the data where the Δ contribution has already been subtracted [16] as our experimental data.

In Fig. 4 we show the result obtained using the quark-model derived $NN \to NN^*(1440)$ potential and its comparison to data. As can be seen, the predicted cross section is smaller that the model-dependent experimental data. If we choose a small value for the width of the $N^*(1440)$, the results come closer to the experimental data. Let us notice that the bigger disagreement with the extracted data corresponds to the region where the error bars are larger, in other words, to the region where the uncertainties related to the theoretical method used to subtract the Δ contribution and interference term are important. For the sake of clarity, let us note that the subtraction of the Δ contribution is proportional to the square of the $\pi N \Delta$ coupling constant. This coupling constant is different as used in baryonic processes, $f_{\pi N \Delta}^2/4\pi = 0.35$, as the one used in our quark model, $f_{\pi N \Delta}^2/4\pi = 0.22$ [18]. This value is crucial when trying to reproduce the $^1S_0$ NN phase shift through the tensor coupling to the $^5D_0$ $N\Delta$. Using the baryonic coupling one would obtain much bigger attraction than observed experimentally. As a consequence, the baryonic calculation of the Δ contribution could be underestimating the region above the peak overestimating in this way the $N^*(1440)$ contribution. The way to wipe out those uncertainties would be to calculate the Δ contribution together with the interference term making use of quark-model baryonic potentials.

It is also worth wile to compare our results to the ones obtained from the baryonic calculation of the RET diagram [16]. To make more clear the comparison we plot the dependence of these results on the value chosen for the σ mass (within the allowed experimental interval), Fig. 5, and on the baryonic cut-off mass needed, Fig. 6. As can be seen the smaller m_σ the bigger the cross section and the smaller the cut-off the smaller the cross section. The significant dependence on the cut-off mass points out the need of having a good description of the scalar short-range part of the interaction. In our quark model framework this scalar piece is not uncertainly dependent on any free parameter but determined by quark antisymmetry plus the dynamics, the OPE and OGE giving most of the cross section, see Fig. 7. It is then clear that the results obtained with the quark-model derived interactions are qualitatively quite different to the ones reported using baryonic degrees of freedom. In fact, the baryonic form-factor could be hiding the effects of the quark substructure that we find in our quark-model treatment through the contributions to the scalar channel from every term in the quark-quark Hamiltonian.

4 Summary

We have carried out a test of a quark-model based $NN \to NN^*(1440)$ potential derived from an universal qq interaction. The consideration of the long-range tail of the potential as compared to the baryonic parametrization allows

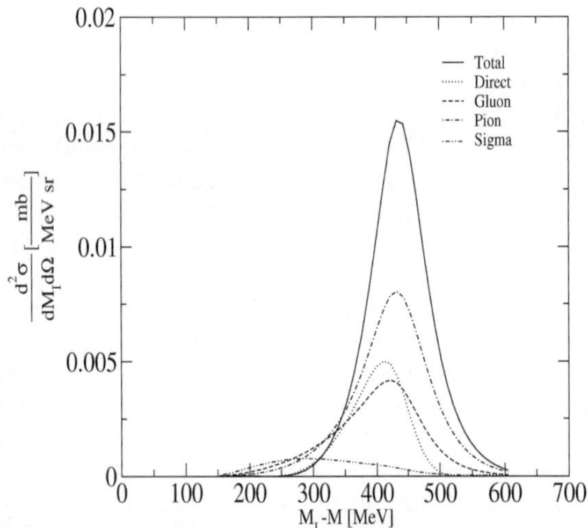

Fig. 7. Detailed contributions to the $p(d, d')$ cross section coming from the different interactions at the quark level, neglecting the interference terms. We denote by direct the result obtained neglecting quark-exchange diagrams. M_I is the invariant mass of the target system

the extraction of the $\pi NN^*(1440)$ and $\sigma NN^*(1440)$ coupling constants. On the other hand the consideration of the physical mechanisms involved in the reactions $p(\alpha, \alpha')$ and $p(d, d')$, in particular the RET, allows to test the scalar short-range part of the interaction. The results we get are quite encouraging in spite of the lack of a full quark model calculation. To pursue this could open a new way to search for effects of the microscopic structure in the mentioned processes.

Acknowledgements. The authors thank Dr. S. Hirenzaki for useful correspondence concerning the calculation of [16]. This work has been partially funded by Ministerio de Ciencia y Tecnología under Contract No. BFM2001-3563, by Junta de Castilla y León under Contract No. SA-109/01, and by EC-RTN (Network ESOP) under Contracts No. HPRN-CT-2000-00130 and HPRN-CT-2002-00311.

References

1. K. Hagiwara et al.: Phys. Rev. D **66**, 010001 (2002)
2. M.T. Peña, D.O. Riska, and A. Stadler: Phys. Rev. C **60**, 045201 (1999)
3. S. Hirenzaki et al.: Phys. Lett. B **378**, 29 (1996); Phys. Rev. C **53**, 277 (1996)
4. L. Alvarez-Ruso: Phys. Lett. B **452**, 207 (1999)
5. B.A. Li, C.M. Ko, and G.Q. Li: Phys. Rev. C **50**, 2675 (1994)
6. S. Huber and J. Aichelin: Nucl. Phys. A **573**, 587 (1994)
7. H. Garcilazo and E. Moya de Guerra: Nucl. Phys. A **562**, 521 (1993)
8. A. Valcarce, F. Fernández, H. Garcilazo, M.T. Peña, and P.U. Sauer: Phys. Rev. C **49**, 1799 (1994)
9. B. Juliá-Díaz, A. Valcarce, P. González, and F. Fernández: Phys. Rev. C **66**, 024005 (2002)
10. D.R. Entem, F. Fernández, and A. Valcarce: Phys. Rev. C **62** 034002 (2000)
11. K. Holinde: Nucl. Phys. A **415**, 477 (1984)
12. D.O. Riska and G.E. Brown: Nucl. Phys. A **679**, 577 (2001)
13. S. Hirenzaki, P. Fernández de Cordoba, and E. Oset: Phys. Rev. C **53**, 277 (1996)
14. H.P. Morsch et al.: Phys. Rev. Lett. **69**, 1336 (1992)
15. P. Fernández de Cordoba et al.: Nucl. Phys. A **586**, 586 (1995)
16. S. Hirenzaki, E. Oset, C. Djalali, and M. Morlet: Phys. Rev. C **61**, 044605 (2000)
17. T.E.O Ericson and W. Weise: *Pions and Nuclei* (Oxford, Clarendon, 1988)
18. A. Valcarce, A. Faessler, and F. Fernández: Phys. Lett. B **345**, 367 (1995)

Hadrons in the nuclear medium

Eur Phys J A (2004) **19**, s01, 105–110
Digital Object Identifier (DOI) 10.1140/epjad/s2004-03-018-8

EPJ A direct
electronic only

Propagation of fast partons in the nuclear medium

Mikkel B. Johnson

Los Alamos National Laboratory, Los Alamos, NM 87544

Received: 30 Aug 2003 / Accepted: 14 Nov 2003 /
Published Online: 6 Feb 2004 – © Società Italiana di Fisica / Springer-Verlag 2004

Abstract. The color dipole approach has been applied in the target rest frame to address the issues of transverse momentum broadening and energy loss of a fast quark propagating in the nuclear medium. A recent application of the theory to the FermiLab E772/E866 experimental data, determining the rate of energy loss of a quark propagating in the medium to be 2 to 3 GeV/fm, will be reviewed. Calculations for the transverse momentum distribution will be presented, and the results will be compared to the E866 data. The theory will be shown to compare favorably to the data, and these results will be shown to suggest that the momentum broadening of a quark is about twice the generally accepted size.

PACS. 13.85.Qk; 24.85+p Drell-Yan processes; medium effects; heavy-ion collisions

1 Introduction

Nucleon-nucleus (or deuteron-nucleus) reactions at high energy are important experimentally because they constitute the conventional background for the less well known dynamics of relativistic nucleus-nucleus collisions, which have been identified as a means for producing the quark-gluon plasma in the labratory at facilities such as the Relativestic Heavy Ion Collider (RHIC) at BNL, and the Large Hadron Collider (LHC) at CERN. To develop sufficient confidence in theory to interpret the more complicated phenomena in nucleus-nucleus collisions, it is vital that the theory be tested against these simpler data first.

Shadowing, quark energy loss, and transverse momentum broadening are three central issues of parton propagation in nucleus-nucleus collisions that may be quantitatively addressed in $p+A$ reactions, specifically in Drell-Yan (DY) reactions. The color dipole approach in the target rest frame provides an attractive means for interpreting these data to learn more about these phenomena. After a brief description of the status of this theory, I will review recent results that are improving our understanding of the physics behind these aspects of parton propagation in the nuclear medium.

Drell-Yan in the target rest frame may be described as a quark of the incident hadron interacting with a target nucleon, radiating a virtual photon $\gamma*$ of mass M that subsequently decays into the observed Drell-Yan dilepton pair $\bar{l}l$. In the target rest frame, the $\gamma*$ is a constituent of projectile fluctions, which are "frozen" by time dilation for a length of time t_c given by the uncertainty relation,

$$t_c = \frac{2E_q}{M_{q\bar{l}l}^2 - m_q^2} \qquad (1)$$

where E_q and m_q refer to the energy and mass of the projectile quark and $M_{q\bar{l}l}^2$ is the square of the effective mass of the fluctuation,

$$M_{q\bar{l}l}^2 = \frac{m_q^2}{\alpha} + \frac{M^2}{1-\alpha} + \frac{k_T^2}{\alpha(1-\alpha)}, \qquad (2)$$

where α is the fraction of the light-cone momentum of the incident quark carried by the lepton pair, and k_T^2 is the square of the transverse momentum of the lepton pair. The fluctuation lifetime t_c is called the coherence time, and the DY reaction occurs when the $\gamma*$ of the fluctuation is released by an interaction between one of the constituents of the fluctuation and a target nucleon. In the color dipole approach, this interaction is mediated by the color dipole cross section, $\sigma_{\bar{q}q}$.

When the scattering takes place on a nucleus, multiple interactions with target nucleons can give rise to various medium effects. Two limiting cases should be distinguished, the short coherence length limit (SCL) reached when the coherence length $\ell_c \equiv c\langle t_c \rangle$ is much smaller than the interparticle spacing d, $\ell_c \ll d$ (in a heavy nucleus, $d \approx 2fm$), and the long coherence length limit (LCL) reached when $\ell_c \gg R_A$, where R_A is the nuclear radius. The coherence length determines the physics and also controls the way that medium effects are taken into account. The theory simplifies these important limits. The intermediate case is generally more difficult to describe; however, if $0 < \ell_c \ll R_A$, the result can be obtained by interpolating between these limits using the square of the longitudinal form factor, $F_A^2(q_c)$, where $q_c = 1/l_c$ is the longitudinal momentum transferred in the reaction. The Green function method [1] was developed to handle the more difficult situations. Our results make use of all these cases.

One important medium effect is shadowing, and a basic consideration is the connection between shadowing and the coherence length. In the SCL there is no shadowing because the duration of the fluctuation is so short that it has no time to interact with the medium. In the LCL, which is applicable to reactions at the LHC and at RHIC under certain kinematic conditions, there is maximal shadowing.

Medium effects also give rise to quark energy loss and to transverse momentum broadening. These both arise in the SCL as the quark from the incident hadron undergoes various additional interactions with nucleons of the nucleus before it radiates the $\gamma*$. For the LCL, the quark may undergo interactions that lead not only to shadowing but also to additional momentum broadening. Nuclear effects giving rise to momentum broadening and shadowing are believed to arise predominantly from the same color dipole cross section $\sigma_{\bar{q}q}(r_T)$ that mediates the DY reaction on a nucleon.

The DY data relevant to shadowing, energy loss, and momentum broadening in which we are interested here is the $p + A$ FermiLab data from the E772/E866 collaboration. This data corresponds to the range of coherent length where the shadowing is weak and the interpolation formula described above applies.

2 Drell-Yan reaction on a nucleon in the target rest frame

In the color dipole approach, the Drell-Yan reaction occurs when a fast projectile quark scatters off the gluonic field of the target, as shown in Fig. 1. The color dipole approach was originally proposed for deep-inelastic scattering (DIS). Extension to Drell-Yan was developed by Kopeliovich [2] and subsequently by Brodsky, et al. [3]. The DY reaction on a nucleon is described by

$$M^2 \frac{d^2\sigma_{DY}}{dM^2dx_1} = \int_{x_1}^{1} dx_q F_q^h(x_q) \int d^2\rho |\Psi(\alpha,\rho)|^2 \sigma_{\bar{q}q}(\alpha\rho),$$

(3)

where $\sigma_{\bar{q}q}$ is the color-dipole cross section, ρ is the transverse distance between the $\gamma*$ and quark in the fluctuation, and Ψ represents the distribution for the incident quark to fluctuate into a quark and the $\gamma*$. In DY, the color dipole consists of the quark before and after the release of the $\gamma*$, whose impact parameters differ by $\alpha\rho$. Here $x_q(x_1)$ is the fraction of light-cone momentum of the incoming hadron h (for us a proton) carried by the quark (lepton pair); the light-cone momentum fraction α is given in terms of these variables, $\alpha = x_1/x_q$; and, F_q^h is the quark distribution function of the incident hadron.

Deep-inelastic scattering from nucleons at HERA at DESY has been used to fix models of the color-dipole cross section. The HERA data suggests saturation, the condition that the color dipole cross section approaches a constant value at large r_T,

$$\sigma_{\bar{q}q}(r_T) = \sigma_0(1 - e^{-r_T^2/R_0^2}).$$

(4)

Models that incorporate this saturation property include the GW [4] and KST [5] models. We see that in both the

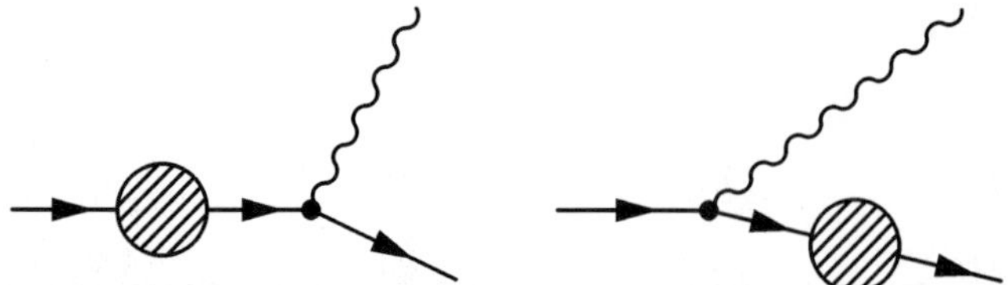

Fig. 1. In the target rest frame, DY dilepton production looks like bremsstrahlung. A quark or an anti-quark from the projectile hadron scatters off the target color field (denoted by the *shaded circles*) and radiates a massive photon, which subsequently decays into the lepton pair. The photon decay is not shown. The photon can be radiated before or after the quark scatters

GW and KST models, for small r_T,

$$\sigma_{\bar{q}q}(r_T) \approx C r_T^2,$$

(5)

where $C = \sigma_0/R_0^2$ is a constant that depends on the quark energy, in the case of the KST model, and on Bjorken x in the case of the GW model. For us, the most important point is that once the color dipole cross section has been fit to experimental DIS data on the nucleon, the color-dipole formalism makes predictions for Drell-Yan both in nucleon-nucleon [6] and nucleon-nucleus scattering with no further adjustment of parameters.

The theory is expected to be valid for small Bjorken x only, and specific calculations [6] have shown it to agree well for $x_2 < 0.1$ in both the magnitude and shape with the next-to-leading order parton model. Although the calculations disagreed with the E772 FermiLab data, they are in excellent agreement with the more recent E866 analysis in [7].

3 Medium effects in DY reactions for p+A collisions in the target rest frame

Medium effects are evident in nuclear ratios of cross sections $R^{A/A'}$. These include analyses of both cross sections themselves, and the analysis of momentum distributions. For each, there are two cases of interest: weak shadowing and maximal shadowing.

As we have stated, shadowing is controled by the coherence length, and explicit calculations in [8] show that for incident quarks of the 800 GeV projectile protons the coherence length is quite a bit less than the nuclear radius, implying weak shadowing. Additionally, these calculations show that the coherence length of the dominant $q-\gamma*$ fluctuation is nearly constant as a function of x_1 and is therefore essentially a function of x_2 alone. For $x_2 > .03$, $\ell_c < 2$ fm, and shadowing can be safely ignored. For smaller x_2 the coherence length exceeds the interparticle spacing at the center of the nucleus, and shadowing therefore begins to become important. The particular admixture of long- and short-coherence length contributions is determined by the longitudinal form factor, as stated in the introduction, and this depends in turn explicitly on the variation of ℓ_c with x_1 and x_2.

For cross section ratios,

$$R^{A/A'} = \frac{d^2\sigma_{DY}^A}{dM^2 dx_1} \Big/ \frac{d^2\sigma_{DY}^{A'}}{dM^2 dx_1}, \tag{6}$$

the E772/E866 FermiLab data at $E_p = 800\ GeV$ has been used for a recent determination of quark energy loss [8]. The DY reaction at RHIC for $x_F = x_1 - x_2 > 0.5$ and at the LHC lies in the regime of long-coherence length physics, and such experiments at these facilities will be exploring a limit that has so far not been accessible experimentally. Predictions of cross sections in $p+A$, $d+A$, and $A+A$ collisions [9] have been made recently in the latter case.

For momentum distribution ratios,

$$R^{A/A'}(p_T) = \frac{d^2\sigma_{DY}^A}{dp_T^2} \Big/ \frac{d^2\sigma_{DY}^{A'}}{dp_T^2}, \tag{7}$$

data relevant to the the case of weak shadowing has been taken at $E_p = 800\ GeV$ by the FermiLab E772/E866 collaboration and was used by them to study quark momentum broadening in nuclei,

$$\delta\langle p_T^2 \rangle = \langle p_T^2 \rangle^A - \langle p_T^2 \rangle^p. \tag{8}$$

This same regime is relevant to $p+A$ collisions at RHIC. The case of maximal shadowing is relevant to the LHC and RHIC, and predictions are made in [9].

3.1 Cross section ratios

Let us consider first the case of weak shadowing for the cross section ratios, as analyzed in [8]. To calculate the cross section, we must calculate the cross section separately in the SCL and LCL and then interpolate between them using using the longitudinal form factor, as explained in the introduction.

The main medium effect in the short coherence length limit $x_2 \gg 0.03$ the $\gamma*$ is the quark energy loss, which occurs before the $\gamma*$ is radiated. The interactions that cause the incident quark to lose energy reduce the quark momentum fraction x_q and hence shift the value of $x_1 = \alpha x_q$. The amount of reduction due to quark energy loss may be calculated as $x_q \approx E_q/E_h \to x_q - \kappa L/E_h$, where L is the path length of the incident quark in the nucleus and κ is its rate of energy loss. Using these kinematic considerations, we find an expression in [8] for the DY cross section $M^2 \frac{d^2\sigma_{DY}^{(SLC)}}{dM^2 dx_1}$ including its dependance on κ. It becomes sensitive to κ for $x_1 \approx 1$.

Mechanisms of energy loss [8] consist of the primary ones of string breaking (SB), for which $\kappa_{SB} \approx 1\ GeV/fm$ (numerically, the string tension), and the independent process of gluon radiation (GR), for which $\kappa_{GR} \approx 3\alpha_s \langle k_T^2 \rangle \approx 0.8\ GeV/fm$, where $\langle k_T^2 \rangle \approx 0.65\ GeV$ is the mean-square momentum of the radiated gluon. The induced energy-loss mechanisms arising from multiple interactions with nucleons are relatively minor for nuclei of ordinary density and can be neglected. We thus expect that $\kappa \approx 1.8\ GeV/fm$.

As we have said, shadowing arises in the LCL through multiple interactions between the quark and the nucleus following the emission of the $\gamma*$. For weak shadowing, Glauber theory to second order in the number of interactions may be used, giving

$$M^2 \frac{d^2\sigma_{DY}^{(LCL)}}{dM^2 dx_1} = \langle \sigma_{\bar{q}q}(\alpha\rho) \rangle (1 - \frac{\sigma_{eff}\langle T_A \rangle}{4}) \tag{9}$$

where the brackets around the color dipole cross section refer to the averages over (ρ, α) in (3), and where $\langle T_A \rangle$ is the average of the thickness function $T_A(b)$ over impact parameter b,

$$< T_A > = \frac{1}{A} \int d^2 b\, T_A^2(b), \tag{10}$$

with $T_A(b) = T_A(b, \infty)$, where

$$T_A(b, z) = \int_{-\infty}^{z} d^2 b\, \rho_A(b, z), \tag{11}$$

and $\rho_A(b, z)$ is the nuclear density. The quantity $\sigma_{eff} = \langle \sigma_{\bar{q}q}^2 \rangle / \langle \sigma_{\bar{q}q} \rangle$ in (9) is obtained by expanding the multiple scattering through second order in $\sigma_{\bar{q}q}$. In the color-dipole approach shadowing is completely determined by the color dipole cross section when shadowing is weak.

Interpolating between the LCL where $x_2 \ll 0.03$ and the SCL where $x_2 \gg 0.03$, we obtain

$$\frac{d^2\sigma_{DY}}{dM^2 dx_1} = \frac{d^2\sigma_{DY}^{(SCL)}}{dM^2 dx_1}(1 - F_A^2(q_c)) + \frac{d^2\sigma_{DY}^{(LCL)}}{dM^2 dx_1} F_A^2(q_c), \tag{12}$$

where $q_c = 1/\ell_c$ and the longitudinal form factor is defined as

$$F_A^2(q) = \frac{1}{< T_A >} \int d^2 b |\int_{-\infty}^{\infty} dz e^{iqz} \rho(b, z)|^2. \tag{13}$$

Note that this interpolation formula has the correct limits for $\ell_c \to 0$ and $\ell_c \to \infty$. It also correctly reproduces the Gribov shadowing formula [10,11] and correctly describes shadowing in DIS [8]. Because the cross section becomes sensitive to both shadowing and κ for $x_1 \approx 1$, it is important that both effects are included in the theory (note that the depencence of shadowing on ℓ_c becomes quite intricate in this limit [12,8]). This theory then provides a quantitative test of the value of the relatively poorly known quark energy loss in the nuclear medium, since shadowing is completely determined by theory.

For comparing the theory to experiment, we have evaluated (12) and compared to the E866/E772 FermiLab data. This data provides the dependence of $R^{A/A'}$ over a range of (A, x_1, M^2); our ability to reproduce these data gives us confidence that we are able to separate the effects of shadowing and energy loss. Typical results are shown in Fig. 2.

The rate of energy loss corresponding to the fit to the the data, as shown in Fig. 2, is

$$\kappa = 2.73 \pm 0.37 GeV/fm \tag{14}$$

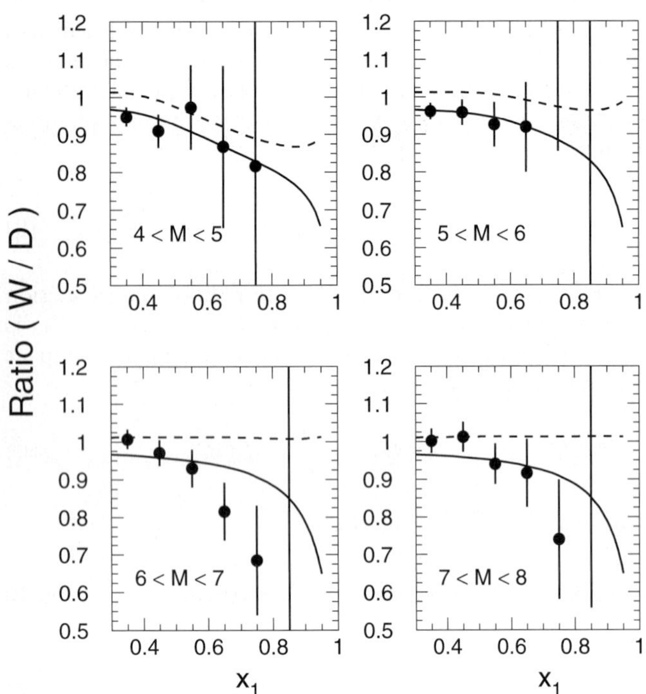

Fig. 2. Examples of ratios of the DY cross sections of tungsten to deuterium as functions of x_1 for various intervals of M. *Dashed curves* correspond to net shadowing contribution, *solid curves* show the full effect including shadowing and energy loss. Data are from [13,14]

Note that the value in (14) is somewhat larger than the theoretical estimate of $1.8\ GeV/fm$ given above; it is also considerably larger than values found in previous analyses of the DY data [15]. We regard our value as more reliable, since the color dipole approach is able to disentangle shadowing from energy loss using a reliable theoretical calculation of the shadowing contribution.

3.2 Momentum distribution ratios [16]

Calculations show that the momentum distributions from E772/E866 experiments with $800 GeV$ protons or measurements at RHIC with $s^{1/2} = 200\ GeV$ protons at $x_F = 0$ correspond to the short coherence length limit, as long as $x_2 > 0.05$. Working in the short coherence length limit simplifies the theory by eliminating the shadowing term in (12). We will restrict our attention to this kinematic regime, which is possible because the data is available binned in x_2 intervals, and thus concentrate on the first (LCL) term in this expression.

It was recognized in [17] that at high energies, the multiple interaction of a quark propagating in the nuclear medium can be eikonalized and exponentiates into a factor containing the color dipole cross section and nuclear thickness function. The physics of momentum broadening of the quark is therefore that of color filtering, or absorption of large dipoles leading to diminished transverse separation with distance. Then, the probability distribution $W^A(k_T)$ that a quark will have acquired transverse momentum k_T

at a position (b,z) in the nucleus A becomes [17]

$$W^A(k_T) = \frac{1}{(2\pi)^2} \int d^2 r_T e^{ik_T \cdot r_T} e^{-\frac{1}{2}\sigma_{\bar{q}q}(r_T)T_A(b,z)}, \quad (15)$$

where $T_A(b,z)$ is defined in (11). We express the cross section, $\sigma_{DY}^A(p_T)$, for a proton to produce a DY pair on a nucleus A with transverse momentum p_T as convolution of the probability $W^A(k_T)$ and the DY momentum distribution on a proton. The resulting expression entails an integral over $\Psi(\alpha, \rho)$ as in (3), and for the results we show below, this integral has been performed numerically. However, for the purpose of discussing the result here, we replace the quantities that depend on α by averages. Then, the ratio $R^{A/p}(p_T)$ may be expressed as in terms of the DY cross section $\sigma_{DY}(p_T)$,

$$R^{A/p}(p_T) = \frac{1}{\sigma_{DY}^p(p_T)} \int d^2 k_T W^A(k_T)\sigma_{DY}^p(p_T - \bar{\alpha}k_T) \tag{16}$$

where $\bar{\alpha}$ arises from the fact that the DY pair carries away a fraction α of the transverse momentum of a quark in the incident proton. We find that $\bar{\alpha} = \langle \alpha^2 \rangle^{1/2} \approx 0.97$. The average over α implicit in (16) also entails an evaluation of the quark energy, $\langle E_q \rangle \approx E_q/3$, since we use the GW model of $\sigma_{\bar{q}q}$, which depends explicitly on this energy.

For our calculations below we have chosen

$$\sigma_{DY}^p(p_T) \propto (1 + p_T^2/\Lambda^2)^{-6} \tag{17}$$

where the value of $\Lambda^2 = 7 GeV^2$ is taken from [18].

Some of the integrals in (15, 16) may be done analytically by expanding out the exponential in (4). We find that to an excellent approximation, the effect of averaging (15) over (b,z) is to replace

$$T_A(b,z) \to \langle T_A \rangle / 2. \tag{18}$$

When evaluating (15), we also take into account gluon shadowing by replacing $\sigma_{\bar{q}q} \to R_G \sigma_{\bar{q}q}$, where R_G is the gluon shadowing function. See [17] for more discussion. Our calculations are however insensitive to R_G.

Our calculation for $R^{W/Be}(p_T)$ is shown in Fig. 3. The data relevant to p_T distributions on nuclei are FermiLab E772/E866 data [15,14,19]. One sees that the calculation is in quite good agreement with experiment. Note that the details of the Cronin effect, the rise above 1 for $p_T \approx 3$ in Fig. 3, are completely explained with no adjustable parameters. We find comparable agreement with $R^{Fe/Be}(p_T)$.

3.3 Transverse momentum broadening

The transverse momentum broadening of a quark propagating in a nucleus is defined in (8) where the mean momentum is defined as

$$\langle p_T^2 \rangle = \frac{\int d_{pT}^2 p_T^2 \sigma_{DY}^A(p_T)}{\int d_{pT}^2 \sigma_{DY}^A(p_T)}. \tag{19}$$

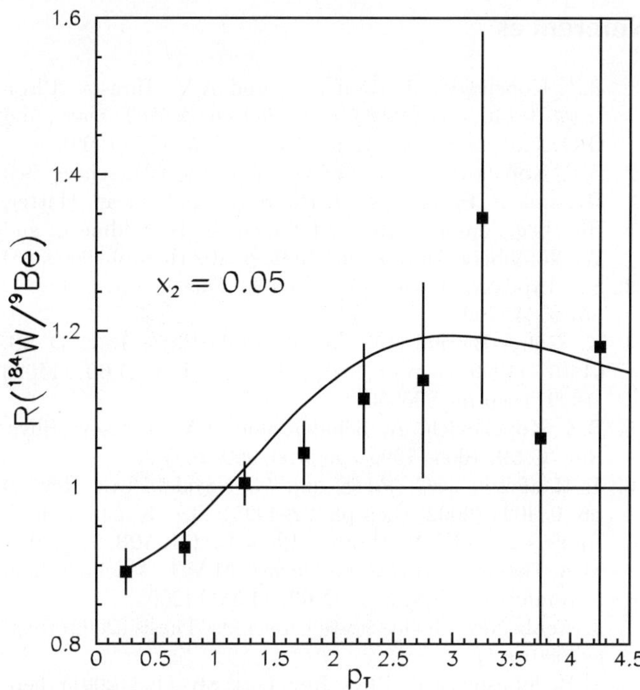

Fig. 3. Comparison of theoretical prediction of $R^{W/Be}(p_T)$ vs. p_T (in GeV/c) to experiment for $x_2 = 0.05$. Data are from the FermiLab E772/E866 collaboration [15,14,19]

The dependence of the quantity $\delta\langle p_T^2 \rangle$ on A has been studied experimentally using the E772 data in [20]. They found that for large A,

$$\delta\langle p_T^2 \rangle^A_{exp} \approx 0.021 A^{1/3} GeV^2. \tag{20}$$

We can make a calculation of the same quantity in the theory just described. From (16) we easily find that

$$\sigma^A_{DY}(p_T) = \frac{1}{\bar{\alpha}^2} \int d^2 r_T e^{i p_T \cdot r} W^A(r_T) \sigma^p_{DY}(r_T) \tag{21}$$

where

$$W^A(r_T) = e^{-\frac{1}{2}\sigma_{\bar{q}q}(r_T)T_A(b,z)} \tag{22}$$

and

$$\sigma^A_{DY}(r_T) = \int d^2 p_T e^{-p_T \cdot r_T} \sigma^A_{DY}(p_T) \tag{23}$$

characterizes how a quark acquires transverse momentum as it propagates through the nucleus. From (19, 21) we find

$$\langle p_T^2 \rangle^A = -\frac{\nabla^2 \sigma^p_{DY}(r_T)}{\sigma^p_{DY}(r_T)}\Big|_{r_T=0} - \frac{\nabla^2 W^A(\bar{\alpha}r_T)}{W^A(\bar{\alpha}r_T)}\Big|_{r_T=0} \tag{24}$$

The first term in (24), the mean-square momentum on a nucleon, is formally infinite. Fortunately, the nucleon contribution gets subtracted to obtain $\delta\langle p_T^2 \rangle$,

$$\delta\langle p_T^2 \rangle = -\bar{\alpha}^2 \frac{\nabla^2 W^A(r_T)}{W^A(r_T)}\Big|_{r_T=0}. \tag{25}$$

Note that the shadowing contribution in (12), which we justified in dropping by going to $x_2 > 0.05$, leads to a

divergent contribution that no longer cancels exactly; in cases where FSI are important other means are needed to define a meaningful quantity [9].

Using (22), along with the observation that $\sigma_{\bar{q}q}(r_T)$ is proportional to r_T^2 at small r_T (see (5)),

$$\delta\langle p_T^2 \rangle = \bar{\alpha}^2 C \langle T_A \rangle \tag{26}$$

where we have used the average of T_A as given in (18).

We may now make estimates to compare with the experimental result in (20). Using values of $C \approx 4-5$ as obtained from the GW or KST models, and taking the sharp-surface model for the nuclear density,

$$\langle T_A \rangle = \frac{3}{2}\rho_0 R_A \tag{27}$$

where $\rho_0 \approx 0.16 fm^{-3}$ is the central density of heavy nuclei and $R_A \approx 1.1 A^{\frac{1}{3}}$, we find from theory that

$$\delta\langle p_T^2 \rangle^A_{Th} \approx 0.047 A^{\frac{1}{3}} \ GeV^2 \tag{28}$$

The theoretical result in (28) is clearly larger than and in apparent disagreement with the experimental result in (20) [17,21]. However, it has been pointed out [22] that large systematic errors must be included in (20) and that when these are taken into account, the result in (28) could be accommodated.

3.4 Predictions for maximal shadowing

As we have stated, the case of maximal shadowing corresponds to $\ell_c \gg R_A$, and experiments in this regime have not yet been possible. The color dipole approach in the target rest frame applies in this limit, and numerous predictions have been presented in [9]. We give one example in Fig. 4, namely the DY ratios as a function of transverse momentum presented separately for longitudinal (L) and transverse + longitudinal (T + L) photon polarization.

In contrast to the weak shadowing limit discussed earlier, transverse momentum broadening here is associated with interactions with the quark following the emission of $\gamma*$. Note that the Cronin peak is much less visible in the LCL, which is a consequence of the gluon shadowing. For evaluating the gluon shadowing here, the Green's function method is required, and we take R_G from the calculations of [23].

4 Summary and conclusions

We have shown that the color dipole approach formulated in the target rest frame has advantages for examining nuclear modifications in $p+A$ collisions because the effects of shadowing may be determined theoretically. In the SCL, $\ell_c \ll R_A$, energy loss and momentum broadening are mediated by multiple interactions of the incident quark with target nucleons before the emission of the $\gamma*$. Analysis of the E772/E866 FermiLab experimental DY data at

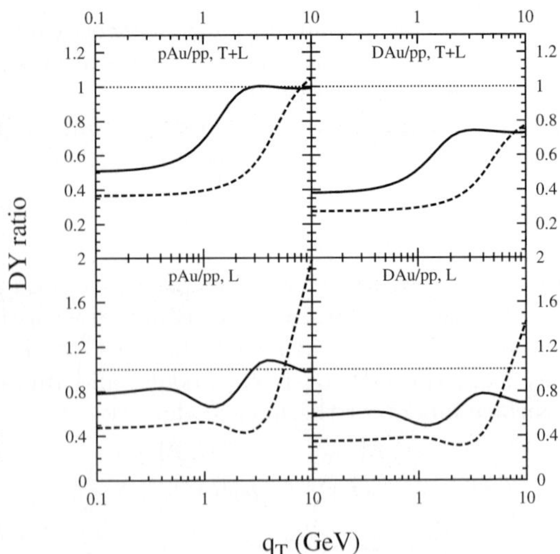

Fig. 4. Nuclear effects on the DY transverse momentum distribution. *Curves* show the DY cross sections for *pAu* (*left*) and deuterium – gold (*right*) collisions per nucleon divided by the DY cross section from *pp* scattering. *Solid curves* are predictions for RHIC ($\sqrt{s} = 200$ GeV) and *dashed* for LHC ($\sqrt{s} = 5.5$ TeV). Calculations are for M=4.5 GeV and $x_F = 0.5$

$E_p = 800\ GeV$ in the color dipole approach has provided improved determinations of these quantities, giving

$$\frac{dE_q}{dz} = -2.73 \pm 0.35\ GeV/fm \qquad (29)$$

and

$$\delta\langle p_T^2\rangle \approx 0.047 A^{\frac{1}{3}}\ GeV^2 \qquad (30)$$

The value of the rate of energy loss is larger than previous determinations, and the value of $\delta\langle p_T^2\rangle$ is about twice as large as the conventional value.

Predictions of the transverse momentum dependence of DY cross sections, for $p + A$ and $A + A$ collisions, have been made in the color dipole approach. These provide tests of the theory in a new physical regime of strong shadowing, where $\ell_c \gg R_A$, that may be explored at the LHC in the future.

Acknowledgements. I would like to acknowledge the essential contributions of my theoretical and experimental collaborators B. Kopeliovich, A. Tarasov, J. Raufeisen J. Moss, M. Leitch, and P. McGaughey. This work was supported in part by U.S. Department of Energy.

References

1. B.Z. Kopeliovich J. Raufeisen, and A.V. Tarasov: Phys. Lett. B **440**, 151 (1998); J. Raufeisen, A.V. Tarasov, and O.O. Voskresenskaya: Eur. Phys. J. A **5**, 173 (1999)
2. B.Z. Kopeliovich: Proc. of the Workshop Hirschegg '95: Dynamical Properties of Hadrons in Nuclear Matter, Hirschegg, January 16–21, 1995, ed. by H. Feldmeyer and W. Nörenberg, Darmstadt, 1995, p. 102 (hep-ph/9609385)
3. S.J. Brodsky, A. Hebecker, and E. Quack: Phys. Rev. D **55**, 2584 (1997)
4. K. Goltec-Biernat and M. Wüstoff: Phys. Rev. D **59**, 014017 (1999) (hep-ph/9807513); Phys. Rev. D **60**, 114023 (1999) (hep-ph/9903358)
5. B.Z. Kopeliovich, A. Schäfer, and A.V. Tarasov: Phys. Rev. C **59**, 1609 (1999) (hep-ph/9908245)
6. J. Raufeisen, J-C. Peng, and G. Nayak: Phys. Rev. D **66**, 034024 (2002) (hep-ph/0204095); B.Z. Kopeliovich, J. Raufeisen, and A.V. Tarasov: Phys. Lett. B **503**, 91 (2001); M.A. Betemps, M.B. Gay Ducati, M.V.T. Machado, and J. Raufeisen: Phys. Rev. D **67**, 114008 (2003)
7. J. Webb: New Mexico State University Thesis (2002) (hep-ex/0301031)
8. M.B. Johnson et al.: Phys. Rev. Lett. **86**, 4483 (2001) (hep-ex/0010051) and M.B. Johnson et al.: Phys. Rev. C **65**, 025203 (2002) (hep-ph/0105195)
9. B.Z. Kopeliovich, J. Raufeisen, A.V. Tarasov, and M.B. Johnson: Phys. Rev. C **67**, 025203 (2003) (hep-ph/0105195)
10. V.N. Gribov: Sov. Phys. JETP **29**, 483 (1969); **30**, 709 (1970)
11. V. Karmanov and L.A. Kondratyuk: Sov. Phys. JETP Lett. **18**, 266 (1973)
12. B.Z. Kopeliovich, J. Raufeisen, and A.V. Tarasov: Phys. Rev. C **62**, 035204 (2000)
13. D.M. Alde et al.: Phys. Rev. Lett. **66**, 133 (1991)
14. Unpublished E772 data
15. M.A. Vasiliev et al.: Phys. Rev. Lett. **83**, 2304 (1999)
16. Results shown in this subsection are preliminary. M.B. Johnson and B.Z. Kopeliovich: unpublished
17. M.B. Johnson, B.Z. Kopeliovich, and A.V. Tarasov: Phys. Rev. C **63**, 035203 (2001) (hep-ph/0006326)
18. Erratum to P. McGaughey et al.: Phys. Rev. D **50**, 3038 (1994)
19. M. Leitch: private communication
20. P.L. McGaughey, J.M. Moss, and J-C. Peng: Annu. rev. Nucl. Part. Sci. **49**, 217 (1999)
21. J. Raufeisen: Phys. Lett. B **557**, 184 (2003)
22. Joel Moss: private communication
23. B.Z. Kopeliovich, A. Schäfer, and A.V. Tarasov: Phys. Rev. D **62**, 054022 (2000) (hep-ph/9908245)

Eur Phys J A (2004) **19**, s01, 111–127
Digital Object Identifier (DOI) 10.1140/epjad/s2004-03-019-7

EPJ A direct

electronic only

Inside-outside hadronization in a medium

B.Z. Kopeliovich[1,2], J. Nemchik[3], E. Predazzi[4], and A. Hayashigaki[2]

[1] Max-Planck Institut für Kernphysik, Postfach 103980, 69029 Heidelberg, Germany
[2] Institut für Theoretische Physik der Universität, 93040 Regensburg, Germany
[3] Institute of Experimental Physics SAV, Watsonova 47, CS-04353 Kosice, Slovakia
[4] Dipartimento di Fisica Teorica, Università di Torino and INFN, Sezione di Torino,
I-10125, Torino, Italy

Received: 30 Oct 2003 / Accepted: 14 Nov 2003 /
Published Online: 6 Feb 2004 – © Società Italiana di Fisica / Springer-Verlag 2004

Abstract. Nuclei are unique analyzers for the early stage of the space-time development of hadronization. DIS at medium energies is especially suitable for this task being sensitive to hadronization dynamics, since the production length is comparable with the nuclear size. This was the driving motivation to propose an experiment at HERMES [1], and provide predictions based on a pQCD model of hadronization. Now the experiment is done and one can compare the predictions with data [2]. The model successfully describes with no adjustment the nuclear effects for various energies, z_h, p_T, and Q^2, for different flavors and different nuclei. It turns out that the main source of nuclear suppression of the hadron production rate is attenuation of colorless pre-hadrons in the medium. An alternative model [11] is based upon an unjustified assumption that the struck parton keeps radiating gluons far beyond the nuclear size and the pre-hadron is produced outside the nucleus. This model has apparent problems attempting to explain certain features of the results from HERMES. A good understanding of the hadronization dynamics is important for proper interpretation of the strong suppression of high-p_T hadrons observed in heavy ion collisions at RHIC. We demonstrate that the production length is even shorter in this case and keeps contracting with rising p_T.

PACS. 24.85.+p Quarks, gluons, and QCD in nuclei and nuclear processes – 25.30.Rw Electroproduction reactions – 25.75.Dw Particle and resonance production

1 Introduction

Recent measurements of inclusive production of hadrons in deep-inelastic scattering (DIS) off nuclei by the HERMES collaboration [2] have provided precious information about the space-time development of hadronization. These measurements were proposed [1] back in 1995 and predictions were made within a model of hadronization based on perturbative QCD. One of the goals was to reach a better understanding of in-medium hadronization in order to provide a more reliable interpretation of high-p_T hadron production in heavy ion collisions considered as a probe for a created dense matter.

Now, when some of the results of these measurements are released, it is proper to compare the predictions with the data and draw conclusions. In present paper we revisit the problem of in-medium hadronization, re-introduce the model, and perform calculations in accordance with the concrete kinematics of HERMES data. We demonstrate that absorption of the produced colorless pre-hadron is the main source of nuclear suppression. Our predictions need no adjustment and are in a good agreement with data.

Older models [3,4,5,6,7] were based on the string model which were quite pedagogical and helpful for intuitive understanding of the space-time development of hadronization. On the other hand, one could not consider such a nonperturbative phenomenology as a realistic scheme for hard reactions like DIS. These models had a very low predictive power, no access to Q^2 and p_T-dependences, missed the effect of color transparency, etc.

The release of data from the HERMES experiment has stimulated a new wave of theoretical models. All of them have low predictive power and are simply fitted to data which are supposed to be explained. This is why all of them agree with data, although they explore quite different physical ideas.

Some of the models [8,9,10] continue developing already known ideas based on the phenomenology of nonperturbative hadronization. Others [11,12], employ the idea of perturbative induced energy loss, but pushing it to the extreme. Namely, they make two strong ad hoc assumptions which have no justification. First, it is assumed that pre-hadrons are produced always outside the nucleus. Second, the effect of induced energy loss is accounted for via a simple shift of the variable in the fragmentation function, what is equivalent to the assumption that hadronization starts only after the leading parton leaves the nucleus.

Since the induced energy loss scenario is a popular explanation for the nuclear suppression of high-p_T events observed at RHIC, we confront some of the predictions of this model with data from HERMES. We sort out those observables which are sensitive to the model assumptions.

1.1 Why nuclear target?

Particles produced in hadronic collisions and detected at macroscopic distances carry limited information about the hadronization dynamics. The most important details are hidden at the early stage. Nuclear targets provide a unique opportunity to look at the early stage of hadronization at distances of few Fermi from the origin. A quark-gluon system originated from DIS propagates through the nuclear medium and interact with other nucleons. Modification of the differential cross section of particle production can bring forth precious information about the structure of the excited system and its space-time development.

We assume in what follows that Bjorken x is sufficiently large, $x \sim 0.1$, therefore the lepton interacts incoherently with only one bound nucleon, no shadowing is possible. Besides, this region is dominated by valence quarks, i.e. one can treat the DIS as electron-quark scattering with energy transfer, ν, to the knocked out quark. At small x dominated by the sea the process of DIS looks differently (in the nuclear rest frame), the virtual photon produces two jets, q and $\bar{q}$, which share the full energy ν. Therefore, the variable z_h, which is the fraction of ν transferred to the detected hadron, cannot reach the kinematic limit $z_h = 1$ for neither of the two jets, unless one of them takes the whole energy of the photon. This fact makes the nuclear ratio to fall at $z_h \to 1$ steeper than at large x. Besides, negative kaons which at large x are produced mainly from gluons, at small x are generated by the same mechanism as positive kaons. This, one should be cautious about the range of x involved into the analysis, and make a proper x-binning of DIS data.

One can also use the process of hadronization as a tool for probing the medium properties. This is the driving idea of so called "jet quenching" probe for creation of dense matter in relativistic heavy ion collisions [13]. This tool, however, can work properly only if the hadronization dynamics is reliably understood, which seems to be still a challenge.

1.2 Vacuum and induced energy losses

Energy loss is a hot topic nowadays, however, related discussions are confusing sometimes, since vacuum and induced energy losses are mixed up. Due to confinement an energetic parton cannot propagate in vacuum as a free particle, but hadronizes and produces a jet of hadrons. The parton shares its energy with the produced hadrons, therefore it is gradually losing energy. We call this *vacuum energy loss*. In the string model [14] the rate of energy

loss should be constant, of the order of the string tension $\kappa = 1\,\mathrm{GeV/fm}$. Therefore, the energy loss rises linearly with time or length of the path,

$$- \Delta E_{vac}(L) = \kappa L \ . \tag{1}$$

If the parton originates from a hard reaction (DIS, high-p_T process, etc), gluon bremsstrahlung should be an additional source of energy loss. Indeed, as a result of a strong kick the parton should shake off a part of its color field with transverse frequencies controlled by the strength of the kick. Apparently, in this case energy loss may be very extensive and may substantially exceed the static value given by the string model. Amazingly, the radiative energy loss turns out to rise linearly with the path length [15] like in the string model,

$$- \Delta E_{vac}(L) = \frac{2}{3\pi}\, \alpha_s(Q^2)\, Q^2\, L \ . \tag{2}$$

This happens because the gluons lose coherence with the source, i.e. are radiated, at different time intervals.

In the case of hadronization inside a medium, an additional source of energy loss is interaction of partons with the medium. Due to multiple collisions the parton increases its transverse momentum squared linearly with the path length, since it performs Braunian motion in the transverse momentum plane. Experimentally, the broadening of transverse momentum is a rather small effect. For instance, the broadening measured in Drell-Yan reaction on a nucleus as heavy as tungsten is only $\Delta p_T^2 \approx 0.1\,\mathrm{GeV}^2$. Apparently, the induced energy loss [17] generated by such a small momentum transfer, is a small addition to the vacuum energy loss originated from a hard process (at least in a cold nuclear matter),

$$- \Delta E_{ind}(L) = \frac{3}{8}\alpha_s\, \Delta p_T^2\, L = \frac{3}{4}\,\alpha_s\, C(E)\, \rho_A\, L^2 \ , \tag{3}$$

where $C(E) = d\sigma_{\bar{q}q}(r_T, E)/dr_T^2\big|_{r_T=0}$ [18,19] is the derivative of the universal phenomenological dipole cross section; ρ_A is the nuclear density.

The intuitive interpretation of these results is rather straightforward. It takes time, called coherent time, to radiate a photon or gluon which have to get incoherent with the source [1],

$$t_c = \frac{2E\alpha(1 - \alpha)}{k_T^2} \ , \tag{4}$$

where k_T and α are the transverse momentum and fraction of energy taken away by the radiated quantum. According to Landau-Pomeranchuk principle radiation at long coherence times does not resolve the details of interaction, whether it was a single kick, or a few of them. What matters is the final transverse momentum of the parton. Since the additional transverse momentum gained by the parton due to multiple interactions in the nuclear medium is very small, the corresponding correction (induced) to the energy loss is small as well.

This is why it is not easy to see in data the effects of induced energy loss. In particular, the analysis of nuclear effects in Drell-Yan reaction performed in [20] provided information about vacuum energy loss. Indeed, according to the kinematics of the E772/E866 experiments (and other experiments at lower energies) most of events correspond to a short coherence time for the Drell-Yan process which occurs nearly momentarily deep inside the nucleus. Prior to that the incoming hadron interacts softly on the face surface of the nucleus. This collision breaks down coherence of the projectile partons which start losing energy on the path from the soft collision point till the point where the Drell-Yan reaction takes place. One should expect the vacuum value for the rate of energy loss which is at least $-dE/dz \approx 1\,\mathrm{GeV/fm}$ or higher, much larger than what one could expect for induced gluon bremsstrahlung. Indeed, the analyses [20] led to $-dE/dz = (2.8 \pm 0.4 \pm 0.5)\,\mathrm{GeV/fm}$, which is a summed effect of vacuum and induced energy losses.

Naively, one might expect that the effects of vacuum energy loss cancel out in the nucleus to nucleon ratio, since they are identical in the numerator and denominator. This is not correct, the vacuum energy loss controls the time scale of hadron production which has a great impact on nuclear effects (see next section).

1.3 Production and formation times

One should discriminate between formation of the final hadron and production of a colorless state which is not yet an eigen state of the mass matrix and may be projected to various hadronic wave functions. We call such states pre-hadrons and the related time scale *production time*, t_p. The properties of such pre-hadron states and their attenuation in nuclei have been intensively studied over last two decades, both theoretically and experimentally. It has been proven that such pre-hadrons of a reduced size are indeed produced, and that nuclei are indeed more transparent for such states. A clear signal of color transparency in DIS was observed recently in the HERMES experiment [21,22].

To form the hadronic wave function the constituents of the pre-hadron have to circle at least once along their orbits. Dilated by Lorentz transformation this time scale which we call *formation time*, t_f, is proportional to z_h, which is the fraction of the photon energy carried by the hadron. On the other hand, the production time t_p was proven in [23,4] and confirmed within the LUND model in [24] to vanish proportionally to $1-z_h$ at $z_h \to 1$. The latter property follows from energy conservation: the longer the hadronization process is lasting, the more energy is lost by the leading quark, unless its energy falls below the desired hadron energy. In particular, at $z_h = 1$ no energy loss for hadronization is allowed, so the colorless ejectile has to be created instantaneously.

1.4 What are the observables?

Nuclear medium affects the momentum distribution of the produced particles, which is called sometimes a modified fragmentation function $D^A_{eff}(z_h, p_T, Q^2, \nu)$, where p_T is the transverse momentum of the hadron, and ν and Q^2 are the energy and virtuality of the photon respectively. The energy dependence signals that QCD factorization is broken and D^A_{eff} cannot be treated as a fragmentation function, e.g. the evolution equations cannot be applied. Only at very large ν and Q^2 nuclear effects disappear and factorization is restored. In this trivial limit, however, all physical information we are interested in is missed.

Experimental results are usually presented in a simple form, as a ratio of the nucleus-to-nucleon hadron multiplicities,

$$R_A(z_h, p_T, Q^2, \nu) = \frac{dn(\gamma^* A \to hX)/dz_h d^2 p_T}{A\, dn(\gamma^* N \to hX)/dz_h d^2 p_T} \ . \quad (5)$$

Because of limited statistics this four-dimensional ratio is usually presented as one-dimensional ratios integrated over other variables. Some of correlations are, however, very informative. For instance, p_T dependence of the ratio binned in z_h brings forth precious information about the space-time pattern of hadronization (see Sect. 9.3).

Nuclear modification of the hadronic spectra was considered for high-p_T hadron production in [4,25], for DIS in [3,5,7] and for hadroproduction of leading particles on nuclei in [24,26].

1.5 What can we learn?

The questions, which can be answered in such an analysis are:

– *How long does it take to produce a pre-hadron?*
Nuclear suppression of hadron production, especially its z_h and p_T dependences, are sensitive to the production time t_p.

– *How does the produced pre-hadron evolve and attenuate in nuclear matter?*
Since we are interested in production of leading particles, we should forbid inelastic interaction of the produced colorless wave packet at times $t > t_p$. It would lead to a new hadronization process which ends up with a smaller value of z_h. The corresponding correction is easy to estimate and it turns out to be small. The condition of no interaction results in an attenuation of the produced wave packet in nuclear matter, but the absorption cross section of the pre-hadron may be different from the hadronic one. Particularly, in DIS with high photon virtuality Q^2 the produced wave packet may have a smaller transverse size, similar to what is known for exclusive particle production. Correspondingly, the color transparency effects [27,28] should be important and we take care of that in what follows.

– *Does a fast quark attenuate in nuclear matter?*
At first glance an energetic quark should not attenuate since it cannot be stopped or absorbed in the medium. Multiple soft interactions can only rotate the quark in color space, while the energy loss is a small fraction of the quark energy. However, saying "attenuation" of a quark,

we mean the suppression of the production rate of the final hadron produced with a certain momentum, related to the quark interactions in the nuclear medium. In the string model any color-exchange interaction leads to formation of a new string, but with reduced initial energy. This fact results in a reduction of hadron production at large $z_h \to 1$ [6]. Moreover, the leading quark may pick up new strings due to presence of higher Fock components, giving rise to induced energy loss in this model [20]. If one tries to treat the reduced production rate of hadrons as an effective attenuation of the quark in the medium, the effective quark absorption cross section cannot be treated as an universal parameter. Its value essentially varies with geometry and kinematics [6]. In the present paper we include induced energy loss explicitly.

2 Lessons of the string model

Apparently, the nonperturbative string model cannot be a realistic approach to such a hard process as DIS. The model does not have any access to Q^2-dependence, color transparency, etc., the phenomena crucial for understanding the properties of in-medium hadronization. At the same time, the model is simple, very intuitive, and helps to understand the general features of the space-time development of hadronization.

As it was already mentioned, we assumed that data are taken at large Bjorken $x > 0.1$ to make sure that valence quarks in the target nucleon give the main contribution. A color string[1] formed between the quark and the debris of the target nucleon, is slowing down the former and speeding up the latter. Naively, one may think about a long string stretched across the nucleus. Instead, a very short object is propagating through the nucleus. Indeed, a simple kinematics shows that the maximal possible length of the string in the rest frame of the nucleus is,

$$L_{max} = \frac{m}{\kappa} \, , \tag{6}$$

where m is the mass of the debris of the target nucleon (e.g. a diquark) attached to the slow end of the string. The string tension $\kappa = 1/(2\pi\alpha'_R) \approx 1\,\mathrm{GeV/fm}$ is related to the slope α'_R of the Regge trajectories [14]. Since the nucleon debris are probably lighter than the nucleon, $L_{max} < 1$ fm. In reality the string should be even much shorter due to the Schwinger phenomenon, spontaneous $q\bar{q}$ pair production from vacuum breaking the string to shorter pieces.

Thus, a heavy but short string with the effective mass $M \approx \sqrt{s} = \sqrt{m_N^2 - Q^2 + 2m_N\nu}$ propagates through the nucleus. Due to $q\bar{q}$ pair production in the color field of the string light pieces of the string are chipped off, while the leading heavy part of the string keeps propagating on. Apparently, the leading quark is slowing down losing energy with a constant rate,

$$\frac{dE}{dt} = -\kappa \, , \tag{7}$$

where $E(t)$ is the quark energy and $t = z$ are the time and longitudinal coordinate. Note that the rate of energy loss (7) is a constant which is independent of the quark flavor, energy and virtuality, and is invariant relative to longitudinal Lorentz boosts. Correspondingly, the total energy loss rises linearly with the length Δz of the path, $\Delta E = \kappa\Delta z$. The lost energy goes to acceleration of the target debris and production of new hadrons. The process of hadronization completes when the mass of the leading piece of the string is reduced by many decays down to the hadronic mass scale and the leading pre-hadron (or a cluster) is produced.

How does the production time depend on the initial quark energy ν and its fraction z_h carried by the produced hadron? For low energy hadrons it is short, but rises with their energy, $t_p \approx \nu z_h/2\kappa$. One might think that such a linear z_h-dependence is an unquestionable truth since it is dictated by Lorentz time dilation. It was claimed, however, in [23,4] and confirmed within Lund model [24], that there is an opposite trend for leading hadrons, namely, the production time vanishes like $t_p = (1 - z_h)\nu/\kappa$ at $z_h \to 1$. Although it was first derived in the string model [23,4], this is a general property, since is dictated by conservation of energy. Indeed, unless color neutralization happens, i.e. a colorless system with energy $z_h\nu$ is produced, the leading quark keeps losing energy. Therefore, the smaller $(1 - z_h)$ is, the faster this process must be completed, otherwise the energy needed for the leading hadron production will be wasted.

Note, that the produced colorless object should be treated as a pre-meson. It may take a long time proportional to $z_h\nu$ to form the final hadron wave function. We call this formation time, and it is always longer than the production time, $t_f > t_p$.

Thus, we arrive at two different types of end-point behavior of the production time [23,4,24],

$$t_p = \begin{cases} \frac{\nu}{2\kappa} z_h & z_h \ll 1 \\ \frac{\nu}{\kappa} (1 - z_h) & 1 - z_h \ll 1 \end{cases} \tag{8}$$

The concept of production time is crucial in the case of nuclear target: a colorless pre-hadron produced inside the nucleus should not interact (inelastically) on the way out of the nucleus, otherwise hadronization will be triggered, and the final hadron will emerge with a substantially reduced energy. Apparently, this condition leads to an attenuation which depends on the interaction cross section of the pre-hadron (see below). For this reason different processes on nuclei are suppressed near the kinematic limit $z_h \to 1$, for example, back-to-back high-p_T dihadron production, DIS, forward particle production in hadron-nucleus interactions, etc.

As was mentioned above, the production time (8) corresponds to creation of a fast colorless piece of the string, a pre-hadron, rather than the final hadron. The latter is supposed to have a wave function describing the specific distribution of quark momenta inside the hadron, different from the quark momentum ordering in the string. The wave function of the hadron is developing over a longer period of time (formation time), which is proportional to

---

[1] Of course a one-dimensional string is an idealization. We neglect the transverse size of the color-flux tube [14], unless otherwise specified.

the hadron energy, $z_h \nu$, rather than to $(1 - z_h)\nu$. These two time scales are mixed up sometimes. However, nuclear attenuation is controlled mainly by the the production time (8), rather than by the formation time which has a moderate influence on nuclear effects. Nevertheless, we will include both effects in our calculations.

Another question which may be answered within the string model is how multiple interactions of the hadronizing quark at earlier stage, $t < t_p$ affects the hadron production rate? Sometimes [5,2] this attenuation is treated as absorption with an effective quark-nucleon cross section treated as a universal fitted parameter. However, the physical origin of such an attenuation discussed in [6] shows that it cannot be described in terms of a universal parameter like a quark absorption cross section. Indeed, every interaction of the leading quark leads to a new hadronization process starting from the very beginning, but with a reduced initial energy $E_q(z) = \nu - \kappa z$, where z is the distance covered by the quark. Since the energy of the final hadron is fixed, the variable of the fragmentation function should be redefined as, $z_h \Rightarrow z_h/(1 - \kappa z/\nu)$. Apparently, the increase of z_h should lead to a suppression which depends on ν and z_h and cannot be treated as a universal parameter [6].

Besides, multiple quark interactions at $t < t_p$ induce an additional energy loss with a rate rising linearly with the path length z [20],

$$\frac{dE}{dz} = -\kappa \left[1 + \langle n(z)\rangle\right] , \tag{9}$$

where $\langle n(z)\rangle = \sigma_{in}^{qN} \rho_A z$ is the mean number of collisions experienced by the quark over the distance z; σ_{in}^{qN} is the model dependent effective quark-nucleon inelastic cross section. The first term in (9), same as in (7), is the usual retarding force produced by the string, while the second term is due to additional strings created by multiple interaction of the quark. These extra terms correspond to higher Fock component of the quark and can be motivated within the dual parton model [29] (more details can be found in [20]).

Thus, the total energy loss on a distance L contains a correction $\propto L^2$.

$$\Delta E(L) = \kappa \, L \left[1 + \frac{1}{2}\sigma_{in}^{qN}\rho_A \, L\right] . \tag{10}$$

In the additive quark model $\sigma_{in}^{qN} \sim 10\,\mathrm{mb}$, which is the same value as follows from the dual parton model [29]. Note, that this value is twice as large as in the more realistic model [30], which demonstrates that the dominant part of the hadronic cross section does not obey quark additivity.

It is questionable whether one can apply this phenomenology to DIS. At this point we feel that our unjustified assumption that soft interaction phenomenology, like the string model, is relevant for final state interaction in DIS has been pushed to the extreme, and we should consider a more realistic approach based on perturbative QCD.

3 Radiative energy loss: Hadronization in vacuum

3.1 Vacuum energy loss

When a high-momentum parton is produced in a hard reaction, it triggers gluon bremsstrahlung, which lasts until the moment of color neutralization. Gluons are not radiated instantaneously, but it takes time to lose coherence with the parent parton. This quantum-mechanical uncertainty called coherence time can be introduced differently. A strict definition comes within precise calculations and is related to interference between quanta radiated at different positions. For instance, two amplitudes of gluon radiation by a quark at different coordinates separated by the longitudinal distance Δz get phase shift $\Delta z q_L$, where $q_L = (M_{Gq}^2 - m_q^2)/2E_q$ is a difference between the longitudinal momenta of the quark and the produced quark-gluon final state which has effective mass squared,

$$M_{Gq}^2 = \frac{k_T^2 + \alpha m_q^2}{\alpha(1 - \alpha)} . \tag{11}$$

Here $\mathbf{k}_T$ and α are the transverse momentum and fraction of the light-cone momentum of the quark carried by the gluon; $E_q = \nu$ is the initial quark energy (we assume that Bjorken x is large and the quark gets the whole energy transfer).

Apparently, only those gluons can interfere, which are radiated within the coherence time interval, or coherence length (we assume $z = t$ and neglect the quark mass hereafter),

$$t_c \sim \frac{2\nu}{k_T^2}\alpha(1 - \alpha) . \tag{12}$$

This time may be also interpreted as the lifetime of the quark-gluon fluctuation, or as the formation time of radiation. At any rate, only those gluons may be treated as radiated during time t, which have coherence time $t_c < t$, otherwise they are still coherent with the quark and should be considered as a part of its color field.

Correspondingly, one can estimate the energy deposited into the radiation as

$$\Delta E(t) = \nu \int\limits_{\lambda^2}^{Q^2} dk_T^2 \int\limits_0^1 d\alpha \, \alpha \, \frac{dn_G}{dk_T^2 d\alpha} \, \Theta(t - t_c) \tag{13}$$

where λ is an infrared cutoff. The number of radiated gluons $dn_G/d\alpha dk_T^2 = \gamma/\alpha k_T^2$, where $\gamma = 4\alpha_s(k_T^2)/3\pi$ [31,15]. As usual, we employ the approximation of soft radiation, $\alpha \ll 1$.

Equation (13) leads to a linear time dependence of energy loss,

$$\Delta E(t) = \frac{\gamma}{2}\left(Q^2 - \lambda^2\right) t \tag{14}$$

i.e. the rate of energy loss is constant, like it is known for strings. This interesting observation was done in [15].

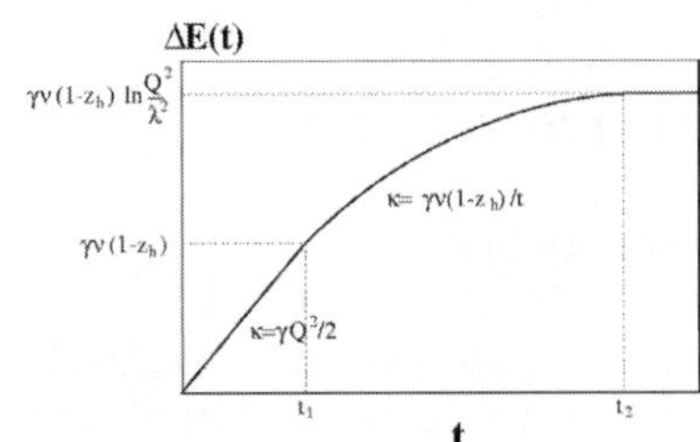

Fig. 1. Time-dependence of energy loss corresponding to (15). The notations are explained in the text

The radiated energy is distributed among produced hadrons, but we are interested in the most energetic one, leading hadron, which carries the largest fraction z_h of the initial momentum of the quark. This condition and energy conservation impose additional restrictions on the integration in (13), resulting in a different time-development of the hadronization and the energy loss.

If the leading hadron is produced with large z_h and includes the leading quark, none of the radiated gluon may leave with a fraction of the initial quark momentum exceeding $\alpha > 1 - z_h$. For instance, in the limiting case $z_h \to 1$ any radiation is forbidden, what leads to a so called Sudakov's suppression of the cross section. With decreasing z_h, this restriction gradually softens, but still should not be forgotten.

In order to estimate the time-dependence of radiative energy loss specified for the case of the leading hadron production we introduce energy conservation in (13) via step-function $\Theta(1-z_h-\alpha)$. The result of integration reads,

$$\Delta E(t) = \frac{\gamma}{2}\, t\,(Q^2 - \lambda^2)\,\Theta(t_1 - t)$$

$$+ \left\{ \nu\gamma(1 - z_h)\left[1 + \ln\left(\frac{t}{t_1}\right)\right] - \frac{\gamma}{2}\lambda^2 t\right\}\Theta(t - t_1)\Theta(t_2 - t)$$

$$+ \nu\gamma(1 - z_h)\ln\left(\frac{Q^2}{\lambda^2}\right)\Theta(t - t_2) \qquad (15)$$

where

$$t_1 = \frac{2\nu}{Q^2}(1 - z_h) = \frac{1 - z_h}{x\, m_N}\,, \qquad (16)$$

$$t_2 = \frac{Q^2}{\lambda^2}t_1\,, \qquad (17)$$

and $x = Q^2/2m_N\nu$ is Bjorken variable.

The time dependence of radiative energy loss is illustrated in Fig. 1 assuming $Q^2 \gg \lambda^2$. During the time interval $t < t_1$, the rate of energy loss is constant, $-dE/dt = \gamma Q^2/2$, exactly as in the case of (13), (14) with no restriction on the radiated energy.

Over a longer time interval, more energetic gluons can be emitted in accordance with (12) and the restriction $\alpha < 1 - z_h$ becomes effective. As a result, the loss of energy slows down to a logarithmic t-dependence. This implies that the rate of energy loss decreases as function of time as $dE/dt = -\gamma\nu(1 - z_h)/t$.

According to (15) at much longer times either very energetic gluons can be emitted (what may be forbidden by the condition $\alpha < 1 - z_h$), or gluons with very small

k_T. The latter are cut off in (13). This completely eliminates any energy loss at long times $t > t_2$ as one can see from (15) (see also Fig. 1). This sounds puzzling, but this is suppressed by a Sudakov-type factor. This suppression becomes active $t > t_1$, when the production of leading hadron with large z_h restricts the spectrum of radiation.

3.2 Leading hadron production

Summing up all the nonradiated gluons we get the following Sudakov's suppression

$$S(t, z_h, Q^2, \nu) = \exp\left[-\tilde{n}_G(t, z_h, Q^2, \nu)\right]\,, \qquad (18)$$

where

$$\tilde{n}_G(t, z_h, Q^2, \nu) = \gamma \int\limits_{1-z_h}^{1}\frac{d\alpha}{\alpha}\int\limits_{\lambda^2}^{Q^2}\frac{dk_T^2}{k_T^2}[1 - \exp(-t/t_c)] \qquad (19)$$

is the number of gluons which have sufficiently short coherence time to be radiated during time interval t, but are conflicting with a leading hadron production since are too energetic, $\alpha > 1 - z_h$.

To be more realistic we should smooth out the step-function in (15). We assume that the radiation rate of gluons with coherence time t_c has a decay rate distribution,

$$\frac{dP(t)}{dt} = \frac{1}{t_c}\,e^{-t/t_c}\,. \qquad (20)$$

Then the probability to radiate a gluon during time interval t is,

$$P(t) = 1 - e^{-t/t_c}\,, \qquad (21)$$

which is a replacement for the step-function $\Theta(t - t_c)$ in (15) and (19).

The Sudakov formfactor, (18), substantially reduces the mean time interval of hadronization and the production time of the leading pre-hadron.

The estimate (1) for energy loss was obtained within the string model, which has no access to Q^2 dependence and assumes that the final hadron is immediately produced once the quark energy degraded down to $z_h\nu$. The perturbative QCD gives opportunity to develop a more detailed and profound model for hadronization and color neutralization. In the large N_c limit each radiated gluon is equivalent to a $q\bar{q}$ pair, and the products of gluon bremsstrahlung can be seen as a system of colorless dipoles (compare with [32,33]). It is natural to assume that the leading hadron originates from the color dipole, which includes the leading quark and the antiquark from the last radiated gluon. In the $1/N_c$ approximation they are automatically tuned in color, i.e. this pair is colorless. This $q\bar{q}$ dipole is to be projected on the hadron wave function, $\Psi_h(\beta, l_T)$, where β is the fractions of the hadron light-cone momentum carried by one of the quarks, and l_T is the transverse momentum of the quarks in the hadron.

Let us evaluate the distribution function $W(t, z_h, Q^2, \nu)$ for production time t of a pre-meson,

$$W(z_h, t, Q^2, \nu) = N \int\limits_0^1 \frac{d\alpha}{\alpha}\, \delta\left[z_h - \left(1 - \frac{\alpha}{2}\right)\frac{E_q(t)}{\nu}\right]$$

$$\times \int\limits_{\Lambda^2}^{Q^2} \frac{dk_T^2}{k_T^2}\, \frac{\exp(-t/t_c)}{t_c} \int dl_t^2\, \delta\left[l_T^2 - \frac{9}{16}k_T^2\right]$$

$$\times \int\limits_0^1 d\beta\, \delta\left[\beta - \frac{\alpha}{2-\alpha}\right] |\Psi_h(\beta, l_T)|^2\, S(z_h, t, Q^2, \nu)\ . \quad (22)$$

Here α is the fraction of the quark light-cone momentum carried by the gluon emitted at the time t by the quark of energy $E_q(t) = \nu - \Delta E(t)$. The mean radiation time of this gluon t_c is given by (12). Assuming that the radiated gluon splits to a $\bar{q}q$ pair which shares the gluon momentum in equal parts, we conclude that the leading colorless $q\bar{q}$ pair is produced with an energy $E_q(t)(1-\alpha/2)$. On the other hand, this must be the observed energy $z_h\nu$ of the produced hadron, the condition controlled by the first δ-functions in (22). The transverse momentum l_T of the quark and the antiquark within the colorless dipole is related to k_T as $l_T = 3\mathbf{k}_T/4$, while the fraction of the hadron light-cone momentum carried by the quark is $\beta(t) = \alpha(t)/[2 - \alpha(t)]$. This is why we need the last two δ-functions in (22). Normalization factor N in (22) cancels in ratio (5), so we do not specify it.

Note that (22) has a probabilistic structure, instead of projecting the production amplitude on the hadron wave function, we convolute the squares of them. This simplification is related to to our need to include the time development of hadronization and make use of the coherence time. To do it correctly one should work with amplitudes taking care of interferences at all stages. In principle, this can be done within the path integral approach (partially applied below), however, technically it is so complicated that exact solution of this problem does not look feasible at present. Nevertheless, we will see in what follows that the approximation (22) does a good job describing data.

We use the parameterization for the hadronic wave function in (22) in the form of the asymptotic light-cone meson wave function,

$$\Psi_h(\beta, l_T^2) \propto \frac{\beta(1-\beta)}{\beta(1-\beta) + a_0} \exp\left[-\frac{R_h^2\, l_T^2/8}{\beta(1-\beta) + a_0}\right], \quad (23)$$

where $R_h^2 = 8\langle r_h^2\rangle_{em}/3$ is related to the mean hadronic electromagnetic radius squared, $\langle r_h^2\rangle_{em}$. Normalization is not important since it can be absorbed into the factor N in (22). The parameter a_0 in (23) incorporates confinement for highly asymmetric $q\bar{q}$ configurations with $\beta \to 0, 1$. We fix this parameter demanding the mean charge radius of the hadron to be correctly reproduced,

$$\langle \rho^2\rangle = \frac{\int \frac{d\beta}{\beta\,(1-\beta)} \int d\rho^2 |\Psi_h(\beta, \rho)|^2\, \rho^2}{\int \frac{d\beta}{\beta\,(1-\beta)} \int d\rho^2 |\Psi_h(\beta, \rho)|^2} = \frac{8}{3}\langle r^2\rangle_{em}\ , \quad (24)$$

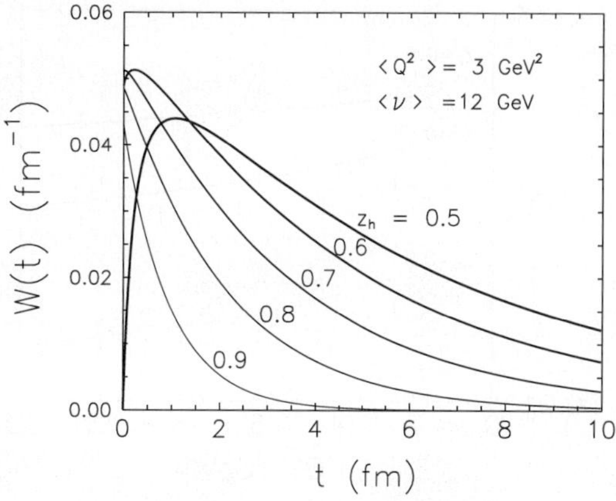

Fig. 2. Distribution of the hadron production time at $\nu = 12\,\mathrm{GeV}$, $Q^2 = 3\,\mathrm{GeV}^2$ and $z_h = 0.5$-0.9

where $\Psi_h(\beta, \rho)$ is the hadronic wave function in coordinate representation, i.e. the Fourier transformation of (23). For production of pions with $\langle r^2\rangle_{em} = 0.44\,fm^2$ [34] we found $a_0 = 1/12$.

Examples for the production-time distribution $W(t)$ at $<\nu> = 12\,\mathrm{GeV}$, $<Q^2> = 3\,\mathrm{GeV}^2$ and different z_h are presented in Fig. 2 [2]. We see that as function of z_h the mean production time decreases similar to the expectations of the string model (5) (but with different value of κ), what is not a surprise, since this behavior reflects conservation of energy.

There are no data, of course, for $W(t, z_h)$ to compare with, but the integral over t at fixed z_h is the probability to produce the hadron with energy $z_h\nu$. This probability is the fragmentation function which is rather well known phenomenologically.

$$\tilde{D}_{h/q}(z_h, Q^2) = \int\limits_0^\infty dt\, W(t, z_h, Q^2)\ . \quad (25)$$

Tilde here means that this function describes fragmentation of a quark only to a (leading) hadron which includes the parent quark.

Now we can fix the normalization factor N in (22). In fact, the normalization of $\tilde{D}_{h/q}$ is different from that for the conventional fragmentation function which is normalized to the particle multiplicity, $\int_0^1 dz_h\, D_{h/q}(z_h, Q^2) = \langle n_h\rangle$. Apparently, only one hadron in the produced jet contains the original quark, therefore,

$$\int\limits_0^1 dz_h\, \tilde{D}_{h/q}(z_h, Q^2) = 1. \quad (26)$$

Now we are in a position to compare the modeled fragmentation function with a realistic phenomenological one. In view of the specifics of our model which is designed to reproduce production of leading hadrons, we may expect

---

[2] Note that a function, $t\,W(t)$ was plotted in Fig. 1 of [1], but factor t was missed in the notations.

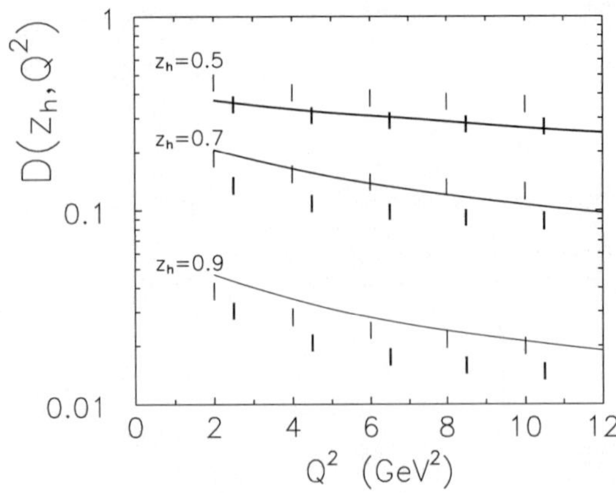

Fig. 3. The model prediction for the fragmentation function $\tilde{D}(z_h, Q^2)$ in comparison with parametrizations fitted to data, [35] (*thin vertical bars*) and [36] (*thick bars*)

agreement only at large z_h, probably $z_h > 0.5$. Our calculations for Q^2 dependence of $\tilde{D}_{h/q}(z_h, Q^2)$ at different z_h are depicted in Fig. 3, and are compared with the popular parametrizations fitted to data, [35] and [36] shown by thin and thick bars respectively. The difference between the parametrizations may be treated as a systematic uncertainty. The model reproduces the phenomenological fragmentation function rather well. This encourages us to apply this approach to nuclear targets as well.

4 Evolution and attenuation of the pre-hadron in nuclear environment

Production of a leading colorless pre-hadron with the detected momentum completes the process of hadronization. The further propagation of the pre-hadron in vacuum and formation of the hadronic wave function have no observable effect, since leads only to a phase shift. This changes drastically in presence of an absorptive medium. In this case the production time t_p is of a special importance. Indeed, when color neutralization happens and a pre-hadron with the desired energy, $z_h\nu$, is produced, any subsequent inelastic interaction in the medium gives rise to a new hadronization process, and the quark energy substantially degrades down to a lower value of z_h. This means that the pre-hadron should have been produced with a larger $z_h' > z_h$, i.e. with a smaller probability. Such a suppression of the hadron production rate turns out to be so strong that we can neglect these double-step processes.

If the energy of the pre-hadron ($\bar{q}q$ dipole) is sufficiently high to freeze the transverse $\bar{q}q$ separation (due to Lorentz time dilation) for the time of propagation through the medium, then the survival probability of the pre-hadron, which we also call nuclear transparency, gets the simple eikonal form,

$$Tr = \left| \frac{\langle \Psi_h(r_T) | \exp\left[-\frac{1}{2}\sigma_{\bar{q}q}^N(r_T) T_A\right] | \Psi_{\bar{q}q}(r_T)\rangle}{\langle \Psi_h(r_T) | \Psi_{\bar{q}q}(r_T)\rangle} \right|^2, \quad (27)$$

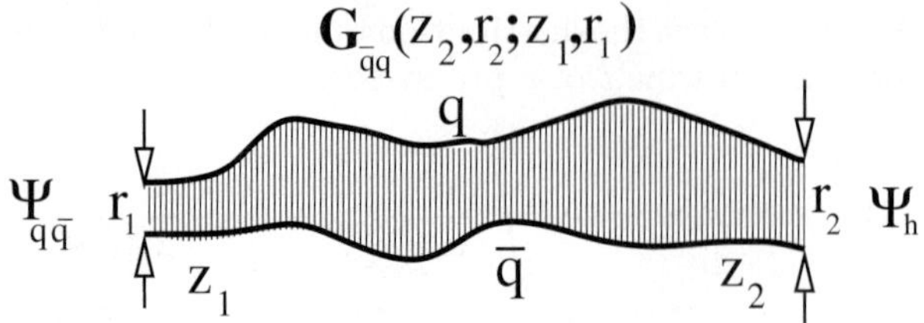

Fig. 4. The light-cone Green function $G_{\bar{q}q}(z_2, \mathbf{r}_2; z_1, \mathbf{r}_1)$ describing propagation of a $\bar{q}q$ pair over all possible paths in a medium between points with coordinates z_1 and z_2

where T_A is already introduced nuclear thickness, given in this case by integral of the medium density along the path of the dipole, $T_A(z_1, z_2) = \int_{z_1}^{z_2} dz\, \rho_A(z)$. The universal dipole-nucleon cross section, $\sigma_{\bar{q}q}^N(r_T)$, was introduced earlier. $\Psi_{\bar{q}q}(r_T)$ and $\Psi_h(r_T)$ are the light-cone wave functions of the initial $\bar{q}q$-dipole (pre-hadron) and the final hadron respectively.

In the case of medium energies (HERMES, Jefferson Lab) the dipole size may substantially fluctuate during propagation through the nucleus and nuclear transparency might substantially differ from (27). The rigorous quantum-mechanical approach incorporating all effects of fluctuations should include all possible trajectories of the quark and antiquark, rather than just fixed quasi-classical ones. This path integral technique leads to an expression for the survival probability of a dipole in a medium similar to (27). However, the exponential attenuation factor in the numerator must be replaced by the light-cone Green function, $G_{\bar{q}q}(z_2, \mathbf{r}_2; z_1, \mathbf{r}_1)$, which describes propagation of the $\bar{q}q$ pair with initial and final separations $\mathbf{r}_1$ and $\mathbf{r}_2$, between points with longitudinal coordinates z_1 and z_2 [37,38,22], as is illustrated in Fig. 4. Then the nuclear transparency (27) takes the form,

$$Tr(z_1, z_2) = \left| \frac{\int d^2 r_1 d^2 r_2 \Psi_h^*(r_2) G_{\bar{q}q}(z_2, \mathbf{r}_2; z_1, \mathbf{r}_1) \Psi_{\bar{q}q}(r_1)}{\int d^2 r\, \Psi_h^*(r)\, \Psi_{\bar{q}q}(r)} \right|^2. \quad (28)$$

The Green function satisfies the two-dimensional light-cone Schrödinger equation,

$$i\frac{d}{dz_2} G_{\bar{q}q}(z_2, \mathbf{r}_2; z_1, \mathbf{r}_1) = \left[\frac{m_q^2 - \Delta_{r_2}}{2 E_h \beta (1 - \beta)} + V_{\bar{q}q}(z_2, \mathbf{r}_2, \beta) \right]$$
$$\times\, G_{\bar{q}q}(z_2, \mathbf{r}_2; z_1, \mathbf{r}_1)\,, \quad (29)$$

with the boundary condition $G_{\bar{q}q}(z_2, \mathbf{r}_2; z_1, \mathbf{r}_1)|_{z_2=z_1} = \delta^{(2)}(\mathbf{r}_1 - \mathbf{r}_2)$. Here $E_h = z_h \nu$ is the hadron energy, and β is the fraction of the hadron light-cone momentum carried by the quark. The first, kinetic, term in squared brackets is responsible for the phase shifts which are related to quark masses, m_q, and their transverse motion described by Laplacian Δ_r acting on coordinate r.

The imaginary part of the light-cone potential in (29),

$$\mathrm{Im}\, V_{\bar{q}q}(z, \mathbf{r}, \beta) = -\frac{1}{2}\, \sigma_{\bar{q}q}^N(r)\, \rho_A(z)\,, \quad (30)$$

is responsible for attenuation of the $\bar{q}q$ in the medium, while the real part represents the interaction between the

q and $\bar{q}$. The latter is supposed to provide the correct light-cone wave function of the final meson. For the sake of simplicity we use the oscillator form of the potential,

$$\operatorname{Re} V_{\bar{q}q}(z, \mathbf{r}, \beta) = \frac{a^4(\beta)\, \mathbf{r}^{\,2}}{2\, E_h\, \beta(1-\beta)}\,, \tag{31}$$

which leads to a Gaussian r-dependence of the light-cone wave function of the meson ground state. The shape of the function $a(\beta)$ was discussed in [39], and fixed by photoproduction data $a^2(\beta) = a_0^2 + a_1^2\beta(1-\beta)$, where $a_0^2 = v^{1.15} \times (112\,\mathrm{MeV})^2$; $a_1^2 = (1-v)^{1.15} \times (165\,\mathrm{MeV})^2$, and v takes any value in the interval $0 < v < 1$.

If we assume also that the small-r dependence for the dipole cross section, $\sigma_{\bar{q}q}^N(r) = C(E_h)\, r^2$, then (29) has an analytical solution [40],

$$G_{\bar{q}q}(z_2, \mathbf{r}_2; z_1, \mathbf{r}_1) = \frac{a^2(\beta)}{2\,\pi\, i\, \sin(\xi\, \Delta z)}$$

$$\times \exp\left\{\frac{i\, a^2(\beta)}{\sin(\xi\, \Delta z)}\left[(r_1^2 + r_2^2)\cos(\xi\,\Delta z) - 2\,\mathbf{r}_1 \cdot \mathbf{r}_2\right]\right\}$$

$$\times \exp\left[-\frac{i\, m_q^2\, \Delta z}{2\, E_h\, \beta\,(1-\beta)}\right]\,, \tag{32}$$

where $\Delta z = z_2 - z_1$ and

$$\xi = \frac{\sqrt{a^4(\beta) - i\beta(1-\beta)E_h\, C(E_h)\, \rho_A}}{E_h\, \beta(1-\beta)}\,, \tag{33}$$

This solution is valid only for a medium with a constant density. However, using the recurrent relations between Green functions for different space intervals one can solve analytically the problem for a varying density as well [37].

5 Nuclear attenuation caused by absorption

We assume that the colorless $\bar{q}q$ pair (pre-hadron) produced at distance l_p from the DIS point has Gaussian distribution of transverse separations with the mean transverse size related to the inverse value of the mean transverse momentum of the quarks,

$$\langle r_1^2 \rangle = \frac{4}{\langle l_T^2(t_p) \rangle} = \frac{64}{9\,\langle k_T^2(t_p) \rangle} \tag{34}$$

where

$$\langle k_T^2(t) \rangle = \frac{\int_{\lambda^2}^{Q^2} dk_T^2 \left[dW(t, z_h, k_T^2, \nu)/d\ln(k_T^2)\right]}{\int_{\lambda^2}^{Q^2} dk_T^2 \left[dW(t, z_h, k_T^2, \nu)/dk_T^2\right]}\,. \tag{35}$$

A $\bar{q}q$ dipole with starting value of transverse size $r_1(t_p)$ experiences evolution, forming the hadronic wave function and getting through the filtering process in the nuclear medium which absorbs more effectively dipoles of larger size. If it were in vacuum, it would take formation time t_f which can be estimated as

$$t_f \approx \frac{2 z_h \nu}{m_{h'}^2 - m_h^2}\,, \tag{36}$$

where h' is the first radial excitation of the hadron h. As was mentioned in the introduction, the formation time linearly rises with z_h, while the production time vanishes at $z_h \to 1$.

The effective fragmentation function modified by the survival probability (transparency) of the pre-hadron in the nuclear matter reads,

$$D_{eff}^A(z_h, Q^2, \nu) = \int d^2\mathbf{b} \int_{-\infty}^{\infty} dz\, \rho_A(\mathbf{b}, z)$$

$$\times \int_0^{\infty} dt\, W(t, z_h, Q^2, \nu)\, Tr(b, z+t, \infty)\,. \tag{37}$$

In the case of vanishing absorption, $\sigma_{\bar{q}q}^N \to 0$, this fragmentation function is identical to one on a free nucleon, except is A times larger. Ratio of the effective fragmentation function on a nuclear target to one on a free nucleon is identical to the experimental definition (5).

As was mentioned, we use the dipole approximation, $\sigma_{\bar{q}q}^N(r) = C\, r^2$. Factor C is proportional to the inverse mean hadron radius squared related to the parameters of the oscillatory potential,

$$C = \frac{1}{2}\, m_q\, \omega\, \sigma_{tot}^{hN}\,, \tag{38}$$

where $\omega = (m_{h'} - m_h)/2$ is the oscillatory frequency. For production of leading pions we took $\omega = 0.3\,\mathrm{GeV}$ and $m_q = 0.2\,\mathrm{GeV}$ which is constrained by the mean charge pion radius,

$$\langle r_T^2 \rangle_h = \frac{2}{m_q\, \omega} = \frac{8}{3} < r^2 >_{ch}\,, \tag{39}$$

and nuclear shadowing data [41]. Using $\sigma_{tot}^{\pi N} = 25\,\mathrm{mb}$ we arrive at $C \sim 1.9$.

6 Comparison with data

Unfortunately low statistics of HERMES data does not allow to perform a multi-dimensional comparison with the theory. One should integrate over all variables except one to get data with reasonable errors.

6.1 Energy dependence of nuclear effects

The results of numerical calculations for energy dependence of the ratio of nuclear-to-nucleon cross sections for leading pion production integrated over z_h, p_T and Q^2 are depicted by dashed curve in Fig. 5 in comparison with with HERMES data. We ignored the induced energy loss in these calculations and integrated over $z_h > 0.5$. Since the data are integrated over $z_h > 0.2$, this may cause some difference.

It is clear why nuclear suppression is maximal at low energies and has a tendency to vanish with increasing energy (data from EMC experiment [42] at much higher energies indeed demonstrate no nuclear effects). As we discussed in previous sections, hadronization process is completed by production of the leading colorless wave packet,

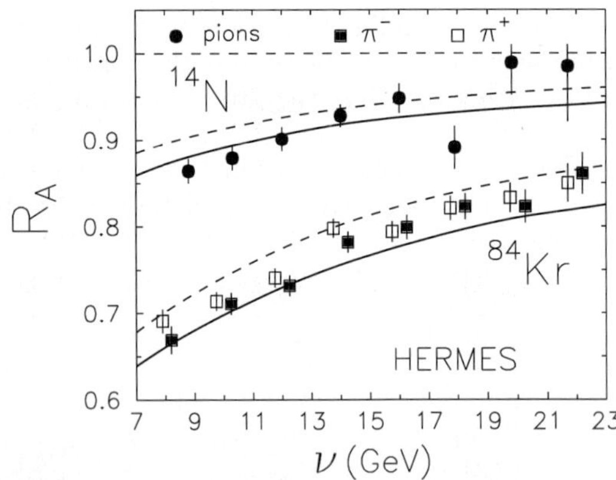

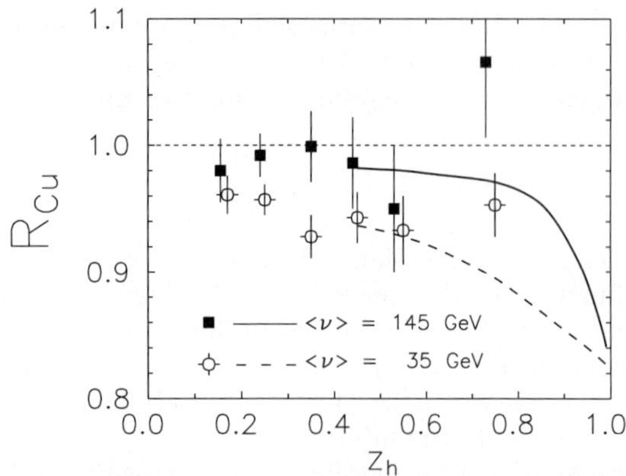

Fig. 5. Comparison of the model calculations for the ν- dependence of nuclear suppression R_{N14} (5) integrated over z_h with the HERMES data [2]. The solid and dashed curve are the model predictions for nuclear suppression with and without induced radiation to be taken into account, respectively

Fig. 6. Comparison of the model calculations for the z_h-dependence of nuclear suppression R_{Cu} (5) at the energy $< \nu >= 35\,\mathrm{GeV}$ (*dashed curve*) and $< \nu >= 145\,\mathrm{GeV}$ (*solid curve*) with the EMC data [42,44]. Calculations are done at $Q^2 = 11\,\mathrm{GeV}^2$

which is then evolving into the final hadron over the formation time t_f given by (36). If color neutralization happens outside the nucleus ($t_p > R_A$), the evolution leads only to a phase shift, resulting in no nuclear effects, as long as we neglect the induced energy loss.

On the other hand, the evolution may cause a strong attenuation if the wave packet is produced inside the nuclear target, as it happens at low energies where $t_p < R_A$. At higher energies t_p rises leaving less room for absorption. Besides, the formation time t_f rises too, and the initial small size, $r_T \sim 1/Q$, of the pre-hadron is evolving slower, leading to a lesser attenuation due to color transparency. Both effects lead to a stronger nuclear suppression at small than at high energies in accordance with data from HERMES [2] (Fig. 5), SLAC [43] and EMC [42, 44] experiments.

6.2 z_h-dependence of nuclear effects

Even at high energies the production time shrinks at large $z_h \to 1$ according to (8) and Fig. 2. Thus, at large z_h the path available for absorption of the pre-hadron reaches maximal length resulting in a strongest nuclear suppression. Such a shape of nuclear ratio falling toward $z_h = 1$ has been expected since 30 years ago [4,24], but high energy experiments could not reach sufficiently large values of z_h. Our predictions [1] are compared with EMC data in Fig. 6. We see that the interval of z_h where l_p is short absorption is at work, is rather short and squeezed towards $z_h = 1$. This is because the energy is too high, but absorption is at work only if $l_p \propto \nu(1 - z_h)$ is shorter than the nuclear size, what needs quite a small $(1 - z_h) \propto 1/\nu$.

This is why measurements at much lower energies of HERMES were proposed in citeknp in order to detect this effect. Fig. 7 demonstrates HERMES data for nitrogen and krypton in comparison with our predictions shown by dashed curves for the case of pure absorption. This is the first experimental confirmation for the longstanding prediction of vanishing production length at $z_h \to 1$.

Note, however, that even at $z_h \to 1$ nuclear suppression vanishes if both ν and Q^2 are sufficiently large. This is demanded in QCD by k_T-factorization and is correctly reproduced by our model. Indeed, at high Q^2 the initial size of the pre-hadron is very small, and it remains small long enough if the energy is high. Then color transparency eliminates nuclear effects.

6.3 Q^2 dependence of nuclear effects

At least two effects are sensitive to a variation of Q^2, color transparency and the production length. Apparently, the initial size of the pre-hadron shrinks at larger Q^2 and the nuclear matter becomes more transparent, i.e. the survival probability for the pre-hadron increases. At the same time, the production length l_p contracts because of rising vacuum energy loss, and provides a longer path available for absorption. This effect leads to a stronger nuclear suppres-

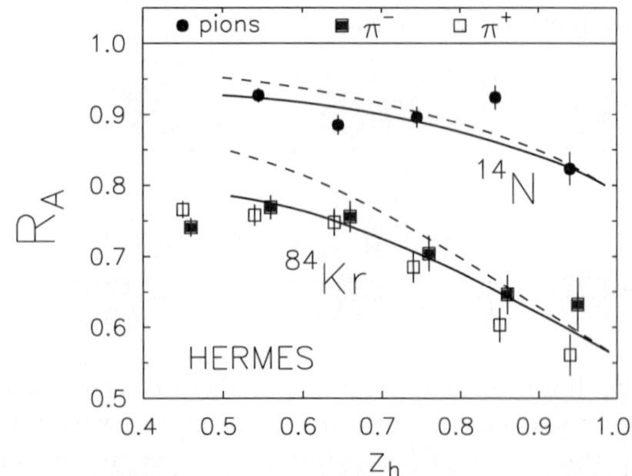

Fig. 7. Model predictions for the z_h- dependence of nuclear ratios R_{14N} (5) and R_{84Kr} calculated at values of $< \nu >$ and Q^2 corresponding to the measurements at HERMES. The *solid and dashed curve* are the model predictions with and without inclusion of induced radiation, respectively

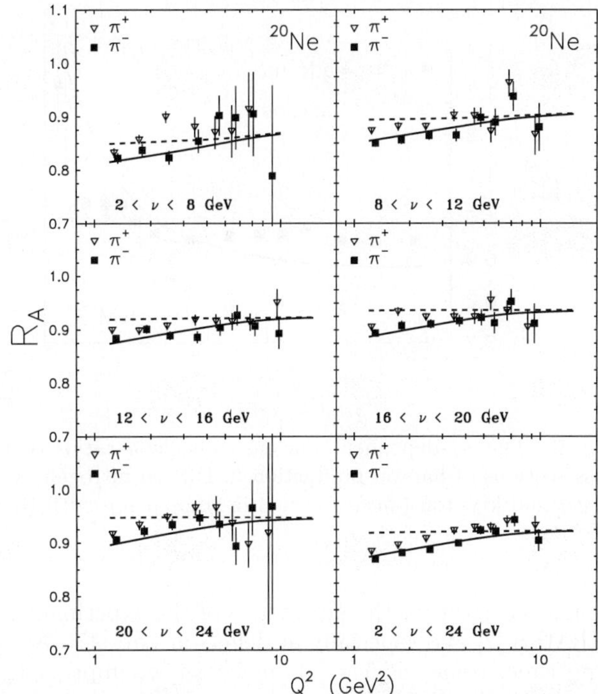

Fig. 8. Comparison of the model calculations for the Q^2- dependence of the ratio R_{Ne} of cross sections integrated over z_h with preliminary data from the HERMES experiment [45]. The *solid and dashed curve* are the model predictions with and without corrections for induced gluon radiation respectively

sion at larger Q^2. Thus, the two effects work in opposite directions and may partially or fully compensate each other. Indeed, it nearly happens: our predictions for pure effect of absorption depicted by dashed curves in Fig. 8 for neon demonstrate a rather weak Q^2-dependence. Although the Q^2-dependence demonstrated by the HERMES data is not steep either, one can see some difference. We will come to this point later after inclusion of corrections related to induced energy loss.

7 Corrections for induced gluon radiation

Multiple interactions of the quark during the time interval between the DIS point and production of the colorless pre-hadron cannot stop or absorb the quark, which keeps propagating through the medium rotating in the color space. Nevertheless, the additional soft kicks gained by the quark enforce it to shake off more gluons. This additional gluon radiation induced by the medium increases the loss of energy compared to vacuum. We calculate the induced energy loss in accordance with (3) where Δp_T^2 can be taken from data on p_T-broadening in Drell-Yan reaction, or from calculations (see next section) which agree with data. Then one should add this induced energy loss to the vacuum one, i.e. to replace $\Delta E_{vac}(t) \Rightarrow \Delta E_{vac}(t) + \Delta E_{ind}(t)$ for the time-dependent quark energy, $E_q(t) \equiv \nu - \Delta E(t)$ in (22).

Another modification of our calculations caused by p_T-broadening in the medium is the modification of the hard scale. Namely, one should replace $Q^2 \Rightarrow Q^2 + \Delta p_T^2(t)$ in (18) for the Sudakov factor and in (22).

Both corrections enhance nuclear suppression, i.e. diminish the nuclear ratio. The results are shown by solid curves in Figs. 5, 7, 8 and 12. Apparently, these induced energy corrections vanish if $l_p \to 0$. This happens at $z_h \to 1$ and at maximal Q^2. Indeed, the dashed and solid curves in Figs. 7, 8 and 12 coincide in these limits.

8 Nuclear broadening of the transverse momentum distribution

According to the perturbative QCD treatment of induced energy loss, it is caused by the broadening of transverse momentum of the quark initiating a jet during the hadronization process (at $t < t_p$). At longer time intervals, $t > t_p$, the produced pre-hadron carries undisturbed information about the quark transverse momentum (since it has no interactions) which thus can be directly measured. Comparison with theoretical predictions seems to be of special importance.

A parton propagating through a medium experiences multiple interactions in the medium and performs Braunian motion in transverse momentum plane. As a result, the mean transverse momentum squared of the quark linearly rises with the path length. Quantitatively this process has been successfully described within the light-cone dipole approach [18,19] (otherwise, most of models just fit the data to be explained). If the DIS takes place on a nucleon with coordinates $(\mathbf{b}, z)$, the medium modified transverse momentum distribution is expressed in terms of the universal cross section of a $\bar{q}q$ dipole with a nucleon,

$$\frac{dN_q^A(b,z)}{d^2p_T} = \int d^2r\, d^2r'\, \Omega_q(\mathbf{r}_T, \mathbf{r}'_T)\, \exp\left[i\mathbf{p}_T(\mathbf{r}_T - \mathbf{r}'_T)\right]$$

$$\times \exp\left[-\frac{T_A(b,z,z+l_p)}{2}\,\sigma_{\bar{q}q}^N(\mathbf{r}_T - \mathbf{r}'_T, \nu)\right]. \qquad (40)$$

Here $T_A(b, z, z+l_p) = \int_z^{z+l_p} dz'\, \rho_A(b, z')$ is the nuclear thickness function; $\sigma_{\bar{q}q}^N(r_T, \nu)$ is the phenomenological dipole cross section fitted to data for the proton structure function and photoabsorption cross section. It includes nonperturbative effects, as well as all corrections for gluon radiation (which give rise to the energy dependence of the cross section) also contributing to the broadening of the quark transverse momentum.

The quark density matrix in coordinate representation,

$$\Omega_q(\mathbf{r}_T, \mathbf{r}'_T) = \frac{k_0^2}{\pi}\, \exp\left[-\frac{k_0^2}{2}(r_T^2 + r_T'^2)\right], \qquad (41)$$

describes the initial distribution of the quark in the target nucleon. It controls the transverse momentum distribution of the valence quark emerging from DIS on a free nucleon. Therefore, the mean transverse momentum is of the order of the inverse proton size, i.e. is rather small, $k_0 \sim \Lambda_{QCD}$. Usually data on hard processes need a much larger primordial transverse momentum. This effect is related to the next-to-leading order corrections related to hard gluon emission. Since we explicitly include gluon radiation in our model, additional next-to-leading order corrections would lead to double counting.

Note that our parameter-free approach successfully explained available data for the Cronin effect in pA collisions predicted it for $d - Au$ collisions at RHIC [22]. Data for nuclear broadening in Drell-Yan reaction also are well reproduced [19].

To get the hadron transverse momentum distribution we should convolute the quark p_T-distribution, (40) calculated without the exponential factor, with the fragmentation function $D_{h/q}(z_h, p_T)$,

$$\frac{1}{\sigma_{DIS}^N} \frac{d\sigma_{DIS}^N(\gamma^* N \to hX)}{d^2 p_T} = \int d^2 k_T \, d^2 q_T \, \frac{dN_q^N}{d^2 q_T}$$
$$\times \frac{dD_{h/q}(z_h, k_T^2)}{d^2 k_T} \delta\left(\mathbf{p}_T - \mathbf{q}_T - \frac{\mathbf{k}_T}{2}\right) . \tag{42}$$

Now we can make use of our model for the vacuum fragmentation function $\tilde{D}_{h/q}(z_h, Q^2)$, (26), and calculate the mean momentum squared of the produced meson. It hardly varies with Q^2 within the kinematic range corresponding to HERMES data. We found $\langle p_T^2 \rangle = \langle k_T^2 \rangle/4 + k_0^2 = 0.3\,\mathrm{GeV}^2$, where $k_0 = 0.2\,\mathrm{GeV}$. This result agrees with the p_T-distribution of hadrons measured at SLAC [46] at energies overlapping with the HERMES energy range.

The p_T-dependence of the effective fragmentation function in nuclear medium, according to (37), reads,

$$\frac{A}{\sigma_{DIS}^A} \frac{d\sigma_{DIS}^A(\gamma^* A \to hX)}{d^2 p_T} = \int d^2 \mathbf{b} \int_{-\infty}^{\infty} dz \, \rho_A(\mathbf{b}, z)$$
$$\times \int_0^{\infty} dt \, Tr(b, z + t, \infty) \int d^2 k_T \, d^2 q_T \, \frac{dN_q^A(b, z)}{d^2 q_T}$$
$$\times \frac{dW(t, z_h, k_T^2, \nu)}{d^2 k_T} \delta\left(\mathbf{p}_T - \mathbf{q}_T - \frac{\mathbf{k}_T}{2}\right) \tag{43}$$

We also included the small effect of Fermi motion in the nucleus.

The calculated ratio of the two distributions, nucleus-to-nucleon, is depicted in Fig. 9 for nitrogen and krypton in comparison with data from the HERMES experiment [2]. The model successfully reproduces the data with no adjustment.

9 Disentangling absorption and induced energy loss

The production length is a fundamental characteristics of the early stage of hadronization, and it is important to sort out those observables which are sensitive to l_p.

Such observables could also help to disentangle different models of hadronization. The current estimates of l_p within different models (strings [4,24,2,8], pQCD [1]) are pretty close and show that l_p is rather short compared to

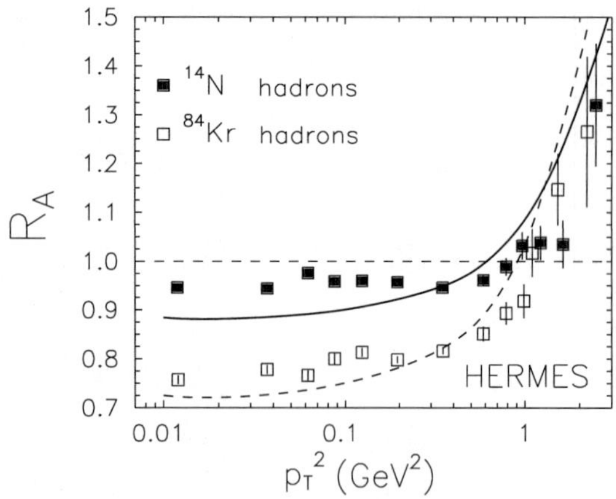

Fig. 9. The p_T-dependent nucleus-to-nucleon ratios of the cross sections of hadron production in DIS on nitrogen (*solid curve*) and krypton (*dashed curve*) in comparison with HERMES data

the nuclear sizes for the kinematics of the experiments at HERMES [2] and especially at Jefferson Lab [47]. At the same time, some models [11] make an assumption that color neutralization happens always outside the nucleus. Although this is an ad hoc assumption with no justification, would be useful to confront it with a direct experimental evidence.

9.1 Flavor dependence

The survival probability for the produced pre-hadron depends on the wave function of the final hadron, as one can see in (28). Also intuition tells us that the projection to a wave function with a larger mean radius should emphasizes larger $\bar{q}q$ separations in the pre-hadron, leading to a stronger absorption. Since pion has a larger radius than the kaon, the former is expected to be produced with a stronger nuclear suppression, than the latter. This may be a good signature for absorptive effects and shortness of l_p. On the contrary, in the case of $l_p \gg R_A$ there should be no sensitivity to the hadronic size.

HERMES experiment has provided recently high quality data for hadron production with particle identification [2]. Here we concentrate only on production of pions and positive kaons, since negative kaons and antiprotons do not contain valence quarks common with the target nucleon, while proton production is much contaminated with the target nucleons.

Our predictions for the production rates of positive kaons and pions integrated over z_h are depicted in Fig. 10 as function of photon energy together with recent data from HERMES. The two curves are different only by the value of the charge radius, $\langle r^2 \rangle_{ch}$ (kaon or pion) used in calculations.

Data confirm the expected identical suppression for positive and negative pions, and less attenuation for kaons. Although our calculations underestimate kaons, this is due to the difference in the bottom limit of integration over z_h, $(z_h)_{min} = 0.2$ in the data, and 0.5 in our calculations. At

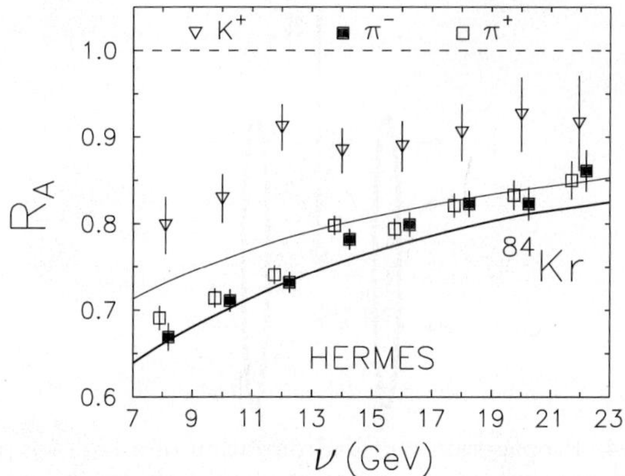

Fig. 10. Comparison of nuclear effects for kaon (*thin curve*) and pion (*thick curve*) productions as function of energy in comparison with HERMES data

small z_h kaons are copiously produced via reaction $\pi p \to K\Lambda$ which has a very low energy threshold. It would make sense if experimental data used a larger value of $(z_h)_{min}$.

Although the alternative model, the energy loss scenario, is not sensitive to the hadronic radius, one may expect a difference in quark fragmentation functions for $q \to \pi$ and $q \to K$, where q is a light quark, u or d. Indeed, the end-point behavior of fragmentation function, at $z_h \to 1$, is dictated by the Regge phenomenology,

$$D_{\pi/q}(z_h) \propto (1 - z_h)^{1-\alpha_\rho(0)} = (1 - z_h)^{0.5} \; ;$$
$$D_{K^+/q}(z_h) \propto (1 - z_h)^{1-\alpha_{K^*}(0)} = (1 - z_h)^{0.8} \; . \quad (44)$$

Since the induced energy loss leads to a shift in $z_h \Rightarrow z_h + \Delta z_h$, it should lead to a stronger suppression kaons which have a steeper fragmentation function. This expectation is in variance with HERES data depicted in Fig. 11.

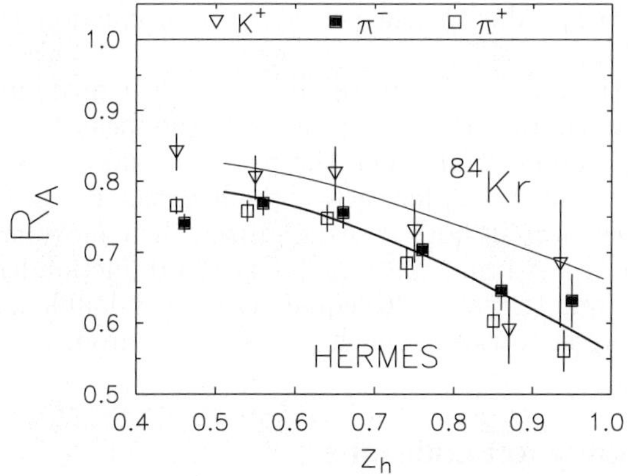

Fig. 11. Nuclear ratios for pions (*thick curve*) and positive kaons (*thin curve*) as function of z_h predicted by our model in comparison with HERMES data

At the same time, the data agree well with our predictions shown by thick and thin curves for pions and positive kaons respectively. This fact confirms our claim that the disagreement with data for kaons seen in Fig. 10 is due to presence of small-z_h events in the data.

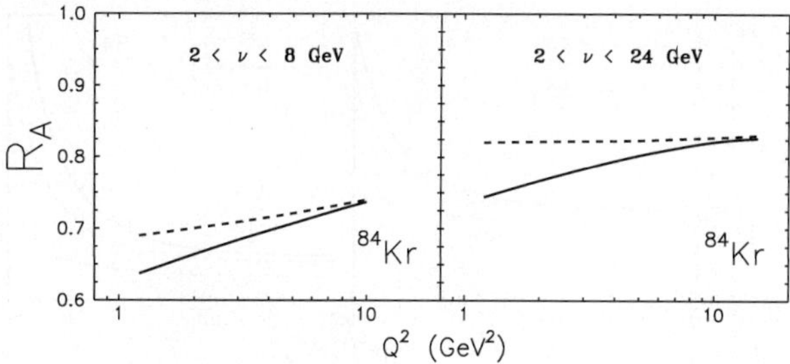

Fig. 12. The same as in Fig. 8, but predicted for krypton

Note that maximum effect of flavor dependence should be expected for shortest l_p, e.g. at $z_h \to 1$. Indeed, our calculations predict quite a large difference between nuclear effects for K^+ and pions at $z_h \to 1$, as is demonstrated in Fig. 11. Unfortunately, the accuracy of the data is not sufficient to see such a variation of the flavor dependence with z_h.

9.2 Q^2-dependence

The fit to HERMES data performed in [11] within the energy loss scenario led to the value of the universal parameter, twist-4 quark-gluon correlation tensor, which is twice as small as extracted from Drell-Yan data and 25 times smaller that follows from data on high-p_T dijet production. Such a steep variation of the parameter which must be process-independent was attributed in [11] to a strong scale dependence. If it were true, the induced energy loss effect would rise with Q^2. Accordingly, the nuclear suppression should become stronger at larger Q^2. This expectation is in obvious contradiction with the HERMES results depicted in Fig. 8.

On the other hand, we expect a more pronounced rise of nuclear ratio with Q^2 for heavy nuclei. Our predictions for krypton are depicted in Fig. 12. Note the sizeable variation of nuclear suppression with Q^2 at low energies. In this case the production time is short at any Q^2 and the variation of nuclear effects versus Q^2 is only due to color transparency. This is why the nuclear ratio steeply rises with Q^2.

9.3 z_h dependence of p_T-broadening

According to discussion in the previous section, nuclear broadening of the transverse momentum distribution might be a most sensitive probe for the production length, and provides a direct measurement of l_p, since $\Delta p_T^2 \propto l_p$. Indeed, only the hadronizing quark at short time interval $t < t_p$ contributes to the broadening. As far as the pre-hadron is created, no further broadening occurs, since inelastic interactions are prohibited for the pre-hadron, and only broadening via elastic rescattering are possible. However, the elastic cross section is so small that even for pions the mean free path in nuclear matter is about 20 fm. It is even longer for a small-size pre-hadron due to

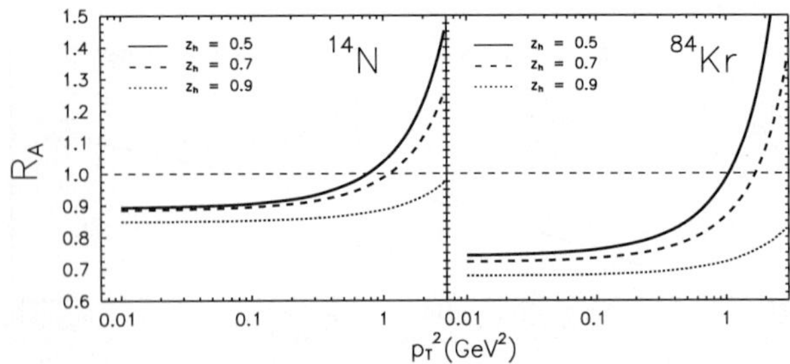

Fig. 13. p_T-dependence of nucleus-to-nucleon ratios binned in z_h for hadroproduction in DIS on nitrogen (*left*) and krypton (*right*)

color transparency. Therefore, we should expect a disappearance of the broadening effect at large $z_h \to 1$ since $t_p \to 0$. In Fig. 13 we show our predictions for the nuclear ratio as function of p_T for different z_h bins.

Note, however, that broadening should be weaker at larger z_h for any model which is able to describe the observed z_h-dependence of the nuclear ratio (see Fig. 7). Indeed, the fact that $R_A(z_h) < 1$ implies that DIS happens only in a part of the nuclear volume, namely its back side. Then, the fact that $R_A(z_h)$ decreases towards $z_h = 1$ means that a smaller part of the nuclear volume is working at larger z_h, i.e. the DIS reaction is pushed towards the back surface of the nucleus. Therefore, even in the energy loss model the path available for broadening (from the DIS point and on) becomes shorter at large z_h. This should also lead to a reduction of the broadening, but rather small, only about 10% for nitrogen and 20% for krypton. At the same time, in our approach the broadening completely vanishes at $z_h \to 1$ since $l_p \to 0$. It would be extremely important for HERMES and the experiment running at Jefferson Lab [47] to provide relevant data.

9.4 Double hadron production

One may think that a process of double hadron production is a sensitive tool which is able to disentangle the inside-outside hadronization. If absorption of the pre-hadrons is the main contributer to nuclear suppression, naively one would expect the nuclear ratio for double-hadron production to be squared, i.e. quite less, compared to inclusive single-hadron channel.

A closer look, however, shows that the situation is quite complicated and such an expectation is oversimplified. Let us assume that the two pre-hadrons are created inside the nuclear medium with production lengths l_1 and l_2 respectively, and $l_2 > l_1$. For the sake of simplicity, just for this consideration, we assume that the nuclear density is constant, both hadrons have the same and constant absorption cross sections σ, and we neglect any additional suppression related to induced energy loss. Then it is straightforward to write the nuclear suppression ratio at given impact parameter b,

$$R_A^{(2h)}(b) = \int\limits_{-\infty}^{\infty} dz\, \rho_A(b,z) \exp\left[-\sigma \int\limits_{z+l_1}^{\infty} dz'\, \rho_A(b,z')\right]$$

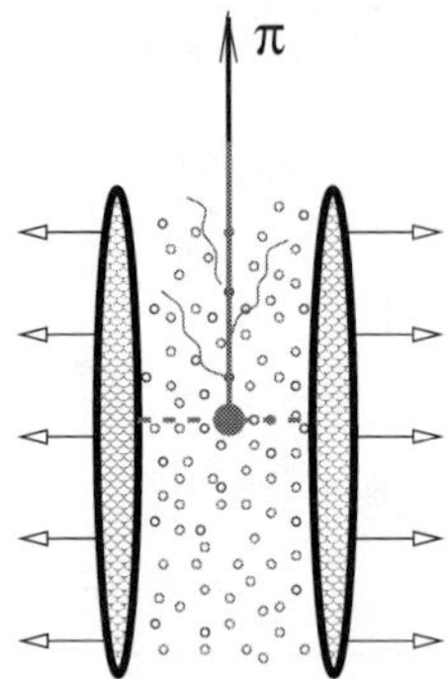

Fig. 14. Propagation and hadronization of a high-k_T parton in the dense partonic matter created right after collision of two nuclei in their c.m. frame

$$\times \exp\left[-\sigma \int\limits_{z+l_2}^{\infty} dz'\, \rho_A(b,z')\right] \tag{45}$$

Neglecting $\exp(-2\sigma\rho_A L) \ll 1$, where $L = \sqrt{R_A^2 - b^2}$, we arrive at the relative, double to single hadron production, nuclear suppressions,

$$\frac{R_A^{(2h)}(b)}{R_A^{(1h)}(b)} = \frac{1 - \frac{1}{2}\exp(-\sigma\rho_A \Delta l) + \sigma\rho_A l_1}{1 + \sigma\rho_A l_p}, \tag{46}$$

where $\Delta l = l_2 - l_1$ and l_p is the single-hadron production length. This ratio varies dependent on the values of l_1, l_2, l_p and the mean free path, $1/\sigma\rho_A$, in the medium. In the case of a very dense matter, e.g. produced in heavy ion collisions, one should expect $l\sigma\rho_A \gg 1$. Contrary to naive expectations, the production rates of single and double hadrons according to (46) should be suppressed equally. The data for high-p_T production at RHIC (see next section) seem to confirm this. This situation may change at very large p_T due to contraction of the production lengths [see (49)].

In the case of DIS in the dilute nuclear medium production lengths are comparable with the mean free path, which is about 2 fm. Then the result should vary dependent on values of z_h for each of the hadrons. For instance, if values of z_h in both cases approach their maximal values, i.e. $z_h \to 1$ and $z_{h1} + z_{h2} \to 1$, all production lengths vanish, and the ratio (46) equals $1/2$, i.e. a double-hadron is suppressed twice as much as a single-hadron.

10 Heavy ion collisions: Energy loss or absorption?

Situation in the process of high-p_T hadron production in heavy ion collisions at very high energies is similar to the process of hadron production in DIS off nuclei at medium energies. Indeed the cartoon in Fig. 14 shows the development of hadronization after the nuclear disks passed through each other. Lots of partons are created from vacuum forming a dense medium of about the transverse size of the nuclei. The high-p_T parton which was created in a hard partonic reaction at the earliest stage of nuclear collision is propagating through this medium. This process ends up with production of a hadron detected at macroscopic distances. The momenta of the hadrons in the c.m. frame of collisions ranges from a few up to 10 GeV (in the available data), what overlaps with kinematics of HERMES [2] and experiment running at Jefferson Lab [47].

Kinematics and interpretation of this process are much less certain than in DIS. In the latter case the initial quark energy is known, as well as z_h, the density and size of the medium. On the contrary, the production cross section of a high-k_T parton is a result of convolution of the initial parton distributions, hard partonic cross section, and fragmentation functions. Therefore the initial parton energy and values z_h are not known. Besides, one has to sum up over different species of partons.

These measurements are aimed at probing the density of the created medium, which is varying in space and time. This brings even more model dependence to the interpretation of data. In view of these uncertainties in interpretation of data on heavy ion collisions, one desperately needs a reliable model of in-medium hadronization tested in more certain situations like DIS.

There is a principal difference between hadronization processes in DIS and high-p_T hadron production at the mid rapidity. The dependence of the production time, l_p, on the jet energy varies substantially dependent on a process. Indeed, our previous experience says that the production time is proportional to the jet energy (the Lorentz factor). In soft interactions,

$$l_p \propto \frac{\nu}{\kappa} . \tag{47}$$

The string tension κ in the denominator provides the correct dimension for l_p.

In the case of DIS the production time is also proportional to energy, but depends on the photon virtuality,

$$l_p \propto \frac{\nu}{Q^2} . \tag{48}$$

Here the right dimension is provided by the photon virtuality, Q^2. The physics is transparent: the higher is Q^2, the stronger is the kick from the photon, the more intensive is gluon radiation and vacuum energy loss by the quark. Correspondingly, the shorter must be the color neutralization time, l_p, if one wants to produce a leading hadron with large z_h.

Now we are coming to the point. In the case of 90° parton scattering, like in the process illustrated in Fig. 14, the jet energy ($\nu \Rightarrow k_T$) and the parton virtuality ($Q^2 \Rightarrow k_T^2$) are controlled by the same parameter, which is the transverse momentum of the parton, k_T. Therefore, in this case,

$$l_p \propto \frac{1}{k_T} . \tag{49}$$

It is clear why the energy dependence of the production time switches to the inverse. The vacuum energy loss with a rate proportional to k_T^2 is so intensive, that in spite of

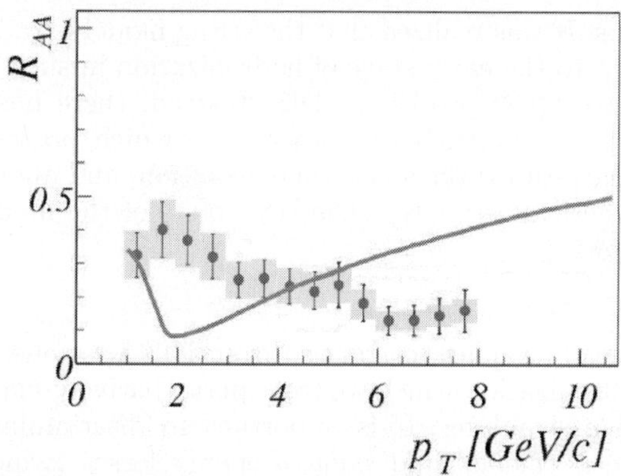

Fig. 15. Data from the PHENIX experiment [48] for pion suppression in 10% central gold-gold collisions at $\sqrt{s} = 130\,\mathrm{GeV}$ in comparison with the prediction from [11]. The figure is borrowed from [48]

the Lorentz factor k_T, the hadronization process must finish shortly after the hard partonic collision, otherwise the parton energy will degrade too much, making impossible production of an energetic hadron.

It is also clear that in pure perturbative QCD calculations the dimensional parameter, Λ_{QCD}, can emerge only under a log. Therefore, (49) presents the unique possibility to provide the correct dimension for l_p. Thus, the higher is the energy of the jet, the faster the leading pre-hadron is produced. This outrageous conclusion may contradict a simple intuition based on the experience with the string model.

The energy dependence of the production length suggested by (49) for high-p_T processes is inverse to what one has for DIS, (48). Correspondingly, energy dependence of nuclear suppression should be opposite to DIS. As long as the nuclear ratio in DIS rises with energy (Figs. 5–10), we should expect a falling p_T-dependence of the nuclear ratio for heavy ion collisions. This unusual effect was indeed observed at RHIC. Data from the PHENIX experiment depicted in Fig. 15 demonstrate quite a strong fall of the ratio at $p_T > 2\,\mathrm{GeV}$. Of course at high p_T one should expect saturation ($l_p = 0$) and following rise due to color transparency.

On the contrary, in the energy loss scenario, there is no principal difference between the two cases. As long as the nuclear ratio rises with energy in DIS, it must rise with p_T in heavy ion collisions (unless one makes different models of hadronization in the two reactions). Such a behavior was indeed predicted in [11] as is shown in Fig. 15. That was a true prediction before the data at high p_T were released. Of course, once data are known, one can find explanation introducing new exotic physical assumptions.

11 Conclusions and discussion

The model of in-medium hadronization [1] developed long time ago (last century) has successfully predicted the nuclear effects for inclusive electroproduction of leading hadrons. It was realized that the string model is rather irrelevant to the early stage of hadronization in such a perturbative QCD process as DIS. Instead, there has been developed a perturbative description which possessed a clear pattern for the space-time evolution, and allowed to do numerical estimates. The key points of the model are as follows.

- A leading quark originated from DIS loses energy for hadronization which we treat perturbatively via gluon bremsstrahlung. It is important to discriminate between *vacuum* and *induced* energy losses (which are frequently mixed up). The former is always present, even on a free nucleon target, and is the main source of energy loss. In the case of in-medium hadronization the quark radiates more gluons due to multiple soft interaction with the medium. This correction is usually much smaller than the vacuum energy loss.
- It is important to discriminate between *production* and *formation* times (which are frequently mixed up). Energy loss (both vacuum and induced) stops when a color neutralization happens, i.e. the leading quark picks up an antiquark and produces a pre-hadron, which is a colorless $\bar{q}q$ dipole with no stationary wave function. The corresponding time interval is called *production time*, or production length. Energy conservation enforces the production time to vanish at the kinematic limit of production of hadrons with maximal possible energy ($z_h \to 1$).
- The pre-hadron attenuates on its way out of the nucleus with an absorptive cross section which is controlled by the varying dipole size. Color transparency is an important ingredient of this dynamics. Eventually, on a longer time scale, called *formation time* (or length) the dipole develops the hadronic wave function. We employ a rigorous quantum-mechanical description of this process within the light-cone Green function approach. Contrary to the production length vanishing at $z_h \to 1$, the formation length reaches a maximal value in this limit.
- All these effects important for nuclear modification of the hadron production rate can be evaluated. Then the model is able to predict in a parameter free way nuclear effects in DIS as function of energy, z_h, Q^2 and p_T. Data from HERMES nicely confirm the predictions.
- Searching for observables which are able to disentangle different models, we concluded that the energy loss scenario should have problems explaining already available data for flavor and Q^2 dependence of nuclear suppression. This model is insensitive to the hadronic size, but the difference in fragmentation function at $z_h \to 1$ leads to more suppression for kaons than for pions. This is in variance with HERMES data. Also comparison of data on different reactions at different hard scales led to a conclusion that p_T broadening and energy loss rise with Q^2. Such an expectation also contradicts HERMES data. It would be very informative to have data on variation of p_T-distribution with z_h, which is a direct way to measure the production length.

- The new feature of hadronization dynamics in high-p_T processes is an inverse dependence of the production time on the jet energy compared to what is known for DIS. Correspondingly, the p_T-dependence of nuclear suppression in heavy ion collisions should be enhanced at larger p_T, contrary to DIS where nuclear effects vanish at high energy. Such a puzzling behavior was indeed disclosed by recent measurements at RHIC. Nevertheless, at higher p_T we expect the effect of color transparency to take over, then the k_T-factorization will be restored.

Acknowledgements. We are grateful for inspiring and helpful discussions to Alberto Accardi, Claudio Ciofi, Miklos Gyulassy, Mikkel Johnson, Larry McLerran, Valeria Muccifora, Pasquale Di Nezza, Hans-Jürgen Pirner, Andreas Schäfer, Ivan Schmidt. This work is supported by the grant from the Gesellschaft für Schwerionenforschung Darmstadt (GSI), grant No. GSI-OR-SCH, and by the grant INTAS-97-OPEN-31696.

References

1. B.Z. Kopeliovich, J. Nemchik, and E. Predazzi: Proceedings of the workshop on Future Physics at HERA, ed. by G. Ingelman, A. De Roeck and R. Klanner, DESY 1995/1996, v. 2, 1038 (nucl-th/9607036); Proceedings of the ELFE Summer School on Confinement physics, ed. by S.D. Bass and P.A.M. Guichon, Cambridge 1995, Editions Frontieres, p. 391 (hep-ph/9511214)
2. HERMES Collaboration, A. Airapetian et al.: Eur. Phys. J. C **20**, 479 (2001) (hep-ex/0012049); V. Miccifora et al.: Nucl. Phys. A **715**, 506 (2003) (hep-ex/0106088); A. Airapetian et al.: hep-ex/0307023
3. A. Bialas: Acta Phys. Polon. B **11**, 475 (1980)
4. B.Z. Kopeliovich and F. Niedermayer: Sov. J. Nucl. Phys. **42**, 504 (1985); Yad. Fiz. **42**, 797 (1985)
5. A. Bialas and J. Czyzewski: Phys. Lett. B **222**, 132 (1989)
6. B.Z. Kopeliovich: Phys. Lett. B **243**, 141 (1990)
7. J. Czyzewski and P. Sawicki: Z. Phys. C **56**, 493 (1992)
8. A. Accardi, V. Muccifora, and H.J. Pirner: Nucl. Phys. A **720**, 131 (2003)
9. T. Falter, W. Cassing, K. Gallmeister, and U. Mosel: nucl-th/0303011
10. N.Z. Akopov, G.M. Elbakian, and L.A. Grigoryan: hep-ph/0205123
11. E. Wang and X.-N. Wang: Phys. Rev. Lett. **89**, 162301 (2002)
12. F. Arleo: Eur. Phys. J. C **30**, 213 (2003)
13. M. Gyulassy and M. Plümer: Nucl. Phys. B **346**, 1 (1989)
14. A. Casher, H. Neubereger, and S. Nussinov: Phys. Rev. D **20**, 179 (1979)
15. F. Niedermayer: Phys. Rev. D **34**, 3494 (1986)
16. S.J. Brodsky and P. Hoyer: Phys. Lett. **298B**, 165 (1993)
17. R. Baier et al.: Nucl. Phys. B **484**, 265 (1997)
18. J. Dolejsi, J. Hüfner, and B.Z. Kopeliovich: Phys. Lett. B **312**, 235 (1993)
19. M.B. Johnson, B.Z. Kopeliovich, and A.V. Tarasov: Phys. Rev. C **63**, 035203 (1991)
20. M.B. Johnson et al.: Phys. Rev. Lett. **86**, 4483 (2001); Phys. Rev. C **65**, 025203 (2002)

21. HERMES Collaboration, A. Airapetian et al.: Phys. Rev. Lett. **90**, 052501 (2003)
22. B.Z. Kopeliovich, J. Nemchik, A. Schäfer, and A.V. Tarasov: Phys. Rev. C **65**, 035201 (2002)
23. B.Z. Kopeliovich and L.I. Lapidus: in Proc. of the VI Balaton Conference on Nuclear Physics, Balatonfuered, Hungary, 1983, p. 73
24. A. Bialas and M. Gyulassy: Nucl. Phys. B **291**, 793 (1987)
25. V.T. Kim and B.Z. Kopeliovich: JINR preprint, **E2-89-727**, Dubna, 1989
26. B.Z. Kopeliovich, L.B. Litov, and J. Nemchik: J. Mod. Phys. E **4**, 767 (1993)
27. B.Z. Kopeliovich, L.I. Lapidus, and A.B. Zamolodchikov: JETP Lett. **33**, 595 (1981)
28. J. Bertch, S.J. Brodsky, A.S. Goldhaber, and J.G. Gunion: Phys. Rev. Lett. **47**, 297 (1981)
29. A. Capella et al.: Phys. Rep. **236**, 225 (1994); A.B. Kaidalov: JETP Lett. **32**, 474 (1980); Sov. J. Nucl. Phys. **33**, 733 (1981); Phys. Lett. **116B**, 459 (1982)
30. B.Z. Kopeliovich, I.K. Potashnikova, B. Povh, and E. Predazzi: Phys. Rev. Lett. **85**, 507 (2000); Phys. Rev. D **63**, 054001 (2001)
31. J.F. Gunion and G. Bertsch: Phys. Rev. D **25**, 746 (1982)
32. A.H. Mueller: Nucl. Phys. B **415**, 373 (1994)
33. B. Andersson et al.: Z. Phys. C **49**, 79 (1991)
34. S.R. Amendolia et al.: Phys. Lett. B **178**, 435 (1986)
35. J. Binnewies, B.A. Kniehl, and G. Kramer: Z. Phys. C **65**, 471 (1995)
36. B.A. Kniehl, G. Kramer, and B. Potter: Nucl. Phys. B **597**, 337 (2001)
37. B.Z. Kopeliovich and B.G. Zakharov: Phys. Rev. D **44**, 3466 (1991)
38. B.Z. Kopeliovich, A. Schäfer, and A.V. Tarasov: Phys. Rev. C **59**, 1609 (1999) (extended version in hep-ph/9808378)
39. B.Z. Kopeliovich, A. Schäfer, and A.V. Tarasov: Phys. Rev. D **62**, 054022 (2000)
40. R.P. Feynman and A.R. Gibbs: Quantum Mechanics and Path Integrals, McGRAW–HILL Book Company, New York 1965
41. B.Z. Kopeliovich, J. Raufeisen and A.V. Tarasov: Phys. Lett. B **440**, 151 (1998); Phys. Rev. C **62**, 035204 (2000)
42. EMC Collaboration, J. Ashman et al.: Z. Phys. C **52**, 1 (1991)
43. L.S. Osborne et al.: Phys. Rev. Lett. **40**, 1624 (1978)
44. N. Pavel: Ph. D. Thesis, **WUB 89-24** (1989) Wuppertal
45. HERMES Collaboration: preliminary, in http://www-hermes.desy.de/notes/pub/trans-public-subject.html#HADRON-ATTENUATION
46. J.F. Martin et al.: Phys. Rev. D **20**, 5 (1979); J.T. Dakin et al.: Phys. Rev. D **10**, 1401 (1974)
47. W. Brooks: 'Jefferson Lab experiment E02-104', talk at the 2d International Conference on Nuclear and Particle Physics by CEBAF at JLAB, May 26-31, 2003, Dubrovnic, Croatia
48. S. Mioduszewski: High-p_T measurements from Phenix, talk at Quark Matter 2002, Long Island, New York, USA (http://www.phenix.bnl.gov/conf_2002.html)
49. B.Z. Kopeliovich and J. Nemchik: JINR preprint **E2-91-150** (1991) Dubna
50. B.Z. Kopeliovich and J. Nemchik: Preprint SANITA **INFN-ISS 91/3** Rome, 1991

Eur Phys J A (2004) **19**, s01, 129–132
Digital Object Identifier (DOI) 10.1140/epjad/s2004-03-020-2

EPJ A direct
electronic only

J/Ψ and heavy-quark production in E866/FNAL and PHENIX

M.J. Leitch[a]

P-25, MS H846, Los Alamos National Laboratory, Los Alamos NM 87545

Received: 18 Aug 2003 / Accepted: 14 Nov 2003 /
Published Online: 6 Feb 2004 – © Società Italiana di Fisica / Springer-Verlag 2004

Abstract. The production of heavy quarks in nuclei is modified from that for a free nucleon by a number of nuclear effects including shadowing of the nuclear gluon distributions, energy loss of the incident gluon, and, in the case of the J/Ψ, disassociation of the $c\bar{c}$ pair (absorption) as it exits the nucleus. Measurements in the E866/NuSea 800 GeV fixed target experiment show a large suppression of the closed-charm yield with strong kinematical dependencies and with a slightly stronger suppression for the Ψ' than for the J/Ψ near $x_F = 0$. On the other hand, a measurement of the D meson nuclear dependence near $x_F = 0$ shows no suppression. At RHIC the J/Ψ is thought to be a key signature for the creation of a quark-gluon plasma (QGP) in heavy-ion collisions, but the non-QGP suppression already seen in p-A collisions at lower energies shows that we must first understand these non-QGP effects on the J/Psi in order to gain a clear understanding of its production in nucleus-nucleus collisions. The most recent run at RHIC included deuteron-gold collisions and will serve as a baseline for these cold nuclear matter effects at RHIC energy. Here I report on the first results for the J/Ψ and for open-charm at PHENIX including the d-Au results. The present knowledge of gluon shadowing is very uncertain, especially in the large rapidity region of the PHENIX muon arms, and these measurements should help us determine how strong it is and perhaps even its dependence on impact parameter.

PACS. 24.85.+p Quarks, gluons, and QCD in nuclei and nuclear processes – 25.75.-9 Relativistic heavy-ion collisions – 25.75.Dw Particle and resonance production – 14.65.Dw Charm quarks – 14.40.Lb Charm mesons

1 Introduction

Here I will review results and prospects for J/Ψ and heavy-quark production from the E866/NuSea experiment at Fermilab and from the PHENIX experiment at RHIC. The topics I will cover include 1) gluon shadowing and energy loss in nuclei, 2) J/Ψ production mechanisms and absorption in nuclei, 3) the Cronin effect, and 4) early results from PHENIX for the just completed d-Au run at RHIC.

2 Shadowing

Shadowing is the depletion of small momentum-fraction (x) gluons in nuclei relative to the nucleon. One way to understand the origin of shadowing is that very low-momentum-fraction partons, by the uncertainty princible, have large size, overlap more with their neighbors and fuse together thus enhancing their population at higher momenta at the expense of lower momenta. Another picture involves coherence between different scattering centers where for large coherence length, larger than the typ-

ical intra-nucleon distance, can result in destructive interference and an effective "shadowing" of the partons deeper inside a nucleus. Some theorists have told me that these two pictures are equivalent, but this is beyond the scope of this talk.

An example of shadowing is seen in the nuclear suppression of the Drell-Yan process in our E866/E772 data is shown in Fig. 1, where one sees a clear depletion of the anti-quarks at target-parton momentum fractions(x_2) below about 0.06.

In Fig. 2 is shown the ratio between lead and hydrogen gluon structure functions from the parameterization of Eskola et al [3]. The production of heavy-quarks goes mainly through gluon-fusion, so the shadowing shown here is relevant for charm production. In this phenomenological approach the main information comes from deep-inelastic muon scattering and the gluon information comes mainly from the Q^2 evolution of these structure functions. In addition to the shadowing at small x values one also sees an enhancement called anti-shadowing which results from the conservation of momentum and balances the depletion at small x. The ranges in x sampled for the measurements I will discuss are shown. The fixed target measurements from Fermilab and CERN both lie in the anti-shadowing

[a] for the E866/NuSea and PHENIX Collaborations
`leitch@lanl.gov`

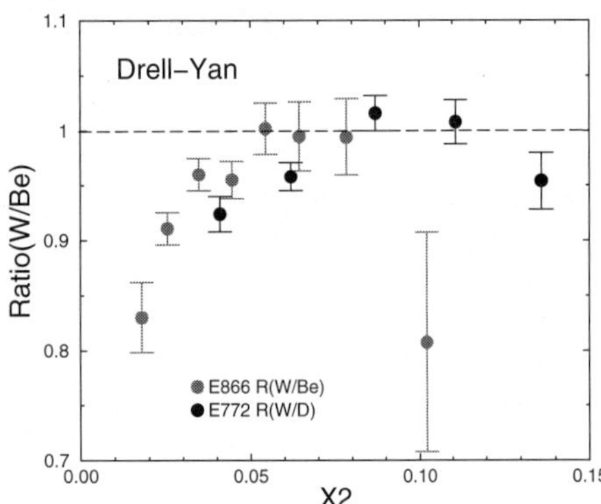

Fig. 1. Ratio of per nucleon cross sections between W and Be versus x_2 for Drell-Yan dimuon production by 800 GeV/c protons from E866 [1] and E772 [2]

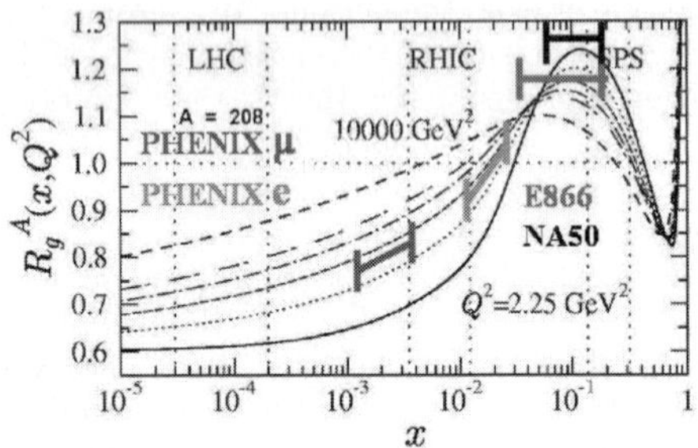

Fig. 2. Gluon shadowing from Eskola et al [3]. Marked in colored bars are shown the approximate x ranges covered by PHENIX $\mu^+\mu^-$ (*Blue*), PHENIX e^+e^- (*Green*), E866 (*Red*) and NA50 (*Black*)

region, while the PHENIX measurements lie in the shadowing region.

The main results from E866/NuSea are shown in Fig. 3 where the nuclear suppression is represented by α, where $\sigma_A = \sigma_p A^\alpha$. The horizontal axis here is the fraction of the maximum allowed momentum that the J/Ψ or $c\bar{c}$ have, or roughly speaking how fast they are going. One sees a substantial suppression which grows much stronger at larger x_F. The J/Ψ and Ψ' look similar at large x_F where they both correspond to a $c\bar{c}$ traversing the nucleus (since they would not hadronize at these x_F values until they are far outside of the nucleus). But at small x_F these resonances may be beginning to hadronize in the nuclear medium and the Ψ', which is larger and more loosely bound, is absorbed a little more strongly than the J/Ψ. Also shown is a single point for the neutral D-meson from E789 [6]. The lack of suppression for the D supports the idea that the suppression of the closed-charm resonances is due to absorption (or dissociation of the $c\bar{c}$) which cannot occur for the D. The stronger suppression of the resonances at large x_F is thought to be caused by a combination of shadowing (since $x_F = x_1 - x_2$ and large x_F corresponds to small x_2) and energy-loss of the incident gluon before the hard interaction. However, if one compares these results versus x_2 to those from NA3 [7] at lower (200 GeV/c) energy the bulk of the observed suppression does not scale with x_2 as

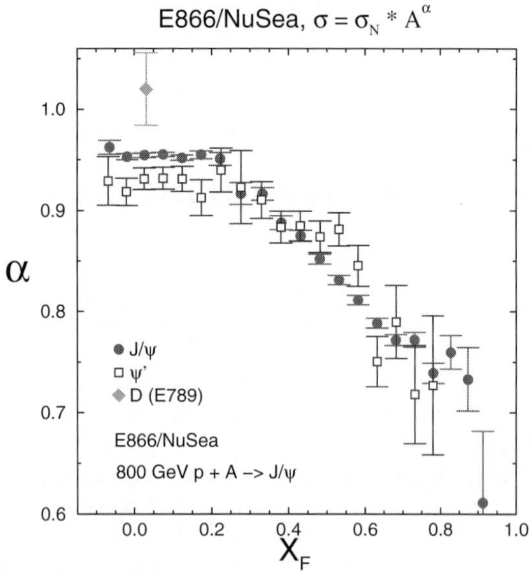

Fig. 3. Nuclear dependence of J/Ψ and Ψ production and of open charm versus x_F for 800 GeV/c protons from E866 [5] and E789 [6]

would be expected if it were all shadowing. This remains a puzzle.

The production of J/Ψ and Ψ' is dominated by gluon-gluon fusion and for the small x values sampled at forward rapidity at RHIC the shadowing of these gluons is quite important, but largely unknown. Different models for gluon shadowing give answers that vary by a factor of two or more. For example the shadowing from Eskola [3], as shown in Fig. 2, for the forward di-muon acceptance in PHENIX gives a reduction of the gluons of about 20%. On the other hand, the model of Kopeliovich [8] gives over a 50% reduction. The effect of these two shadowed gluon structure functions is shown in Fig. 4 for dimuon measurements in PHENIX and gives a huge difference in the cross section at forward rapidity. Clearly we must measure the shadowing at RHIC to resolve this uncertainty.

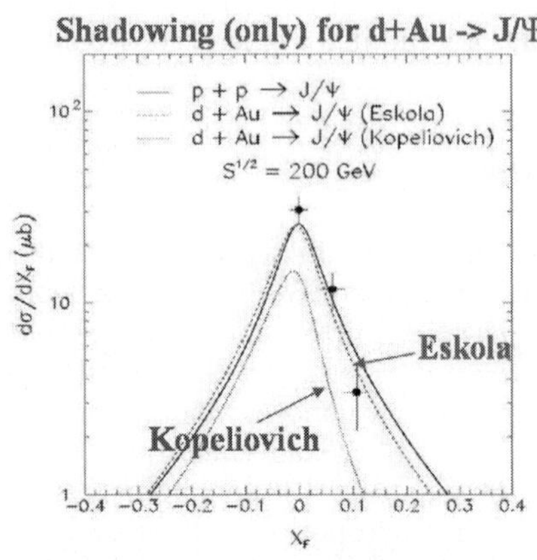

Fig. 4. Rapidity distribution for $J/\Psi \rightarrow \mu^+\mu^-$ in PHENIX for two different gluon shadowing models from Eskola et al [3] and Kopeliovich [8]

Since J/Ψ suppression is one of the leading signatures for QGP in heavy-ion Collisions, it is clearly critical to understand the suppression for normal nuclear matter. This is often expressed in terms of the atomic mass dependence of the nuclear cross section in terms of α, as in $\sigma_A = \sigma_p * A^\alpha$. For early measurements in fixed target experiments α values of 0.92 (E772 [9]) and 0.919 ± 0.015(NA50 [11]) were found. More recently E866 determined 0.954 ± 0.003 at mid-rapidity [5], where this result is larger than that found in the earlier E772 experiment using the same spectrometer because of biases caused by a narrow p_T acceptance in the earlier measurements which is much smaller and corrected for in E866. In addition E866 found a small but significantly larger absorption for Ψ which was discussed earlier. Finally, the most recent result from the NA50 CERN experiment [12], with more p-A data included in the analysis, indicates a larger value of α for the J/Ψ of 0.934 ± 0.014 which is consistent with the E866 result. This latest NA50 result corresponds to a J/Ψ nuclear cross section in a simple absorption model of 4.4 mb compared to the earlier result of 6.2 mb. Since these results may be sampling the gluon distributions in the anti-shadowing region (see Fig. 2), they should be interpreted with care when considering RHIC energies where measurements are not in the anti-shadowing region.

2.1 Parton energy loss and the cronin effect

Another important nuclear effect is energy loss of the gluons in nuclear matter before the hard interaction. A calculation from Kopeliovich [13] illustrates how this causes a general increase in nuclear suppression for larger values of x_F as shown in Fig. 5. However, this effect is thought to be negligible at RHIC energies, since the energy loss is thought to be fixed and is therefore small compared to the typical energies of partons at higher energies.

The Cronin effect, or p_T broadening, is a general phenomena [14] attributed to multiple scattering of the incident parton that is seen for various produced particles and is also seen for J/Ψ production. The systematics for measurements at several energies compared to the broadening

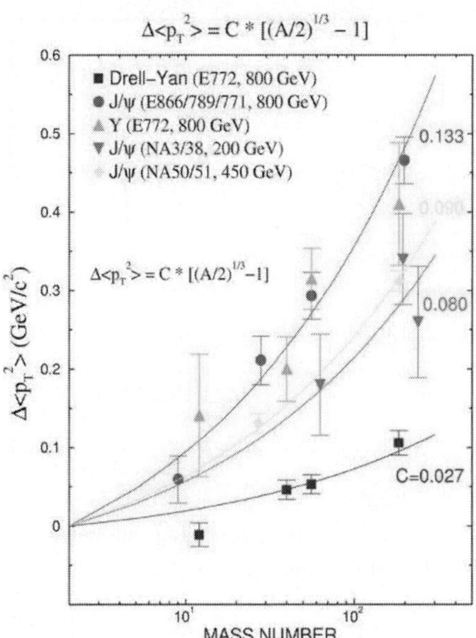

Fig. 6. Broadening of p_T in nuclear production for heavy-vector mesons and for the Drell-Yan process

for the Drell-Yan process is shown in Fig. 6 in terms of $\Delta < p_T^2 >$, where $\Delta < p_T^2 >$ is the difference in the average p_T^2 between a heavy nucleus and deuterium. Broadening for the heavy vector mesons is substantially larger than that for the Drell-Yan process and grows with increasing $\sqrt{s}$.

It is also interesting to note, as seen in some preliminary results from E866, shown in Figs. 7 and 8, that even in deuterium there appear to be some small but non-zero nuclear effects similar to those seen in heavier nuclei. Using a parameterization that works well for the heavy nuclei, but with an effective A that is smaller than 2 appears to explain the trends seen in this data as shown in the figure.

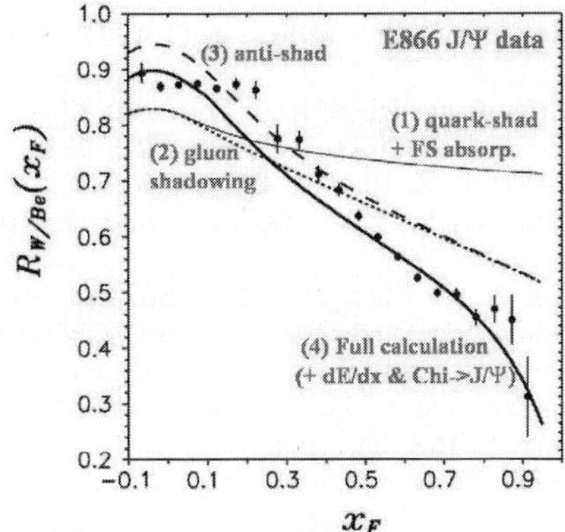

Fig. 5. The effect of energy loss (*difference between solid and long-dashed curves*) on J/Ψ suppression at fixed target energies from the model of Kopeliovich [13]

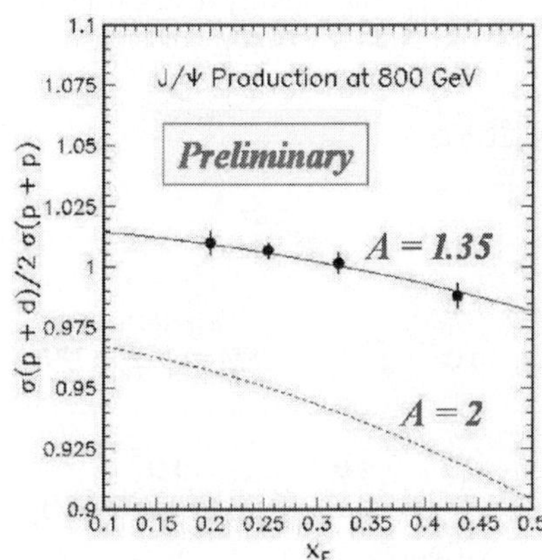

Fig. 7. Preliminary results from E866 showing small nuclear effects for deuterium versus x_F

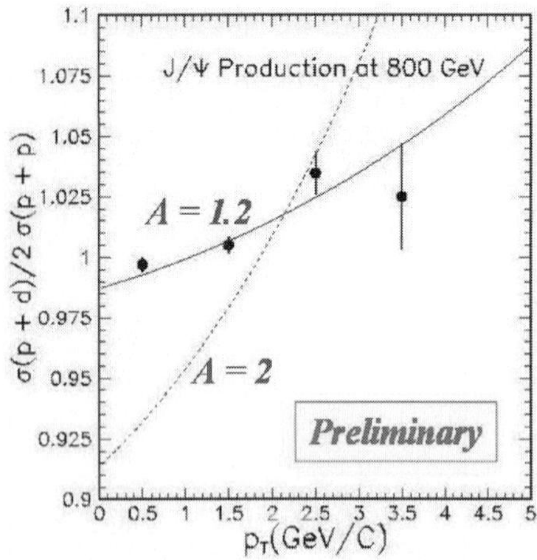

Fig. 8. Preliminary results from E866 showing small nuclear effects for deuterium versus p_T

2.2 Early PHENIX J/Ψ measurements

At RHIC the program for charm physics is just beginning, and the first measurements of the J/Ψ in p-p collisions have just come out from the 2002 run [15], as shown in Figs. 9 and 10. These results are limited by the low luminosity and J/Ψ statistics (65 counts for the $\mu^+\mu^-$ channel) achieved in this RHIC run.

In 2003 new measurements of the J/Ψ for d-Au and p-p collisions at PHENIX are just being made. Mass peaks from an early partial analysis of the data are shown in Fig. 11. A substantial increase in the number of J/Ψ's

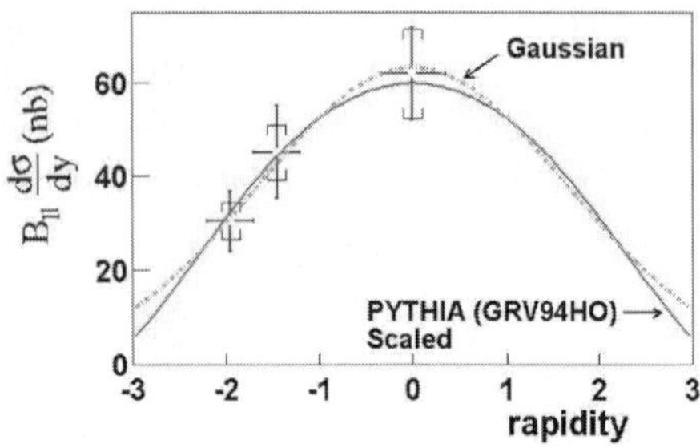

Fig. 9. First results for the rapidity distribution from PHENIX at RHIC for J/Ψ in 200 GeV/c p-p collisions

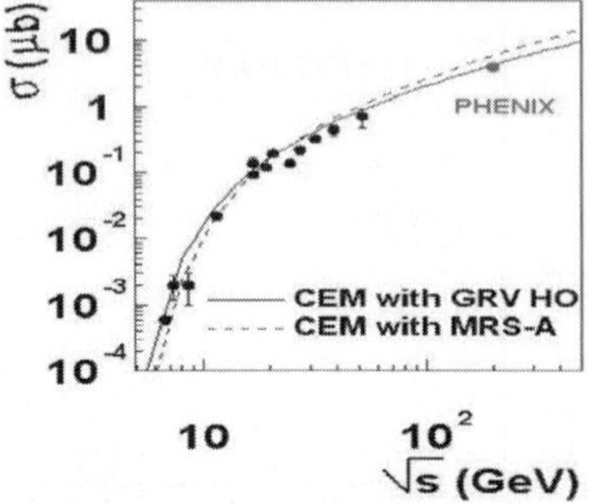

Fig. 10. The PHENIX total cross section for J/Ψ at 200 GeV/c p-p collisions compared to measurements at lower energies

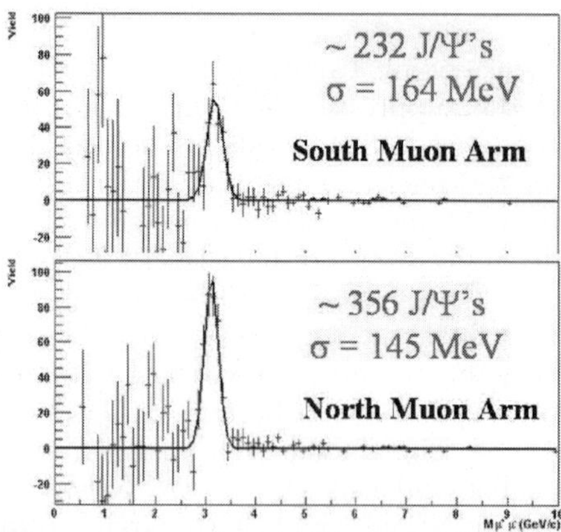

Fig. 11. Early results for the J/Ψ in 200 GeV/c d-Au for the PHENIX south and north muon arms showing a good mass resolution and a substantial number of counts from a portion of the 2003 runs data

obtained in this years run compared to last year is seen and excellent mass resolutions of the two muon arms in PHENIX, approaching the design values, is seen. We estimate that the entire data sample, when analyzed, will have around 1000 J/Ψ's in each muon arm (negative and positive rapidity) and several hundred from the central rapidity measurement using e^+e^- pairs. Yields perhaps a factor of two below this are expected from this year's p-p run. From these data we should be able to provide quantitative constraints on the nuclear shadowing of gluons, its centrality dependence, and on the p_T broadening. This will also serve as a firm basis for the long Au-Au run that we expect to have next year. In that run we expect to finally be able to address the question of J/Ψ suppression in the possible presence of a QGP.

References

1. M.A. Vasiliev et al. (E866/NuSea Collaboration): Phys. Rev. Lett. **83**, 2304 (1999)
2. D.M. Alde et al.: Phys. Rev. Lett **64**, 2479 (1990)
3. K.J. Eskola, V.J. Kolhinen, and R. Vogt: hep-ph/0104124
4. K.J. Eskola, V.J. Kolhinen, and P.V. Ruuskanen: Nucl. Phys. B **535**, 351 (1998) and K.J. Eskola, V.J. Kolhinen, and C.A. Salgado: Eur. Phys. J. C **9**, 61 (1999) and hep-ph/9807297
5. M.J. Leitch et al.: Phys. Rev. Lett. **84**, 3256 (2000)
6. M.J. Leitch et al.: Phys. Rev. Lett. **72**, 2542 (1994)
7. J. Badier et al.: Z. Phys. C **20**, 101 (1982)
8. B.Z. Kopeliovich, A.V. Tarasov, and J. Hufner: hep-ph/0104256
9. D.M. Alde et al.: Phys. Rev. Lett **66**, 133 (1991)
10. D.M. Alde et al.: Phys. Rev. Lett **66**, 2285 (1991)
11. M.C. Abreu et al.: Phys. Lett. B **444**, 516 (1998)
12. E. Scomparin, Quark Matter 2001 and M.C. Abreu et al.: Phys. Lett. B **438**, 35 (1998)
13. Kopeliovich, Tarasov, and Hufner: Nucl. Phys. A **696**, 669 (2001) and hep-ph/0104256
14. D. Antreasyn et al.: Phys. Rev. D **19**, 764 (1979)
15. S. Adler et al. (PHENIX collaboration): submitted to Phys. Rev. Lett. and hep-ex/0307019

Eur Phys J A (2004) **19**, s01, 133–137

Digital Object Identifier (DOI) 10.1140/epjad/s2004-03-021-1

EPJ A direct

electronic only

Hadronization and final state interaction effects in semi-exclusive deep inelastic scattering off nuclei

C. Ciofi degli Atti[1], L.P. Kaptari[1,2], and B.Z. Kopeliovich[3]

[1] Department of Physics, University of Perugia and INFN, Sezione di Perugia, via A. Pascoli, Perugia, I-06100, Italy
[2] Bogoliubov Laboratory of Theoretical Physics JINR, Dubna, Russia
[3] Max Planck Institut für Kernphysik Postfach 103980, 69029 Heidelberg, Germany

Received: 30 Oct 2003 / Accepted: 14 Nov 2003 /
Published Online: 6 Feb 2004 – © Società Italiana di Fisica / Springer-Verlag 2004

Abstract. Recent calculations of the effects of hadronization and final state interaction (FSI) in semi-exclusive deep-inelastic scattering (DIS) $A(e, e'(A-1))X$ processes are reviewed. The basic ingredient underlying these calculations, *viz* the time-dependent effective debris-nucleon cross section is illustrated, and some relevant results on complex nuclei and the deuteron are presented. In the latter case, particular attention is paid to the choice of the kinematics, for such a choice would in principle allow one to investigate both the structure function of a bound nucleon as well as the hadronization mechanisms. It is stressed that a planned experiment at Jlab on the process $D(e, e'p)X$ could be very useful in that respect.

PACS. 24.85.+p Quarks, gluons, and QCD in nuclei and nuclear processes – 13.60.-r Photon and charged-lepton interaction with hadrons

1 Introduction

To date, information on hadronization mechanisms comes mainly from the measurement of the multiplicity ratio of the lepto-produced hadrons in semi inclusive $A(e, e'h)X$ processes [1], whose interpretation on the basis of the quark re-interaction model of [2] appears to be very convincing. However, it should also be pointed out that the more exclusive processes of $A(e, e'(A-1))X$ (the *semi-exclusive* process) proposed in [3], though difficult to perform, could provide more direct information on quark re-interaction and hadronization mechanisms, as shown recently in [4]. The semi-exclusive reaction $A(e, e'(A-1))X$ represents the process in which an incoming electron undergoes a DIS process off a low-momentum bound nucleon, with the scattered electron and the recoiling $(A-1)$ nucleus detected in coincidence in the final state. Much theoretical attention has been devoted to such process on a deuteron target, i.e. when $A = D$ and $(A-1) = p$ or n (see e.g. [5]-[7]). These calculations (except [6], to be discussed later on) are based upon the *plane wave impulse approximation (PWIA)*, according to which: i) X results from DIS off one of the two nucleons in the deuteron, ii) the second nucleon N recoils without interacting with X and is detected in coincidence with the scattered electron. The process $A(e, e'(A-1))X$ on a complex nucleus was thoroughly investigated in [3] within the PWIA, i.e. by assuming that the nucleon debris created by the virtual photon propagates without re-interacting with the spec-

tator nucleus, which, therefore, always remains intact, an assumption which, at first sight, might appear unjustified.

As a matter of fact, as pointed out in [4], DIS off a bound nucleon results in the production of a multi-particle final state with an effective mass squared equal to

$$s' \simeq m_N^2 - Q^2 + 2 m_N \nu - 2 \sqrt{\nu^2 - Q^2}\, p_L$$
$$= Q^2\left(\frac{1}{x} - 1\right) + m_N^2 - 2|\mathbf{q}|p_L \tag{1}$$

where $Q^2 = \mathbf{q}^2 - \nu^2$ is the four-momentum transfer, ν the virtual photon energy in the rest frame of the nucleus, p_L the longitudinal Fermi momentum of the nucleon relative to the direction of the virtual photon ($\mathbf{q} \parallel z$), m_N the nucleon mass, and $x = \dfrac{Q^2}{2m_N\nu}$ the Bjorken scaling variable (we neglect here the binding energy of the nucleon). At high energies and far from the quasi-elastic region ($x \approx 1$), the effective mass is large, $\sqrt{s'} \gg m_N$, and one expects the production of many particles, which can interact traveling through the nucleus. This would substantially suppress the probability for the spectator nucleus to remain intact. However, the process of multi-particle production has a specific space-time development, and it turns out that not so many particles have a chance to be created inside the nucleus. In order to check such an expectation, in [4] the propagation of the struck nucleon debris and its re-interaction with the nuclear medium, occurring in the semi-exclusive process on complex nuclei $A(e, e'(A-1))X$, has been considered with the aim of also clarifying if and to which extent the process is sensitive to the details of quark hadronization in nuclear environment. To this end an effective time-dependent cross section σ_{eff} (to be called the *debris-nucleon* cross section) has been obtained, which describes the interaction with the nuclear medium of the hadronization products of a highly virtual quark created in a DIS process off a bound nucleon. The main features of the *debris-nucleon* cross section are discussed in the next section.

2 The debris-nucleon cross section

In the derivation of the effective debris-nucleon cross section of [4] use has been made of the color string model [8] and the gluon radiation mechanism [9,2], namely an approach has been adopted which takes into account both the production of hadrons due to the breaking of the color string which is formed after a quark is knocked out off a bound nucleon, as well as the production of hadrons originating from gluon radiation.

According to QCD, DIS is the process in which an incident electron interacts with a target quark by exchanging a gauge boson, making the quark highly virtual. The formation of the final, detectable hadrons, occurs after the space-time propagation of the created nucleon debris, with a sequence of soft and hard production processes. The theoretical description of these processes, which generally cannot be treated within perturbative QCD, requires the use of model approaches. Most of them are based upon the quark color string model [8], according to which at world interval of the order of $\simeq 1 fm$, the string, which is formed by the highly virtual leading quark and the remnant target quarks, breaks into a hadron and another, less stretchy string. Further, at longer space-time intervals, this decay process iterates unless the energy of the string is too low for hadron production and all the final hadrons are formed. However, since the hadronization process can also be accompanied by gluon perturbative bremsstrahlung [9], the string model itself is not sufficient for a consistent treatment of hadronization. A reliable model must incorporate both the perturbative and the non-perturbative aspects of the hadron formation process. Note, that the hadronization process starts at extremely short space-time intervals, hence a direct experimental study of these intervals is difficult to undertake in DIS off a free nucleon. As a matter of fact, the final hadrons do not carry much information about their early stage of hadronization, and therefore only nuclear targets, which consist of a number of scattering centers, allow one to probe short times after the hadronization has started. The time-dependent cross section σ_{eff} describing the scattering of the nucleon debris with the surrounding medium turns out to read as follows [4]

$$\sigma_{eff}(t) = \sigma_{tot}^{NN} + \sigma_{tot}^{\pi N}\left[n_M(t) + n_G(t)\right], \qquad (2)$$

where σ_{tot}^{NN} ($\sigma_{tot}^{\pi N}$) are the total cross sections describing nucleon-nucleon (meson-nucleon) interactions, and $n_M(t)$

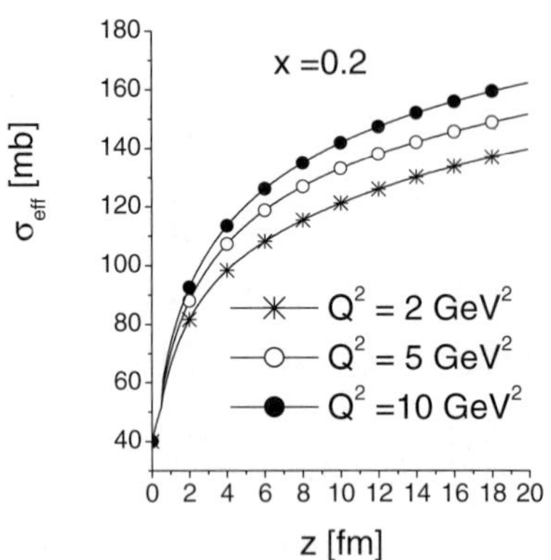

Fig. 1. The debris-nucleon effective cross section (2) plotted vs the distance z for a fixed value of the Bjorken scaling variable x and various values of the four-momentum transfer Q^2 (after [4])

and $n_G(t)$ are the effective numbers of created mesons and radiated gluons, respectively, and are explicitly given in [4]. It should be pointed out that in obtaining (2) the color-dipole picture was employed by replacing each radiated gluon by a color-octet $q\bar{q}$ pair.

The cross section (2) exhibits a rather complex Q^2- and x-dependence, which, however, asymptotically tends to a simple logarithmical behavior. This is illustrated in Fig. 1, where the dependence of σ_{eff} upon the coordinate z along the propagation direction is shown for different values of the four momentum transfer Q^2 and a fixed value of the Bjorken scaling variable x.

It should be stressed that in [4] it has been shown that the effective cross Sect. (2) can reasonably explain the Fermilab data on the production of protons in semi inclusive Deep Inelastic muon-Xenon scattering at 490 GeV [11]. In this experiment, at a given value of x, *grey tracks* were observed, which have been interpreted as protons in the momentum range $200 < p < 600 \ MeV/c$. Using (2) the average number $< n_g >$ of *grey tracks vs* Q^2 has been calculated, obtaining a satisfactory explanation of the experimental data of [11]; this makes us confident in the correctness of the calculation of FSI effects in semi-exclusive processes off nuclear targets, to be discussed in the next section.

3 The process $A(e, e'(A-1))X$ in complex nuclei

In a nucleus, at each hadronization point one expects re-interactions of the produced hadrons with the nuclear constituents, so that the multiplicity of final particles is predicted to be reduced relative to the case of nucleon targets. Thus, by comparing the same DIS process off a single nucleon and off nuclear targets, information on the space-time structure of the hadronization process could be obtained. As already pointed out, the theoretical model of hadronization developed in [2], proved to be very effective for the explanation of the leading hadron multiplicity ratios (nucleus to nucleon) measured at HERMES [1] in semi-inclusive processes. It should however be pointed out

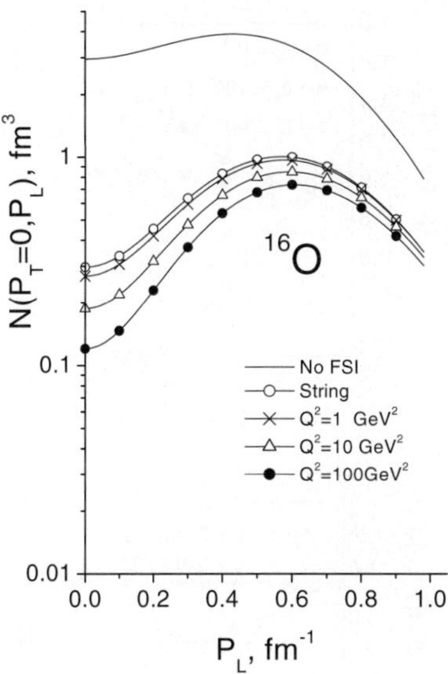

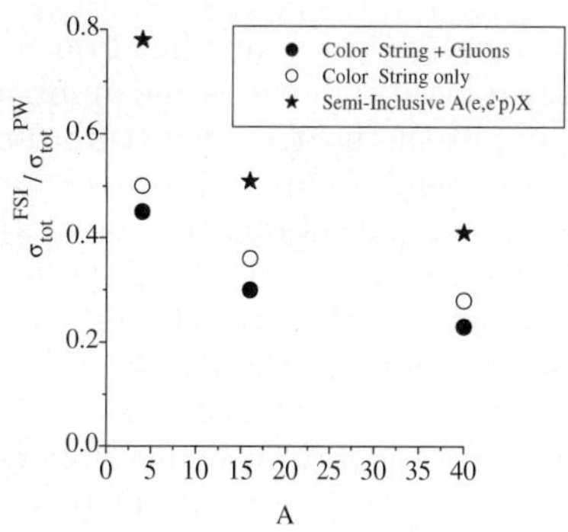

Fig. 2. The proton Momentum Distribution for ^{16}O ((6) – No FSI) compared with the Distorted Momentum Distribution $N(\mathbf{P}_{A-1})$ (3) plotted vs P_L for $P_T = 0$. The curve labeled by open dots has been obtained using the effective cross section for the nucleon debris corresponding to the color string model, whereas the other curves correspond to the cross section which includes also the gluon bremsstrahlung. The stars represent the distorted *proton* momentum distributions calculated in [13] for the semi-inclusive quasi-elastic process $^{16}O(e, e'p)X$ (After [4])

Fig. 3. The nuclear transparency i.e. the ratio between the total cross section σ_{tot}^{FSI}, obtained by integrating (3) over $\mathbf{P}_{A-1}$) and the PWIA total cross section σ_{tot}^{PW}, obtained in the same way but disregarding the Final State Interaction ($S(\mathbf{b}, z) = 0$, $\sigma_{tot} \equiv \sigma_{tot}^{PW}$). The *open dots* correspond to the debris-nucleon effective cross section given by the color string model, whereas the *full dots* correspond to the cross section where the gluon bremsstrahlung has also been considered [(2)] at $Q^2 = 100 GeV^2$. The stars represent the *proton* transparency calculated in [12,13] for the reaction $A(e, e'p)X$ (After [4])

that the initial stage of hadronization is difficult to investigate by semi-inclusive processes, where the non leading hadrons are strongly affected by subsequent cascade processes and therefore do not carry information on their formation mechanism.

In [4] it has been shown that the deep inelastic semi-exclusive process $A(e, e', (A-1))X$, where the nucleus $(A-1)$ is detected in coincidence with the scattered electron, could be an effective tool to study the mechanisms and the initial stage of hadronization. The approach of [4] is based on a traditional Glauber-type nuclear structure approach in which the ground state wave function of the initial nucleus $\Psi_A^0(\mathbf{r}, \mathbf{r}_2 \ldots \mathbf{r}_A)$ is written as a product of a function ϕ, describing the motion of the hit nucleon and the wave function $\Psi_{A-1}^f(\mathbf{r}_2 \ldots \mathbf{r}_A)$ of the spectator; moreover $|\Psi_{A-1}^f(\mathbf{r}_2 \ldots \mathbf{r}_A)|^2$ is, as usually, approximated by a product of single particle densities. Within these assumptions, the cross section can be shown to be governed by the *distorted momentum distributions*

$$n_A^{FSI}(\mathbf{P}_{A-1}) \equiv N(\mathbf{P}_{A-1}) = \left| F_{A,A-1}^{FSI}(\mathbf{P}_{A-1}) \right|^2 \quad (3)$$

where

$$F_{A,A-1}^{FSI} \simeq \int e^{i\mathbf{P}_{A-1}\mathbf{r}} \phi(\mathbf{r}) \left[1 - \frac{S(\mathbf{b}, z)}{2(A-1)} \right]^{A-1} d\mathbf{r} \quad (4)$$

and

$$S(\mathbf{b}, z) = \int_z^{\infty} dz' \, \rho_A(\mathbf{b}, z') \, \sigma_{eff}(z' - z) \quad (5)$$

In these equations $\rho_A(\mathbf{b}, z)$ is the nuclear density ($\int d\mathbf{r}\rho_A(\mathbf{r}) = A$), and $\sigma_{eff}(z' - z)$ is given by (2). Note that when FSI are absent, i.e. when $\sigma_{eff} = 0$, the usual nucleon momentum distribution in the nucleus is recovered

$$n_A(|\mathbf{P}_{A-1}|) \simeq \left| \int e^{i\mathbf{P}_{A-1}\mathbf{r}} \phi(\mathbf{r}) \, d\mathbf{r} \right|^2 \quad (6)$$

The distorted momentum distributions for ^{16}O are shown in Fig. 2, whereas the nuclear transparency for the nucleon debris created in a DIS process and for a nucleon knocked out in a quasi-elastic scattering semi-inclusive $A(e, e'p)X$ process, is exhibited in Fig. 3.

4 The process $D(e, e'p)X$

To perform an experiments on semi-exclusive processes $A(e, e'(A-1))X$ off complex nuclei is not easy task; however, in the case of a deuteron target the experiment appears to be feasible [14]. As a matter of fact the process

$$e + D = e' + X + N \quad (7)$$

has been the object of many theoretical calculations, mainly aimed at studying the neutron structure function [5]-[7], and its experimental investigation is planned to be performed at JLab [14].

Process (7) has many attractive features with respect to the inclusive process $^2H(e, e')X$, both for extracting information on the nucleon structure functions and for the investigation of hadronization mechanism. Indeed, in spite of the fact that inclusive DIS processes have provided us in the past with fairly precise knowledge of parton distributions in hadrons, conclusive information about the origin of the EMC effect is still lacking; moreover, important details on the neutron structure function are unknown, which is mostly due the difficulties and ambiguities related to the unfolding of the neutron structure functions from nuclear data [15]. Semi-exclusive processes could provide, on the contrary, unique information on both the origin of the EMC effect, and the details of the neutron structure function; moreover, they can also be used as a unique tool to investigate hadronization processes. Obviously, a reliable treatment of semi-exclusive processes requires a careful treatment of the FSI of the nucleon debris X with the final nuclear system $(A-1)$. Intuitively, one expects, on one hand, that if the proton is detected in the backward hemisphere, FSI effects should not play a relevant role, so that the process could be used to investigate the structure functions of bound nucleons; on the other hand, the effects from FSI are expected to be relevant in the process when the recoiling nucleon is detected in the direction perpendicular to the three-momentum transfer, in which case information on the hadronization mechanism could be obtained. In view of the planned experiments at JLab [14], a detailed quantitative calculation of FSI effects in process (7) has been carried out in [16]. There it has been found that the central quantity describing the process is the distorted momentum distribution

$$n_D^{FSI}(\mathbf{p}_s, \mathbf{q}) = \tag{8}$$

$$\frac{1}{3}\frac{1}{(2\pi)^3}\sum_{\mathcal{M}_D}\left|\int d\mathbf{r}\Psi_{1,\mathcal{M}_D}(\mathbf{r})S(\mathbf{r},\mathbf{q})\chi_f^+\exp(-i\mathbf{p}_s\mathbf{r})\right|^2$$

where χ_f is the spin function of the spectator nucleon and $S(\mathbf{r},\mathbf{q})$ the S-matrix describing the final state interaction between the debris and the spectator, $viz.$

$$S(\mathbf{r},\mathbf{q}) = \tag{9}$$

$$1 - \theta(z)\frac{\sigma_{eff}(z,Q^2,x)(1-i\alpha)}{4\pi b_0^2}\exp(-b^2/2b_0^2).$$

When FSI are absent ($\sigma_{eff} = 0$) the usual deuteron momentum distributions are recovered, $viz.$

$$n_D(|\mathbf{p}_s|,) = \tag{10}$$

$$\frac{1}{3}\frac{1}{(2\pi)^3}\sum_{\mathcal{M}_D}\left|\int d\mathbf{r}\Psi_{1,\mathcal{M}_D}(\mathbf{r})\chi_f^+\exp(-i\mathbf{p}_s\mathbf{r})\right|^2$$

In what follows, the momentum distributions are expressed in terms of the light cone variable α_s, $p_\parallel$ and p_T, defined as

$$\alpha_s = \frac{E_s - p_\parallel}{m}, \quad p_\parallel = |\mathbf{p}_s|\cos\theta_s, \quad p_T = |\mathbf{p}_s|\sin\theta_s \tag{11}$$

where θ_s is the angle between $\mathbf{p}_s$ and $\mathbf{q}$, and the spectator four-momentum is $p_s \equiv (E_s, \mathbf{p}_s)$ (note, that in the DIS kinematics the light cone z-axis is directed opposite to the vector $\mathbf{q}$). In [16] the problem was addressed of finding proper kinematics which would allow the investigation of both the nucleon structure function and the hadronization mechanism; in the first case FSI effects should be minimized, whereas in the second case they have to be

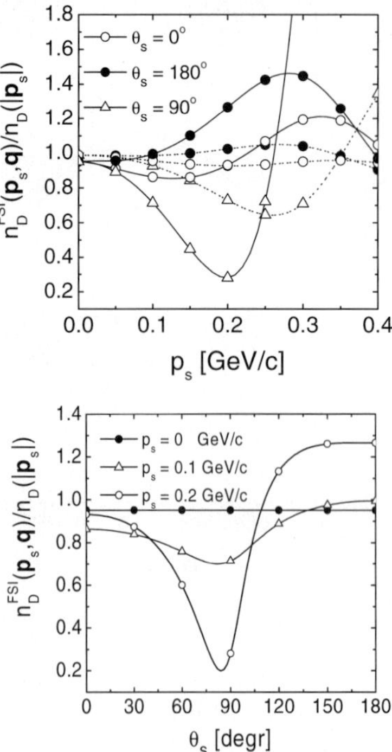

Fig. 4. *Upper panel*: the Deep Inelastic Scattering ratio n_D^{FSI}/n_D for the process $D(e,e'p)X$ with n_D^{FSI} and n_D (given, respectively, by (8) and (10)), calculated vs the momentum $p_s \equiv |\mathbf{p}_s|$ of the spectator nucleon emitted at different angles θ_s. The *full lines* correspond to the Q^2- and z-dependent debris-nucleon effective cross section σ_{eff} shown in Fig. 1, whereas the *dashed lines* correspond to a constant cross section $\sigma_{eff} = 20$ mb. *Lower panel*: the same as in the upper panel, vs the emission angle θ_s of the spectator, for different values of the spectator momentum. Calculations have been performed at $Q^2 = 5$ $(GeV/c)^2$ and $x = 0.2$ (After [16])

maximized. Since all of the FSI effects are contained in the distorted momentum distribution $n_D^{FSI}(\mathbf{p}_s, \mathbf{q})$, its deviation from the nucleon momentum distributions $n_D(|\mathbf{p}_s|)$ could provides clear signature of FSI effects. This is illustrated in Fig. 4, which shows the ratio n_D^{FSI}/n_D, calculated using two different models for the effective cross section σ_{eff}, representing its upper and lower limits, viz the time- and Q^2- dependent cross section of [4] (solid lines), and a constant cross section $\sigma_{eff} = 20$ mb (dashed lines) considered in [6]. In our numerical calculations the parameters entering (10), viz the slope b_0 and the ratio α of the real to the imaginary parts of the forward amplitude, have been taken from NN scattering data at high energies. It should be pointed out, in this respect, that our results, in the range of considered momenta, are not very sensitive to the value of α; whereas the values of the latter is known in the case of *nucleon-nucleon* scattering, the value for *debris-nucleon* scattering should rely on some theoretical models. This point is under investigations and the results will be presented elsewhere. It can be seen from Fig. 4 that the predictions given by the two different models of the effective cross section ere rather different, particularly when the recoiling proton is emitted perpendicularly to $\mathbf{q}$,

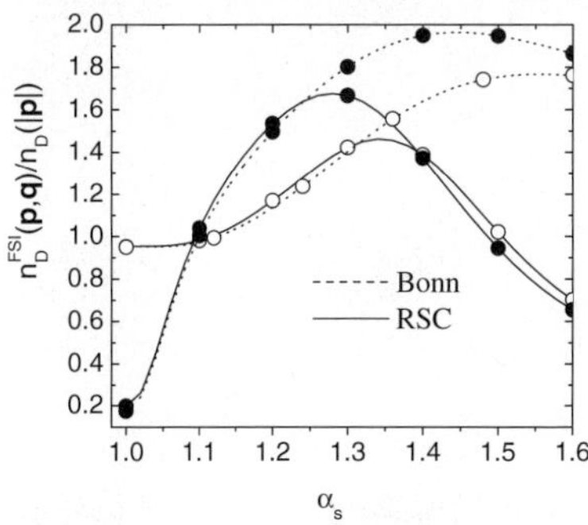

Fig. 5. The Deep Inelastic Scattering ratio n_D^{FSI}/n_D for the process $D(e, e'p)X$ *vs* the light cone variable α_s, calculated at $Q^2 = 5\ GeV^2/c^2$ and $x = 0.2$ using two different deuteron wave functions: the *solid lines* correspond to the RSC potential and the *dashed lines* to the Bonn potential. Calculations have performed in correspondence of two values of the transverse momentum p_T of the recoiling nucleon (cf. 11), namely $p_T = 0\ GeV/c$ (*open dots*) and $p_T = 0.2\ GeV/c$ (*full dots*). The results correspond to the debris-nucleon cross section shown in Fig. 1 (After [16])

i.e. at $\theta_s \sim 90^o$, and with large values of the momentum ($p_s \sim 0.2\ GeV/c$). Thus, by investigating this kinematical region one could, in principle, obtain unique information about the magnitude of σ_{eff} and, consequently, about the hadronization mechanism.

The results exhibited in Figs. 4 also demonstrate that FSI effects are essentially reduced in parallel kinematics ($\theta = 0^o, 180^o$) and also at small values of p_s. It can be shown that in this region the cross section provides direct information on the structure function F_2 of a bound nucleon, so that a reliable investigation of off-mass shell effects [18,19] in DIS could be possible. The results exhibited in Figs. 4 have been obtained by using the deuteron wave function corresponding to the Reid Soft Core (RSC) potential [20]. Calculations have been repeated with the Bonn interaction [21]. As shown in Fig. 5, at small values of α_s different potential models yield basically the same results; however at moderate and large values of α_s the predictions by different deuteron wave functions appreciably differ.

To sum up, from the analysis exhibited in [16], it can be concluded that FSI effects in the semi-exclusive process (7) are negligible in the backward kinematics and slow momenta of the detected nucleon, which would allow one to investigate the nucleon structure function of bound nucleons, in particular the neutron one; if, on the contrary, the spectator nucleon is emitted perpendicularly to the momentum transfer, FSI effects are enhanced and different models for the hadronization process could be investigated.

5 Summary and conclusions

We have reviewed recent calculations of FSI effects which occur in the semi-exclusive deep inelastic $A(e, e'(A-1))X$ process. The FSI is generated by the rescattering with the medium of the nucleon debris formed by the deep inelas-

tic scattering of the virtual photon. These calculations are based upon the effective time-dependent debris-nucleon cross section obtained in [4] on the basis of the color string [8] and the gluon radiation [9] models. The detailed calculation [16] of the reaction $D(e, e'p)X$, shows that the planned experiment at Jlab [14], aimed at investigating such a reaction, may provide unique information on the hadronization mechanisms.

Acknowledgements. This work was partially supported by the Ministero della Istruzione, Università e Ricerca (MIUR), through the funds COFIN01. L.P.K. is indebted to the University of Perugia and INFN, Sezione di Perugia, for warm hospitality and financial support.

References

1. The HERMES Collab: "Hadron formation in deepinrelastic scattering", Eur. Phys. J. C **20**, 479 (2001)
2. B. Kopeliovich, J. Nemchik, and E. Predazzi: Workshop on Future Physics at HERA, Hamburg, Germany, 30-31 May 1996. In *Future physics at HERA* 1038-1042, Amsterdam NIKHEF - NIKHEF-96-026 (96/09,rec.Oct.), 14-18; nucl-th/9607036
   B. Kopeliovich, J. Nemchik, and E. Predazzi, in Proceedings of the *ELFE Summer School on Confinement Physics*, edited by S.D. Bass and P.A.M. Guichon: Editions Frontieres, 1995, p. 391, Gif-sur-Yvette (hep-ph/9511214)
3. C. Ciofi degli Atti, L.P. Kaptari, and S. Scopetta: Eur. Phys. J. A **5**, 191 (1999)
4. C. Ciofi degli Atti and B.Z. Kopeliovich: Eur. Phys. J. A **17**, 133 (2003)
5. S. Simula: Phys. Lett. B **387**, 245 (1996)
6. W. Melnitchouk, M. Sargsian, and M.I. Strikman: Z. Phys. A **359**, 99 (1997)
7. A.E.L. Dieperink and S.I. Nagorny: Nucl. Phys. A **629**, 290c (1998)
8. F.E. Low: Phys. Rev. D **12**, 163 (1975); A. Casher, H. Neuberger, and S. Nussinov: Phys. Rev. D **20**, 179 (1979)
9. B. Kopeliovich and F. Niedermayer: Phys. Lett. B **117**, 101 (1982)
10. C. Ciofi degli Atti and B.Z. Kopeliovich: to appear
11. M.R. Adams et al. The Fermilab E665 Collaboration: Z. Physik C **65**, 225 (1995)
12. H. Morita, C. Ciofi degli Atti, and D. Treleani: Phys. Rev. C **60**, 034603 (1999)
13. M. Alvioli, C. Ciofi degli Atti, and H. Morita: nucl-th/0309086
14. C. Keppel, S. Kuhn, and W. Melnitchouk: Spokespersons, *The Structure of Free Neutron Via Spectator Tagging*, TJNAF BoNuS Collab E03-012
15. A.Y. Umnikov, F.C. Khanna, and L.P. Kaptari: Z. Phys. A **348**, 211-215 (1994)
16. C. Ciofi degli Atti, L.P. Kaptari, and B. Kopeliovich: Eur. J. Phys. A, in press; nucl-th/0307052
17. C. Ciofi degli Atti, L.P. Kaptari, and D. Treleani: Phys. Rev C **63**, 044601 (2001)
18. A.Y. Umnikov, F.C. Khanna, and L.P. Kaptari: Phys. Rev. C **53**, 377 (1996)
19. W. Melnitchouk, A.W. Schreiber, and A.W. Thomas: Phys. Rev. D **49**, 1199 (1994)
20. R.V. Reid Jr.: Ann. Phys. **50**, 411 (1968)
21. R. Machleid, K. Holinde, and C. Elster: Phys. Rep. **149**, 1 (1987)

Eur Phys J A (2004) **19**, s01, 139–142
Digital Object Identifier (DOI) 10.1140/epjad/s2004-03-022-0

EPJ A direct
electronic only

Charmonium production in high-energy nuclear collisions

Alberto Polleri

[1] Physik Department, Technische Universität München, D-85747 Garching, Germany
[2] ECT*, Villa Tambosi, I-38050 Villazzano (Trento), Italy

Received: 20 Oct 2003 / Accepted: 14 Nov 2003 /
Published Online: 6 Feb 2004 – © Società Italiana di Fisica / Springer-Verlag 2004

Abstract. We discuss some issues concerning the evolution of charmonia as they interact with the constituents of the fireball produced in high-energy nucleus-nucleus collisions. We study in detail the evolution in different regimes, controlled by collision energy, kinematics and geometry.

PACS. 25.75.-q, 25.75.Nq, 24.85.+p, 14.40.Gx

The initial proposal of studying charmonium production in heavy-ion collisions [1] generated a lively dabate on whether or not the experimentally observed anomalous suppression can be attributed to QGP formation. Various effects can potentially account for the observed data because they provide the necessary non-linear dependence on the number of participating particles which becomes significant only when going beyond pA collisions.

In view of the uncertainties inherent in the construction of models which consider such effects, we attempt an approach to the problem which maximizes the number of independent physical constraints, making use of the knowledge of the medium evolution as inferred by a variety of other observables. We then explore whether the same description is consistent with the observed J/ψ measurements. We stress that this is possible because the specific time evolution of the medium is not constructed or fitted in order to reproduce the J/ψ suppression effect, but it is constrained beforehand.

The medium itself is constituted, in the QGP phase, by quark and gluon quasiparticles, with an equation of state in accordance with lattice calculations [2]. It evolves in time following a fireball construction as explained in [3]. Charmonium production is treated as a two-step process, factorizing direct production from the subsequent evolution in the medium. The first part includes the conventional description [4] of nuclear effects within the Glauber model framework, with values of the absorption cross section either extrapolated from the suppression observed experimentally already in pA collisions, or fitted to reproduce the results of more sophisticated computations [5, 6]. The second part is a description within kinetic theory of the interaction of J/ψs with the different degrees of freedom that populate the evolving medium at different times.

In the following we will assume that J/ψs are fully formed hadrons by the time the thermalized medium is produced, and they will subsequently interact with its degrees of freedom. Since the medium itself consists of a QGP for a significantly long time, we begin discussing how J/ψ interacts with quark and gluon quasi-particles.

It is clear that collisions of Ψ with either quarks or gluons will lead to dissociation of the bound state. On the other hand, a quark can interact only via gluon exchange. Within the quasi-particle model, the latter process is effectively already included in the definition of the temperature dependent gluon mass. In other words, Ψs only see gluonic quasi-particles in the plasma. We concentrate now on the process $\Psi(p) + g(k) \rightarrow c(q_1) + \bar{c}(q_2)$, and label momenta as indicated in parentheses. We then come to the problem of computing cross sections involving a relativistic bound state. In the present case one can argue that the $c\bar{c}$ system is, to first approximation, non-relativistic, greatly simplifying the treatment. Moreover, as was done originally by Bhanot and Peskin [7] one can argue that the lowest-lying quarkonium levels can be approximately described using the Coulomb part of the potential. Then, with operator product expansion methods or more recent non-relativistic factorization techniques [8], it is possible to obtain an analytic expression for the cross section for the cases of $J/\psi, \psi'$ and χ_c. The calculated cross sections are not necessarily reliable since the Coulomb potential approximation is used without rigorous justification. This problem is serious especially for the ψ' and χ_c states, which have much too large dissociation cross sections. Within the adopted framework we choose to re-scale the ψ' and χ_c cross sections by a multiplicative factor κ such that the obtained J/ψ suppression pattern at SPS energies is in agreement with experiment. The required value is $\kappa = 0.2$

We now consider the possibility of $c\bar{c}$ coalescence in the QGP, a manner for production of Ψ which has been recently considered by several authors [9,10,11,12]. We do this by means of a cross section, applying detailed balance to the reaction $\Psi g \leftrightarrow c\bar{c}$, and use the cross section calculated above for Ψ dissociation by gluons.

Coming now to the case of hadronic dissociation, we recall that many approaches have been developed in the literature to compute mesonic dissociation cross sections of Ψ. In our approach the medium evolution is constrained by the freeze-out analysis, leaving no room for adjustments. Since the particle densities in the hadronic phase are almost two orders of magnitude lower than in the QGP phase, unless one employs exceptionally large cross sections, it seems unlikely that hadronic dissociation can be at all relevant. Given the inherent uncertainties we will not consider the possibility of hadronic dissociation of Ψ.

The natural framework in which to study the time evolution of Ψ is that of kinetic theory. We use a semi-classical treatment, setting up a relativistic kinetic equation for the Ψ phase-space distribution. The general problem of kinetic evolution of bound states in a strongly interacting medium was studied recently [13]. Here we make use of those results, adapting the treatment to our needs. For details see [14]. We then assume that the kinetic equation describing the time evolution of the phase-space density f_Ψ of Ψ is

$$p^\mu \partial_\mu f_\Psi(p,x) = C_F^\Psi(p,x) - C_D^\Psi(p,x)\, f_\Psi(p,x)\,, \quad (1)$$

consisting of the drift term $p^\mu \partial_\mu f_\Psi$ on the l.h.s. and of a collision term on the r.h.s. The collision term is made of a dissociation (loss) part

$$C_D^\Psi(p,x) = \frac{1}{2} \int d\Phi_3(k,q_1,q_2)\,(2\pi)^4\delta^4(p+k-q_1-q_2)$$
$$\times\, \overline{W}_{\Psi g \to c\bar{c}}\, f_g(k,x) \quad (2)$$

describing Ψ dissociation by quasi-particle gluons, studied in [15,16] to address the problem of charmonium suppression, and a formation (gain) part

$$C_F^\Psi(p,x) = \frac{1}{2} \int d\Phi_3(k,q_1,q_2)\,(2\pi)^4\delta^4(p+k-q_1-q_2)$$
$$\times\, \overline{W}_{c\bar{c} \to \Psi g}\, f_c(q_1,x)\, f_{\bar{c}}(q_2,x) \quad (3)$$

describing Ψ formation by $c\bar{c}$ fusion, introduced in a simplified manner in [10]. In the equations above f_c, $f_{\bar{c}}$ and f_g are the phase-space distributions of the degrees of freedom participating in the collisions. The Lorentz invariant 3-body the phase-space integration measure is $d\Phi_3$, and the transition probabilities $\overline{W}$ are averaged over the initial color and spin polarizations and summed over the final ones. They are related to the dissociation and formation cross sections discussed before, which can be expressed as

$$\sigma_D^\Psi(s) = \frac{1}{4F_{\Psi g}} \int d\Phi_2(q_1,q_2)(2\pi)^4\delta^4(p+k-q_1-q_2)\, \overline{W}_{\Psi g \to c\bar{c}}$$
$$(4)$$

for dissociation and

$$\sigma_F^\Psi(s) = \frac{1}{4F_{c\bar{c}}} \int d\Phi_2(p,k)\,(2\pi)^4\delta^4(p+k-q_1-q_2)\, \overline{W}_{c\bar{c} \to \Psi g}$$
$$(5)$$

for formation. Here $d\Phi_2$ is the 2-body phase-space integration measure

Concerning the definitions of the phase-space distribution of gluons, we adopt the expression

$$f_g(k,T) = \nu_g\, C(T) \left\{ \exp\left[E_g(k,T)/T\right] \pm 1 \right\}^{-1}, \quad (6)$$

where ν_g is the number of gluon degrees of freedom, $C(T)$ is the confinement factor [2] and E_g is the gluon quasiparticle energy. The time evolution of the gluon distribution is all contained in the proper time dependence of the temperature.

For c quarks, as they are produced in the hard initial collision, we take their spectrum as computed with pQCD. We assume that they do not interact significantly with the medium, approximately moving on straight lines according to the free streaming equation. Then, we define the phase-space density of charm quarks as

$$f_c(q,x) = \frac{(2\pi)^3}{\tau\, m_\perp^c}\, \frac{dN_{AB}^c}{dy_c d^2\mathbf{q}_\perp}\, \delta(y_c - \eta)\, \rho_\perp(r_\perp(\tau))\, \theta(T(\tau) - T_c)$$
$$(7)$$

and analogously for $\bar{c}$ quarks. The step function θ is introduced to account for the fact that below T_c c and $\bar{c}$ quarks have hadronized into D mesons and are no more available to coalesce into Ψ. For simplicity in numerical computations, the transverse position density is taken as a box of the same radius of the fireball. The δ-function, arising from the assumption that c quarks are produced in a very narrow longitudinal region, strongly correlates the momentum rapidity y_c and the space-time rapidity η.

With several simplifications [14] it is possible to reduce (1) to a simple first order differential equation for the rapidity distribution of Ψ as function of proper time as

$$\frac{\partial}{\partial\tau}\frac{dN_\Psi}{dy}(\tau) = \lambda_F^\Psi(y,\tau) - \lambda_D^\Psi(y,\tau)\frac{dN_\Psi}{dy}(\tau)\,, \quad (8)$$

whose solution, valid at $y = 0$, is obtained with few elementary steps and provides the final Ψ rapidity distribution at $y = 0$ as

$$\frac{dN_\Psi^f}{dy} = \left\{ \frac{dN_\Psi^0}{dy} \exp\left[-\int_{\tau_0}^{\tau_f} d\tau'\, \lambda_D^\Psi(y,\tau')\right] \right.$$
$$\left. + \int_{\tau_0}^{\tau_f} d\tau'\, \lambda_F^\Psi(y,\tau') \exp\left[-\int_{\tau'}^{\tau_f} d\tau''\, \lambda_D^\Psi(y,\tau'')\right] \right\}. \quad (9)$$

Again, the solution found holds for the different charmonia J/ψ, ψ' and χ_c. The rates λ_D^Ψ and λ_F^Ψ are given respectively by

$$\lambda_D^\Psi(\tau) = \int \frac{d^3k}{(2\pi)^3}\, V_{\Psi g}\, \sigma_D^\Psi(s)\, f_g(k,T(\tau)) \quad (10)$$

and

$$\lambda_F^\Psi(\tau) = \frac{1}{\tau} \int d^2p_\perp d^2q_\perp^{\,1} d^2q_\perp^{\,2}\, K\, \sigma_F^\Psi(s)\, S_c(\bar{y}_c, q_\perp^1) S_c(\bar{y}_c, q_\perp^2)\,, \quad (11)$$

with K representing a collection of factors [14]. The structure of this solution is self-evident. The first term describes the dissociation of Ψs initially produced in the hard collision, with the usual exponential suppression acting at all times from τ_0 to τ_f, while the second term describes formation of Ψs from $c\bar{c}$ in the QGP, from the initial time τ_0 up to an intermediate value τ' and their subsequent suppression from τ' to τ_f, integrated over all values of τ'. This last term may become important as soon as the number of charmed quarks is large enough.

Using the elements of the calculation as discussed previously, we can now compute the time dependence of dissociation and formation rates. The dissociation rate depends on the fireball temperature which, in turn, depends on time. How T depends in detail on τ, was evaluated for different impact parameters and collision energies [3, 14] Dissociation and formation rates are then combined together according to (9) to give the observed Ψ rapidity distribution. Since we are interested in the final J/ψ yield, we need to take into account decays into the charmonium ground state of ψ' and χ_c. Experimentally is has been observed that measured J/ψs come in a fraction of 60 % from direct production, while a fraction of 10 % comes from ψ' decays and a fraction of 30 % is from χ_c.

We now look at the case of $Pb+Pb$ collision at $\sqrt{s} = 17.4$ GeV and compute the J/ψ spectrum as function of the impact parameter b. Then we construct the $J/\psi/DY$ ratio

$$R_{J/\psi/DY}(b) = N_0 \, \frac{dN_{J/\psi}(b)/dy}{dN_{\Psi}^{pp}/dy \, N_{coll}(b)}, \qquad (12)$$

where we assume that the Drell-Yan spectrum in $Pb+Pb$ collisions scales with the number of collisions N_{coll}. The overall normalization is fixed at $N_0 = 53.5$. The ratio is then plotted in Fig. 1 (solid line) as function of the mean transverse energy

$$E_T(b) = \epsilon_T \, N_p(b) \,. \qquad (13)$$

The quantity $\epsilon_T = 0.274$ is the amount of produced transverse energy per participant. As reference, we also plot the curve obtained by considering only nuclear effects (dotted line) and neglecting the contribution of the produced medium. The agreement with the NA50 data [17] is quite remarkable, in particular the slope of the curve, considering that fireball parameters have not at all been tuned to this particular observable.

We now proceed to examine central collisions at RHIC energy. In order to compare with data from the PHENIX experiment [18], obtained at mid-rapidity from Ψ decays into e^+e^- pairs, we construct the quantity

$$N_{J/\psi}^*(b) = B_{e^+e^-} \frac{dN_{J/\psi}(b)/dy}{N_{coll}(b)} \qquad (14)$$

and plot it in Fig. 2 as function of the number of participants. The quantity $B_{e^+e^-}$ is the branching of J/ψ into e^+e^- pairs, while the overall normalization is fixed at $N_{J/\psi}^*(N_p = 0) = 0.15$. Although a comparison with the data is, at present, premature, we see that our results lie within experimental errors. Again, the contribution from

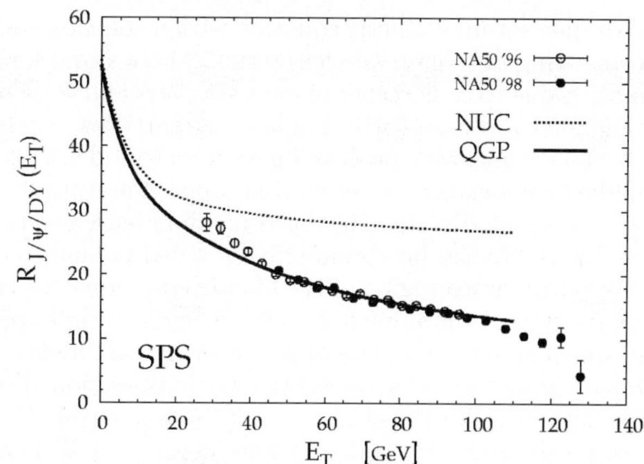

Fig. 1. Result at SPS energy for the $J/\psi/DY$ ratio as function of the transverse energy. The *dotted curve*, labelled "*NUC*", includes only nuclear effects, while the *solid line*, labelled "*QGP*", is the complete result including dissociation by collisions with gluonic quasi-particles

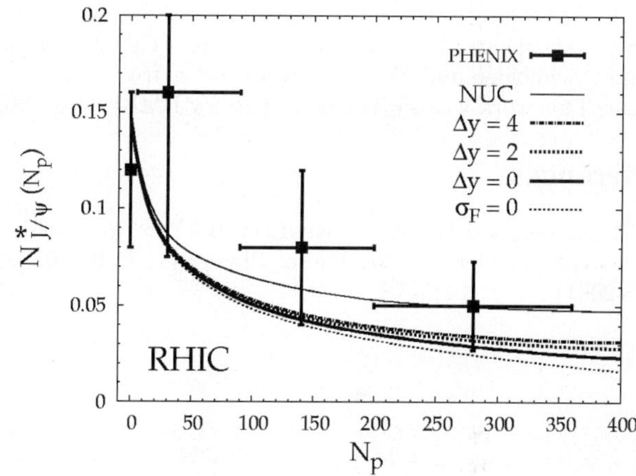

Fig. 2. Result at RHIC energy for the J/ψ yield, scaled by the number of binary collisions, as function of the number of participants

$c\bar{c}$ coalescence into J/ψ is significant, although not dramatic. All the obtained curves are monotonically decreasing and we do not find any inversion of this tendency at any large value of N_p. In other words, we do not find a net enhancement of J/ψ, but this result needs to be confirmed by more accurate calculations.

Summarizing, we have calculated J/ψ production from an expanding fireball created in relativistic heavy-ion collisions over a wide range of centralities and beam energies, from SPS at $\sqrt{s} = 17.3$ GeV ($E_{lab} = 158$ GeV) to RHIC at $\sqrt{s} = 200$ GeV. The produced medium was described assuming QGP formation and thermalization. A phenomenological quasi-particle model for quarks and gluons, in accordance with lattice QCD thermodynamics, was applied to model the partonic phase. This provided a realistic EoS which we then used to drive the expansion dynamics of the medium by means of a fireball model, characterized by time dependent temperature and volume.

We then set up a kinetic equation with a collision term incorporating both gain and loss terms. These stand for Ψ formation due to coalescence of $c\bar{c}$ quark pairs and Ψ dissociation due to collisions with gluon quasi-particles, respectively. The elementary process $\Psi g \to c\bar{c}$ was modeled by a simplified dissociation cross section, approximating Ψ as a Coulomb bound state. The corresponding back reaction $c\bar{c} \to \Psi g$ could then be obtained by detailed balance.

With proper averaging of the kinetic equation over the spatial extent of the fireball, we then found a simple solution, which allowed direct comparison with experiment. At SPS energy we were able to describe the suppression effect in the data, without the need to invoke hadronic comovers. These results support the hypothesis that the QGP is actually produced, at a transient stage, in $Pb+Pb$ collisions at $\sqrt{s} = 17.3$ GeV. We also considered extrapolations up to RHIC energies where, despite the more extreme conditions as compared to SPS, a sizable fraction of primordial J/ψs still survives. Although a clear trend towards more copious J/ψ production in by $c\bar{c}$ coalescence was found, no net J/ψ enhancement was present in the end.

Acknowledgements. I would like to thank Thorsten Renk, Roland Schneider and Wolfram Weise for a fruitful collaboration. This work was supported in part by BMBF and GSI.

References

1. T. Matsui and H. Satz: Phys. Lett. B **178**, 416 (1986)
2. R.A. Schneider and W. Weise: Phys. Rev. C **64**, 055201 (2001)
3. T. Renk, R. Schneider, and W. Weise: Phys. Rev. C **66**, 014902 (2002)
4. D. Kharzeev, C. Lourenco, M. Nardi, and H. Satz: Z. Phys. C **74**, 307 (1997)
5. B. Kopeliovich, A. Tarasov, and J. Hüfner: Nucl. Phys. A **696**, 669 (2001)
6. A. Capella: Coherence Effects in Charmonium Production off Nuclei: Consequences for J/ψ Suppression, nucl-th/0207049
7. M.E. Peskin: Nucl. Phys. B **156**, 365 (1979); G. Bhanot and M.E. Peskin: Nucl. Phys. B **156**, 391 (1979)
8. Y. Oh, S. Kim, and S.H. Lee: Phys. Rev. C **65**, 067901 (2002)
9. P. Braun-Munzinger and J. Stachel: Phys. Lett. B **490**, 196 (2000)
10. R.L. Thews, M. Schroedter, and J. Rafelski: Phys. Rev. C **63**, 054905 (2001)
11. L. Grandchamp and R. Rapp: Phys. Lett. B **523**, 60 (2001); Nucl. Phys. A **709**, 415 (2002)
12. M.I. Gorenstein et al.: Phys. Lett. B **524**, 265 (2002)
13. A. Polleri: *Bound State Kinetics in High-Energy Nuclear Collisions*, nucl-th/0303065
14. A. Polleri, T. Renk, R. Schneider, and W. Weise: *Kinetic description of charmonium production in high-energy nuclear collisions*, nucl-th/0306025
15. J.P. Blaizot and J.Y. Ollitrault: Phys. Rev. D **39**, 232 (1989)
16. X.M. Xu, D. Kharzeev, H. Satz, and X.N. Wang: Phys. Rev. C **53**, 3051 (1996)
17. M.C. Abreu et al. [NA50]: Phys. Lett. B **477**, 28 (2000)
18. S.S. Adler et al. [PHENIX]: nucl-ex/0305030

Eur Phys J A (2004) **19**, s01, 143–146
Digital Object Identifier (DOI) 10.1140/epjad/s2004-03-023-y

EPJ A direct

electronic only

Rescattering and finite formation time effects in electro-disintegration of the deuteron in the cumulative region

M.A. Braun[1], C. Ciofi degli Atti[2], and L.P. Kaptari[2,3]

[1] Department of High Energy Physics, S. Petersburg University, 198904 S. Petersburg, Russia
[2] Department of Physics, University of Perugia and INFN, Sezione di Perugia, via A. Pascoli, Perugia, I-06100, Italy
[3] Bogoliubov Laboratory of Theoretical Physics, JINR, Dubna, Russia

Received: 30 Oct 2003 / Accepted: 14 Nov 2003 /
Published Online: 6 Feb 2004 – © Società Italiana di Fisica / Springer-Verlag 2004

Abstract. The role of rescattering due to the final state interaction (FSI) and the influence of the finite formation time (FFT) on the inclusive $D(e,e')X$ and exclusive $D(e,e'p)n$ electro-disintegration of the deuteron are studied in the cumulative kinematical region $x > 1$ and moderate values of the 4-momentum transfer $Q^2 = 2 \div 10$ $(\mathrm{GeV/c})^2$. The spins are averaged out. It is found that in the inclusive process the relative magnitude of rescattering steadily grows with x and that at $x = 1.7$ it has the same order as the plane wave impulse approximation (PWIA) contribution, with the finite formation time effects decreasing the rescattering contribution by $\sim 30\%$. In the exclusive process, with increasing momentum transfer, FFT substantially reduces the effects from FSI, although the latter are still appreciable in the region of momentum transfer investigated.

PACS. 24.85.+p Quarks, gluons, and QCD in nuclei and nuclear processes – 13.60.-r Photon and charged-lepton interaction with hadrons

1 Introduction

High-energy electro-disintegration of the deuteron is a powerful tool to investigate, first, how and when at larger energies and momentum transfers the description in terms of hadrons (nucleons and mesons) transforms into the one in terms of quarks and gluons and, second, the Colour Transparency (CT) effects predicted by QCD. In terms of relevant Feynman diagrams, one expects that whereas at comparatively low energies virtual nucleons and mesons with standard propagators can be used, at large enough virtualities such a description gradually becomes invalid. To clearly see this phenomenon one has to be able to reach high enough virtualities in the process, which can be achieved by choosing the kinematics forbidden for free (on-mass-shell) nucleons, i.e. the so-called *cumulative* kinematics, which corresponds to values of the Bjorken scaling variable larger than one (note that the importance of studying the x-dependence of CT was first stressed in [1]). In the PWIA the cross-sections are directly related to the behaviour of the deuteron wave function. However such a direct relation is broken by FSI, which naively are expected to be large in the cumulative region; it is here that CT effects are to be taken into account. CT predicts that at high Q^2 FSI become small, so that, in principle, choosing both x and Q^2 large enough one can neglect FSI and have access, via the PWIA, to the high-momentum behaviour of the nuclear wave function. It is of particular im-

portance that for high values of x and Q^2 the rescattering energy may be quite small [2], so that only two-nucleon intermediate states can be considered and the simplest Feynman diagrams with only nucleon lines (although at high virtuality) have to be evaluated. In such a kinematical situation CT effects can originate from the virtual excitation of the ejectile.

In this study we investigate the relative role of FSI and CT effects in the deuteron electro-disintegration, using an approximate picture in which the spins are altogether averaged out. Taking the relativistic spin into account, apart from purely technical difficulties, inevitably introduces a variety of unknown off-shell components of both the electromagnetic production vertex and the rescattering amplitude, which deprives the results of any predictive power. To relate our absolute cross-sections to the experimental data we introduce an effective electromagnetic form factor, chosen to reproduce the PWIA results with full relativistic spins [2]. In our approach CT is introduced on the hadronic level using its equivalence to the final formation time (FFT) effect [3].

2 Basic formulas

Separating the leptonic part, the process we are going to study is

$$\gamma^*(q) + d(2p) \rightarrow N(p_1) + N(p_2), \tag{1}$$

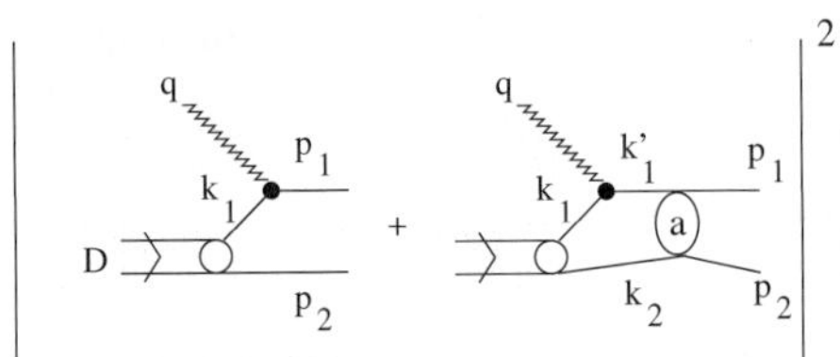

Fig. 1. The square of the sum of the PWIA and rescattering amplitudes proportional to the cross sections

where 4-momenta are denoted in brackets. The total c.m. energy squared in the process is

$$s = (2p + q)^2 = Q^2 \frac{2 - x}{x} + M^2, \quad x = \frac{Q^2}{2qp}, \qquad (2)$$

where M is the deuteron mass. The cumulative region we are going to study corresponds to $1 < x < 2$. From (2) one concludes that at x close to 2 the c.m. energy remains close to the threshold even if Q^2 is large, so that only the lowest two-nucleon intermediate states can be considered. The calculation of the cross-section for the process (1) then reduces to the evaluation of the square modulus of the two standard diagrams shown in Fig. 1, which correspond to the PWIA and rescattering contributions, respectively.

Our final expression for the inclusive process

$$e(l) + d(2p) \to e(l') + X \qquad (3)$$

can be standardly expressed through the hadronic tensor $W_{\mu\nu}$, which is just the square of the two amplitudes shown in Fig.1 integrated over the momenta of the two final nucleons. It can be shown that in order to obtain the exclusive cross-section for the lepto-disintegration, averaged over the azimuthal angles, it is sufficient to drop this integration.

The PWIA contribution to the amplitude is just the hadronic matrix element of the electromagnetic current

$$J_{\mu}^{PWIA} = 2\gamma(Q^2) K_\mu G(k_1^2), \qquad (4)$$

where γ is the effective electromagnetic form-factor of the (scalar) nucleon, the vector K is chosen to guarantee conservation of the electromagnetic current, and $k_1 = 2p - p_2 = p_1 - q$ is the momentum of the active nucleon before the interaction. The quantity $G(k_1^2)$ is the relativistic deuteron wave function, which in principle can be sought as a solution of the Bethe-Salpeter equation for the deuteron. However the relativistic potential for this equation is unknown at high momentum transfers and virtualities, so rather than use forms for this potential fitted to comparatively low energy and momentum transfer scattering, we directly approximate G using as a guide its non-relativistic limit. To calculate the rescattering contribution we choose a system with $q_\perp = 0$ ("lab" system for reaction (1)) which simplifies the integrations over the intermediate nucleon momenta. In this system the integration over k_{2-} puts either the spectator or the active nucleon on the mass-shell, depending on whether k_{2+} is

lower or higher than $P_+ + q_+$. In both cases, as in the PWIA contribution, the integrand involves only the relativistic deuteron wave function with one of the nucleons on its mass shell.

As a result we find the rescattering contribution to the matrix element of the hadronic current in the form

$$J_{\mu}^{resc} = 2\gamma(Q^2) \int dV(k_1, k_2) K_\mu G(v) \frac{a(k_2|p_2)}{m^2 - (k_1 + q)^2}, \qquad (5)$$

where $k_1 + k_2 = 2p$, $K = k_1 + q(k_1 q)/Q^2$ and a is the rescattering amplitude, with dV and v having different forms in the two mentioned regions of integrations. All invariant arguments entering the integrand in (5) have to be expressed through the light-cone integration variables, taking into account that either $k_2^2 = m^2$, or $k_1^2 = m^2$.

The total hadronic tensor is obtained as

$$W_{\mu\nu} = \int dV(p_1, p_2)(J_{\mu}^{PWIA} + J_{\mu}^{resc})(J_{\nu}^{PWIA} + J_{\nu}^{resc})^*, \qquad (6)$$

where $dV(p_1, p_2)$ is the standard invariant phase volume for the produced nucleons. As far as CT is concerned, we introduce it via the FFT of the hit hadron, which manifests itself through a dependence of the scattering amplitudes and verteces on the virtuality of the hit hadron after $\gamma*$ absorption (see [3]). We assume that the effect of this dependence can be modelled by a monopole form-factor, which is equivalent to changing the ejectile propagator as follows

$$\frac{1}{m^2 - k_1'^2} \to \frac{1}{m^2 - k_1'^2} - \frac{1}{m^{*2} - k_1'^2}. \qquad (7)$$

This substitution is equivalent to assuming that there are two different ejectile states with masses m and m^* whose contribution to rescattering cancels out in the limit of high momentum, in agreement with the underlying ideas of CT [3]. Following [3], [4], we choose for the mass m^* the value 1.8 GeV.

3 Numerical calculations

We parametrize the relativistic deuteron wave function $G(k^2)$ using as a guide the nucleon density in the deuteron at comparatively low momenta. In the non-relativistic limit one obtains

$$G^2(k^2) = 2M(2\pi)^3 |\Psi(\mathbf{k}^2)|, \quad \mathbf{k}^2 = \frac{1}{2}(m^2 - k^2 - M\epsilon), \qquad (8)$$

where ϵ is the deuteron binding energy. For $|\Psi|^2$ we have taken the form which corresponds to the AV14 interaction [5]. As for the rescattering amplitude, it was chosen in the form

$$a(s, t) = (\alpha + i)\sigma^{tot}(s)\sqrt{s(s - 4m^2)}e^{bt}, \qquad (9)$$

with the values of the parameters σ^{tot}, α and b taken from [6].

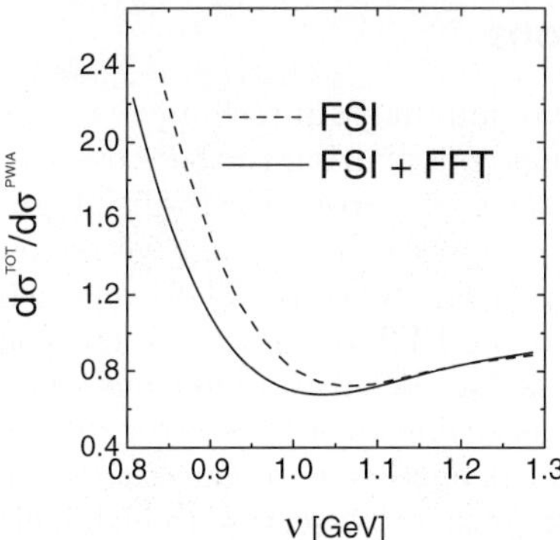

Fig. 2. Ratio of the total inclusive cross section $d\sigma^{TOT} \equiv d\sigma^{TOT}/dE'd\Omega'$ to the PWIA cross section $d\sigma^{PWIA} \equiv d\sigma^{PWIA}/dE'd\Omega'$. In the *dashed curve* $d\sigma^{TOT}$ includes only rescattering effects, whereas in the *full curve* both rescattering and FFT effects are present

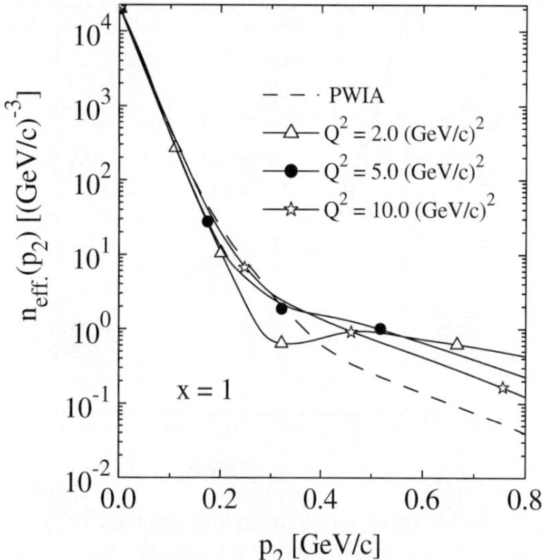

Fig. 3. The effective momentum distribution (11) *vs.* the neutron recoil momentum $|\mathbf{p}_2| \equiv p_2$ at $x = 1$. The *dot-dashed curve* represents the PWIA result, whereas the *other curves* include also rescattering and FFT effects at various values of Q^2

As already mentioned, our main goal was to estimate the magnitude of the FSI and the influence of CT (or, equivalently, FFT) in the cumulative region. Accordingly, our basic quantities to be calculated are the ratios of the rescattering to the PWIA contributions. We expect that the error due to neglecting spins is reduced in these ratios. Still, to have a clearer physical picture and to be able to compare with the experimental data, we also tried to calculate absolute values of the cross-sections. To this end we introduced an effective electromagnetic form-factor for the active nucleon, chosen to approximately take into account magnetic interaction of both the proton and neutron. Comparison of our impulse approximation with the results obtained with a full account of spins [2] leads to the choice

$$\gamma^2(Q^2) = \gamma_D^2(Q^2)\frac{2 + \tau(\mu_p^2 + \mu_n^2)}{1 + \tau}, \qquad (10)$$

where γ_D is the standard dipole form-factor, $\tau = Q^2/(4m^2)$, and $\mu_{p,n}$ are the anomalous magnetic moments of the proton and neutron. With this choice our PWIA results practically coincide with the ones with full relativistic spins taken into account.

The inclusive cross-sections for the process (3) have been calculated at points corresponding to the experimental data of [7], with the initial electron energy $E = 9.761$ GeV and scattering angle $\theta = 10^o$. We have considered values for the final electron energy which cover the region of x in the interval $1.0 < x < 1.71$. Some relevant kinematical values characterizing the chosen points are listed in Table 1.

It can be seen, that at high values of x (high cumulativity) p_{lab} is small, well below the threshold energy for pion production. This justifies our approach based on the assumption that only nucleon degrees of freedom are relevant; however at smaller x p_{lab} grows and such an assumption becomes of disputable validity.

In Fig. 2 the ratios of the full cross section (which includes FSI with or without FFT) to the PWIA cross section is shown. It can be seen that the rescattering contribution, which is very small at $x \sim 1$, steadily grows with

Table 1. Some kinematical variables in the inclusive electro-disintegration cross section corresponding to the kinematics of of [7], *viz* incident electron energy $E = 9.761 GeV$ and scattering angle $\theta = 10^o$. ν is the energy transfer, x the Bjorken scaling variable, Q^2 the square 4-momentum transfer, s the produced invariant mass, and p_{lab} the the momentum of the struck nucleon in the system where the spectator is at rest p_{lab}. Note that the inelastic threshold corresponds to $s \simeq 4\ GeV^2$ ($p_{lab} \simeq 0.8\ GeV$) (cf. [2])

ν, GeV	x	Q^2, $(GeV/c)^2$	s, GeV^2	p_{lab}, GeV/c
0.826	1.71	2.65	3.96	0.73
0.872	1.61	2.64	4.14	0.88
0.930	1.50	2.62	4.38	1.06
0.987	1.41	2.60	4.61	1.21
1.056	1.30	2.58	4.89	1.40
1.137	1.20	2.56	5.22	1.61
1.228	1.10	2.53	5.58	1.82
1.332	1.00	2.50	6.00	2.07

x, reaching an order of about 50% already at $x \sim 1.3$. The relative role of FFT (or CT) also rises with x. At the maximum value of x studied, $x = 1.71$, FFT effects decrease FSI by $\sim 30\%$. This however does not make FSI smaller than the PWIA contribution, so that they cannot be neglected at all.

Let us now discuss the exclusive cross-section $d(e, e')pn$. In order to minimize the error of neglecting spins, we consider the reduced cross-section

$$n_{eff}(|\mathbf{p}_2|) \equiv n_{eff}(|\mathbf{p}_2|, x, Q^2) = |\Psi(|\mathbf{p}_2|)|^2\frac{\sigma_{excl}}{\sigma_{excl}^{PWIA}}, \quad (11)$$

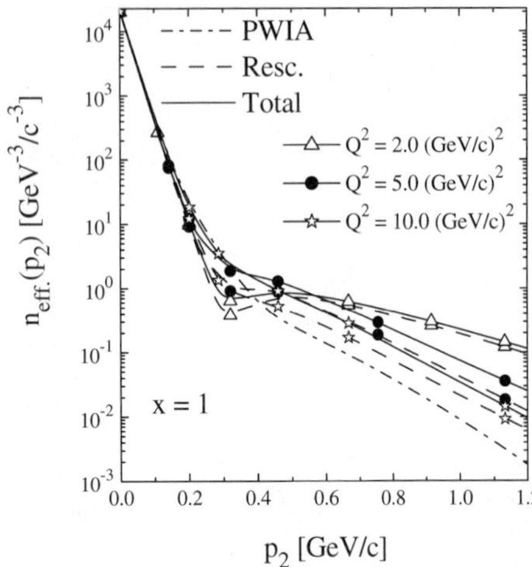

Fig. 4. The effective momentum distribution (11) at $x = 1$ which includes rescattering and FFT effects (*full*), and rescattering effects only (*dashed*); the difference between the *full and dashed curves* is due to FFT effects; the *dot-dashed curve* represents the PWIA result

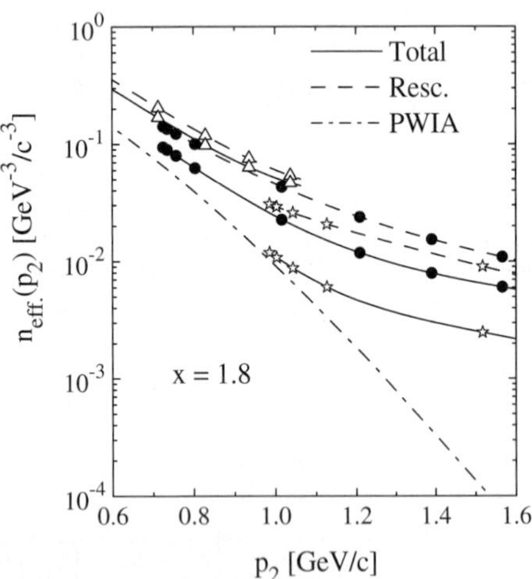

Fig. 5. The same as in Fig. 4 at $x = 1.8$

which in PWIA reduces to the input nucleon momentum distribution in the deuteron. Here p_2 is the momentum of the unobserved nucleon (the missing momentum). Figure 3 shows the effective momentum distributions *vs* the missing momentum at $x = 1$, calculated taking rescattering and FFT into account. Calculations have been performed at fixed values of x and Q^2 and different missing momenta $\mathbf{p_2}$. Due to energy conservation, this corresponds to different values of the angle between $|\mathbf{q}|$ and $|\mathbf{p_2}|$.

In Figs. 4 and 5 the effects of rescattering and FFT are shown separately at $x = 1$, and in the deep kinematical region, $x = 1.8$. It can be seen from Fig.4 that at $x = 1$ FFT effects, as expected, increase with Q^2, in agreement with the results obtained for the process $^4He(e, e'p)^3H$ ([8]). At $x = 1.8$ (Fig. 5) one observes that with the growth of the missing momentum, FSI appreciably decrease due to FFT, which makes the distorted cross-section more similar to the PWIA one.

4 Conclusions

Using the Feynman diagram technique we have calculated the rescattering contribution to the cross-section of inclusive and exclusive deuteron electro-disintegration, paying particular attention to the cumulative region $x > 1$. Our main goal has been to study CT effects, which we have introduced via the FFT of the struck nucleon. In our calculations spins has been averaged out and all particles were treated as scalar ones. The error introduced by this approximation is expected to cancel to a large extent in the ratio of the total cross-section (which includes FSI and FFT effects), with and without FSI and FFT effects, to the PWIA cross section. To be able to compare absolute magnitudes of the cross-sections with the experimental data we used an effective electromagnetic vertex for the active nucleon, chosen to describe the PWIA results with spins.

The results of our calculations show that the relative magnitude of the FSI steadily grows with x. At $Q^2 \sim 2$ (GeV/c)^2 and $x \sim 1.7$, the rescattering contribution raises the cross-section nearly by a factor of 2.5, whereas the introduction of FFT decreases back the cross section by a factor of 1.6, which is evidently not enough to disregard FSI altogether. As expected, at large values of x the influence of FFT grows up: it can be seen from Fig. 5 that at $Q^2 = 10$ (GeV/c)^2 and $|\mathbf{p_2}| \sim 1 \div 1.5$ GeV/c, FFT decreases the pure FSI results by a factor of 4.

In conclusion, we have found that the CT or FFT effects are clearly visible in the electro-disintegration of the deuteron in the cumulative region, which may serve as a tool for their experimental study. We have also found that although FFT effects decrease the effects of the FSI rather substantially at $Q^2 = 2 \div 10 (GeV/c)^2$, this is by far not sufficient to neglect FSI altogether.

Acknowledgements. This work was partially supported by the Ministero dell'Istruzione, Università e Ricerca (MIUR), through the funds COFIN01. M. B. and L.P.K. are indebted to the University of Perugia and INFN, Sezione di Perugia, for warm hospitality and financial support.

References

1. B.K. Jennings and B.Z. Kopeliovich: Phys. Rev. Lett. **70**, 3384 (1993)
2. C. Ciofi degli Atti, L.P. Kaptari, and D. Treleani: Phys. Rev. C **63**, 044601 (2001)
3. M.A. Braun, C. Ciofi degli Atti, and D. Treleani: Phys. Rev. C**62**, 034606 (2000)
4. L.L. Frankfurt, W.R. Greenberg, G.A. Miller, and M.I. Strikman: Phys. Rev. C **46**, 2547 (1992)
5. R.B. Wiringa, R.A. Smith and T.L. Ainsworth: Phys. Rev. C **29**, 1207 (1984)
6. R.A. Arndt et al.: Partial-Wave Analysis Facility (SAID), http:/said.phys.vt.edu
7. S. Rock et al: Phys. Rev. Lett. **49**, 1139 (1982); R.G. Arnold et al: ibid. **61**, 806 (1988)
8. H. Morita, M.A. Braun, C. Ciofi degli Atti, and D. Treleani: Nucl. Phys. A **699**, 328c (2002)

Eur Phys J A (2004) **19**, s01, 147–151
Digital Object Identifier (DOI) 10.1140/epjad/s2004-03-024-x

Hadron multiplicity in semi-inclusive lepton-nucleon and lepton-nucleus scattering

O. Benhar[1], S. Fantoni[2], G.I. Lykasov[3], U. Sukhatme[4], and V.V. Uzhinsky[3]

[1] INFN and Physics Department, Università "La Sapienza". I-00185 Roma, Italy
[2] International School for Advanced Studies (SISSA). I-34014 Trieste, Italy
[3] Joint Institute for Nuclear Research. Dubna, Moscow Region, 141980 Russia
[4] State University of New York at Buffalo. Buffalo, NY 14260-4600, USA

Received: 30 Sep 2003 / Accepted: 14 Nov 2003 /
Published Online: 6 Feb 2004 – © Società Italiana di Fisica / Springer-Verlag 2004

Abstract. We discuss multi-hadron production in both inelastic neutrino-nucleon interactions in the current fragmentation region and neutrino-nucleus collisions in the target fragmentation region. Our analysis, carried out within the framework of the quark-gluon string model, is mainly focused on the difference between these two processes. We show that the Q^2 dependence of hadron multiplicity in the current and target fragmentation regions is indeed completely different. The study of inelastic $\nu - A$ scattering in the target fragmentation region also provides new information on nuclear structure at small $N - N$ distances. The results of the proposed approach are in satisfactory agreement with the data recently obtained at CERN by the NOMAD Collaboration.

PACS. 13.60.Hb Total and inclusive cross sections – 13.15.+g Neutrino interactions

1 Introduction

Deep-inelastic particle interactions play a decisive role in the development of the modern theory of fundamental interactions. For instance, the investigation of multi-hadron production in inelastic scattering of leptons off nucleons and nuclei is a tool to study both the dynamics of such processes and the quark structure of the target. Of particular interest, in this context, is the analysis of the Q^2 dependence over a wide range of values. The difficulties associated with the increase of the QCD coupling constant $\alpha_s(Q^2)$ at moderate and low Q^2 have led to the development of alternative approaches to analyze soft hadron interactions. A well known example is the the $1/N$ expansion of the scattering amplitude, N being the number of colors or quark flavors, suggested by t'Hooft [1] and Veneziano [2]. Based on this approach, the Quark-Gluon String Model (QGSM) and the Dual Parton Model (DPM) have been developed in [3] and [4,5], respectively, to analyze hadronic reactions. The first application of the QGSM to inelastic lepton-nucleon processes at moderate and low Q^2 has been carried out in [6].

In this paper we investigate inelastic neutrino scattering from both protons and nuclei within the QGSM, focusing mainly on the differences between these reactions. In Sect. 2 we outline the main features of the theoretical approach, whereas Sect. 3 is devoted to a comparison between our results and the data recently obtained at CERN

by the NOMAD collaboration. Finally, in Sect. 4 we summarize our findings and state the conclusions.

2 Theoretical framework

In the QGSM [3,6], hadron production in the reactions $\nu(\bar{\nu}) + p \rightarrow \mu^-(\mu^+) + h + X$ is described in terms of planar and cylindrical graphs, as shown in Fig. 1. The planar graph of panel (a) describes neutrino scattering off a valence quark, corresponding to one Reggeon exchange in the t-channel [6], whereas the cylindrical graph of panel (b) describes neutrino scattering off sea quarks, corresponding to one-Pomeron exchange in the t-channel [6]. The figure also shows the occurrence of hadronization in the colorless quark-antiquark and quark-diquark strings.

The relativistic invariant distribution of hadrons produced in the process $\nu(\bar{\nu}) + p \rightarrow \mu^-(\mu^+) + h + X$ is defined as

$$\rho_{\nu(\bar{\nu})+p\rightarrow\mu^-(\mu^+)+h+X} = E_h \frac{dN}{d^3 p_h d\Omega dE'} , \qquad (1)$$

where E_h and $\mathbf{p}_h$ are the total energy and three momentum of the produced hadron, respectively, whereas E' and Ω are the energy and the solid angle of the final state muon. The right hand side of (1) can be written in the following general form [7,8]:

$$\rho_{\nu(\bar{\nu})+p\rightarrow\mu^-(\mu^+)+h+X} = \Phi(Q^2) \big\{ F_P(x, Q^2; z, p_{ht})$$
$$+ F_C(x, Q^2; z, p_{ht}) \big\} , \quad (2)$$

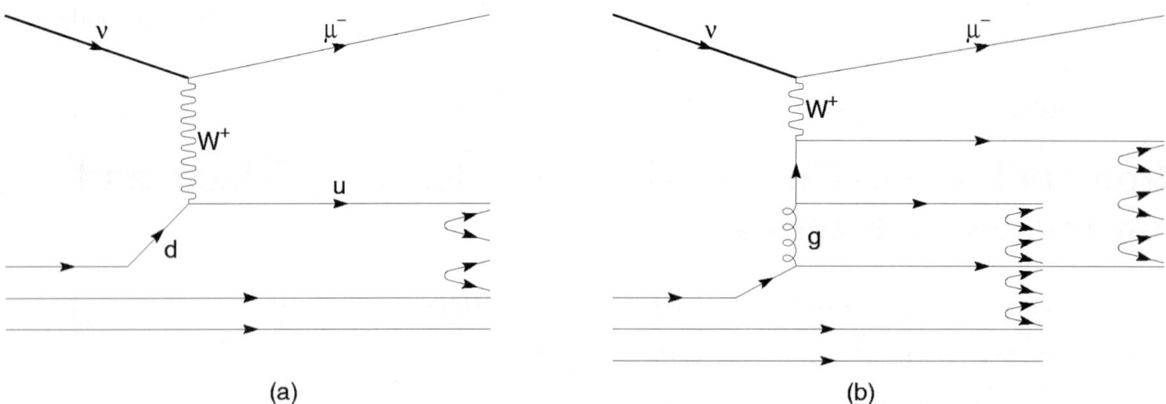

Fig. 1. Planar, one-Reggeon exchange (*panel* **a**), and cylindrical, one-Pomeron exchange (*panel* **b**), graphs (*see text*)

with

$$\Phi(Q^2) \;=\; mE\frac{G^2}{\pi}\frac{m_W^2}{Q^2+m_W^2}\;, \qquad (3)$$

where G is the Fermi weak coupling constant, E is the energy of the incoming neutrino, m and m_W are the nucleon and W-boson masses, respectively, $x = Q^2/2(p_\nu \cdot k)$ is the Bjorken variable and p_ν and k are the four momenta of the initial neutrino and nucleon. In general, z is the light cone variable defined as $z = (E_h + p_{hz})/(E + p_z)$, where p_{hz} is the component of the hadron momentum parallel to the momentum of the incoming neutrino, p_z, (see, e.g., [3,6] and [4,5]). At large energies of the final hadron it reduces to the longitudinal momentum fraction of the hadron, with respect to the neutrino, in the rest frame of the target proton. The variable z can also be treated as the Feynman variable $x_F = 2p_L^*/W_X$, defined as the longitudinal momentum fraction in the hadronic center of mass system (HCMS). Here p_L^* denotes the longitudinal hadron momentum in the HCMS and W_X is the mass of the hadrons produced in the reaction. Finally, p_{ht} is the transverse momentum of the produced hadron with respect to the current (hadronic jet) direction, (see, e.g. [9]).

The probabilty distributions of hadron production associated with the planar and cylindrical graphs of Fig. 1 are given by $F_P(x, Q^2; z, p_{ht})$ and $F_C(x, Q^2; z, p_{ht})$, respectively. They can be computed either analytically, as in [7], or using the Monte Carlo (MC) approach to generate all quark-antiquark and quark-diquark strings shown in Fig. 1, as done in [8].

The main ingredients for the calculations of observables in the reaction under discussion are the quark distributions in a nucleon and their fragmentation functions to hadrons. In addition to the dependence upon x, quark distributions also depend on Q^2 and the transverse momentum k_t. Following [7], we use a factorized form for these distributions:

$$q_f(x, Q^2; k_t) \;=\; q_f(x, Q^2)g_q(k_t)\;, \qquad (4)$$

with the function g_q chosen in the form

$$g_q(k_t) \;=\; \frac{B^2}{2\pi}e^{-Bk_t}\;, \qquad (5)$$

where $B = 1/\langle k_t \rangle \simeq 4(GeV/c)^{-1}$, while $\langle k_t \rangle \simeq 0.25$ GeV/c is the average transverse momentum of a quark in a nucleon. As for the function $q_f(x, Q^2)$, we use the fit suggested in [10], including true Regge x-asymptotic at $x \to 0, x \to 1$ and small Q^2, and its QCD prediction at large Q^2.

In general, the fragmentation functions (FF) of quarks (diquarks) into hadrons, $D_{q(qq)}^h$, depend on the hadron momentum fraction z_1 and the hadron transverse momentum with respect to a quark (diquark) momentum direction, $\widetilde{p}_{ht}$. Here we choose again the factorized form

$$D_{q(qq)}^h(z_1, \widetilde{p}_{ht}) \;=\; D_{q(qq)}^h(z_1)g_q(\widetilde{p}_{ht})\;, \qquad (6)$$

with the function g_q defined as in (5).

The functions $D_{q(qq)}^h(z_1)$ are obtained, according to the recursive cascade model procedure suggested in [11], from the integral equation

$$D_{q(qq)}^h(z_1) \;=\; f(z_1) \;+\; \int_{z_1}^1 \frac{dx}{x}f(x)D_{q(qq)}^h\left(\frac{z_1}{x}\right)\;, \qquad (7)$$

the function $f(x)$ being chosen in the form

$$f(x) \;=\; x^\beta(1-x)^\gamma\;. \qquad (8)$$

According to the main assumptions underlying the QGSM, the fragmentation functions $D_{q(qq)}^h(z_1)$ should satisfy true Regge asymptotic at $z_1 \to 1$ and $z_1 \to 0$ [3]. These constraints determine the values of the parameters. The detailed procedure is presented in [12]. In general the FF depend not only on z_1 and p_{ht}, but also on Q^2. At low Q^2 they have to reproduce the true Regge asymptotic [3], while at large Q^2 they have to describe e^+e^- annihilation data. As we mainly analyze inelastic $\nu - N$ interactions at moderate Q^2, one can assume that the Q^2-dependence of the FF be negligibly weak.

Let us now consider pion production in the target fragmentation region in $\nu - A$ interactions. Our study of these processes focuses on the kinematical region of large pion momentum ($p_\pi > 0.3$ GeV/c), where the effects of final state interactions (FSI), leading to pion absorption associated with production of baryon resonances, are expected to be small [13,14] and the impulse approximation (IA) cane be safely used.

Within the framework of the IA the relativistic invariant semi-inclusive spectrum of hadrons, in particular pions, produced by the semi-inclusive inelastic process $\ell + A \to \ell' + \pi + X$,

$$\rho_{\ell+A\to\ell'+\pi+X} \equiv E_h d\frac{\sigma}{d^3 p_h d\Omega dE'} \ , \qquad (9)$$

can be written in the convolution form [7]

$$\rho_{\ell+A\to\ell'+\pi+X}(x, Q^2; z, p_t) = \int_{z\leq y} dy d^2 k_t f_A(y, Q^2, k_t)$$

$$\times \left[\frac{Z}{A} \rho_{\ell+p\to\ell'+\pi+X}(x/y, Q^2; z/y, p_t - k_t) \right. \qquad (10)$$

$$\left. + \frac{N}{A} \rho_{\ell+n\to\ell'+\pi+X}(x/y, Q^2; z/y, p_t - k_t) \right] \ ,$$

where Z, N and A are the numbers of nucleons, protons and neutrons in the nucleus, $z = (pp_\nu)/(P_A p_\nu) M_A/m$, p is the four-momentum of the produced pion and M_A and m are the nucleus and nucleon masses, respectively. $\rho_{\ell+p(n)\to\ell'+\pi+X}$ denotes the semi-inclusive spectra of pions produced by interactions of the lepton ℓ with a quasi-free proton (neutron). The nucleon distribution function $f_A(y, k_t)$ is defined as [15]

$$f_A(y, k_t) = \int dk_0 dk_z S(k) y \delta\left(y - \frac{M_A}{m}\frac{(kq)}{(P_A q)} \right) \ , \quad (11)$$

where q is the four-momentum transferred by the lepton, $q^2 = -Q^2$, $S(k)$ is the relativistic invariant function describing the nuclear vertex with an outgoing virtual nucleon, $y = (M_A/m)(kp_\nu)/(P_A p_\nu)$ and P_A, k and p_ν are the four-momenta of the nucleus, nucleon and initial neutrino, respectively.

The distribution function $f_A(y)$, defined as the integral of $f_A(y, k_t)$ over $d^2 k_t$, can be calculated within nuclear many-body theory approximating $S(k)$ with the nonrelativistic spectral function $P(k, E)$, yielding the probability of finding a nucleon with momentum $\mathbf{k}$ and removal energy $E = m - k_0$ [16]. However, due to the limited range of momentum and removal energy covered by nonrelativistic calculations of $P(k, E)$ (typically $|\mathbf{k}| < k_{min} \sim 0.7 - 0.8$ GeV/c and $(m - k_0) < 0.6$ GeV, see e.g. [16]), this procedure can only be used in the region $y < y_0 \sim 1.7 - 1.85$. An alternative approach to obtain $f_A(y)$ at larger y, based on the calculation of the overlap of the relativistic invariant phase-space available to quarks belonging to strongly correlated nucleons, has been proposed in [15,17].

The convolution form of (11) implies that to analyze the production of pions emitted in the whole backward hemisphere by $\nu - A$ interaction one needs to know the dependence of the nucleon distribution upon y and k_t. We assume the following factorized form:

$$f_A(y, k_t) = f_A(y) g_A(k_t) \ , \qquad (12)$$

where the function g_A is chosen to be a Gaussion

$$g_A(k_t) = \frac{1}{\pi \langle k_t^2 \rangle} \exp(-k_t^2/\langle k_t^2 \rangle) \ , \qquad (13)$$

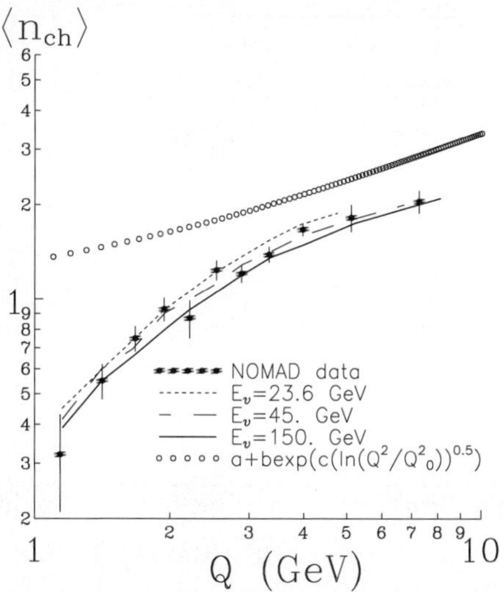

Fig. 2. Mean multiplicity of charged hadrons in the current fragmentation region for the process $\nu + p \to \mu^- + h + X$, as a function of the momentum transfer Q. The *open circles* correspond to the QCD fit given by (14), while the *solid, long-dash* and *short-dash lines* correspond to our calculations at $E_\nu = 150$, 45 and 23.6 GeV, respectively. The experimental data are taken from [18]

$\langle k_t^2 \rangle$ being the average value of the squared transverse momentum of a nucleon in the target nucleus.

3 Comparison to data

The results of our calculation of different observables of the reactions $\nu + N \to \mu^- + h + X$ and $\nu + C^{12} \to \mu^- + \pi^- + X$ are presented in Figs. 2–4.

The mean charged multiplicity in the current region is shown in Fig. 2. The multiplicity measured by the NOMAD collaboration [18], $\langle n_{ch} \rangle$, turns out to be close to $\langle n_{ch} \rangle/2$ measured in $e^+ e^-$ experiment at energy $E = \sqrt{s}$ and to $\langle n_{ch} \rangle$ obtained from ep and $\bar{\nu}_\mu p$ at $E = Q$. In Fig. 2 we compare our results to $\langle n_{ch}^{QCD} \rangle/2$ obtained from a QCD calculation of the charged multiplicity in $e^+ e^-$. The QCD results, represented by open circles, correspond to evolution of parton distributions in the leading log approximation, yielding the following fit for $\langle n_{ch}^{QCD} \rangle$ (see [18]):

$$n_{ch}^{QCD} = a + b \exp\left(c\sqrt{\log(Q^2/Q_0^2)} \right) \ , \qquad (14)$$

where $a = 2.257$, $b = 0.094$, $c = 1.775$ and $Q_0 = 1$ GeV/c. The solid, long-dash and short-dash lines in Fig. 2 correspond to our Monte Carlo calculations at initial neutrino energies $E_\nu = 150$ GeV, $E_\nu = 45$ GeV, $E_\nu = 23.6$ GeV, respectively. The NOMAD experimental data are averaged over the initial energy [18].

Figure 2 shows that the QCD fit, while failing to reproduce the NOMAD data on the charged multiplicity at $Q < 5$ GeV/c, provides a reasonable description of the behavior at large Q. As discussed in [18] the same fit also

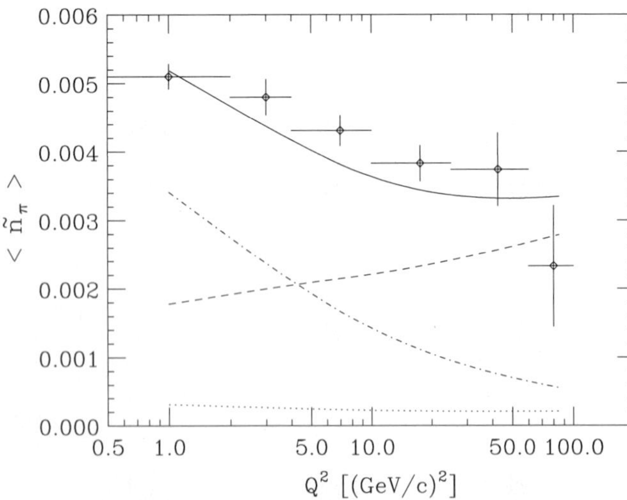

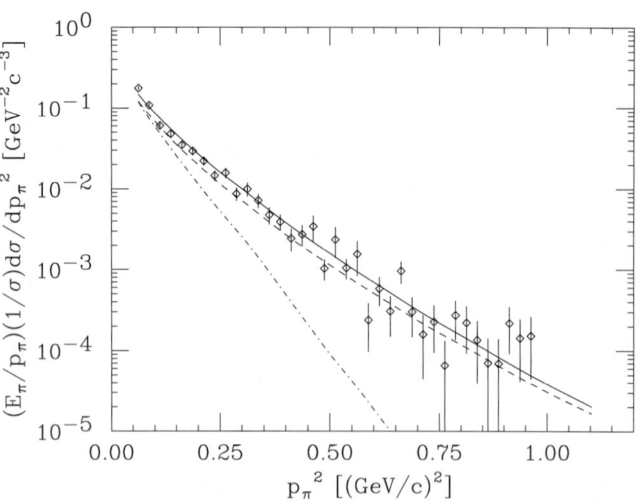

Fig. 3. Mean multiplicity of charged pions produced in the backward hemisphere in $\nu - C$ interaction as a function of Q^2. The *solid line* shows the results of the full calculation, including both graphs of Fig. 1. The *dashed* and *dash-dot lines* correspond to the separated contributions of the cylindrical graph of Fig. 1b and the planar graph of Fig. 1a, respectively. The *dots* show the results obtained setting $P(\mathbf{k}, E) \equiv 0$ in the high energy-momentum domain, not covered by the nonrelativistic calculation of [16]. The experimental data are taken from [9]

Fig. 4. p_π^2-dependence of the spectrum of backward pions produced in inelastic $\nu - C$ interaction. The *solid curve* corresponds to the full calculation whereas the *dashed line* has been obtained including only the cylindrical graph of Fig. 1b. The *dash-dot line* shows the results obtained setting $P(\mathbf{k}, E) \equiv 0$ in the high energy-momentum domain, not covered by the nonrelativistic calculation of [16]. The experimental data are taken from [9]

satisfactorily describes $\langle n_{ch} \rangle$ obtained from $e^+ e^-$, ep and $\bar{\nu}_\mu p$ experiments. On the other hand, the approach proposed in [8] describes the NOMAD data at $Q < 5\,\mathrm{GeV}/c$ rather well.

The multiplicity of pions, normalized to the cross section of the process $\nu + A \rightarrow \mu + X$, σ, is defined as

$$\langle \widetilde{n}_\pi \rangle \equiv \frac{\langle n_\pi \rangle}{\sigma} = \frac{1}{\sigma} \int_{x_{min}}^{x_{max}} dx$$
$$\times \int_{z_{min}}^{z_{max}} \frac{dz}{z} \int_0^{p_{max}} d^2 p \; \rho_{\nu + A \rightarrow \mu + \pi + X} \; . \quad (15)$$

It can be split in two parts, corresponding to the planar, or one-Reggeon exchange, diagram (Fig. 1 (a)) and cylindrical, or one-Pomeron exchange graph (Fig. 1 (b)).

The NOMAD collaboration carried out a study of inelastic $\nu - C$ interactions in which the negative pions emitted in the whole backward hemisphere, with respect to the incoming neutrino beam, were detected [9]. The data of Fig. 3 show the nultiplicity of negative pions carrying momenta $0.35 < p_\pi < 0.8\,\mathrm{GeV}/c$, measured in a kinematical setup in which W_X increases as Q^2 increases. Theoretical calculations have been performed applying the same kinematical conditions.

As shown in Fig. 3, the planar graph provides the main contribution at small Q^2, while the cylindrical one dominates at $Q^2 > 10\,(\mathrm{GeV}/c)^2$. It clearly appears that both diagrams have to be included to explain the observed Q^2 dependence. The dotted line in Fig. 3 has been obtained setting $P(\mathbf{k}, E) = 0$ in the domain of large energy and large momentum, not covered by the nonrelativistic calculations of ref.[16]. Comparison between the solid and dot-

ted line shows that the dominant contribution to $< \widetilde{n}_\pi >$ does indeed come from the high momentum tail of the nucleon distribution.

In Fig. 4 we compare the p_π^2 dependence of the experimental spectrum to the one resulting from our approach. The contribution of the cylindrical graph dominates the spectrum and the inclusion of the high momentum tail of the nucleon distribution, corresponding to $p > 0.4\,\mathrm{GeV}/c$, is clearly needed to describe the data.

4 Summary and conclusions

In this paper we have analyzed inelastic $\nu - N$ and $\nu - A$ processes at low and moderate Q^2 within the QGSM. The main conclusions can be summarized as follows. The dynamics of hadron production in the current fragmentation and target fragmentation regions is different. At low Q^2 the contribution of the cylindrical graph to the hadron multiplicity is much more prominent in the first kinematical region than it is in the second one. There is some analogy between pion production in the backward hemisphere from target fragmentation and soft $h - N$ processes at $x_F \rightarrow -1$, where the one-Pomeron exchange graph dominates at large initial energy (see, e.g., [6]). In conclusion, the contributions of the graphs of Figs. 1 (a) and (b) have different Q^2 dependences in the different kinematical regions corresponding to current and target fragmentation.

The conventional perturbative QCD calculations of hadron multiplicities, both in $\nu - N$ and $\nu - A$ inelastic interactions, do not reproduce the NOMAD data, whereas the application of the QGSM allows for a rather satisfactory description. As for the role of nuclear structure, the dominant contribution to the spectrum of pions emitted in the backward hemisphere in $\nu - A$ inelastic scattering is clearly coming from the high momentum tail of the nucleon distribution, which can be described in terms of overlaps of distributions of three-quark colorless objects [19].

References

1. G. t'Hooft: Nucl. Phys. B **72**, 461 (1974)
2. G. Veneziano: Phys. Lett. B **52**, 200 (1974)
3. A.B. Kaidalov: Phys. Lett. B **116**, 459 (1982); Sov. J. Nucl. Phys. **45**, 902 (1987)
4. A. Capella, U. Sukhatme, C.I. Tan, and J. Tran Thanh Van: Phys. Lett. B **81**, 68 (1979); Z. Phys. **63**, 68 (1979)
5. A. Capella, U. Sukhatme, C.I. Tan, and J. Tran Thanh Van: Phys. Rep. **236**, 223 (1994)
6. A.B. Kaidalov: Sov. J. Nucl. Phys. **33**, 733 (1981)
7. O. Benhar, S. Fantoni, G.I. Lykasov, and U. Sukhatme: Phys. Lett. B **527**, 73 (2002)
8. G.I. Lykasov, U. Sukhatme, and V.V. Uzhinsky: Phys. Lett. B **553**, 217 (2003)
9. NOMAD Collaboration, P. Altegoer et al: Phys. Lett. B **455**, 439 (1999)
10. A. Capella, A. Kaidalov, C. Merino, and J. Tran Thanh Van: Phys. Lett. B **337**, 358 (1994); Phys. Lett. B **343**, 403 (1995); A. Kaidalov and C. Merino: Eur. Phys. J. C **10**, 153 (1999)
11. R.D. Field and R.P. Feynman: Nucl. Phys. B **136**, 1 (1978); U. Sukhatme: Phys. Lett. B **73**, 478 (1978); Zeit. Phys. C **2**, 321 (1979); U. Sukhatme, K. Lassila, and R. Orava: Phys. Rev. D **25**, 2975 (1982)
12. V.V. Uzhinsky and S.Y. Shmakov: Sov. J. Nucl. Phys. **53**, 1034 (1991)
13. L.L. Frankfurt and M.I. Strikman: Phys. Rep. **160**, 236 (1988)
14. N.S. Amelin and G.I. Lykasov: Sov. J. Nucl. Phys. **33**, 100 (1981)
15. O. Benhar, S. Fantoni, G.I. Lykasov, and N.V. Slavin: Phys. Rev. C **57**, 1532 (1998)
16. O. Benhar, A. Fabrocini, and S. Fantoni: Nucl. Phys. A **505**, 267 (1989)
17. O. Benhar, S. Fantoni, and G.I. Lykasov: Eur. Phys. J. A **7**, 415 (2000)
18. NOMAD Collaboration, J. Altegoer et al: Phys. Lett. B **445**, 439 (1999)
19. O. Benhar, S. Fantoni, G.I. Lykasov, and N.V. Slavin: Phys. Rev. C **55**, 244 (1997)

Eur Phys J A (2004) **19**, s01, 153–158

Digital Object Identifier (DOI) 10.1140/epjad/s2004-03-025-9

EPJ A direct

electronic only

Medium modification of the proton form factor

Steffen Strauch for the Jefferson Lab E93-049 collaboration

Department of Physics, The George Washington University, Washington, DC 20052, USA

Received: 31 Jul 2003 / Accepted: 14 Nov 2003 /
Published Online: 6 Feb 2004 – © Società Italiana di Fisica / Springer-Verlag 2004

Abstract. I argue that the double ratio of proton-recoil polarization-transfer coefficients, P'_x and P'_z, of the quasielastic $^4\mathrm{He}(\mathbf{e}, e'\mathbf{p})^3\mathrm{H}$ reaction with respect to the elastic $^1\mathrm{H}(\mathbf{e}, e'\mathbf{p})$ reaction is sensitive to possible medium modifications of the proton form factor in $^4\mathrm{He}$. Recent measurements at both Mainz and Jefferson Lab of this double ratio at four-momentum transfers squared between between 0.4 $(\mathrm{GeV}/\mathrm{c})^2$ and 2.6 $(\mathrm{GeV}/\mathrm{c})^2$ are discussed. I show that the data challenge state-of-the-art conventional meson-nucleon calculations, as these are unable to describe the results. The data hint at the need to include medium modifications of the proton form factor, as predicted by a quark-meson-coupling model, in the calculations. A recently approved follow-up experiment at a Q^2 of 0.8 $(\mathrm{GeV}/\mathrm{c})^2$ and 1.3 $(\mathrm{GeV}/\mathrm{c})^2$ with unprecedented precision will provide one of the most stringent tests of the applicability of various calculations.

PACS. 25.30.Dh Inelastic electron scattering to specific states – 24.70.+s Polarization phenomena in reactions – 14.20.Dh Protons and neutrons – 13.40.Gp Electromagnetic form factors – 13.88.+e Polarization in interactions and scattering – 27.10.+h $A \leq 5$

1 Introduction

The standard nuclear-physics model describes nuclei as clusters of protons and neutrons held together by a strong force mediated by meson exchange. The success of the nuclear shell model demonstrates that this approximation has been highly effective. A recent example is the very good description of the differential cross section and left-right asymmetry in the $^{16}\mathrm{O}(e, e'p)$ reaction [1] with a fully relativistic distorted-wave impulse approximation (RDWIA).

However, protons and neutrons are not the fundamental entities of the underlying theory, quantum chromodynamics. At nuclear densities of about 0.17 nucleons/fm^3, nucleon wave functions have significant overlap. In the chiral limit, one expects nucleons to lose their identity altogether and nuclei to make a transition to a quark-gluon plasma. For that reason, one may expect that, for some observables, the free-nucleon approximation is a highly uneconomical approach. The use of medium-modified nucleons as quasiparticles might be a better choice. To demonstrate that the interpretation of the experimental data is more efficient in terms of medium modifications of nucleon form factors, it is required to have excellent control of the inherent many-body effects, such as meson-exchange currents (MEC) and isobar configurations (IC). In addition, when probing nuclear structure, one has to deal with final-state interactions (FSI).

In unpolarized $\mathrm{A}(e, e'p)$ experiments involving light- and medium-mass nuclei, deviations were observed in the longitudinal/transverse character of the nuclear response

compared with the free-proton case [2,3,4]. Below the two-nucleon-emission threshold, these deviations were originally interpreted as changes in the nucleon form factors within the nuclear medium. However, strong-interaction effects on the ejected proton (final-state interactions) later also succeeded in explaining the observed effect [5]. Still, tantalizing hints of medium effects remain for unpolarized longitudinal/transverse separations in the $^4\mathrm{He}(e,e'p)^3\mathrm{H}$ reaction [6,7].

A reaction which is believed to have minimal sensitivity to reaction-mechanism effects is polarization transfer in quasielastic nucleon knockout (see, e.g., [8]). The polarization-transfer observables are sensitive to the properties of the nucleon in the nuclear medium, including possible modification of the nucleon form factor and/or spinor. This can be seen from free electron-nucleon scattering, where the ratio of the electric to magnetic Sachs form factors, G_E and G_M, is given by [9]:

$$\frac{G_E}{G_M} = -\frac{P'_x}{P'_z} \cdot \frac{E_e + E_{e'}}{2m_p} \tan(\theta_e/2). \qquad (1)$$

Here, P'_x and P'_z are the transferred polarizations, transverse and longitudinal to the proton momentum (see [10]). The beam energy is E_e, the energy (angle) of the scattered electron is $E_{e'}$ (θ_e), and m_p is the proton mass. This relation was used to extract G_E/G_M for the proton (most recently in [11,12,13]).

In the following, the results of recent $^4\mathrm{He}(\mathbf{e}, e'\mathbf{p})^3\mathrm{H}$ experiments to study the proton knock-out process will be discussed. In these experiments $^4\mathrm{He}$ was chosen as target because of its high density and relatively simple structure, which allows for RDWIA and microscopic calculations. The kinematics of these experiments, quasielastic scattering at low missing momentum with symmetry about the three-momentum-transfer direction, minimize conventional many-body effects, as will be demonstrated below.

2 Experiments

The first ^{4}He polarization-transfer measurement was performed at the Mainz microtron (MAMI) at a four-momentum squared $Q^2 = 0.4$ (GeV/c)^2 [14]. More recently these measurements were extended to $Q^2 = 0.5$, 1.0, 1.6, and 2.6 (GeV/c)^2 in experiment E93-049 at Jefferson Lab Hall A [15,16]. Since these experiments were designed to detect differences between the in-medium polarizations compared to the free values, both ^{4}He and ^{1}H targets were employed (due to beam-time constraints, only ^{4}He data were acquired at $Q^2 = 2.6$ (GeV/c)^2). The induced polarization, P_y, was also measured in the JLab experiment [16].

In each experiment, two high-resolution spectrometers were used, one to detect the scattered electron and one to detect the recoiling proton. The latter was equipped with a focal-plane polarimeter (FPP). The proton-recoil-polarization observables were extracted by means of the maximum-likelihood technique, utilizing the azimuthal distribution of protons scattered off the graphite analyzer in the FPP.

The results of the measurements are expressed in terms of the polarization-transfer double ratio, in which the helium polarization ratio is normalized to the hydrogen polarization ratio measured in the identical setting:

$$R = \frac{(P'_x/P'_z)_{^4\mathrm{He}}}{(P'_x/P'_z)_{^1\mathrm{H}}}. \tag{2}$$

As a cross-check, the hydrogen results were also used to extract the free-proton form-factor ratio G_E/G_M, which was found to be in excellent agreement with previous data [11, 12].

Nearly all systematic uncertainties cancel in R: the polarization-transfer observables are to first order independent of instrumental asymmetries in the FPP, and their ratio is independent of the electron-beam polarization and carbon analyzing power. Small systematic uncertainties are predominantly due to uncertainties in the spin transport through the proton spectrometer.

3 Results

Figure 1 shows the results for R at $Q^2 = 1.0$ (GeV/c)^2 as a function of missing momentum. The five data points correspond to five bins in the acceptance within the single experimental setting. Negative values of missing momentum correspond to the recoiling nuclei having a momentum component antiparallel to the direction of the three-momentum transfer.

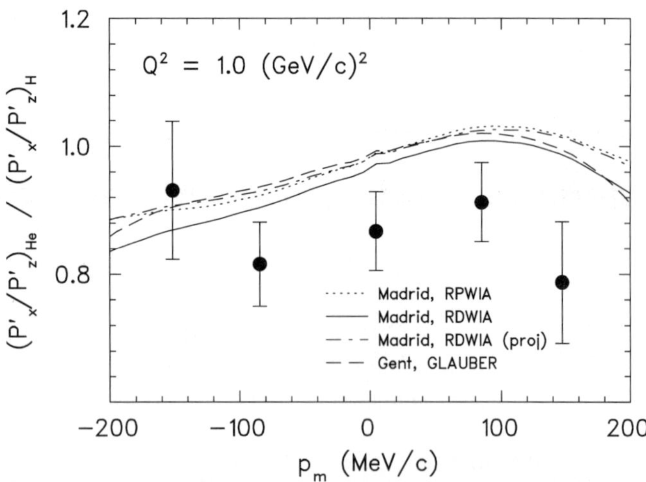

Fig. 1. Polarization-transfer double ratio in ^{4}He$(\mathbf{e}, e'\mathbf{p})^3$H at $Q^2 = 1.0$ (GeV/c)^2 as a function of missing momentum. The data [16] are compared to calculations by the Madrid [17] and Gent [18,19] groups. Error bars are statistical only; systematic uncertainties are much smaller

The data are compared with several calculations, which were averaged over the experimental acceptance. In the relativistic calculations by the Madrid group [17], the Coulomb gauge and the current operator $cc1$ [20] were used. The dotted line is a result of a calculation in plane-wave impulse approximation (RPWIA). This calculation predicts $R \approx 1$ at $p_m = 0$. It overpredicts the data by about 10%, although the missing-mass distribution of the data is reasonably well described. The relativistic distorted-wave impulse approximation (RDWIA) gives a smaller value of R ($\approx 3\%$) than the RPWIA but still overpredicts the data (solid curve). The calculation uses the MRW optical potential of [21]. The main difference in the results between this full calculation and the RPWIA is due to the enhancement of the negative energy components of the relativistic bound and scattered proton wave functions. This can be demonstrated by projecting the wave functions over positive-energy states (dash-dotted curve) resulting in essentially the same polarization-transfer ratio as given by RPWIA. Also, a Glauber calculation by the Gent group [18,19] (long-dashed line) is unable to account for the data.

The induced polarization P_y in the $(e, e'\mathbf{p})$ reaction is identically zero, in the absence of FSI effects (in the one-photon-exchange approximation). This observable thus constitutes a stringent test of various FSI calculations. Figure 2 shows the results, corrected for (small) false asymmetries, as a function of Q^2. One sees that the induced polarizations are small for all measured Q^2 values.

The dashed and dot-dashed curves constitute RDWIA calculations by the Madrid group [17] with the MRW [21] and RLF [22] relativistic optical potentials. For the induced-polarization case, the RDWIA curves do not depend on the nucleon form-factor model chosen in the calculations. Figure 2 confirms the expected smallness of the induced polarizations, and seems to indicate a reasonable agreement with the RDWIA calculation.

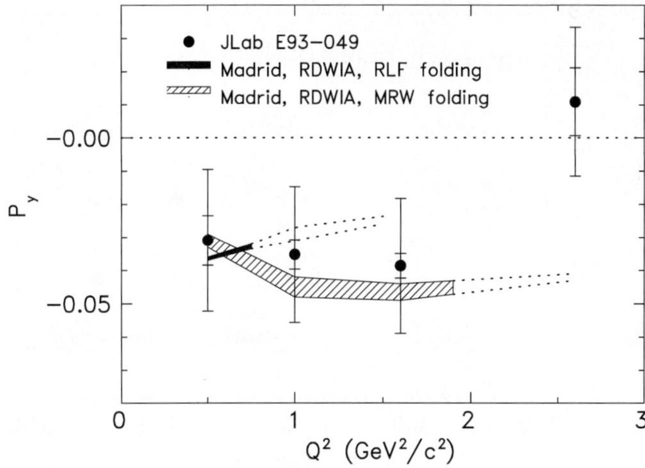

Fig. 2. Measured values of the induced polarizations for the $^4\text{He}(e, e'\mathbf{p})^3\text{H}$ reaction [16]. The inner uncertainty is statistical only, the total uncertainty includes a systematic uncertainty of ± 0.02, due to the imperfect knowledge of the false asymmetries. The *hatched areas* show the results for different RDWIA calculations [17] with the MRW [21] and RLF [22] relativistic optical potentials. All theoretical curves are averaged over the experimental acceptance. Dotted curves indicate calculations in which the optical potentials were extrapolated beyond the region of validity

However, due to the relatively large systematic uncertainties caused by possible false asymmetries in the FPP, the test is not conclusive. A far more detailed study of the induced polarization can be made with data expected from JLab experiment E03-104 [23].

4 Discussion

The disagreement between the data and the RDWIA calculations for the polarization double ratio R is puzzling, since, as mentioned earlier, these relativistic calculations provide good descriptions of, *e.g.*, the induced polarizations measured at Bates in the $^{12}\text{C}(e, e'\mathbf{p})$ reaction [24] and of A_{TL} in $^{16}\text{O}(e, e'p)$ as previously measured at JLab [1]. To find a possible explanation for that disagreement, a detailed study of model dependencies was performed.

First, the model input in the calculations of the Madrid group was studied. Figure 3(a) shows again the $Q^2 = 1.0$ (GeV/c)^2 polarization-transfer double-ratio data. The hatched area indicates the range of different RDWIA calculations. For these calculations, various bound-state wave functions, current operators ($cc1$ and $cc2$), and optical potentials were chosen. The optical potentials include the MRW folding parameters [21] and the RLF folding parameters [22]. The latter are valid only up to $T_{\text{lab}} \approx 400$ MeV, and thus are already beyond its regime of validity for Q^2 of 1.0 (GeV/c)^2. Different optical potentials were constructed for each parameter set based upon the experimental ^3H density, based upon a simple Woods-Saxon density for ^3H with the same root-mean-square radius as the experimental one, and, finally, one with an 30% increased

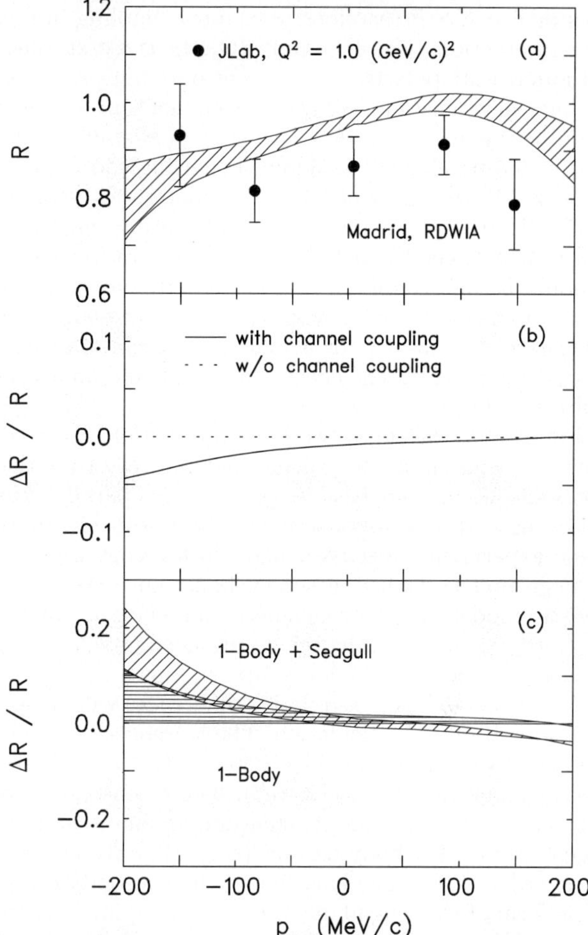

Fig. 3. Study of model sensitivities to the polarization-transfer double ratio in the $^4\text{He}(\mathbf{e}, e'\mathbf{p})^3\text{H}$ reaction at $Q^2 = 1.0$ (GeV/c)^2 . **a** E93-049 polarization transfer double ratio at $Q^2 = 1.0$ (GeV/c)^2 along with a PWIA and different RDWIA calculations of the Madrid group [17]. **b** Relative difference between the polarization transfer ratio including channel coupling and without channel coupling. The calculation follows the approach of [25]. **c** Differences in the polarization double ratios relative to R_{cc1} in the $^4\text{He}(e, e'\mathbf{p})^3\text{H}$ reaction as a function of missing momentum (perpendicular kinematics). Plotted are results for different de Forest current operators with and without the MEC seagull diagram. The 1-body $cc1$ result serves as a baseline. Calculation courtesy of Meucci [26]

root-mean-square radius, resulting in an unrealistic potential. Different parametrizations of the free-nucleon form factors were also used. To a very good approximation, the ratio R is independent of such variations, if the *same* parametrization is used in the ^4He and ^1H calculations. It is evident from Fig. 3(a) that there is hardly any sensitivity in the polarization-transfer ratio at low missing momentum (within the models examined) to the different and partly extreme choices of optical potential, one-body current operator, and bound-state wave function.

Next, the effect of channel coupling on R for the $^4\text{He}(\mathbf{e}, e'\mathbf{p})^3\text{H}$ reaction was estimated. Recently Kelly [25] investigated the sensitivity of recoil-polarization observables in $A(\mathbf{e}, e'\mathbf{p})B$ reactions to channel coupling in final-state interactions (for ^{12}C and ^{16}O). In these studies it was found that polarization-transfer observables for proton knockout with modest missing momentum appear to be quite insensitive to details of the final-state interaction, including channel coupling. Following the approach of Kelly [25], the effect of channel coupling was estimated for the ^{4}He$(\mathbf{e}, e'\mathbf{p})^3$H reaction by calculating the relative difference between the polarization transfer ratio with and without channel coupling. Couplings between the proton $1s_{1/2}$ and neutron $1s_{1/2}$ states were considered. Figure 3(b) shows the results for a four-momentum transfer of $Q^2 = 1.0$ (GeV/c)^2. The effect is minimal, and on average is of the order of $1\% - 2\%$.

Finally, the effects of two-body interactions were considered. Available model calculations indicate that these contributions are smallest in quasielastic, parallel kinematics and at low missing momentum; the kinematics of the experiments discussed here. In a recent work, the Pavia group has studied meson-exchange currents in a relativistic model for electromagnetic one-nucleon emission [26] (for ^{12}C and ^{16}O). Meucci has provided us with similar calculations for ^{4}He; these results are shown in Fig. 3(c) for $Q^2 = 1.0$ (GeV/c)^2 as relative differences with respect to the one-body $cc1$ calculation. The horizontally hatched band covers the range of calculations with three different one-body currents ($cc1$, $cc2$, $cc3$). The calculations confirm that, at low missing momentum, ambiguities due to the choice of the 1-body current ($cc1$, $cc2$, $cc3$) are small, of the order of 3%. The inclusion of the two-body current in the form of the seagull diagram with one-pion exchange (diagonally hatched band) has an asymmetric effect on the polarization ratio about $\mathbf{p}_m = 0$; the effect of MEC is also small. It reduces the polarization transfer ratio by about 2% on average. In addition, these calculations predict the effect of MEC to decrease with increasing four-momentum transfer (not shown).

Also the results of the microscopic calculations of Laget [8] for $Q^2 = 0.4$ (GeV/c)^2 and 0.5 (GeV/c)^2 were found to be nearly identical to the RPWIA results (see Fig. 5) indicating that reaction mechanisms like MEC, IC, or charge exchange do not contribute significantly to R in the present kinematics.

Does the present failure of state-of-the-art calculations to describe R indicates a breakdown of standard meson-nucleon calculations? Probably not. It is likely that more complete and more complicated many-body calculations eventually will account for the data. Similarly, the observed cross-section scaling in deuteron photodisintegration at high energies (see e.g. [27]) does not "prove" a breakdown of meson-nucleon calculations, but rather indicates the onset of a regime where quark-gluon-inspired models may give a more effective basis for the interpretation of the data. In the same way, the observed reduction of R in the ^{4}He$(\mathbf{e}, e'\mathbf{p})^3$H reaction motivates the use of a new approach in the interpretation of the data.

Indeed, a calculation by Lu et al. [28], using a quark-meson-coupling (QMC) model, suggests a measurable deviation from the free-space form-factor ratio over the Q^2

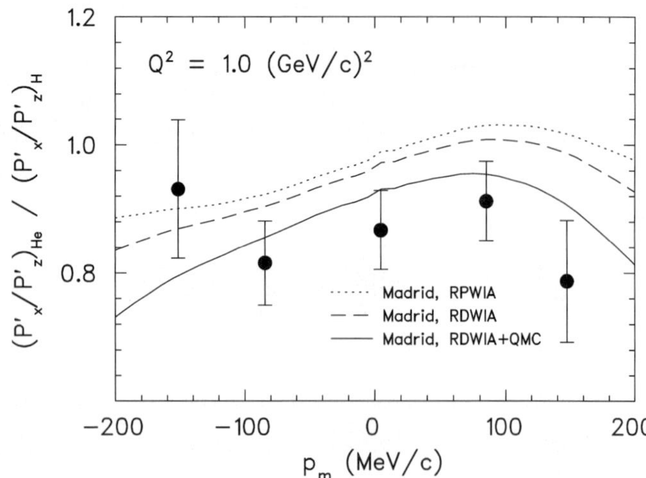

Fig. 4. E93-049 polarization transfer double ratio at $Q^2 = 1.0$ (GeV/c)^2 along with a PWIA and different RDWIA calculations of Udias [17]

range $0.0 < Q^2 < 2.5$ (GeV/c)^2. Note that the calculation is consistent with present constraints on possible medium modifications for both the electric form factor (from the Coulomb Sum Rule, with $Q^2 < 0.5$ (GeV/c)^2 [29,30,31]), and the magnetic form factor (from a y-scaling analysis [32], for $Q^2 > 1$ (GeV/c)^2), and limits on the scaling of nucleon magnetic moments in nuclei [33]. Similar measurable effects have been calculated in the light-front-constituent quark model of Frank et al. [34]. Recently, Yakshiev et al. [35] investigated possible modifications to the electromagnetic form factors of the nucleons in the ^{4}He nucleus in the framework of a modified Skyrme model, up to $Q^2 = 0.6$ (GeV/c)^2. Furthermore, the group of Miller is presently studying medium effects in the framework of a chiral soliton model for the proton [36].

However, the notion of medium modification of single-particle properties like, e.g., the electromagnetic form factors of a nucleon, in a nuclear environment is a purely theoretical concept [37]. Thus, distinguishing possible changes in the spatial structure of nucleons embedded in a nucleus from more conventional many-body effects is only possible within the context of a model; presently the RDWIA calculation of the Madrid group.

The RDWIA results are brought into excellent agreement with the data by replacing the free-nucleon form factor with a density dependent, medium-modified form factor based on the QMC model [28] (solid curve in Fig. 4).

A summary of all available ^{4}He$(\mathbf{e}, e'\mathbf{p})^3$H data [14,16] is shown in Fig. 5 along with different acceptance-averaged calculations. The polarization-transfer double ratio is plotted as a function of Q^2. In order not to obscure the result by small kinematical differences between the individual ^{1}H and ^{4}He measurements, data and calculations are shown with R_{PWIA} as a baseline. The Mainz data point at $Q^2 = 0.4$ (GeV/c)^2 closely coincides with the more recent results at $Q^2 = 0.5$ (GeV/c)^2 of JLab.

On average, the RPWIA calculation (baseline) overestimates the data by $\approx 10\%$. The dotted line shows

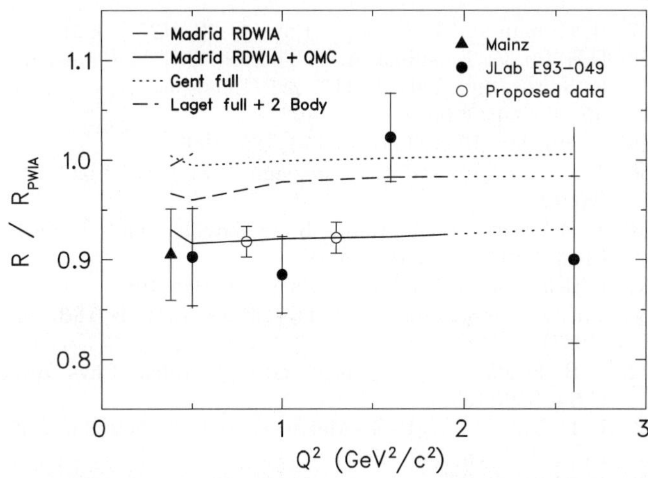

Fig. 5. Superratio R/R_{PWIA} as a function of Q^2 (*closed circles*). The *dashed line* shows the results of the RDWIA calculation of the Madrid group [17]. The *dotted line* shows the results of the full nonrelativistic model by the Gent group [18, 19]. The dash-dotted line shows the results of the full calculation, including two-body currents, of Laget [8]. The *solid line* indicates the full relativistic calculation of the Madrid group [17] including medium modifications as predicted by a quark-meson-coupling model [28]. For $Q^2 > 1.8$ $(\mathrm{GeV/c})^2$ the Udias calculations maintain a constant relativistic optical potential and are indicated as *short-dashed lines*. The lines connecting the acceptance-averaged theory calculations are to guide the eye only. The open circles indicate the projected data for experiment E03-104 [23]

the results of the full nonrelativistic model of the Gent group [18,19]. The full calculation, including two-body currents, of Laget [8] is only available for the lower two Q^2 values (dash-dotted). Both calculations are comparable to the results of RPWIA. The RDWIA calculation (dashed curve) gives a slightly smaller value of R but still overpredicts the data.

The inclusion of medium-modified form factors as predicted by Lu *et al.* [28] into the RDWIA calculation gives a significantly improved agreement with data, with the exception of the datum at $Q^2 = 1.6$ $(\mathrm{GeV/c})^2$ (χ^2 per degree of freedom of 1.3 for the five data points), in contrast to the use of free form factors (χ^2 per degree of freedom of 2.2). All calculations by the Madrid group [17] use the Coulomb gauge, the $cc1$ current operator as defined in [20], and the MRW optical potential of [21]. The $cc2$ current operator gives higher values of R, worsening agreement with the data. As discussed above, various choices for, spinor distortions, current operators, and relativistic corrections affect the theoretical predictions by $\leq 3\%$, and presently cannot explain the disagreement between the data and the RDWIA calculations.

Also shown in Fig. 5 are the projected data points at $Q^2 = 0.8$ $(\mathrm{GeV/c})^2$ and 1.3 $(\mathrm{GeV/c})^2$ for the newly approved JLab experiment E03-104 [23]. This experiment will reduce the statistical uncertainties in the double polarization ratio at each Q^2 by over a factor of two compared to the previous measurements. These two Q^2 values

were selected since they lie in a region where theoretical calculations are expected to be reliable. This measurement will provide one of the most stringent tests to date of the applicability of conventional meson-nucleon calculations.

Any medium modification of bound-nucleon form factors carries implications for the nuclear EMC effect [36]. For example strong constraints are placed on models of the nuclear EMC effect by model-independent relations derived on the basis of quark-hadron duality, which relate the medium modification of the electromagnetic form factors to the modification of the deep-inelastic structure function of a bound proton [38].

5 Summary

Polarization transfer in the quasielastic $(e, e'p)$ reaction is sensitive to possible medium modifications of the bound-nucleon form factor, while at the same time largely insensitive to other reaction mechanisms. The ${}^4\mathrm{He}(e, e'\mathbf{p}){}^3\mathrm{H}$ polarization-transfer double-ratio data are not well described by modern relativistic RDWIA calculations using free nucleons as quasiparticles. The data can be efficiently described if medium-modified nucleons are used in the calculations. However, the evidence from the present data is still limited. New high-precision data expected from JLab Hall A should provide a more stringent test of conventional meson-nucleon calculations.

This work was supported in part by the U.S. Department of Energy under grant DE–FG02–95ER40901. Southeastern Universities Research Association (SURA) operates the Thomas Jefferson National Accelerator Facility under U.S. Department of Energy contract DE–AC05–84ER40150.

References

1. J. Gao et al.: Phys. Rev. Lett. **84**, 3265 (2000)
2. G. van der Steenhoven et al.: Phys. Rev. Lett. **57**, 182 (1986); **58**, 1727 (1987)
3. P. Ulmer et al.: Phys. Rev. Lett. **59**, 2259 (1987)
4. D. Reffay-Pikeroen et al.: Phys. Rev. Lett. **60**, 776 (1988)
5. T.D. Cohen, J.W. Van Orden, and A. Picklesimer: Phys. Rev. Lett. **59**, 1267 (1987)
6. A. Magnon et al.: Phys. Lett. B **222**, 352 (1989)
7. J.E. Ducret et al.: Nucl. Phys. A **553**, 697c (1993)
8. J.M. Laget: Nucl. Phys. A **579**, 333 (1994)
9. A.I. Akhiezer and M.P. Rekalo: Sov. J. Part. Nucl. **3**, 277 (1974); R. Arnold, C. Carlson, and F. Gross: Phys. Rev. C **23**, 363 (1981)
10. With the initial and final electron momentum $\mathbf{k}_i$ and $\mathbf{k}_f$, the coordinate system is given by the unit vectors $\hat{z} = (\mathbf{k}_i - \mathbf{k}_f)/|\mathbf{k}_i - \mathbf{k}_f|$, $\hat{y} = (\mathbf{k}_i \times \mathbf{k}_f)/|\mathbf{k}_i \times \mathbf{k}_f|$, and $\hat{x} = \hat{y} \times \hat{z}$
11. M.K. Jones et al.: Phys. Rev. Lett. **84**, 1389 (2000)
12. O. Gayou et al.: Phys. Rev. C **64**, 038202 (2001)
13. O. Gayou et al.: Phys. Rev. Lett. **88**, 092301 (2002)
14. S. Dieterich et al.: Phys. Lett. B **500**, 47 (2001)
15. Jefferson Lab experiment E93-049 Polarization Transfer in the Reaction ${}^4\mathrm{He}(e, e'p){}^3\mathrm{H}$ in the Quasi-elastic Scattering Region, R. Ent and P. Ulmer, spokespersons

16. S. Strauch et al.: Phys. Rev. Lett. **91**, 052301 (2003)
17. J.M. Udias et al.: Phys. Rev. Lett. **83**, 5451 (1991); J.A. Caballero, T.W. Donnelly, E. Moya de Guerra, and J.M. Udias: Nucl. Phys. A **632**, 323 (1998); J.M. Udias and J.R. Vignote: Phys. Rev. C **62**, 034302 (2000)
18. J. Ryckebusch, D. Debruyne, W. Van Nespen, and S. Janssen: Phys. Rev. C **60**, 034604 (1999)
19. D. Debruyne: Ph.D. thesis, University of Gent (2001)
20. T. de Forest: Nucl. Phys. A **392**, 232 (1983)
21. J.A. McNeil, L. Ray, and S.J. Wallace: Phys. Rev. C **27**, 2123 (1983)
22. C.J. Horowitz: Phys. Rev. C **31**, 1340 (1985)
23. Jefferson Lab experiment E03-104 *Probing the Limits of the Standard Model of Nuclear Physics with the* $^4He(e,e'p)^3H$ *Reaction*, R. Ent, R. Ransome, S. Strauch, and P. Ulmer: spokespersons
24. R.J. Woo et al.: Phys. Rev. Lett. **80**, 456 (1998)
25. J.J. Kelly: Phys. Rev. C **59**, 3256 (1999)
26. A. Meucci, C. Giusti, and F.D. Pacati: Phys. Rev. C **66**, 034610 (2002), and A. Meucci private communication
27. R. Gilman and F. Gross: J. Phys. G **28**, R37 (2002)
28. D.H. Lu, K. Tsushima, A.W. Thomas, A.G. Williams, and K. Saito: Phys. Lett. B **417**, 217 (1998) and Phys. Rev. C **60**, 068201 (1999)
29. J. Jourdan: Phys. Lett. B **353**, 189 (1995)
30. J. Morgenstern and Z.-E. Meziani: Phys. Lett. B **515**, 269 (2001)
31. J. Carlson, J. Jourdan, R. Schiavilla, and I. Sick: Phys. Lett. B**553**, 191 (2003)
32. I. Sick: Comm. Nucl. Part. Phys. **18**, 109 (1988)
33. T.E.O. Ericson and A. Richter: Phys. Lett. B **183**, 249 (1987)
34. M.R. Frank, B.K. Jennings, and G.A. Miller: Phys. Rev. C **54**, 920 (1996)
35. U.T. Yakhshiev, U-G. Meissner, and A. Wirzba: Eur. Phys. J. A **16**, 569 (2003)
36. G. Miller: private communication (2003)
37. H. Arenhövel: private communication (2003)
38. W. Melnitchouk, K. Tsushima, and A.W. Thomas: Eur. Phys. J. A **14**, 105 (2002)

Cold and hot nuclear matter

Eur Phys J A (2004) **19**, s01, 159–164
Digital Object Identifier (DOI) 10.1140/epjad/s2004-03-026-8

EPJ A direct

electronic only

Few-nuclon physics based on chiral dynamics

W. Glöckle[1], E. Epelbaum[1], H. Kamada[2], U.-G. Meißner[3], A. Nogga[4], and H. Witała[5]

[1] Ruhr-Universität Bochum, Institut für Theoretische Physik II, D-44870 Bochum, Germany
[2] Dpt. of Physics, Faculty of Engineering, Kyushu Institute of Technology, 1-1 Sensuicho, Tobata, Kitakyushu 804-8550, Japan
[3] Universität Bonn, Helmholtz-Institut für Strahlen- und Kernphysik (Theorie), Nußallee 14-16, D-53115 Bonn
[4] Department of Physics, University of Arizona, Tucson, Arizona 85721, USA
[5] Jagiellonian University, Institute of Physics, Reymonta 4, 30-059 Cracow, Poland

Received: 4 Aug 2003 / Accepted: 14 Nov 2003 /
Published Online: 6 Feb 2004 – © Società Italiana di Fisica / Springer-Verlag 2004

Abstract. After a brief review of results in few-nucleon physics based on conventional nuclear forces the steps towards nuclear forces in the effective field theory approach constrained by chiral symmetry are indicated. Several low-energy constants are adjusted to nucleon-nucleon phase shift values and in the three-nucleon system to the triton binding energy and the doublet nd scattering length. Applications based on NNLO nuclear forces to the three- and four-nucleon systems are very promising. Finally, a new regularization scheme is described which leads to a nice convergence of the chiral expansion.

PACS. 21.30.Cb Nuclear forces in vacuum – 21.45.+v Few-body systems

1 Introduction

In recent years so called high precision NN forces have been developed [1]. They are constructed with the aim to describe the rich set of NN data up to the pion threshold very well. These forces typically depend on about 45 fit parameters and are not founded on a systematic theoretical approach. The accurate description is considered to be a prerequisite for the use of those forces in few-nucleon sytems with $A > 2$. Over the years accurate methods have also been developed to solve the Schroedinger equation for few-nucleon bound states [2]. The result is that NN forces alone underbind light nuclei as documented in Table 1. In view of the composite nature of nucleons and a relatively low lying excited state of the nucleon, the delta, a three-nucleon force caused by an intermediate delta excitation appears as a natural candidate to fill that gap of underbinding.

Thus three-nucleon force models have been developed around that picture of an intermediate delta excitation generated by two-pion exchanges [6]. Adding them to the NN forces and adjusting their parameters one can easily achieve the correct binding energy of ^{3}H (^{3}He), the first nuclei, where 3N forces are acting. Thus an enriched Hamiltonian consisting of NN and 3N forces can now be applied to predict 3N scattering observables, the binding energy of light nuclei etc. Table 2 displays the prediction [7] for the α particle binding energy using different NN and 3N force combinations. This looks promising.

Also promising are predictions for some 3N scattering observables shown in Figs. 1,2. On the other hand for cer-

Table 1. Theoretical ground state binding energies for various light nuclei and various NN forces in comparison to experiment. The results for $A = 6$ and 8 are from [5]

	Nijm	CD Bonn	AV18	Exp
^{3}H	−7.74	−8.01	−7.6	−8.48
^{4}He	−24.98	−26.26	−24.1	−28.30
^{6}He			−23.9	−29.3
^{6}Li			−26.9	−32.0
^{8}Li			−31.8	−41.3
^{8}Be			−45.6	−56.5

Table 2. Theoretical predictions for the α-particle binding energy for various NN and 3N force contributions

2N+3N forces	^{3}H	^{4}He
CD Bonn + TM [6]	−8.48	−28.4
AV18 + TM [6]	−8.45	−28.36
AV18 + URB [6]	−8.48	−28.50
Exp	−8.48	−28.30

tain other spin observables the addition of these 3N forces does not improve the description, as shown in Fig. 3. This reveals that the spin and momentum dependencies of 3N forces are yet by far not understood. Certainly the variety of meson exchange processes contributing to the NN forces should also show up in 3N forces, which points to a rich set of 3N forces to be explored. Recently steps in that direction have been undertaken [9] by considering three-pion

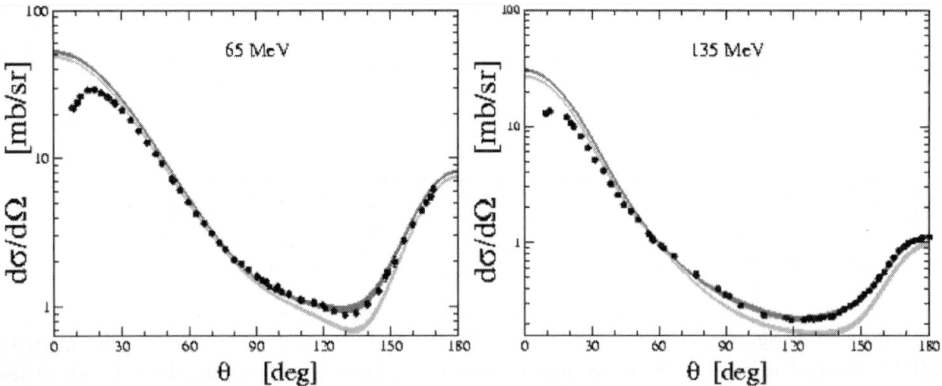

Fig. 1. Differential cross section for elastic nd scattering at $E_{lab} = 65$ und 135 MeV. The light shaded band refers to NN force predictions only, the dark shaded band includes 3N forces. Data from [3], [4]

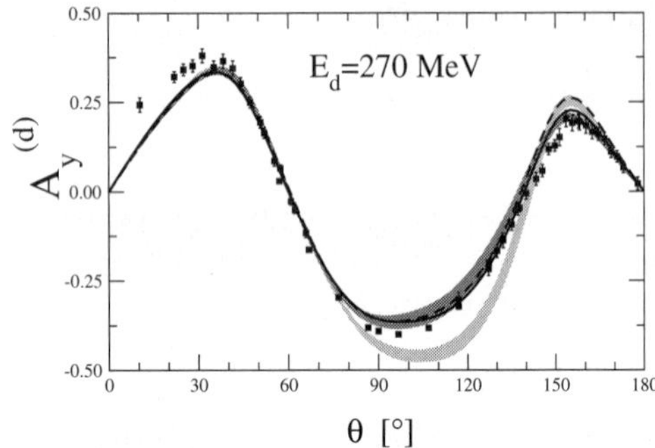

Fig. 2. Deuteron vector analysing power A_y at $E_{lab} = 135$ MeV. The solid (*dashed*) lines refer to AV18 + URBANA IX (CD Bonn + TM') nuclear forces. Data from [8]. For remaining notation see Fig. 1

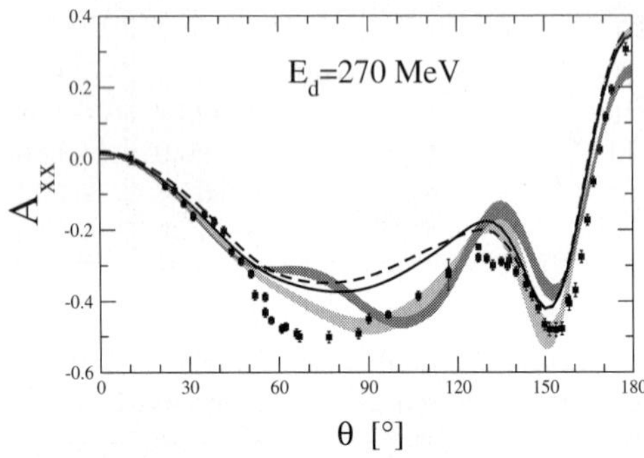

Fig. 3. Tensor analysing power A_y at $E_{lab} = 135$ MeV. For notation see Fig. 2

exchanges with intermediate deltas, which improve significantly the description of low lying states up to $A = 10$, see [10]. Nevertheless what has been and is still missing in all that work is a systematic path towards nuclear forces. This appears to change now with the effective field theory approach constrained by chiral symmetry.

2 Effective field theory approach

It is well known that the QCD Lagrangian for massless up and down quarks is invariant under global $SU(2)_L \times$

$SU(2)_R$ transformations in the flavour space. Further there is ample evidence that this symmetry is spontaneously broken down to the isospin group $SU(2)_V$ and as consequence three massless Goldstone bosons exist, which can be identified with massless pions. The actual nonvanishing masses of the up and down quarks introduce an explicit symmetry breaking which leads to the nonvanishing pion masses. That spontaneously and explicitly broken symmetry governs also the interactions between pion and nucleon fields in an effective field theory set up. There it has to appear as a nonlinear realisation of the chiral group as worked out in [11]. In that effective field theory framework constrained by chiral symmetry the ordering scheme for the most general Lagrangian appears naturally in a low momentum regime, where the momenta stay below a certain mass scale. A convenient formulation to control low momenta for nucleons is the heavy baryon formalism, see e.g. [12]. One arrives at an effective Lagrangian for pion and nucleon fields suitable for low energy nuclear physics, which is arranged according to increasing values of the chiral dimension

$$\Delta = d + \frac{1}{2}n - 2 \,. \tag{1}$$

Here d is the number of derivatives or M_π-insertions and n the number of nucleon fields for each coupling term. Chiral symmetry enforces that $\Delta \geq 0$. Besides coupling terms among pions and nucleons there occur also coupling terms among pions themselves and nucleon themselves.

To arrive at nuclear forces old-fashioned time-ordered perturbation theory has originally been applied [13,14], which, however, leads to energy-dependent forces. They are not suitable for applications to nuclear systems with > 2 nucleons. We apply a method of unitary transformations [15], where the field theoretical Hamiltonian acting in the Fock space of pions and nucleons is block-diagonalised such that the space of pure nucleon states is decoupled from the rest space (which includes pions). This decoupling and thus the derivation of the effective nucleonic Hamiltonian can be organised according to the ratio of generic external nucleonic momenta and the previously mentioned mass scale [16]. The mass scale is of the order of the ρ-mass where new physics beyond pion

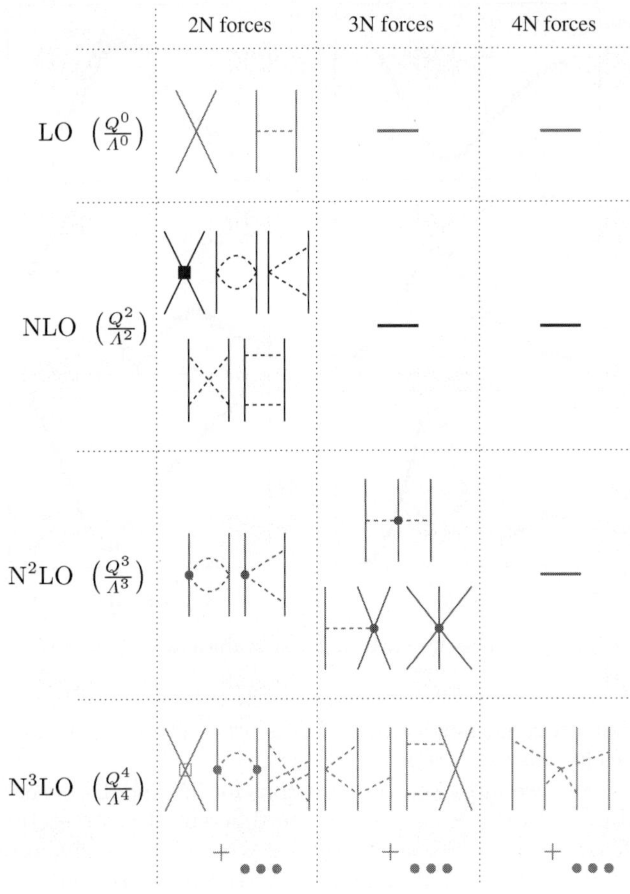

Fig. 4. Hierarchy of the nuclear forces

but no new contact forces. Note that all multi-pion exchanges are parameter free in the sense that the constants at the vertices, called low-energy constants (LEC's), can be taken from the analysis of the πN system. The constants going with the NN contact forces, however, have to be adjusted in the NN system. At NNLO there occur for the first time nonvanishing three-nucleon forces of three topologies: a two-pion exchange, a one-pion exchange between a NN contact interaction and the third nucleon and a pure three-nucleon contact force. The one strength parameter of the latter one has to be adjusted in the 3N system, whereas the one strength parameter at the NNNNπ vertex could in principle be determined from pion production in the NN system. Since this has not yet been worked out sufficiently well we will adjust this constant also in the 3N system (see below). Then, at NNNLO, additional two- and three-pion exchange processes and more NN contact forces contribute in the NN system. Very interestingly, at this order, a whole host of new three-nucleon forces arises, which do not include new 3N contact interactions and therefore can be expected to provide a very interesting predictive power for systems with $A \geq 3$. Finally, in that order, for the first time four-nucleon forces occur. The overview displayed in Fig. 4 nicely shows the ordering scheme that NN forces are stronger than 3N forces and 3N forces are stronger than 4N forces etc. Of course all that is restricted to a low momentum regime where the generic nucleon momenta Q are smaller than the chiral symmetry breaking scale.

The loop diagrams have to be regularized. In [18,19] we used an infinite cut-off, which is in that case equivalent to dimensional regularization, and combined the diverging terms with the contact forces. As a result at NNLO rather strong intermediate and short range attractive forces arise, which lead to a strong cut-off dependence in the D-waves and to deeply bound spurious NN states in low partial waves. Though these states lie outside the range of validity of the low-momentum theory and do not harm the low energy NN observables, they cause technical difficulties in $A > 2$ systems. For instance in the Faddeev–Yakubovsky scheme NN forces are summed up into NN t-matrix, which has poles at NN bound state energies. These spurious NN bound states appearing as additional poles have to be handled with great care which is technically not trivial [18]. In addition, they also lower the 3N binding energy because of orthogonality arguments. As a consequence the 3N forces have to provide more attraction than one is used to have in the conventional framework of nuclear forces [20]. In our first approach [18] we avoided that dynamical scenario by lowering artificially the values of the LEC's c_3 and c_4 at the NπN vertices below the strengths found in the πN system. This leads to NN forces which in [18] and in the following will be denoted by NNLO* forces and which no longer support spurious NN bound states. We refer to [18] for a discussion of that procedure. Now we introduced a different regularisation scheme [21], which will be described below and restore the consistency of the LEC's c_3 and c_4 with the πN system.

exchanges occurs. The resulting nuclear forces are of two types: multi-pion exchanges and a sequence of nucleonic contact forces. To each nuclear force diagram a mass dimension can be associated which is given as

$$\nu = -4 + 2N + 2L + \sum_i V_i \Delta_i, \qquad (2)$$

where N is the number of nucleons entering (or leaving) the diagram, L is the number of loops, V_i the number of vertices of type i and Δ_i the chiral dimension mentioned before. Because of the property $\Delta \geq 0$, ν can not be negative and starts for $N \geq 2$ at $\nu = 0$. Thus the nuclear forces can be classified according to increasing values of ν. This is schematically displayed in Fig. 4.

We see for $\nu = 0$ the one-pion exchange acompanied by NN contact forces. The contributions with $\nu = 0$ are usually called the leading order (LO) contributions. $\nu = 1$-terms do not appear due to parity conservation. Thus the next-to-leading order (NLO) terms go with $\nu = 2$ and comprise various types of two-pion exchanges and several contact forces quadratic in the momenta and M_π. At $\nu = 3$, the next-to-next-to-leading order (NNLO), there occur additional two-pion exchanges with higher order vertices,

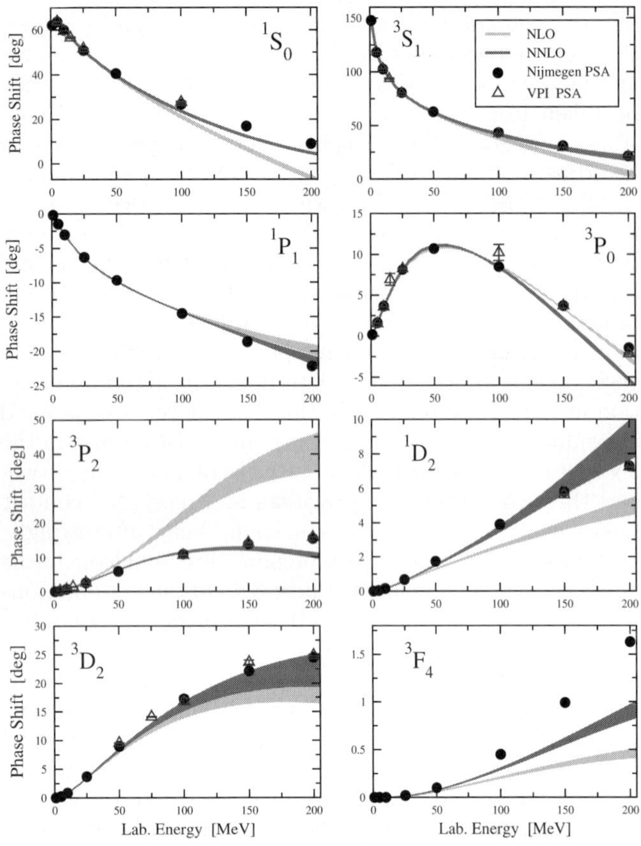

Fig. 5. NN phase shifts at NLO and NNLO

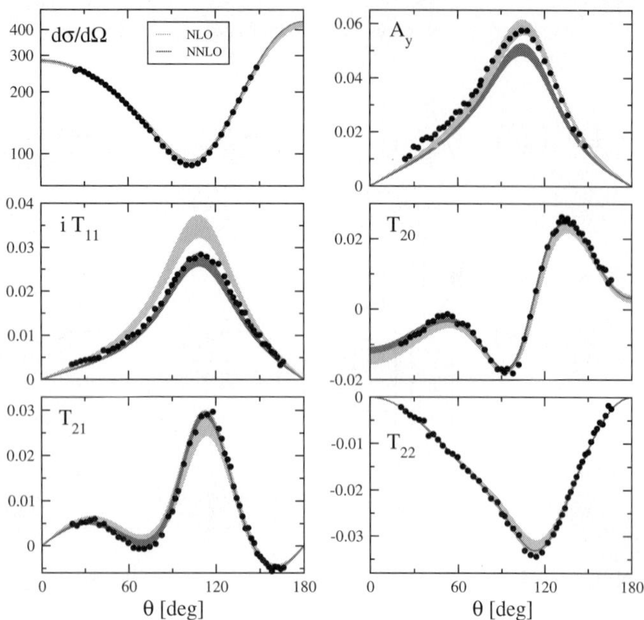

Fig. 6. nd elastic scattering observables at $E_{\text{lab}} = 3$ MeV

As a last step one has to adjust the forces to the low momentum regime where the theory is valid. We enforce that by introducing smooth cut off functions in the form

$$V_\Lambda^{\text{reg}}(\mathbf{p}', \mathbf{p}) = e^{-p'^4/\Lambda^4}\, V(\mathbf{p}', \mathbf{p})e^{-p^4/\Lambda^4} \tag{3}$$

It turns out that

$$500 \text{ MeV} \le \Lambda \le 600 \text{ MeV} \tag{4}$$

is a good choice. The cut-off should be not too small in order not to cut off the pion exchange physics, and not too large in order not to enter into the domain of short range physics, which is not controlled in EFT. One expects [22] that the dependence will get weaker with increasing order in ν.

As a first step in the application the LEC's for the NN and 3N forces have to be adjusted. At LO and NLO there are 2 +7 LEC's going with the NN contact forces. They are adjusted [23] to the S- and P-wave NN phase shifts and to ϵ_1 which are known from NN phase shift analysis. The two LEC's going with the 3N forces are adjusted to the ^{3}H binding energy and the doublet nd scattering length $^2a_{nd}$ [19]. Then up to NNLO all parameters are fixed and the Hamiltonian including NN and 3N forces can be applied to predict other 3N and $A \ge 4$ observables.

In Fig. 5 we display some NN phase shifts at NLO and NNLO. We see a nice improvement in going to NNLO, which comes about by additional two-pion exchanges. In

that order no new contact forces arise. Deuteron properties are also well reproduced [18].

Now we move on to predictions for 3N and 4N observables. In Figs. 6,7 we show the differential cross section in elastic nd scattering together with vector and tensor analysing powers at NLO and NNLO (this still refers to the choice NNLO*, mentioned above and will be changed in the near future). We see for both energies a nice improvement at NNLO in relation to NLO. The so called low energy A_y-puzzle [24], however, remains still open at this order and we expect that only higher order 3N forces and possibly relativistic corrections will solve that long standing problem. Altogether these results are very promising. This is also the case regarding the α-particle binding energy shown in Table 3. While at NLO there is still a rather large Λ-dependence it shrinks at NNLO and one ends up rather close at the experimental value (Please note that at this level we restricted ourselves to np forces only and had to correct for that, see [19]). Of course that shrinkage is also connected to the fact that at NNLO the ^{3}H binding energy has been adjusted. The strong correlation between ^{3}H and ^{4}He binding energies, known from investigations with conventional forces [25,7], enforces that the ^{4}He binding energy can not be far off the experimental value. There are also interesting and promising results for the $p + d \rightarrow N + N + N$ break up process and we refer the reader to [19].

Table 3. Chiral predictions for the triton- and α-particle binding energies at NLO and NNLO* compared to the experimental values, which are corrected for np forces

	NLO	NNLO	"Exp"
^{3}H	$-7.53\ldots-8.54$	-8.68	-8.68
^{4}He	$-23.87\ldots-29.57$	$-29.51\ldots-29.98$	-29.6

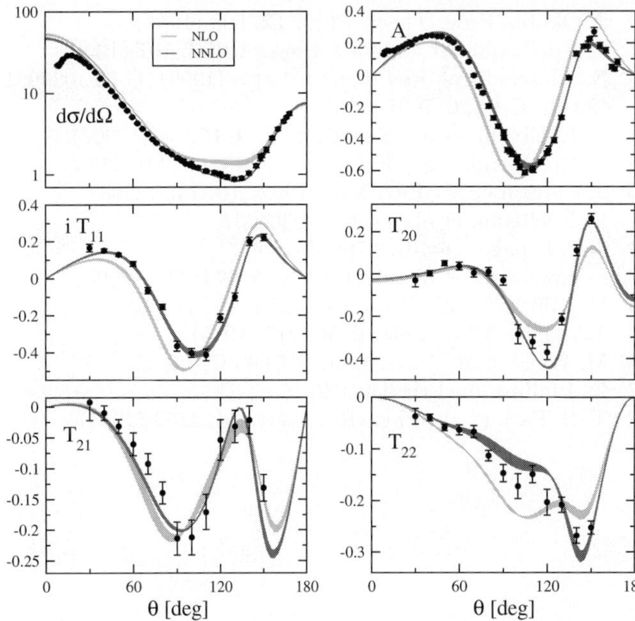

Fig. 7. nd elastic scattering observables at $E_{\mathrm{lab}} = 65$ MeV

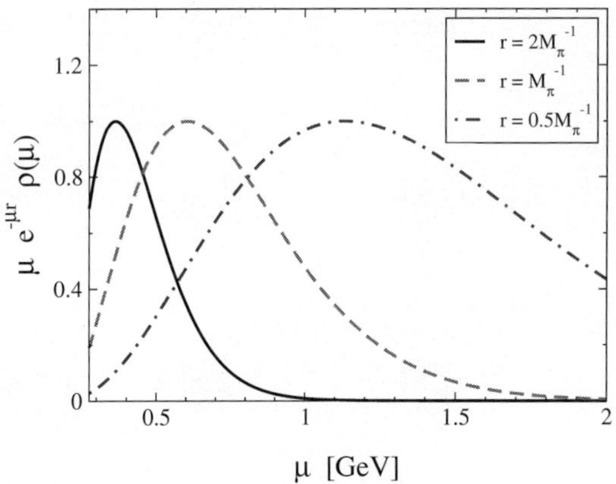

Fig. 8. The integrand of (6) for various r-values

The last step forward to be reported in this overview are recent investigations [21] on the spectral function regularisation (SFR). The two-pion exchange forces can be written as

$$V(q) = \frac{2}{\pi} \int_{2M_\pi}^{\infty} d\mu\, \mu\, \frac{\rho(\mu)}{\mu^2 + q^2} \,, \tag{5}$$

modulo subtractions. This is obviously a superposition of Yukawa interactions, where the spectral function $\rho(\mu)$ is known analytically and simply related [17] to the analytical expressions of the potential. In configuration space it results

$$V(r) = \frac{1}{2\pi^2 r} \int_{2M_\pi}^{\infty} d\mu\, \mu\, e^{-\mu r}\, \rho(\mu) \tag{6}$$

It is instructive to regard the integrand in (6) as a function of μ, the mass exchanged between the two nucleons, and this for different pair distances r. This is shown in Fig. 8 for the isoscalar central part of the subleading (i.e. NNLO) two-pion exchange potential. We see at small r-

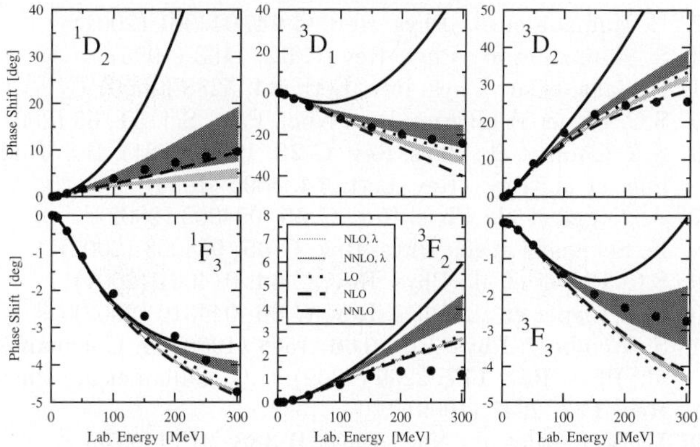

Fig. 9. Peripheral NN phase shifts at LO (*dotted line*), NLO and NNLO in dimensional regularization (*dashed and solid lines*) compared to predictions for SFR (*light and dark shaded bands*)

values large μ-components contribute in contrast to large r's where only small μ-components contribute substantially. Now the exchange of masses μ larger than say Λ belong to short range physics parametrised in this effective field theory approach by contact forces. Thus it is mandatory to cut-off the integral over μ. We do it by a sharp cut off, which introduces another parameter λ, which can be chosen in a similar range as Λ. This "long-distance" regularisation cures now the insufficient results achieved in dimensional regularisation (or infinite cut-off regularisation) at NNLO and at higher orders. We illustrate that in Fig. 9 in the case of peripheral NN phase shifts. We see the results where the loops are evaluated in dimensional regularisation, deviating drastically from the NN phase shift values, whereas using that new regularisation scheme one achieves a nice convergence in going from NLO to NNLO.

This regularization scheme is presently applied also to NNNLO where additional short range forces allow for a rather good description of the NN phase shift values up to about 200–250 MeV nucleon laboratory energy. In that order also a rich group of 3N forces occur, which is under investigation.

Summarising, this new effective field theory approach constrained by chiral symmetry is a systematic way to generate nuclear forces where NN and 3N forces are consistent. Because of the low-momentum cut-off it also allows to incorporate in a well converging manner relativistic corrections.

Moreover, since that approach is based on a Lagrangian the coupling of the photon to the pion-nucleon system is well defined and nucleonic electromagnetic current operators can be constructed which are consistent to nuclear forces, a requirement which is not sufficiently well taken care of in the conventional approach up to now. First steps in that direction have already been done [26, 27,28].

References

1. R.B. Wiringa et al.: Phys. Rev. C **51**, 38 (1995); V.G.J. Stoks et al.: Phys. Rev. C **49**, 2950 (1994); R. Machleidt: Phys. Rev. C **63**, 024001 (2001)

2. H. Kamada et al.: Phys. Rev. C **64**, 044001 (2001)
3. S. Shimizu et al.: Phys. Rev. C **52**, 1193 (1995)
4. H. Sakai et al.: Phys. Rev. Lett. **84**, 5288 (2000)
5. S.C. Pieper et al: Ann. Rev. Nucl. Part. Sci. **51**, 53 (2001)
6. S.A. Coon et al.: Phys. Rev. C **23**, 1970 (1981); B.S. Pudliner et al.: Phys. Rev. Lett. **74**, 4396 (1995)
7. A. Nogga et al.: Phys. Rev. C **65**, 054003 (2002)
8. K. Sekiguchi et al.: Phys. Rev. C **65**, 034003 (2002)
9. S.C. Pieper et al.: Phys. Rev. C **64**, 014001 (2001)
10. S.C. Pieper et al.: Phys. Rev. C **66**, 044310 (2002)
11. S. Weinberg: Phys. Rev. **166**, 1568 (1968); S. Coleman et al.: Phys. Rev. **177**, 2239 (1969); C.G. Callan et al.: Phys. Rev. **177**, 2247 (1969)
12. V. Bernard et al.: Nucl. Phys. B **388**, 315 (1992)
13. S. Weinberg. Nucl. Phys. B **363**, 3 (1991)
14. C. Ordóñez et al.: Phys. Rev. Lett. **72**, 1982 (1994)
15. S. Okubo: Prog. Theor. Phys. **12**, 603 (1954)
16. E. Epelbaoum et al.: Nucl. Phys. A **637**, 107 (1998)
17. N. Kaiser: Phys. Rev. C **61**, 014003 (1999); C **62**, 024001 (2000); C **64**, 057001 (2001)
18. E. Epelbaum et al.: Eur. Phys. J. A **15**, 543 (2002)
19. E. Epelbaum et al.: Phys. Rev. C **66**, 064001 (2002)
20. J. Carlson et al.: Rev. Mod. Phys. **70**, 743 (1998)
21. E. Epelbaum et al.: nucl–th/0304037
22. G.P. Lepage: nucl-th/9706029
23. E. Epelbaum et al.: Nucl. Phys. A **671**, 295 (2000)
24. W. Glöckle et al.: Phys. Rep. **274**, 107 (1996)
25. J.A. Tjon: Phys. Lett. B **56**, 217 (1975)
26. M. Walzl et al.: Phys. Lett. B **513**, 37 (2001)
27. D. Phillips: nucl-th/0304046
28. T.-S. Park et al.: Phys. Rev. Lett. **74**, 4153 (1995)

Eur Phys J A (2004) **19**, s01, 165–169
Digital Object Identifier (DOI) 10.1140/epjad/s2004-03-027-7

EPJ A direct

electronic only

The ^{4}He(e,e'p)^{3}H reaction at JLab

Bodo Reitz, for the Jefferson Lab Hall A Collaboration

Thomas Jefferson National Accelerator Facility, Newport News, Virginia 23606, USA

Received: 5 Sep 2003 / Accepted: 14 Nov 2003 /
Published Online: 6 Feb 2004 – © Società Italiana di Fisica / Springer-Verlag 2004

Abstract. In the recent experiment E97-111 at Jefferson Lab the unseparated cross section for the (e,e'p) reaction on ^{4}He was measured at recoil momenta up to 530 MeV/c. In the plane-wave impulse approximation, many calculations predict a sharp minimum in the cross section for recoil momenta around 450 MeV/c and show that its location is sensitive to the short-range part of the internucleon potential. However, reaction dynamic effects such as final-state interactions and meson-exchange currents can obscure such a minimum. To distinguish and study these effects data were taken at several different kinematic points. The preliminary results of the experiment are presented and compared to recent model calculations.

1 Introduction

Studying few-body nuclear targets via the (e,e'p) reaction is a powerful method to investigate specific aspects of the nucleus. The ^{4}He nucleus is an especially interesting target since it already has many of the ingredients of a complex, heavy nucleus, while as an A=4 system, microscopic calculations are still feasible. Measurements of the cross section for the two-body breakup reaction ^{4}He(e,e'p)^{3}H and extracting the spectral function for ^{4}He $\rightarrow t + p$ allow a study of the effective nucleon momentum distributions in this nucleus. Those momentum distributions are sensitive to ground-state short-range correlations, as well as to reaction dynamic effects such as final-state interactions, meson-exchange contributions, Δ-excitations, and relativistic effects.

The higher beam energies available at Jefferson Lab (JLab) as compared to other facilities such as MAMI, Mainz or NIKHEF, Amsterdam, provide more flexibility in the selection of kinematics and allow an extension of the measurements to higher momentum transfers. That way it is possible at JLab to obtain the spectral function at high recoil momentum in parallel kinematics. This gives access to regions where short-range correlations and possibly the internal structure of the nucleons become important. In the classical nuclear-physics literature nuclei are mostly described in terms of the independent-particle model, which itself can be derived from Hartree-Fock type calculations using effective interactions. These effective interactions have to be constructed by a procedure which has to take the short-range part as well as the long-range part of the NN-interaction into account. Whereas the long range part of the interaction is well understood in terms of the exchange of physical mesons, at shorter range only a phenomenological description is available. This presumably reflects a breakdown of the meson-exchange picture

at small separations. Measuring the short-range component of the NN-interaction will teach us about the transition from mesonic to chromodynamic degrees of freedom.

An especially promising example for studying short-range correlations is the ^{4}He $\rightarrow t + p$ spectral function, as discussed in [1]. Using an harmonic-oscillator nuclear model, these authors find a spectral function which monotonically drops as the recoil momentum of the struck proton increases. However, if they use realistic NN-interactions which include two-body currents via the ATMS method, the resulting spectral function is shaped like a classic diffraction pattern with a minimum at recoil momenta around 450 MeV/c. Furthermore, the structure of the spectral function at high momenta is sensitive to the short-range part of the NN-interaction. It is fortuitous that it is not possible to couple a proton and a triton in a relative $L = 2$ state to the $J^\pi = 0^+$ ^{4}He ground state. Otherwise the longer-ranged tensor correlations which are strong in the d-wave channel could severely obscure the minimum, as it happens e.g. for the two-body breakup of ^{3}He $\rightarrow d + p$. Therefore, it was concluded in [1], that an observation of a minimum in the ^{4}He $\rightarrow t + p$ spectral function could be directly linked to the s-wave correlation in the ground state. Other calculations of the spectral function for this channel [2,3] yield a similar result. The position of the minimum, as well as the position of the second maximum, again depend on the details of the interaction used.

The spectral function can be determined by measuring the cross section for the (e,e'p) reaction, which is to first order proportional to the spectral function

$$\frac{d^6\sigma}{d\Omega_{e'}dE_{e'}d\Omega_{p'}dE_{p'}} = K\sigma_{eN}\frac{1}{\eta}\alpha_{SF}S\left(\mathbf{p_r}, \epsilon_m\right) \quad (1)$$

where K is a kinematical factor; η is the recoil factor; α_{SF} a spectroscopic factor; σ_{eN} the elementary off-shell

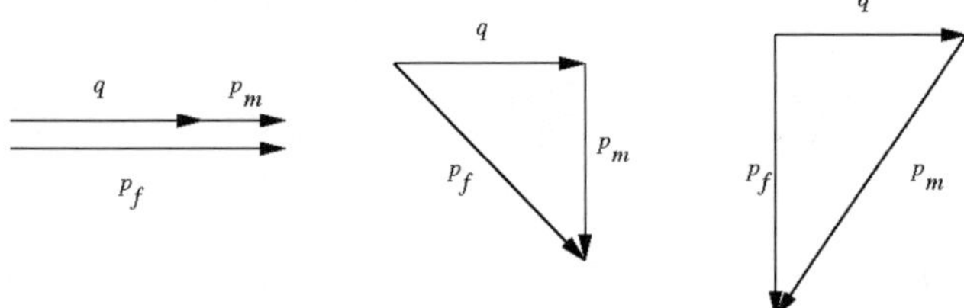

Fig. 1. Various values of p_m for fixed $(\omega, \mathbf{q})$

eN cross section, which includes all the dependence on the polarization of the virtual photon; and S is the spectral function, which gives the joint probability to find a nucleon inside the nucleus with a momentum $-\mathbf{p_r}$ and a separation energy ϵ_m. In this first order approximation for the (e,e'p) reaction the separation energy is given by the missing energy, and the recoil momentum $|\mathbf{p_r}|$ is equal to the missing momentum p_m. Thus a measurement of the (e,e'p) cross section in the appropriate kinematical region is a direct probe of short-range correlations in ^{4}He. However, this first-order approximation is not perfect and the plane-wave impulse approximation (PWIA) does not give a complete description of such experimental data. Reaction dynamic effects such as final-state interactions (FSI), meson-exchange currents (MEC), Δ-excitations (IC), and relativistic effects are relatively enhanced in the region of interest, since the cross section goes through a minimum.

Two experiments[4,5] have previously measured the cross section for ^{4}He(e,e'p)^{3}H at recoil momenta covering the range of interest. Neither of these experiments observed a significant signature of the dip near 450 MeV/c. According to [6,7,8,9], the absence of a minimum is due to the combined effects of FSI and MEC. Figure 1 shows how these measurements were done: the electron kinematics were fixed to an electron energy transfer ω of 215 MeV and to a momentum transfer $|\mathbf{q}|$ of 400 MeV/c. The missing momentum was varied by changing the detection angle for the recoiled proton. At a recoil momentum of 450 MeV the angle θ_{pq} between $\mathbf{q}$ and the outgoing proton was therefore about 50°.

Many ideas have been formulated about how to suppress contamination from these reaction dynamic processes in experiments. They usually require more kinematical flexibility than previously possible at the facilities in Mainz and at NIKHEF, due to the limited beam energy available there, which puts correlated constraints on ω, q, and p_m. To reach the high missing momenta p_m at NIKHEF it was necessary to detect the knock-out proton at a large angle with respect to $\mathbf{q}$. Final-state interactions can seriously distort measurements in this region, since at the same electron kinematics processes involving FSI are possible, where the primarily knocked out proton had a smaller recoil momentum, but due to FSI scattered to larger angles and therefore is reconstructed at larger p_m. Since the cross section at lower missing momentum can be several orders of magnitude higher, those rescattered events can significantly contribute and even dominate the cross section in the dip region. This effect can be partly avoided or at least minimized utilizing parallel kinematics,

where the recoiled proton is detected along the q-axis. In terms of the y parameter, where y is the minimum momentum a struck proton could have originally had while still satisfying the measured (e,e') kinematics, parallel kinematics coincide with $|y| = p_m$. However, there are two possibilities to fulfill this condition, positive and negative y, corresponding to parallel (the struck proton moves in the same direction as the virtual photon) and antiparallel alignment (the struck proton moves in the opposite direction). This parameter also provides a relation between the chosen electron kinematics and the quasi-free peak position. Positive y indicates an excess of energy transfer relative to the momentum transfer (high-energy side of the quasi-elastic peak). Although this would be undesirable for inclusive experiments, it is thought to be advantageous for the special situation of (e,e'p) in parallel kinematics, since both $\mathbf{q}$ and $\mathbf{p}_s$ must line up to the final ejected-proton momentum, suppressing contaminating or multistep processes. Negative y would be favorable for inclusive measurements, since due to the smaller value of the energy transfer MEC and IC effects are smaller. These qualitative arguments for utilizing parallel kinematics are supported by calculations [10,11]. Those calculations also show that although both positive and negative y can help to suppress reaction dynamic effects, the positive y option is superior for (e,e'p).

The higher beam energies available at JLab allow a substantial variation in the four momentum transfer Q^2 for a given ϵ_m, p_m region. Those variations are helpful in two respects: to help discriminate between one- and two-body currents contributing to the cross section and to suppress the contaminant two-body currents. The one-body direct knockout process of interest only depends on Q^2 through the electron-proton cross section σ_{ep}, while MEC and IC contributions are expected to have a very different Q^2 behavior. Higher values of Q^2 will help to suppress MEC and IC contributions due to the additional $1/Q^2$ dependences of the meson propagators $N\pi$ and $N\rho$, and of the $NN\pi$ ($NN\rho$) form factors.

Close to quasi-elastic kinematics the momentum transfer essentially determines the momentum of the outgoing proton. FSI are a strong function of the proton energy. From proton scattering experiments it is known that they are lowest at proton momenta of about 700 MeV/c. Above this momentum, absorption effects begin to increase, but the elastic rescattering continues to decrease. However, for the case of the two-body breakup the latter effect is more important; therefore higher momentum transfer appears to be favorable in terms of suppressing FSI.

2 The experiment

The E97-111 experiment ran in the fall of 2000 at Jefferson Lab Hall A, using the standard equipment available there: two high resolution spectrometers (HRS) with standard detector packages and the cryo-target system for a high pressure gasous ^{4}He target [12]. All ideas about suppression of reaction dynamic effects described in the previous section were accommodated in this experiment.

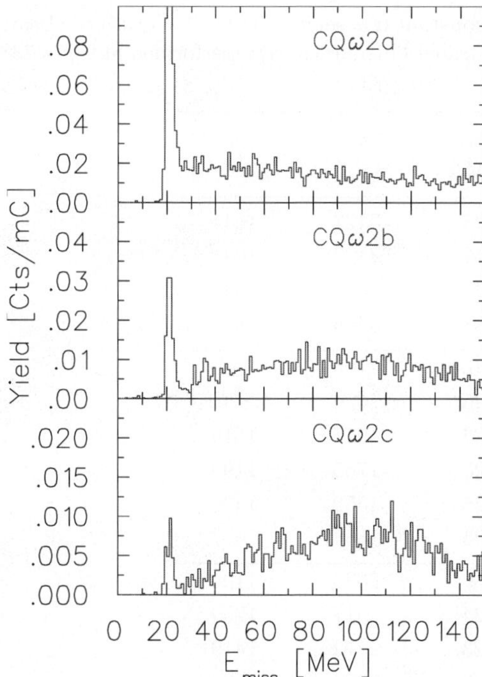

Fig. 2. Missing-energy spectra for the E97-111 experiment. Shown are the spectra for a beam energy of 3952 MeV, an electron scattering angle of 20.90°, and an average missing momentum of 395 MeV/c (*top*), 446 MeV/c (*middle*), and 495 MeV/c (*bottom*)

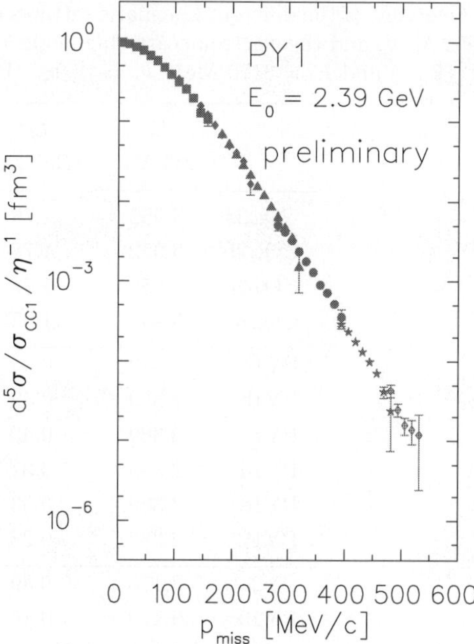

Fig. 3. Preliminary results for the reduced cross section for the six PY1 parallel kinematics at a beam energy of 2390 MeV. The momentum transfer Q^2 ranged from 0.44 $(\text{GeV/c})^2$ for the lowest missing momenta to 0.28 $(\text{GeV/c})^2$ for the highest missing momenta. The error bars only show the statistical uncertainty

The main emphasis was on measuring the ^{4}He(e,e'p)^{3}H cross section at recoil momenta up to 530 MeV/c in parallel kinematics at two different beam energies 2.389 GeV and 3.170 GeV/c. Additional data were taken in two quasi-perpendicular kinematics, with ω fixed to 525 and 487 MeV and Q^2 values of 1.78 and 1.82 $(\text{GeV/c})^2$. Each of these kinematic points required several settings of the two Hall A HRS spectrometers, summarized in Table 1.

The exclusiveness of the two-body breakup channel is guaranteed by means of cuts on the missing energy ϵ_m. Since the residual nucleus is a triton, which has no excited states, the ^{4}He(e,e'p)^{3}H reaction will only occur at $\epsilon_m = 19.81$MeV/c. The continuum is well separated, with a threshold for the three-body breakup of 26.1 MeV and 28.3 MeV for the four-body breakup. Figure 2 shows the missing energy spectra for the three settings in the so-called CQω2 configuration, the peak at roughly 20 MeV corresponds to two-body breakup events.

3 Preliminary results

Preliminary results for the reduced cross sections in the PY1 kinematics are shown in Fig. 3. The error bars show the statistical error only. The cross section is divided by the elementary e-p off-shell cross section σ_{CC1}, using the description of [13] and the recoil factor η, to remove the basic kinematical dependence on the polarization of the virtual photon. At this beam energy recoil momenta from 0 to 530 MeV/c were covered. The reduced cross section

falls monotonically within this momenta range. Figure 4 shows a similar plot of the reduced cross section for the PY2 kinematics at a higher beam energy of 3.17 GeV. The experimental data again cover recoil momenta up to 530 MeV/c. Two sets of theoretical predictions are also shown. The first set from J.M. Laget [14] consists of a PWIA calculation (dashed), a calculation including FSI (dotted), and the full calculation (solid), including FSI as well as MEC and IC. For the FSI at lower energies the phase shift description of [6] was used, which describes elastic NN scattering. At higher energies the high energy parameterization of the NN scattering amplitude of [15] was used, its imaginary part representing the absorptive part of the NN interaction.

The second group of calculations is from H. Morita and C. Ciofi degli Atti [16]. The first curve (long dashes) is a Glauber type calculation (labeled G), and the dashed-dotted curve (labeled G+FFT) additionally includes finite formation time (FFT) effects [17]. Although data were taken in parallel kinematics and at high momentum transfers, the reduced cross section still falls monotonically in the investigated region, with no sign of a minimum or a change in the slope. This feature appears in all but the PWIA calculations. Neither of the full calculations preserves the minimum in the spectral function at this kinematical setting. Laget's calculations indicate, that this is mainly due to FSI, whereas the inclusion of MEC and IC has only a small effect on the predicted cross section. These calculations also show that below 280 MeV/c the PWIA cross section is larger than the one of the full calculation, above that value the PWIA cross section is smaller.

Table 1. Overview of the different kinematic settings of E97-111. For the constant $\mathbf{q}$-ω settings (CQω2, CQω3) the beam energy E_i was 3952 MeV, and the electron scattering angle θ_e was 20.90°. The parallel kinematics were performed at $E_i = 2389$ MeV, $\theta_e = 16.9^\circ$ (PY1) and $E_i = 3170$ MeV, $\theta_e = 18.98^\circ$ (PY2)

	E_0 (GeV)	Q^2 ((GeV/c)^2)	ω (MeV)	p_m (MeV/c)	p_e (MeV/c)	p_p (MeV/c)
CQω2a	3.952	1.78	525	395	3427	1041
CQω2b	3.952	1.78	525	446	3427	1041
CQω2c	3.952	1.78	525	495	3427	1041
CQω3	3.952	1.82	487	468	3465	960
PY1a	2.389	0.44	284	26	2105	744
PY1b	2.389	0.42	369	126	2020	870
PY1c	2.389	0.40	478	226	1911	1015
PY1d	2.389	0.37	624	325	1765	1193
PY1e	2.389	0.33	830	425	1559	1431
PY1f	2.389	0.28	1035	495	1354	1657
PY2a	3.170	0.89	537	24	2633	1105
PY2b	3.170	0.85	653	124	2517	1250
PY2c	3.170	0.80	798	223	2372	1419
PY2d	3.170	0.73	985	323	2185	1627
PY2e	3.170	0.65	1239	423	1931	1900
PY2f	3.170	0.57	1481	493	1689	2154

This indicates that the FSI tends to shift cross section from low p_m (where the spectral function is high) to the dip region (where the PWIA cross section is tiny). The FFT effects, which are believed to restore the minimum at very high momentum transfers, are in these kinematics not of great importance. At lower recoil momenta the full calculations of both groups give a reasonable description of the data. Starting at around 350 MeV/c they start to differ from the data and among themselves. Whereas the Laget calculation overpredicts the cross section, the Ciofi calculation underpredicts it at high missing momenta.

Figure 5 shows the preliminary results for the reduced cross section in perpendicular kinematics CQω2 at a beam energy E_i of 3952 MeV and a Q^2 value of 1.78 (GeV/c)^2. The experimental data cover missing momenta from 300 to 530 MeV/c. The experimental data are again compared to the calculations from J.M. Laget. In the investigated region of missing momenta, the experimental data fall monotonically, there is no visible minimum around p_m=450 MeV/c and no sign of a second maximum at higher values of p_m, although the data can not exclude the possibility that the reduced cross section flattens out at missing momenta above 500 MeV/c. The calculations from Laget show again that in the framework of PWIA a minimum at 460 MeV/c should appear, but that it is mainly filled due to reaction dynamic effects. In contrast to the PY2 settings, both FSI and MEC/IC effects contribute to the cross section in the dip. However, at this higher momentum transfer it is predicted that although the dip is filled significantly, there is still some structure left in the reduced cross section, with the reduced cross

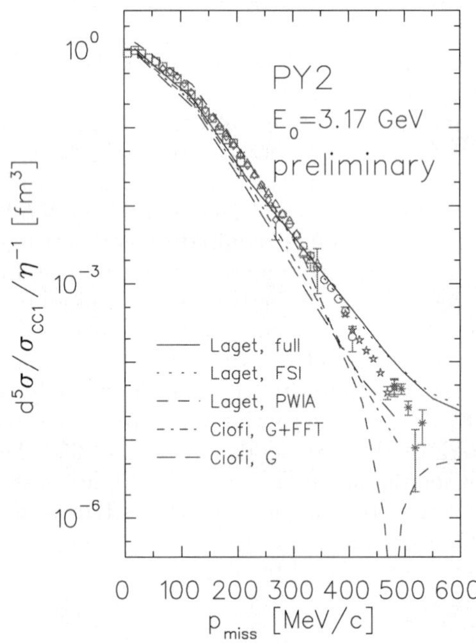

Fig. 4. Preliminary results for the reduced cross sections of the PY2 parallel kinematics at a beam energy of 3170 MeV. The momentum transfer Q^2 ranged between 0.89 (GeV/c)^2 and 0.57 (GeV/c)^2. The error bars on the experimental data points only show the statistical uncertainty. The *dashed line (short dashes)* shows the theoretical prediction by Laget in PWIA, the solid line depicts the full calculation, the *dotted line* only includes FSI. The *dashed curve (long dashes)* is a Glauber calculation by Ciofi and Morita, the *dash-dotted curve* additionally includes finite formation time effects

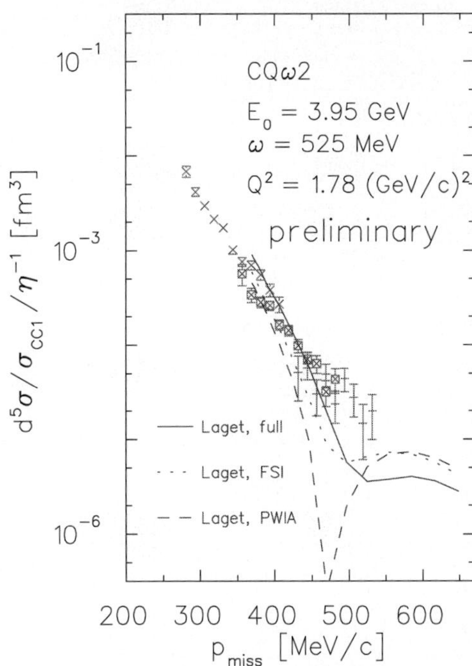

Fig. 5. Preliminary results for the reduced cross section in the CQω2 kinematics. The value of Q^2 was fixed to 1.78 (GeV/c)^2 at a beam energy of 3952 MeV. As in the previous figure, error bars only include the statistical uncertainty. The *dashed line* shows the theoretical prediction by Laget in PWIA, the *solid (dotted) line* is his full (FSI) calculation

section flattening out at missing momenta above 480 MeV At missing momenta above 550 MeV/c, which is outside the scope of the E97-111 experiment, the PWIA and the FSI calculations mainly agree, and only MEC/IC contributions lead to a reduction of the cross section.

4 Conclusions

We have studied the ^{4}He(e,e'p) proton knockout for the two-body breakup at missing momenta up to 530 MeV/c in several different kinematics, two settings utilizing parallel kinematics, and two in perpendicular kinematics. All of these data sets show a monotonically decreasing reduced cross section as the missing momentum increases. The minimum, which is predicted by most of the available PWIA calculations at a recoil momentum around 450 MeV/c, is not observed. However, the preliminary data are in reasonable agreement with the predictions from the latest full calculations, which include FSI, MEC, IC and relativistic effects. To observe the minimum, one would probably have to go to even higher momentum transfers.

Acknowledgements. I would like to thank J.M. Laget, H. Morita and C. Ciofi degli Atti for providing their theoretical calculations and for valuable discussions. This work was supported by DOE contract DE-AC05-84ER40150 under which the Southeastern Universities Research Association (SURA) operates the Thomas Jefferson National Accelerator Facility for the United States Department of Energy.

References

1. S. Tadokoro, T. Katayama, Y. Akaishi, and H. Tanaka: Prog. Theor. Phys. **78**, 732 (1987)
2. H. Morita, Y. Akaishi, and H. Tanaka: Prog. Theor. Phys. **79**, 863 (1988)
3. R. Schiavilla. V. Pandharipande, and R. Wiringa: Nucl. Phys. A **449**, 219 (1986)
4. J.J. van Leeuwe et al.: Phys. Rev. Lett. **80**, 2543 (1998)
5. J.M. LeGoff et al.: Phys. Rev. C **50**, 2278 (1994)
6. J.M. Laget: Nucl. Phys. A **579**, 333 (1994)
7. R. Schiavilla: Phys. Rev. Lett. **65**, 835 (1990)
8. S.I. Nagorny et al.: Sov. J. Nucl. Phys **49**, 465 (1989)
9. S.I. Nagorny et al.: Sov. J. Nucl. Phys **43**, 228 (1991)
10. A. Bianconi and M. Radici: Phys. Lett. B **363**, 24 (1995)
11. L.L. Frankfurt, M.M. Sargsian, and M.I. Strikman: Phys. Rev. C **56**, 1124 (1997)
12. J. Alcorn et al.: submitted to NIM A (2003)
13. T. deForest, Jr.: Nucl. Phys. A **392**, 232 (1983)
14. J.M. Laget: private communication
15. J.M. Laget: nucl-th/0303052 (2003)
16. C. Ciofi degli Atti and H. Morita: private communication
17. H. Morita, M.A. Braun, C. Ciofi degli Atti, and D. Treleani: Nucl. Phys. A **699**, 328c (2002)

Eur Phys J A (2004) **19**, s01, 171–174

Digital Object Identifier (DOI) 10.1140/epjad/s2004-03-028-6

EPJ A direct

electronic only

^{3}He(e,e′p)d and ^{3}He(e,e′p)pn at high momentum transfer

Results from the Jefferson Lab E89-044 experiment

D.W. Higinbotham, for the Jefferson Lab Hall A Collaboration

Jefferson Lab, Newport News, VA 23606 USA

Received: 4 Sep 2003 / Accepted: 14 Nov 2003 /
Published Online: 6 Feb 2004 – © Società Italiana di Fisica / Springer-Verlag 2004

Abstract. The (e,e′p) reaction has proven to be a useful tool for studying specific aspects of the nucleus. The Jefferson Lab continuous electron beam accelerator facility (CEBAF) along with the high resolution spectrometers in the Hall A end station allow this reaction to be studied in kinematics heretofore unaccessible. In this paper the results of the Jefferson Lab Hall A ^{3}He(e,e′p)d and ^{3}He(e,e′p)pn measurements, performed at a fixed $\mathbf{q}$=1.5 GeV/c and ω=840 MeV, will be presented along with theoretical calculations by G. Salme, J.-M. Laget, and J. Udias.

PACS. 21.45.+v few-body systems – 25.70.Bc quasielastic scattering

1 Introduction

The E89-044 proposal [1] to study the ^{3}He(e,e′p) reaction at high momentum transfer was one of the original Jefferson Lab Hall A experiments and drove much of the design of the hall and its two high resolution spectrometers. The 2×10^{-4} momentum resolution of these spectrometers allows to separate the ^{3}He(e,e′p) two-body break-up channel from the continuum, a gap of only 2.2 MeV, at incident beam energies in excess of 4800 MeV.

The E89-044 experiment took data over a broad range of kinematics to study the single-nucleon structure of nuclei and to study the behavior of nucleons embedded in the nuclear medium. In this paper, preliminary results from the constant $\mathbf{q}$=1.5 GeV/c and ω=840 MeV portion of the experiment for both the ^{3}He(e,e′p)d and ^{3}He(e,e′p)pn reactions will be presented along with theory calculations by G. Salme, J.-M. Laget, and J. Udias. New calculations by C. Ciofi degli Atti are presented elsewhere in these proceedings.

2 Kinematics

The kinematics for the (e,e′p) reaction are shown in Fig. 1. The scattering plane is defined by the incoming electron, $e = (E_e, \mathbf{e})$, and the outgoing electron, $e' = (E'_e, \mathbf{e}')$. The four-momentum of the virtual photon is given by $q^{\mu}=(\omega, \mathbf{q})$ and the four-momentum of the outgoing proton is given by $p'^{\mu} = (E_p, \mathbf{p}')$. The four-momentum square, $Q^2 = q^2 - \omega^2$, is defined such that for electron scattering Q^2 is always positive. The missing momentum vector is defined as $\mathbf{p}_m = \mathbf{q} - \mathbf{p}'$ and represents the momentum of the recoiling system.

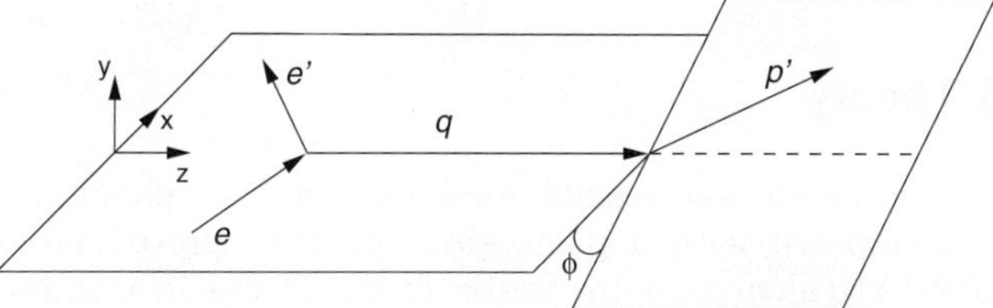

Fig. 1. A schematic of the kinematics for the (e,e′p) reaction

The form of the differential cross section for (e,e′p) reactions in the one-photon exchange approximation without polarization is:

$$\frac{d^6\sigma}{d\Omega_{e'}dE_{e'}d\Omega_{p'}dE_{p'}} = \frac{E_p p'}{(2\pi)^3}\sigma_{Mott}[v_T R_T + v_L R_L + v_{TL} R_{TL}\cos\phi + v_{TT}R_{TT}\cos 2\phi], \quad (1)$$

with ϕ the angle between the plane defined by $\mathbf{e}$ and $\mathbf{e}'$ and the plane defined by $\mathbf{p}'$ and $\mathbf{q}$, and σ_{Mott} the Mott cross section,

$$\sigma_{Mott} = \frac{4\alpha^2 E_{e'}^2}{Q^4}\cos^2\frac{\theta_e}{2}. \quad (2)$$

The kinematical factors $v_L, v_T, v_{TL},$ and v_{TT} are:

$$v_L = \frac{Q^4}{q^4}, \quad (3)$$

$$v_T = \frac{Q^2}{2q^2} + \tan^2(\theta_e/2), \quad (4)$$

$$v_{TL} = \frac{Q^2}{q^2}\left[\frac{Q^2}{q^2} + \tan^2(\theta_e/2)\right]^{1/2}, \text{ and} \quad (5)$$

$$v_{TT} = \frac{Q^2}{2q^2}. \tag{6}$$

For the two-body break-up channel, ^{3}He(e,e'p)d, the proton energy and angle with respect to $\mathbf{q}$ are correlated because the missing energy is fixed. In this case the differential cross is written as

$$\frac{d^5\sigma}{d\Omega_{e'}d\Omega_{p'}dE_{e'}} = \frac{E_p p'}{(2\pi)^3}\sigma_{Mott}f_{rec}^{-1}[v_T R_T + v_L R_L +$$
$$v_{TL}R_{TL}\cos\phi + v_{TT}R_{TT}\cos 2\phi], \tag{7}$$

where f_{rec} is the recoil factor,

$$f_{rec} = \left[1 - \frac{E_{p'}}{E_t}\frac{\mathbf{P_m}\cdot\mathbf{p'}}{p'^2}\right]. \tag{8}$$

The response functions, R_L, R_T, R_{TL}, R_{TT} can be separated by a suitable choice of the kinematic parameters. In perpendicular in-plane kinematics at a fixed $\mathbf{q}$ and ω, one can separate R_T, R_{TL}, and a combination of the R_L and R_{TT} response functions, denoted as R_{L+TT}. One can also measure the cross-section asymmetry A_{TL} for a given $\mathbf{q}$ and ω. This asymmetry is defined as:

$$A_{TL} = \frac{\sigma(\phi = 180°) - \sigma(\phi = 0°)}{\sigma(\phi = 180°) + \sigma(\phi = 0°)}. \tag{9}$$

3 Theory

The cross-section results presented in this paper will be compared with a plane-wave impulse approximation (PWIA) calculation by Salme et $al.$ [2] and the microscopic calculations of J.-M. Laget. The A_{TL} asymmetry will be compared with a relativistic mean-field calculation by J.M. Udias along with Laget's microscopic calculation. The details of the microscopic and relativistic calculations are presented below.

3.1 Microscopic calculation

In the microscopic calculation performed by J.-M. Laget, the reaction amplitude is expanded in diagrams which are computed in momentum space [3]. The code used the Urbana V14 potential and relativistic kinematics. For the PWIA calculation, the possibility of the virtual photon coupling to either the initial state proton or the deuteron is taken into account. In the full calculations of J.-M. Laget, two-body and three-body mechanisms such as final-state interactions and meson-exchange currents are taken into account.

3.2 Relativistic mean-field calculation

In the relativistic mean-field calculations performed by J.M. Udias [5], the one-body nucleon current is calculated with relativistic wave functions for the initial bound and

final outgoing nucleons. The bound-state wave function is represented by a four-spinor. The spinors in a completely relativistic calculation are often referred to as $distorted$ because of the presence of the scalar and vector potentials appearing in the lower component [6,7]. The spinor is obtained by solving the Dirac equation with scalar-vector potentials. Proton distortions are calculated solving the Dirac equation using an optical potential. The Jefferson Lab Hall A ^{16}O(e,e'p)^{15}N response function and A_{TL} asymmetry results [8] were well described by this type of calculation [9,10]. While a mean-field calculation cannot reproduce the ^{3}He(e,e'p) cross section, it will be shown that it does describe the general characteristics of the extracted A_{TL} asymmetry.

4 Results

4.1 Two-body break-up

In order to isolate the two-body break-up channel from the continuum, the (e,e'p) data was plotted as a function of missing energy. Due to the high momentum resolution of the Hall A spectrometers, even with a 4.8 GeV electron beam it was possible to fit the two-body peak and separate it from the continuum which, as a function of missing energy, begins only 2.2 MeV from the two-body peak. The MCEEP code [11] was used to calculate acceptance and to take radiative corrections into account.

In Figs. 2, 3, and 4 the results of the ^{3}He(e,e'p)d cross-section extractions are shown. These results are all for a fixed $\mathbf{q} = 1.5$ Gev/c and $\omega = 840$ MeV, but are were taken at different values of ϕ and ϵ to allow response-function separations. The cross-section in Fig. 2 is for $\phi = 180°$ and $\epsilon = 0.94$ and in Fig. 3 is for $\phi = 0°$ and $\epsilon = 0.94$. Combining these results using 7 the R_{TL} response function can be extracted and by using 9 the A_{TL} asymmetry, as shown in Fig. 5, can be extracted. The cross section results shown in Fig. 4 are for $\phi = 180°$ and ϵ of 0.11. Combining this result with those shown in Fig. 2 will allow the R_T and R_{L+TT} response functions to be extracted.

These results can be divided into three distinct ranges in missing-momentum. The first is the low missing momentum range from 0 to 200 MeV/c where the PWIA models describe the data reasonably well. The second region from 200 to 700 MeV/c requires the inclusion of final-state interactions and meson-exchange currents to reproduce the enhancement in the cross sections and cause oscillations in the A_{TL} asymmetry. The third region, around 1000 MeV/c, presently cannot be explained by any model. The forthcoming response function separations will further test the models and allow a deeper understanding of the reaction mechanisms.

4.2 Continuum

For the ^{3}He(e,e'p)pn data, the focus was to study the effect of two-nucleon correlations. Previous continuum data

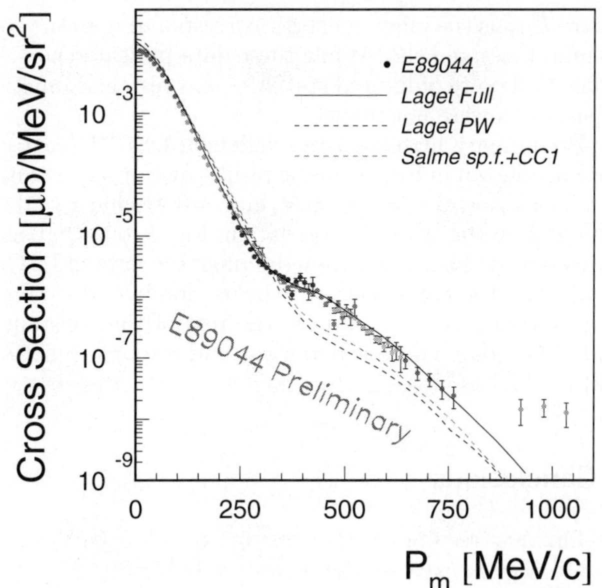

Fig. 2. Shown are the preliminary cross section results for the reaction ^{3}He(e,e'p)d as a function of missing momentum with a beam energy of 4807 MeV and with a fixed $\mathbf{q} = 1500$ MeV/c and $\omega = 840$ MeV [1]. The theory curves show a PWIA calculation along with the latest full calculation of J.M. Laget. The enhancement in the cross section near 300 MeV/c and continuing to larger missing momentum is predominately due to final-state interactions. There is no clear indication yet from theory what causes the enhancement of the cross section near 1000 MeV/c

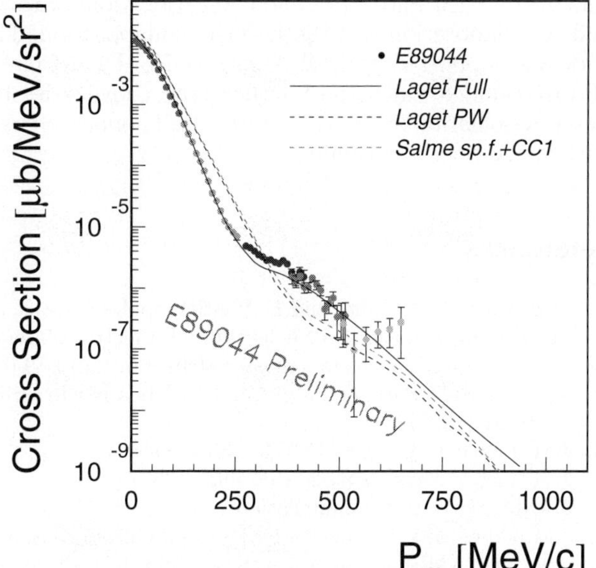

Fig. 3. Shown are the preliminary cross-section results for the reaction ^{3}He(e,e'p)d as a function of missing momentum at a beam energy of 4807 MeV and with a fixed $\mathbf{q} = 1500$ MeV/c and $\omega = 840$ MeV [1]. The theory curves show a PWIA calculation along with the latest full calculation of J.M. Laget

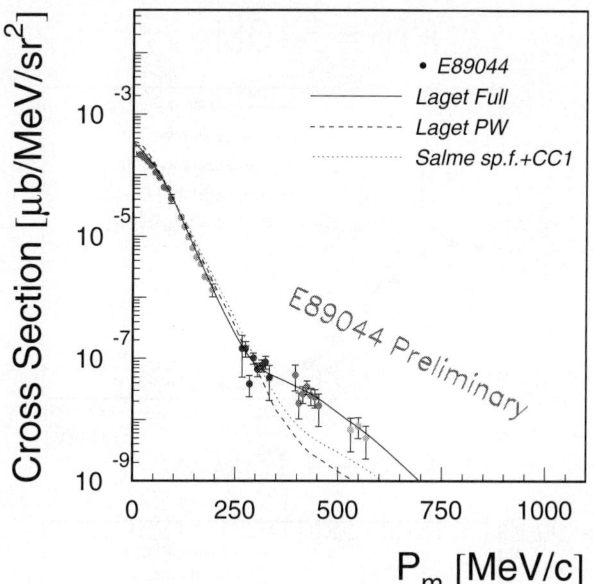

Fig. 4. Shown are the preliminary cross-section results for the reaction ^{3}He(e,e'p)d as a function of missing momentum at a beam energy of 1250 MeV and a fixed $\mathbf{q} = 1500$ MeV/c and $\omega = 840$ MeV [1]. The theory curves show a PWIA calculation along with the latest full calculation of J.M. Laget

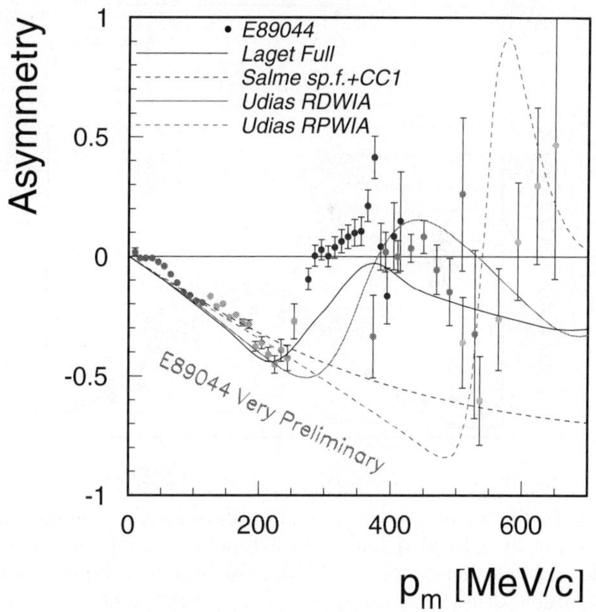

Fig. 5. Shown are the preliminary A_{TL} data. The curves show the latest calculation of J.M. Laget along with a preliminary result of J. Udias. At lower missing momentum the theories are most sensitive to relativistic effects, while at the larger missing momentum the theories become sensitive to final-state interaction effects

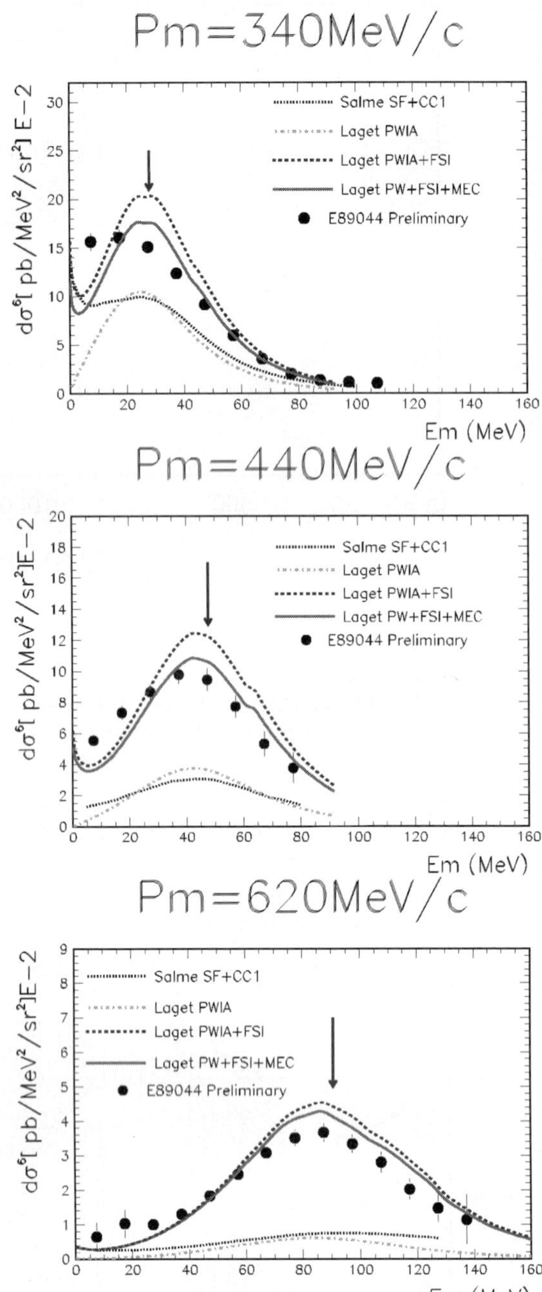

Fig. 6. Shown are the cross-section data for the ^{3}He(e,e'p)pn reaction at a fixed $\mathbf{q}$ and ω as a function of missing energy. The dotted curves show a PWIA prediction of Salme, while the other curves show the theoretical predictions of J.-M. Laget from PWIA to full calculations including meson-exchange current and final-state interaction effects. The arrow shows where one would expect the peak in the cross section due to the photon coupling to a correlated pair of nucleons

on ^{4}He have shown a correlation peak which shifts outward in missing energy as a function of missing momentum as:

$$E_m = E_{thr} + \frac{A-2}{A-1} \times \frac{\mathbf{p_m}^2}{2M_p} \qquad (10)$$

where E_{thr} is the missing energy corresponding to the continuum threshold [12]. While these data indicated correlations, they were of limited statistics and response function separations were not made.

Preliminary high statistics Jefferson Lab ^{3}He(e,e'p)pn data are shown in Fig. 6. These results are for a constant $\mathbf{q}$ = 1.5 GeV/c and ω = 840 MeV, an ϵ = 0.94, and ϕ = 180°. The peak in the cross-section data in Fig. 6 follows 10 as is indicated by the arrows; though unlike the data of Le Goff *et al.* [12], the cross section appears dominated by final-state interactions. Analysis of the rest of the constant $\mathbf{q}$ and ω continuum data is underway and response functions will be forthcoming.

5 Conclusions

Preliminary results of the constant $\mathbf{q}$ = 1.5 GeV and ω = 840 MeV/c part of the Jefferson Lab E89-044 experiment have been presented. As predicted by theory, the cross section is strongly enhanced by final-state interactions for missing momentums greater than 200 MeV/c. At the highest missing momenta, around 1000 MeV/c, the results have yet to be explained by theory.

I would like to acknowledge the hard work and dedication of the doctoral students working on the analysis of the E89-044 experiment: Marat Rvachev with M.I.T., Emilie Penel-Nottaris with Grenoble, and Fatiha Benmokhtar with Rutgers. Also, I would also like to thank the experiment spokespersons, Marty Epstein, Arun Saha, and Eric Voutier, for their hard work and dedication along with the Hall A collaboration and the E89-044 collaboration. This work was supported by DOE Contract No. DE-AC05-84-ER40150 under which the Southeastern Universities Research Association (SURA) operates the Thomas Jefferson National Accelerator Facility.

References

1. M.B. Epstein, A. Saha, and E. Voutier: spokespersons, Selected Studies of the ^{3}He Nuclei through Electrodisintegration at High Momentum Transfer, Jefferson Lab E89-044
2. A. Kievsky, E. Pace, G. Salme, and M. Viviani: Phys. Rev. C **56**, 64 (1997)
3. J.M. Laget: Phys. Rev. D **579**, 333 (1994)
4. J.M. Laget: Phys. Lett. B **199**, 493 (1987)
5. J. Udias: private communication
6. J.M. Udias, J.A. Caballero, E. Moya de Guerra, J.R. Vignote, and A. Escuderos: Phys. Rev. C **64**, 024614 (2001)
7. J.J. Kelly: Phys. Rev C **60**, 044609 (1999)
8. J. Gao et al.: Phys. Rev. Lett. **84**, 3265 (2000)
9. J.M. Udias et al.: Phys. Rev. C **51**, 3246 (1995)
10. J.M. Udias, J.A. Caballero, E. Moya de Guerra, J.E. Amaro, and T.W. Donnelly: Phys. Rev. Lett. **83**, 5451 (1999)
11. P. Ulmer, MCEEP: Monte Carlo for electro-nuclear coincidence experiments, CEBAF-TN-91-101
12. J.M. Le Goff et al.: Phys. Rev. C **50**, 2278 (1994)

Eur Phys J A (2004) **19**, s01, 175–178
Digital Object Identifier (DOI) 10.1140/epjad/s2004-03-029-5

EPJ A direct

electronic only

Two nucleon correlations measured with ^{3}He(e,e'pp)n

L.B. Weinstein and R. Niyazov[a], for the CLAS Collaboration

Old Dominion University, Norfolk, VA 23529, USA

Received: 18 Jul 2003 / Accepted: 14 Nov 2003 /
Published Online: 6 Feb 2004 – © Società Italiana di Fisica / Springer-Verlag 2004

Abstract. Despite tremendous progress in recent years in measuring single nucleon momentum distributions in nuclei, two-nucleon momentum distributions remain largely unmeasured. Here we report on a measurement of ^{3}He(e,e'pp)n that is primarily sensitive to the two-nucleon momentum distribution and relatively uncontaminated by final state rescattering or by two-body currents.

PACS. 21.45.+v – 25.30.Dh

1 Introduction

Single nucleon properties in nuclei have been thoroughly studied, primarily through proton knockout from nuclei, such as O($e, e'p$) [1]. The valence proton momentum distributions and more deeply bound proton momentum and energy distributions have been measured by many experiments [2]. Valence knockout is well described by single nucleon knockout calculations that include the effects of final state interactions. However, the fact that a) we only see about 70% of the 'expected' number of nucleons and b) there is a large cross section at large excitation energies of the residual nucleus indicates that multi-nucleon processes play a significant role. While there have been many $(e, e'p)$ measurements from nuclei, there have been very few $(e, e'pp)$ measurements.

There are two general sources of two-nucleon knockout from nuclei. One source is the interaction of the virtual photon with two nucleons, typically through meson exchange currents or isobar configurations. This is referred to as a two-body current. The second one is the interaction of the virtual photon with one nucleon of a 'correlated pair'. (Note that the distinction between correlations [in the wave function] and currents [in the operator] is blurred by the fact that unitary transformations can transform one into the other.)

A correlated pair is typically a pair of nucleons that have large relative momentum (because they are at short range) and small total momentum. Thus, one signature of correlations is finding two nucleons with large relative momentum and small total momentum in the initial state. Unfortunately, the effects of NN correlations are frequently obscured by the effects of two body currents [3]. In order to disentangle these competing effects, a series of comprehensive measurements are needed.

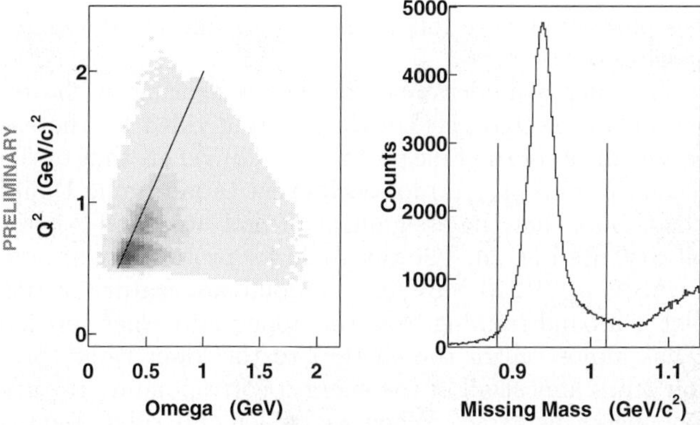

Fig. 1. a Q^2 vs ω for ^{3}He(e,e'pp)n at $E_{beam} = 2.2$ GeV. Note the huge kinematic acceptance. **b** Missing mass for ^{3}He(e,e'pp). We cut at the indicated lines to select (e,e'pp)n events

The seven experiments took data simultaneously in Spring 1999, measuring approximately 500 million $A(e, e'X)$events, using the CEBAF Large Acceptance Spectrometer (CLAS) at Jefferson Lab, a 4π magnetic spectrometer, with 1.16, 2.26 and 4.46 GeV polarized electrons incident on targets from ^{3}He to ^{56}Fe [4]. This paper reports the results from 2.2 and 4.4 GeV electrons on ^{3}He.

2 Measuring ^{3}He(e,e'pp)n

We studied ^{3}He(e,e'pp)n by measuring electron induced two proton knockout reactions from ^{3}He using the CLAS detector and cutting on the missing mass. Figures 1a and b show the electron acceptance and undetected neutron missing mass resolution for $E_{beam} = 2.2$ GeV. The threshold of the CLAS is approximately 250 MeV/c for protons.

Because this is the first time that ^{3}He(e,e'pp)n has been measured using an almost 4π detector, our data anal-

---

[a] *Present address:* Jefferson Lab, Newport News, VA 23606

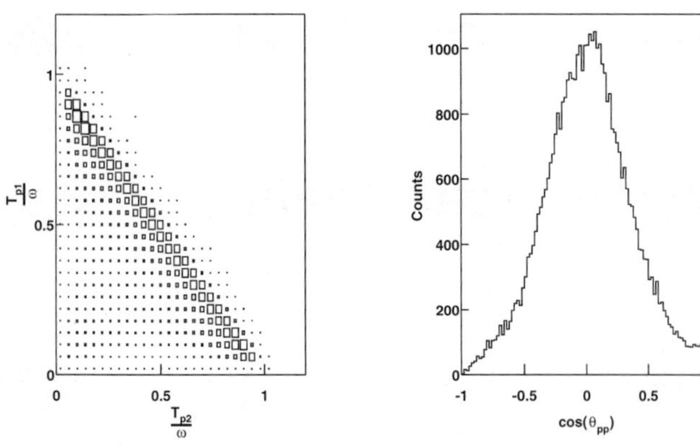

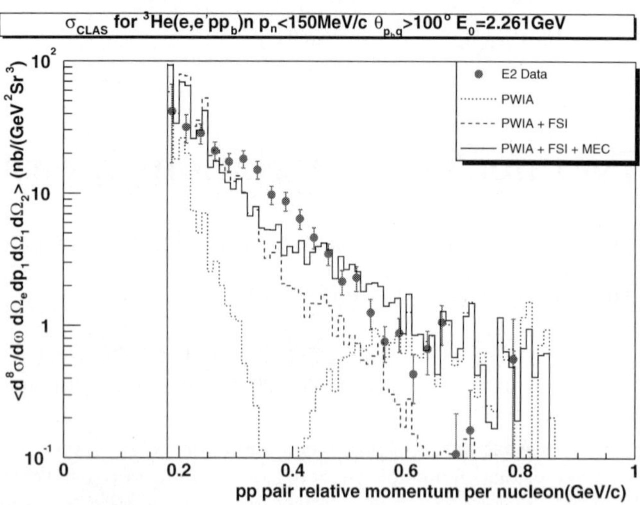

Fig. 2. a Nucleon kinetic energy distribution for 2.2 GeV ^{3}He(e,e'pp)n. The kinetic energy of proton 1 divided by ω (T_{p1}/ω) is plotted versus the same for proton 2. (Note $p_p \geq 250$ MeV/c.) Note the dominant band running from (1,0) to (0,1) corresponding to very low energy neutrons. **b** The cosine of the opening angle between the two protons for events with $p_n < 250$ MeV/c. Note the large peak at 90°

Fig. 3. *Preliminary* cross section vs *pp* relative momentum $\mathbf{p}_{rel} = [(\mathbf{p}_{forward} - \mathbf{q}) - \mathbf{p}_{backward}]/2$ for ^{3}He(e,e'pp)n events with a fast backward proton and a slow neutron. The points show the data, the dotted line is the PWIA calculation, the dashed line also includes FSI and the solid line also includes two-body currents. All calculations are by Laget

ysis philosophy is to follow and understand the dominant features of the data.

In order to understand the energy sharing in the reaction, we plotted the kinetic energy divided by the energy transfer of the first proton (T_{p1}/ω) versus that of the second proton (T_{p2}/ω) for each event (a lab-frame Dalitz plot). Note that the assignment of protons 1 and 2 is arbitrary. See Fig. 2a. The threshold for proton detection in CLAS is $p_p \geq 250$ MeV/c. The dominant feature of this plot is a band running from the upper left corner (proton 1 has almost all of the energy) to the lower right (proton 2 has almost all of the energy) corresponding to very low energy neutrons. When we cut on this ridge, requiring that $p_n < 250$ MeV/c, the opening angle between the two protons peaks at 90° (see Fig. 2b). As we know from teaching introductory physics, when a moving object collides elastically with an identical stationary object, the opening angle is always 90°. Thus, this peak indicates single proton knockout followed by hard *pp* rescattering.

Further evidence for this comes from preliminary data from Zhang [6]. He analyzed the same ^{3}He(e,e'pp)n data with $p_n < 150$ MeV/c, looking at events with a fast backward proton ($\theta_{pq} > 100^o$). See Fig. 3. It was conjectured that these fast backward protons could not have been affected by final state interactions (FSI). However, calculations by Laget show that the cross section is dominated by final state interactions with significant contributions from two-body currents (meson exchange currents and isobar configurations). These results are very consistent with the 90° peak in the *pp* opening angle distribution.

Since we are looking for two-nucleon correlations, we eliminate these FSI-dominated events by setting the same threshold for neutrons as for protons: $p_n \geq 250$ MeV/c. When we look at this Dalitz plot (see Fig. 4) we see three peaks at the three corners of the plot, corresponding to events where two 'fast' nucleons each have less than 20% of the energy transfer and the third 'leading' nucleon has the remainder. We call the two nucleons 'fast' because

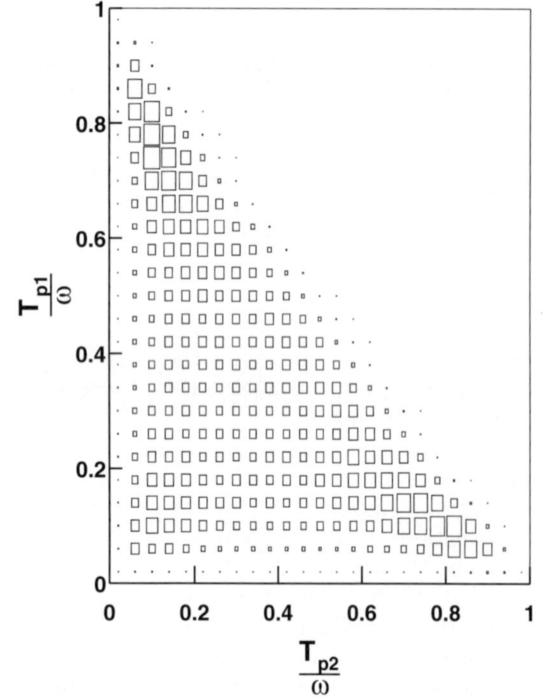

Fig. 4. Nucleon kinetic energy distribution for 2.2 GeV ^{3}He(e,e'pp)n. The kinetic energy of proton 1 divided by ω (T_{p1}/ω) is plotted versus the same for proton 2. (Note $p_p, p_n \geq 250$ MeV/c.) Note the peaks in the corners

$p \gg p_{fermi}$. These peaks are much more pronounced at $E_{beam} = 4.4$ GeV (not shown). We cut on these peaks where the two fast nucleons each have less than 20% of the energy transfer.

Then we looked at the opening angle of the two fast nucleons. Figure 5-a shows the pair opening angle for fast

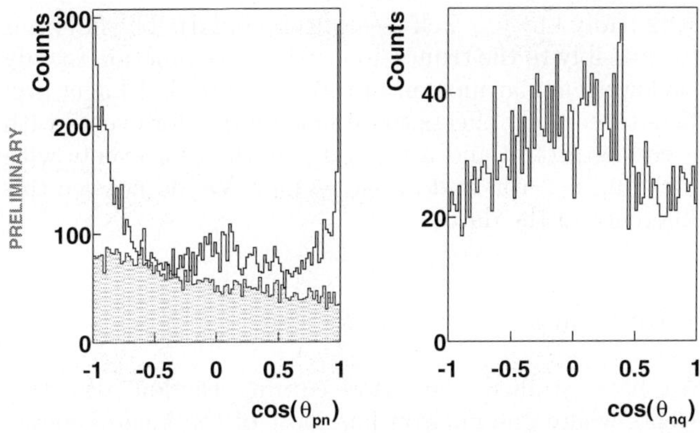

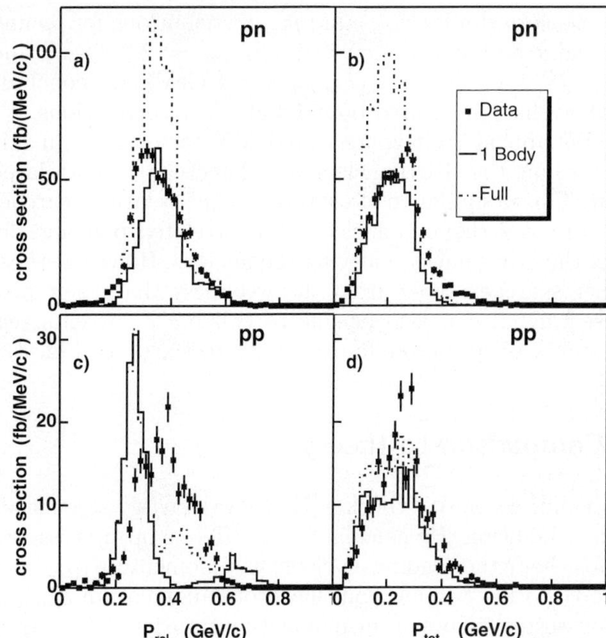

Fig. 7. 2.2 GeV ^{3}He$(e, e'pp)n$ cross section vs. momentum. Points = data, solid histogram = Laget 1-body calculation, dot-dashed histogram = Laget full calculation. **a** leading proton and a fast pn pair: pair relative momentum p_{rel}; **b** leading proton and a fast pn pair: total momentum p_{tot}; **c** leading neutron and a fast pp pair: p_{rel}; **d** leading neutron and a fast pp pair: p_{tot}

Fig. 5. **a** Opening angle of the fast pn pairs for events in the upper left and lower right corners of Fig. 4b. The histogram shows the data, the filled histogram shows the fire ball phase space simulation (with arbitrary normalization). **b** The angle between the neutron in the fast pn pair and **q** where $p_\perp < 300$ MeV/c

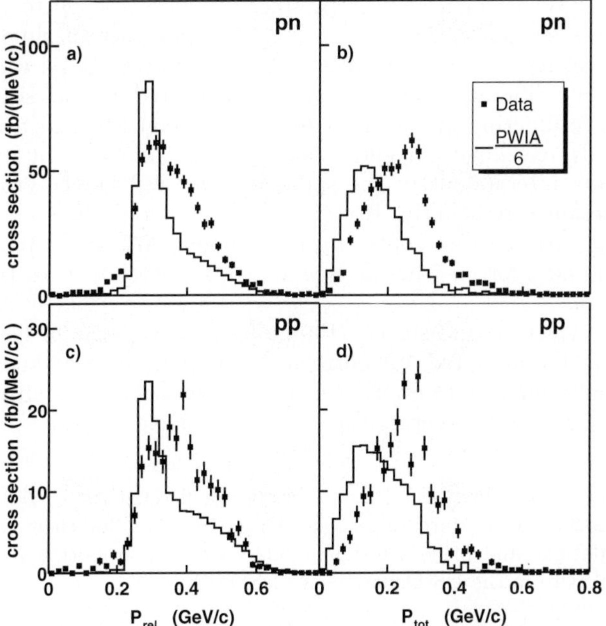

Fig. 6. 2.2 GeV ^{3}He$(e, e'pp)n$ cross section vs. momentum. Points = data, histogram = PWIA decreased by a factor of 5. **a** leading proton and a fast pn pair: pair relative momentum p_{rel}; **b** leading proton and a fast pn pair: total momentum p_{tot}; **c** leading neutron and a fast pp pair: p_{rel}; **d** leading neutron and a fast pp pair: p_{tot}

pn pairs with a leading proton. Note the large peak at 180 degrees ($\cos\theta_{NN} \approx -1$). The distribution for fast pp pairs with a leading proton is identical. The peak is not due to the cuts, since we do not see it in a fire ball phase space simulation assuming three body absorption of the virtual photon and phase space decay. It is also not due to the CLAS acceptance since we see it both for leading protons (which we detect) and leading neutrons (which we infer from missing mass). This back-to-back peak is a strong indication of correlated NN pairs.

3 Studying correlated pairs

Now consider these presumably correlated pairs. Since we believe that we have observed events where the leading nucleon absorbed the virtual photon and the two fast nucleons are emitted back to back, we cut on the perpendicular momentum of the leading nucleon to deemphasize rescattering ($p_\perp < 300$ MeV/c). This cut selects the back-to-back events very cleanly. Unfortunately, there are only 3400 fast pn and 1100 fast pp events remaining in the entire 2.2 GeV data set (and ten times fewer at 4.4 GeV).

If the fast back-to-back NN pairs are really uninvolved in the photon absorption, then they should be distributed isotropically (the angular distribution of the neutrons with respect to **q** is shown in Fig. 5b). Further evidence that the fast NN pair is a spectator comes from the average momentum of the pair along **q**. This is about 0.07 GeV/c for $E_{beam} = 2.2$ GeV and about 0.1 GeV/c for $E_{beam} = 4.4$ GeV, much less than the average momentum transfers of $Q^2 = 0.7$ and 1.4 (GeV/c)^2 respectively.

The 2.2 GeV fast NN pair relative ($p_{rel} = \frac{1}{2}|\mathbf{p}_1 - \mathbf{p}_2|$) and total ($p_{total} = |\mathbf{p}_1 + \mathbf{p}_2|$) momentum distributions are shown in Fig. 6. Note how similar the pp and pn distributions are. The 4.4 GeV distributions (not shown) are similar.

Thus, because when we select a quasifree leading nucleon, the fast NN pairs are back to back, relatively isotropic and have small average momentum along **q**, we conclude that the fast NN pair is a spectator. Because we measure similar p_{total} and p_{rel} distributions for pp and pn pairs and $0.5 < Q^2 < 1$ ($E_{beam} = 2.2$ GeV) and $1 < Q^2 < 2$ (GeV/c)^2 ($E_{beam} = 4.4$ GeV), we conclude that we have measured bound state NN correlations.

We appear to have measured NN correlations in ^{3}He by striking the **third** nucleon and detecting the correlated pair. This is similar to other proposed correlation searches where you strike one nucleon of a correlated pair and detect the other nucleon leaving the nucleus. However, these other searches suffer from the weakness that their proposed signal can also be due to two body currents (eg: photon absorption on an exchanged meson) (see Fig. 3).

4 Comparison to theory

Calculations by W. Glöckle [7] at lower energy strengthen this conclusion. He calculated the ^{3}He(e,e'pp)n cross section where the leading nucleon has momentum $\mathbf{p}_N = \mathbf{q}$ and the other two nucleons have total momentum $p_{total} = 0$ for various values of the momentum transfer, $400 \leq |\mathbf{q}| \leq 600$ MeV/c, and relative momentum. He found that MEC did not contribute, rescattering of the leading nucleon did not contribute, and the continuum state interaction of the outgoing NN pair decreased the cross section by a factor of approximately 10 relative to the PWIA result. Thus, he found that this reaction is a very clean way to measure the overlap integral between the NN continuum state and the same two nucleons in the bound state.

C. Ciofi degli Atti and L. Kaptari also found that the continuum interaction of the outgoing pair significantly decreased the cross Sect. [8].

We compared our results to two other calculations, 1) a Plane Wave Impulse Approximation (PWIA) calculation by M. Sargsian [9] using Glöckle's bound state wave function with no final state interactions, and 2) a diagrammatic calculation by J.-M. Laget [10,11,12] using a Faddeev wave function from P. Sauer and including one-, two-, and three- body mechanisms as well as rescattering terms. We averaged all of the models over the CLAS acceptances and cuts using a monte carlo.

The PWIA calculation of Sargsian has Q^2 vs ω, NN pair opening angle, p_{rel}, and p_{total} distributions that are consistent with the data (see Fig. 6). The momentum distributions peak at smaller momentum than the data. This is probably due to a much stronger NN-pair continuum interaction in the NN s-wave than in the p-wave. This interaction therefore reduces the s-wave strength more than the p-wave, shifting the peaks to higher momenta. The PWIA cross section is a factor of 5 larger than the data which is consistent with the expected effects of the NN continuum state interaction calculated by Glöckle, by Ciofi degli Atti and by Laget. It also predicts the same ratio of pp to pn pairs as seen in the data.

Laget's one-body calculations describe the pn pairs well, both qualitatively and quantitatively (see Fig. 7a and b). However, the full calculation overestimates the data by about 60%. The calculation describes p_{rel} for pp

pairs badly but p_{tot} well (see Fig. 7c and d). The failure is due possibly to the truncation of the wave function to only the lower angular momentum states. Note that Laget predicts three-body effects to be much larger for events with a leading proton and a fast pn pair than for events with a leading neutron and a fast pp pair. We do not see this difference in the data.

5 Summary

We have studied the ^{3}He(e,e'pp)n reaction, selecting events where one nucleon has most of the kinetic energy and has less than 300 MeV/c of momentum perpendicular to $\mathbf{q}$. When we do this, we see isotropic, back-to-back, fast NN pairs with small average momentum along $\mathbf{q}$. We have measured the total and relative momentum distributions of these pairs and found that they do not depend significantly on isospin (pp vs pn pairs) or on momentum transfer.

PWIA calculations reproduce the observed pp to pn cross section ratio, indicating the importance of single-nucleon knockout mechanisms. Calculations by Laget with many different diagrams and a truncated bound state wave function predict that leading-nucleon FSI and two-body exchange currents are negligible, and continuum-state interactions of the spectator pair reduce the cross section significantly. However, the predicted three-body exchange current contributions of about 20% for pp pairs and 50% for pn pairs do not improve agreement with the data.

Thus, by measuring ^{3}He(e,e'pp)n, we appear to have directly measured NN correlations without any significant contamination from other processes by striking the **third** nucleon and detecting the spectator correlated pair.

Acknowledgements. I thank Claudio Ciofi degli Atti, Walter Glockle, Jean-Marc Laget and Misak Sargsian for their calculations and physics insight. This work was supported by a grant from the US Department of Energy.

References

1. J. Gao et al.: Phys. Rev. Lett. **84**, 3265 (2000)
2. J. Kelly: Adv. Nucl. Phys. **23**, 75 (1996)
3. S. Janssen et al.: Nucl. Phys. A **672**, 285 (2000)
4. B. Mecking et al.: Nucl. Inst. and Meth. A **503**, (2003)
5. M. Mestayer et al.: Nucl. Inst. and Meth. A **449**, 81 (2000)
6. B. Zhang: Ph.D. thesis, MIT (2002)
7. W. Glöckle et al.: Acta Phys. Polon. B **32**, 3053 (2001), `nucl-th/0109070`
8. C. Ciofi degli Atti and L. Kaptari: Phys. Rev. C **66**, 044004 (2002), `nucl-th/0203041`
9. M. Sargsian: *Private communication*
10. J. Laget: J. Phys. G **14** (1988)
11. G. Audit et al.: Nucl. Phys. A **614**, 461 (1997)
12. J. Laget: Phys. Rev. C **35** (1987)

Eur Phys J A (2004) **19**, s01, 179–183
Digital Object Identifier (DOI) 10.1140/epjad/s2004-03-030-0

EPJ A direct
electronic only

Experimental search for color transparency at Jefferson Lab

D. Dutta

Duke University, Durham, NC, USA

Received: 16 Sep 2003 / Accepted: 14 Nov 2003 /
Published Online: 6 Feb 2004 – © Società Italiana di Fisica / Springer-Verlag 2004

Abstract. Mapping the transition of the strong interaction from the hadronic degrees of freedom to the partonic degrees of freedom of quantum chromodynamics (QCD), is an important goal in intermediate energy nuclear and particle physics. One of the popular approaches used to study the transition between these two regions involves searching for the onset of various phenomena which are naturally predicted in QCD. One such phenomena is Color Transparency (CT), which refers to the suppression of final (and initial) state interactions of hadrons with the nuclear medium in exclusive processes at high momentum transfers. At Jefferson Lab the search for CT has focused on the (e,e'p) reaction. I will review the status of the experimental search for CT covering these experiments and others spanning over a decade. Some of the future experiments being planned for the 12 GeV upgrade of JLab will also be discussed.

PACS. 14.20.Dh – 13.40.Gp – 21.10.Ft

1 Introduction

Exclusive processes are essential in studies of transitions from non-perturbative to perturbative QCD region. Mapping this transition from the hadronic degrees of freedom to the partonic degrees of freedom is an important goal in intermediate energy nuclear and particle physics. The need for such a mapping arises from the fact that at low energies or long distances the nucleon-meson picture in the standard model of nuclear physics is very successful in describing the overall features of the strong interaction, while at high energies or short distances perturbative QCD (pQCD) with its quark-gluon degrees of freedom allows extremely precise description of the interaction. Unfortunately, there is no clear understanding of how these two regimes are connected. Manifestation of the underlying quark-gluon degrees of freedom of QCD naturally gives rise to a distinct set of phenomena in exclusive processes on nucleons and nuclei. One of the popular methods used to explore the transition region is to look for the onset of such phenomena. One such fundamental predictions of QCD is the phenomena of Color Transparency (CT), it refers to the vanishing of the final (and initial) state interactions of hadrons with the nuclear medium in exclusive processes at high momentum transfer [1]. It is based on the idea that, at sufficiently high momentum transfer, the dominant amplitudes for exclusive reactions involve hadrons of reduced transverse size. These small color singlet objects also known as point like configuration (PLC) have long enough lifetimes to be able to pass undisturbed through the nucleus because of reduced interactions with the medium. This is a novel QCD phenomenon which, if observed, would be a clear manifestation of hadrons fluctu-

ating to a small size in the nucleus. Moreover, it also contradicts the traditional Glauber multiple scattering theory in the domain of its validity. Nuclear transparency defined as the ratio of the cross section per nucleon for a process on a bound nucleon in the nucleus to the cross section for the process on a free nucleon, is the commonly used observable in searches for this phenomena. Therefore, measurements of nuclear transparency have attracted a significant amount of effort over the last two decades. A clear signature for the onset of CT would involve a dramatic rise in the nuclear transparency as a function of momentum transfer involved in the process, i.e. a positive slope with respect to the momentum transfer.

2 The $A(e, e'p)$ reaction

A number of searches for color transparency have been carried out in the last decade in experiments using the $A(p, 2p)$ and $A(e, e'p)$ reactions and coherent and incoherent meson production from nuclei [2] – [8]. The $A(e, e'p)$ reaction was first used to measure nuclear transparency at the MIT-Bates lab [9] and the NE-18 experiment at the Stanford Linear Accelerator Center [3] was the first to use this reaction to look for color transparency. These transparency measurements using the $A(e, e'p)$ reaction have been continued at Jefferson Lab (JLab), where the advantage of the continuous wave electron beam has helped improve the precision of the measurements and extend them to higher momentum transfer squared (Q^2). In these reactions an electron beam is incident on a nuclear target and knocks out a proton from within the nucleus. The scattered electron and the knocked out proton are detected

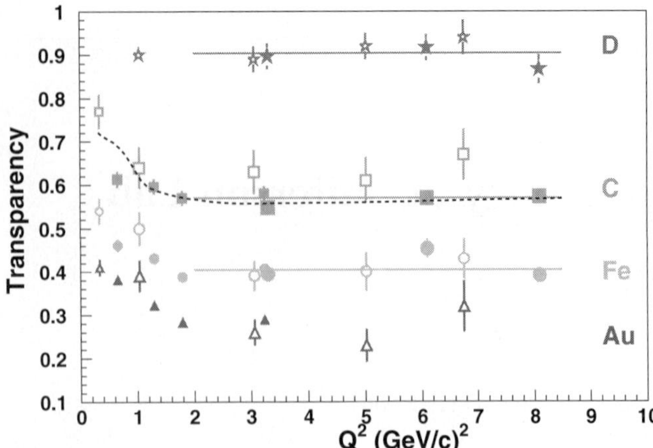

Fig. 1. Nuclear transparency as a function of Q^2, for ^{2}H (*stars*), ^{12}C (*squares*), ^{56}Fe (*circles*) and ^{197}Au (*triangles*). The *small open symbols* are results from MIT-Bates [9], the *large open symbols* are results from the SLAC experiment NE-18 [3], the *small solid symbols* are results from the earlier JLab experiment [4] and the *large solid symbols* are results from the later JLab experiment [5]. The *dashed line* is a Glauber calculation of Pandharipande et al. [10] and the *solid lines* are fit to a straight line of the results for $Q^2 > 2.0$ (GeV/c)^2

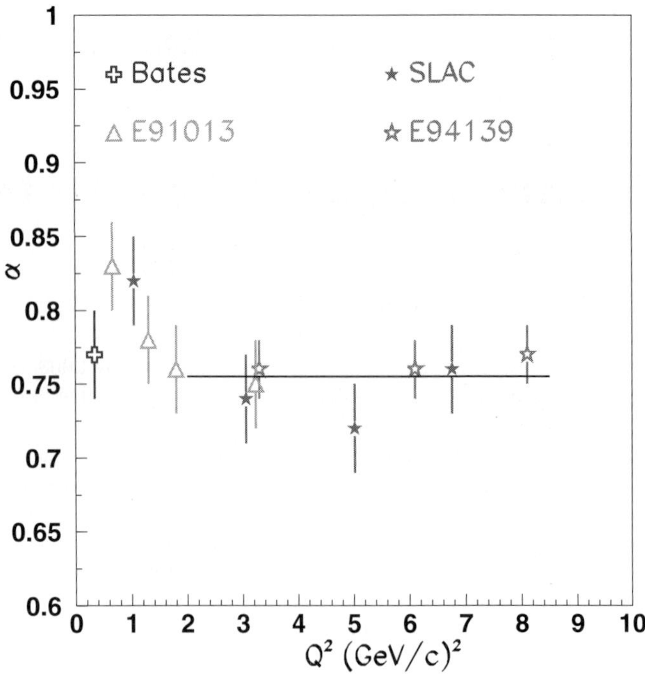

Fig. 2. The parameter α as a function of Q^2. E91013 and E94139 are the two JLab experiments

in coincidence in a pair of magnetic spectrometers. The proton transparency in the target nucleus is extracted by taking the ratio of the experimental yield to the yield from a Plane Wave Impulse Approximation (PWIA) simulation of the experiment. The PWIA calculation does not include any final state interactions, hence the ratio of the experimental yield to the PWIA yield is a measure of the fraction of the protons which are lost on the way out of the nucleus. The CT prediction implies that for fast protons there will a large decrease in these final state interactions and thus one should observe a dramatic increase in the transparency as a function of momentum transfer squared.

At JLab two $A(e, e'p)$ experiments were performed in experimental Hall C, to measure the nuclear transparency of a wide range of nuclei over a large range of Q^2. The first of these experiments was performed on ^{1}H, ^{12}C, ^{56}Fe and ^{197}Au and measured the nuclear transparency up to a Q^2 of 3.2 (GeV/c)^2. This experiment helped establish a baseline for traditional nuclear physics calculation. The later experiment extended the transparency measurements to a high Q^2, in search of a deviation from the expectations of the traditional calculations. The second experiment was performed on ^{1}H, ^{2}H, ^{12}C and ^{56}Fe nuclei and in both experiments the ^{1}H data were used for normalization purposes only. The results of the nuclear transparency measurements with $A(e, e'p)$ experiments from JLab, SLAC and MIT-Bates are shown in Fig. 1. It is clear from Fig. 1 that the transparency is independent of Q^2 up to a Q^2 of 8.1 (GeV/c)^2 and the results agree with the expectations of conventional nuclear physics.

One can parametrize the hadron nucleus cross-section as $\sigma_N = \sigma_0 A^\alpha$, where σ_0 is the N-N cross-section in free space. Nuclear transparency can thus be parametrized as $T = A^{1-\alpha}$. One can then look at the A dependence of the transparency by fitting the transparency at fixed Q^2 to the form $A^{1-\alpha}$. Figure 2 shows a plot of the parameter α as a function of Q^2 for all the JLab transparency results. For $Q^2 > 2$ (GeV/c)^2 the parameter α is equal to 0.76 which agrees very well with the expectations of conventional nuclear physics model based on just geometrical arguments. Thus these results from JLab conclusively rule out the onset of CT phenomena, up to a Q^2 of 8.1 (GeV/c)^2, for protons produced in exclusive reactions in a nuclear medium.

The $(e, e'p)$ experiments searching for the onset of CT will be an integral part of the proposed 12 GeV upgrade of JLab and will be able to extend these measurements up to Q^2 of 18 (GeV/c)^2. The projected results for transparency of ^{12}C from an $(e, e'p)$ experiments with a 11 GeV beam at an upgraded JLab is shown in Fig. 3. It is clear that these future measurements would be able to look for any onset of CT at higher Q^2 and also be able to resolve the wide range of predictions from various models of the CT effect.

3 The $\gamma n \to \pi^- p$ reaction

It is more probable to produce a small transverse size in a $q\bar{q}$ system than in a three quark system, thus one expects an earlier onset of CT for meson production. Support for this comes from recent results on coherent diffractive dissociation of 500 GeV/c pions into di-jets at Fermilab [8], coherent and incoherent rho mesons production from heavy nuclei in the HERMES experiment [7]. However, none of these results can be considered as conclusive evidence for the onset of CT.

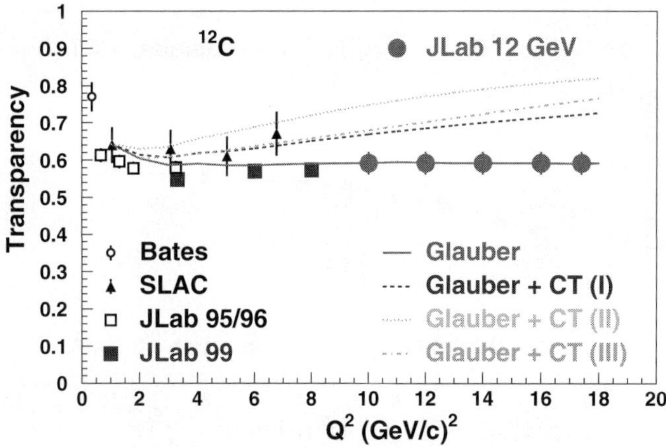

Fig. 3. Projected results from a measurement of transparency of ^{12}C using the $(e, e'p)$ reaction at JLab after the 12 GeV upgrade, shown along with the predictions of various CT calculation, CT(I) [13], CT(II) [11], CT(III) [12] and also a traditional calculation without CT [10]

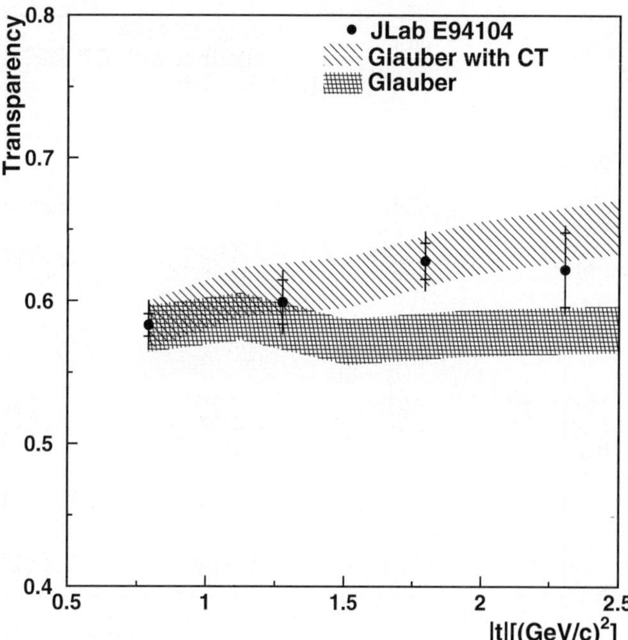

Fig. 4. The nuclear transparency of ^{4}He$(\gamma,\mathrm{p}\,\pi^-)$ at $\theta_{cm}^\pi = 70°$, as a function of momentum transfer square $|t|$. The inner error bars shown are statistical uncertainties only, while the outer error bars are statistical and point-to-point systematic uncertainties (2.7%) added in quadrature. In addition there is a 4% normalization/scale systematic uncertainty which leads to a total systematic uncertainty of 4.8%

A recent experiment at JLab used the $\gamma n \to \pi^- p$ reaction on ^{4}He to measure the transparency of pions and protons in ^{4}He. There are several important advantages to the choice of the ^{4}He nucleus and the $\gamma n \to \pi^- p$ process. Nucleon configurations obtained from the Monte Carlo method based on the exact nuclear ground state wavefunction are available for ^{4}He [14]. These configurations along with the elementary hadron-nucleon cross-sections can be used to carry out precise calculations of the nuclear transparency [15] in the framework of Glauber theory [16]. Therefore, precise measurement of nuclear transparency from ^{4}He nuclei is a benchmark test of these traditional nuclear calculations and can be used to explore where the calculations start to break down. Furthermore, light nuclei such as ^{4}He are predicted to be better for the search of CT phenomenon because of their relatively small nuclear sizes, which are smaller than the length scales over which the hadrons of reduced transverse size revert back to their normal sizes. [17,18].

The experiment was performed in Hall A at JLab. The continuous wave electron beam impinged on a copper radiator to generate an untagged bremsstrahlung photon beam. The combined photon and electron beam was then incident on either helium or liquid deuterium targets. The two High Resolution Spectrometers (HRS) in Hall A, were used to detect the outgoing pions and recoil protons in coincidence. Based on two-body kinematics the incident photon energy is reconstructed for each event using the measured angles and momenta of the π^- and p. In case of ^{4}He we also assume that the residual nucleus is ^{3}He. The experimental yield is obtained from the reconstructed photon energy spectra by integrating over a 100 MeV window starting 25 MeV below the electron beam energy. This ensures that the contributions from multi-pion processes are negligible. A Monte Carlo simulation of the experiment was used to account for the spectrometer acceptances. As per the definition of nuclear transparency one needs the cross section for $\gamma n \to \pi^- p$ reaction in ^{4}He and in free

space to extract transparency. However, since there are no free neutron targets a deuterium target was used and a correction for deuterium transparency was applied. The transparency was thus extracted from a super-ratio of the data and Monte Carlo yields from ^{4}He and ^{2}H targets corrected for the deuterium transparency.

The extracted nuclear transparency for the ^{4}He target along with calculations is shown in Figs. 4 and 5. The Glauber calculation use ^{4}He configurations, which are snapshots of the positions of the nucleons in the nucleus, obtained from the variational wave function of Arriaga *et al.* [14]. These contain correlations generated by the Argonne v_{14} and Urbana VIII models of the two-body and three-body nuclear forces respectively. The classical transparency was calculated from these configurations using the method described in [15]. The hadron-nucleon total cross-sections were taken from [19]. The calculation which includes the CT effect was obtained by repeating the calculation mentioned above with the hadron-nucleon total cross-section modified according to the quantum diffusion model [17]. This procedure is also described in [15].

In Figs. 4 and 5 the traditional nuclear physics calculation appears to deviate from the data at the higher energies. The absolute magnitude of the calculations with CT was normalized to the calculation without CT at the lowest energy point, however, it is the momentum transfer squared ($|t|$) dependence of the transparency which is of greater significance. The $|t|$ dependence is not affected by

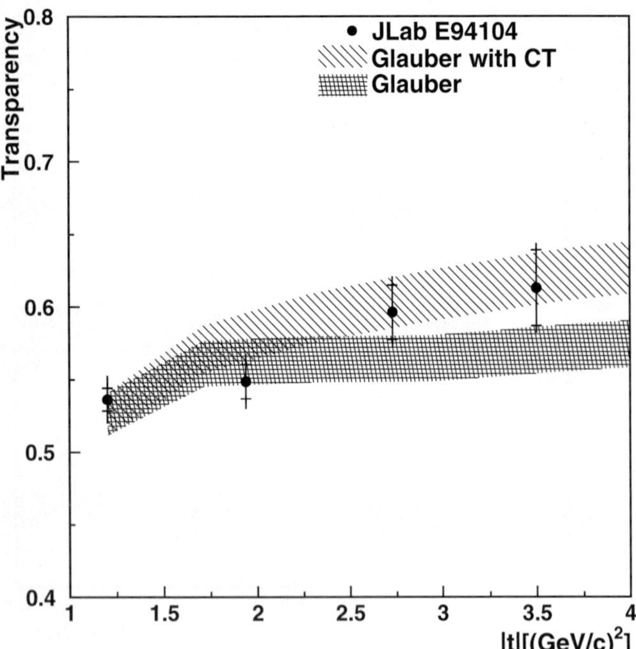

Fig. 5. The nuclear transparency of $^4\mathrm{He}(\gamma,\mathrm{p}\,\pi^-)$ at $\theta^\pi_{cm} = 90°$, as a function of momentum transfer square $|t|$. The inner error bars shown are statistical uncertainties only, while the outer error bars are statistical and point-to-point systematic uncertainties (2.7%) added in quadrature. In addition there is a 4% normalization/scale systematic uncertainty which leads to a total systematic uncertainty of 4.8%

the normalization systematic uncertainties. The deviation from Glauber calculation is larger at $\theta^\pi_{cm} = 90°$, as expected for a CT-like effect, since it is at a higher pion $|t|$. It is also interesting that the results are consistent with the rise expected for CT at the same photon energy at which the onset of scaling behavior was observed in the cross-section for the $\gamma\,\mathrm{n} \to \pi^-\,\mathrm{p}$ and the $\gamma\,\mathrm{p} \to \pi^+\,\mathrm{n}$ processes [20]. Thus, these data suggest the onset of deviation from traditional calculations, but future experiments with significantly improved statistical and systematic precision are essential to put these results on a firmer basis.

4 The $(e, e'\pi)$ reaction

As mentioned in the previous section it is more probable to produce a small transverse size in a $q\bar{q}$ system than in a three quark system, and thus one expects an earlier onset of CT for meson production. There are some hints of experimental support for these claims, however, none of these results can be considered as conclusive evidence for CT. Thus there is an urgent need for a systematic study of pion transparency over a wide range of Q^2 and A. An experiment to measure pion transparency over a Q^2 range of $1-5$ $(\mathrm{GeV/c})^2$ on $^1\mathrm{H}$, $^2\mathrm{H}$, $^{12}\mathrm{C}$, $^{26}\mathrm{Al}$, $^{64}\mathrm{Cu}$ and $^{197}\mathrm{Au}$ has been approved for Hall C at JLab. This experiment will measure the electroproduction of pion from nuclei by detecting the scattered electron and the knocked out pions in coincidence using the HMS and the SOS spectrome-

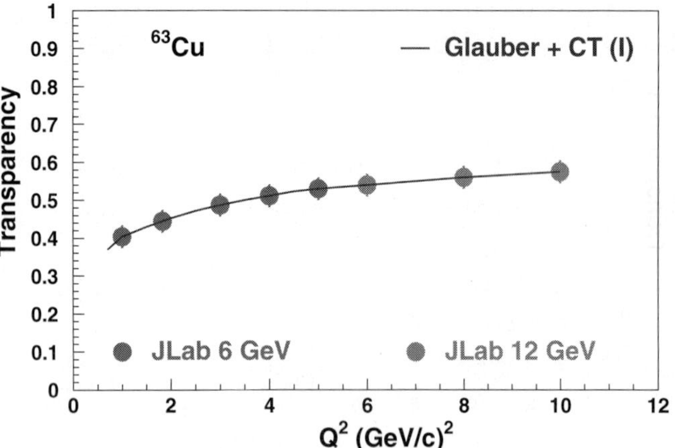

Fig. 6. The projected results for pion transparency in Copper along with the CT calculations of [13]. The statistical and systematic uncertainties have ween added in quadrature. Projections for both JLab upgraded to 12 GeV are also shown. Traditional nuclear physics calculations (Glauber calculations) predict an energy independent transparency

ters in Hall C. The pion transparency will be extracted by comparing the pion production from heavy nuclei to that from hydrogen. The Q^2 interval will be scanned in steps of 1 $(\mathrm{GeV/c})^2$ to provide a baseline for comparison with traditional nuclear physics calculations at the lower Q^2 points, while at the higher Q^2 points it will provide invaluable data on the predicted early onset of CT in $q\bar{q}$ systems. The combined statistical and systematic uncertainty is expected to be 5–7%, while pQCD based calculations predict enhancements as large as ≈ 40 % between Q^2 of 1 to 5 $(\mathrm{GeV/c})^2$ for Gold nuclei [13]. The projected results along with a pQCD based calculation [13] is shown in Fig. 6. This experiment would clearly be able to identify an early onset of CT if it exists.

Experiments measuring the pion transparency will greatly benefit form extension to higher Q^2 range since largest effects are predicted around Q^2 of 10 $(\mathrm{GeV/c})^2$ and thus are an integral part of the JLab 12 GeV upgrade program. The projected results for experiments which will be possible at JLab upgraded to 12 GeV are shown in Fig. 6.

5 Conclusions

Exclusive processes are crucial in studying the transition from the nucleon-meson to the quark-gluon picture. One of the signatures which forms an integral part of such studies is CT. Experiments at JLab has provided some useful clues. They have shown that there is no conclusive evidence for the onset of CT in qqq systems such as protons, up to Q^2 of 8.1 $(\mathrm{GeV/c})^2$. Another recent experiment [21] has shown interesting deviations form traditional calculations, which support claims for an early onset of CT in $q\bar{q}$ systems. However, future experiments with better statistical and systematic precision in this energy range together with improved theoretical calculations are crucial for confirming these results. The upcoming $(e, e'\pi)$

experiment should provide us some valuable information on pion transparency at momentum transfers up to a Q^2 value of 5 $(\text{GeV}/c)^2$. The planned upgrade of JLab to 12 GeV will potentially resolve the issue of the onset of phenomena such as CT by extending the search for CT to very high Q^2.

Acknowledgements. The author is indebted to Drs. R. Ent, H. Gao and R.J. Holt for many fruitful discussions. The author would also like to thank Dr. H. Gao for a careful reading of this manuscript. This work was supported by he U.S. Department of Energy under contract number DE-FG02-03ER41231.

References

1. S.J. Brodsky and A.H. Mueller: Phys. Lett. B **206**, 685 (1988)
2. A.S. Carroll et al.: Phys. Rev. Lett. **61**, 1698 (1988); I. Mardor et al.: Phys. Rev. Lett. **81**, 5085 (1998); A. Leksanov et al.: Phys. Rev. Lett. **87**, 212301-1 (2001)
3. N.C.R. Makins et al.: Phys. Rev. Lett. **72**, 1986 (1994); T.G. O'Neill et al.: Phys. Lett. **351**, 87 (1995)
4. D. Abbott et al.: Phys. Rev. Lett. **80**, 5072 (1998)
5. K. Garrow et al.: Phys. Rev. C **66**, 044613 (2002)
6. M.R. Adams et al.: Phys. Rev. Lett. **74**, 1525 (1995)
7. A. Airapetian et al.: Phys. Rev. Lett. **90**, 052501 (2003)
8. E.M. Aitala et al.: Phys. Rev. Lett. **86**, 4773 (2001)
9. G. Garino et al.: Phys. Rev. C **45**, 780 (1992)
10. V.R. Pandharipande and S.C. Pieper: Phys. Rev. C **45**, 791 (1992); H. Gao, V.R. Pandharipande, and S.C. Pieper: (private communication)
11. L.L. Frankfurt, M.I. Strikman, and M.B. Zhalov: Phys. Rev. C **50**, 2189 (1994)
12. N.N. Nikolaev et al.: Phys. Rev. C **50**, R1296 (1994)
13. B. Kundu et al.: Phys. Rev. D **62**, 113009 (2000)
14. A. Arriaga, V.R. Pandharipande, and R.B. Wiringa: Phys. Rev. C **52**, 2362 (1995)
15. H. Gao, R.J. Holt, and V.R. Pandharipande: Phys. Rev. C **54**, 2779 (1996)
16. R.J. Glauber: in *Lectures in Theoretical Physics*, edited by W. E. Brittin et al.: (Interscience, New York, 1959)
17. G.R. Farrar, H. Liu, L.L. Frankfurt, and M.I. Strikman: Phys. Rev. Lett. **61**, 686 (1988)
18. Nucl. Phys. A **580**, 365 (1994)
19. K. Hagiwara et al.: Phys. Rev. D **66**, 010001 (2002)
20. L.Y. Zhu el al.: Phys. Rev. Lett **91**, 022003 (2003)
21. D. Dutta et al.: to appear in Phys. Rev. C

Eur Phys J A (2004) **19**, s01, 185–188
Digital Object Identifier (DOI) 10.1140/epjad/s2004-03-031-y

EPJ A direct
electronic only

First results on kaonic hydrogen from the DEAR experiment

C. Guaraldo[1], G. Beer[9], A.M. Bragadireanu[1,5], M. Cargnelli[4], C. Curceanu (Petrascu)[1,5], J.-P. Egger[2,3], H. Fuhrman[4], M. Iliescu[1,5], T. Ishiwatari[4], K. Itahashi[7], M. Iwasaki[6], B. Lauss[8], V. Lucherini[1], L. Ludhova[2], J. Marton[4], F. Mulhauser[2], T. Ponta[1,5], L.A. Schaller[2], R. Seki[10,11], D. Sirghi[1,5], F. Sirghi[1], P. Strasser[6], and J. Zmeskal[4]

[1] INFN-Laboratori Nazionali di Frascati, C.P. 13, Via E. Fermi 40, I-00044 Frascati, Italy;
[2] Universite de Fribourg, Institut de Physique, Bd. de Perolles, CH-1700 Fribourg, Switzerland;
[3] Universite de Neuchatel, Institut de Physique, 1 rue A.-L. Breguet, CH-2000 Neuchatel, Switzerland;
[4] Institute for Medium Energy Physics, Boltzmanngasse 3, A-1090 Vienna, Austria;
[5] Institute of Physics and Nuclear Engineering "Horia Hulubei", P.O. Box MG - 6, R-76900, Bucharest, Romania;
[6] Institute of Physical and Chemical Research (RIKEN), 2-1 Hirosawa, Wako, Saitama 351-01, Japan;
[7] Tokyo Institute of Technology, 2-12-1 Ookoyana Meguro, Tokyo 152, Japan;
[8] Physics Department, University of California and Berkeley, Berkeley CA 94720, USA;
[9] University of Victoria, Department of Physics and Astronomy, P.O. Box 3055 Victoria, BC, Canada V8W3P6;
[10] W.E. Kellogg Radiation Laboratory, California Institute of Technology, Pasadena, CA 91125, USA;
[11] Department of Physics and Astrophysics, California State University, Northridge, CA 91330, USA

Received: 29 Jul 2003 / Accepted: 14 Nov 2003 /
Published Online: 6 Feb 2004 – © Società Italiana di Fisica / Springer-Verlag 2004

Abstract. The objective of DEAR (DAΦNE Exotic Atom Research) is the measurement of the K_α line shift and width, due to the strong interaction, in kaonic hydrogen and a similar measurement - the first one - in kaonic deuterium. The aim is a precision determination of the antikaon-nucleon isospin dependent scattering lengths in order to obtain the kaon nucleon sigma terms. In a first phase, dedicated to the calibration and optimization of the setup, the experiment collected data on kaonic nitrogen: for the first time a complex of three transitions ($7 \rightarrow 6$ at 4.6 keV, $6 \rightarrow 5$ at 7.6 keV and $5 \rightarrow 4$ at 14 keV) was measured, and the corresponding yields obtained. The kaonic hydrogen measurement was performed for a total integrated luminosity of about 60 pb^{-1} in the last months of the 2002 year. Data analyses are in progress, preliminary results being hereby presented, together with those on kaonic nitrogen.

PACS. 36.10.-K – 36.10.Gv

1 DEAR scientific program

The objective of DEAR (<u>DA</u>ΦNE <u>E</u>xotic <u>A</u>tom <u>R</u>esearch) [1] is the precise determination of the isospin dependent antikaon-nucleon scattering lengths, through a percent level measurement of the K_α line shift and broadening, due to strong interaction, in kaonic hydrogen, and a similar (the first one) measurement in kaonic deuterium.

A kaonic atom is formed when a negative kaon (coming from the ϕ-decay, produced at DAΦNE) enters the target, loses its kinetic energy through ionization and excitation processes, and is eventually captured, replacing the electron, in an excited orbit ($n \simeq 25$). Via different cascade processes (Auger effect, molecular dissociation, radiative transitions) the kaonic atom deexcites to lower states. When a low-n state with small angular momentum is reached, the strong interaction with the nucleus comes into play. This strong interaction is the reason for the shift in energy of the lowest-lying level from the purely elec-

tromagnetic value and for the finite lifetime of the state corresponding to an increase in the observed level width.

For kaonic hydrogen and kaonic deuterium the K-series transitions are of main experimental interest, since they are the only ones affected by the strong interaction. The K_α lines are clearly separated from the kigher K transitions. The shift ϵ and the width Γ of the 1s state of kaonic hydrogen are related in a fairly model-independent way (Deser-Trueman formula [2]) to the real and imaginary part of the complex s-wave scattering length, a_{K-p}:

$$\epsilon + i\Gamma/2 = 412 \cdot a_{K-p} \; eV \; fm^{-1} \qquad (1)$$

A similar relation applies to the case of kaonic deuterium and to its corresponding scattering length, a_{K-d}:

$$\epsilon + i\Gamma/2 = 601 \cdot a_{K-d} \; eV \; fm^{-1} \qquad (2)$$

The measured scattering lenghts are then related to the isospin-dependent scattering lengths, a_0 and a_1:

$$a_{K-p} = (a_0 + a_1)/2, \quad a_{K-n} = a_1 \qquad (3)$$

The extraction of a_{K-n} from a_{K-d} requires a more complicated analysis than the simple impulse approximation (K^- scattering from each free nucleon): higher order contributions associated with the K^-d three-body interaction have to be taken into account. This means to solve the three-body Faddeev equations by the use of potentials, taking into account the coupling among multichanneled interactions.

An accurate determination of the K^-N isospin dependent scattering lengths will place strong constraints on the low-energy K^-N dynamics, which in turn constraints the SU(3) description of chiral symmetry breaking [3]. Crucial information about the nature of chiral symmetry breaking, and to what extent the chiral symmetry must be broken, is provided by the calculation of the meson-nucleon sigma terms, whose definition is given below.

By considering the scattering process:

$$M_a(q) + N(p) \rightarrow M_b(q') + N(p') \qquad (4)$$

with the four momentum of the particles indicated in parenthesis and a, b denoting the SU(3) indices of the mesons, the meson nucleon sigma-term is defined [4] as the expectation value in the nucleon state of the equal-time double commutator of the chiral symmetry breaking part of the strong-interaction Hamiltonian:

$$\sigma_{MN}^{ba} = i < p|[Q_b^5, [Q_a^5, H_{SB}]]|p >, \qquad (5)$$

where $Q_{a,b}^5$ is the axial-vector charge and H_{SB} is the symmetry breaking part operator of the total Hamiltonian. The sigma term is then a quantity which directly gives the degree of chiral symmetry breaking. Its relation to the scattering amplitude $T_{ba}(\nu, t, q^2, q'^2$, kinematic invariants of the process) is given by the so-called low-energy theorem. Precisely, one has, in the soft-meson limit ($q^2, q'^2 \rightarrow 0$, $\nu = t = 0$):

$$\sigma_{MN}^{ba} = -f_a f_b T_{ba}(0,0,0,0) \qquad (6)$$

where the f's are the meson decay constant.

A phenomenological procedure, which requires an extrapolation of the measured amplitude, via dispersion relations, to a favored point of the (t, ν) plane, the so-called Cheng-Dashen point and then to zero, where, according to (6), sigma terms are defined, allows to extract the latter. Presently only estimates, with 70% uncertainties, exist; a measurement of the K^-N scattering lengths at few percent level should allow the determination of sigma terms with a precision better than 20%.

The sigma terms are also important inputs for the determination of the strangeness content of the proton. The strangeness fraction depends of both kaon-nucleon and pion-nucleon sigma terms, being more sensitive to the first ones [5].

2 The DEAR setup

The schematic drawing of the DEAR setup is shown in Fig. 1. The main elements of the setup are indicated in

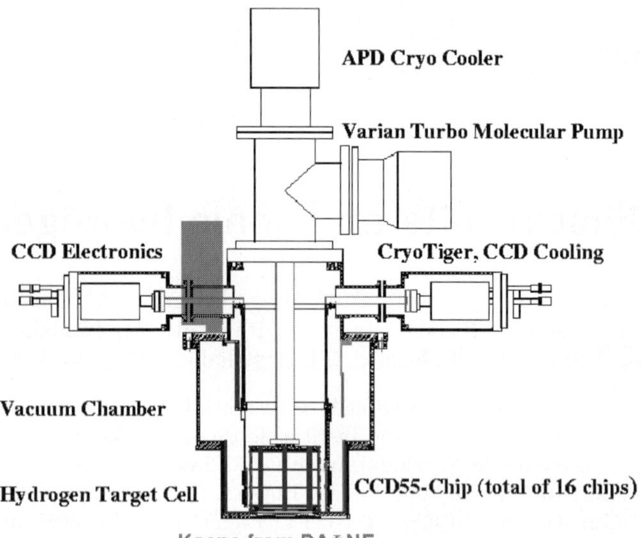

Fig. 1. The DEAR experimental setup

the figure. A pressurized cryogenic target (23 K and 1.82 bar) was used for the kaonic hydrogen measurement, in order to optimize the distribution of the kaons stopped in the target, and to avoid an important loss of the signal due to the Stark effect.

The target cell was done in kapton 75μm thick, cylindrical shaped, with a diameter of 11 cm, reinforced with epoxy-fiberglass bars, in order to have as less material as possible in front of the CCDs, to avoid electronic transitions which could interfere with the line to be measured.

The cryogenic setup was equipped with 16 CCD-55. The following performances were obtained:

- thermal noise: about 15 eV FWHM;
- energy resolution at 5.9 keV: 136 eV;
- linearity of the scale from 1 to 16 keV: 10^{-4};
- stability: fluctuations below 4 eV/month;
- charge transport inefficiency: 10^{-4}.

The setup was installed in one of the two interaction regions of DAΦNE and had periods of data taking starting from December 1999, when first collisions were achieved.

The first periods of data taking were dedicated to background understanding and reduction, by the use of appropriate shielding and machine optics solutions. The first measurement of an exotic atom at DAΦNE, namely kaonic nitrogen, followed in May 2001; this measurement was re-done in October 2002, when, for the first time, three transitions of kaonic nitrogen were clearly seen ($7 \rightarrow 6$ at 4.6 keV, $6 \rightarrow 5$ at 7.6 keV and $5 \rightarrow 4$ at 14 keV) and the corresponding yields extracted. Results of this measurement are presented Sect. 3.

The first measurement by DEAR of kaonic hydrogen was performed in the period November - December 2002. Preliminary results of this measurement are presented in Sect. 4.

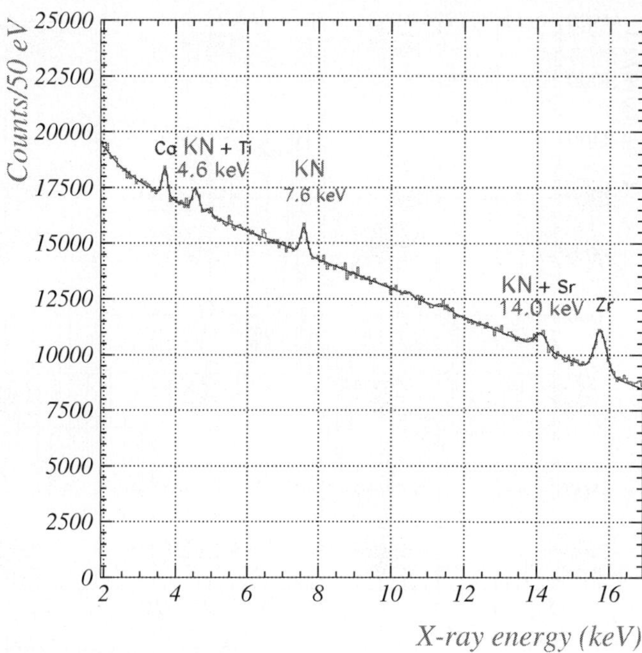

Fig. 2. Kaonic nitrogen spectrum obtained in the October 2002 measurement

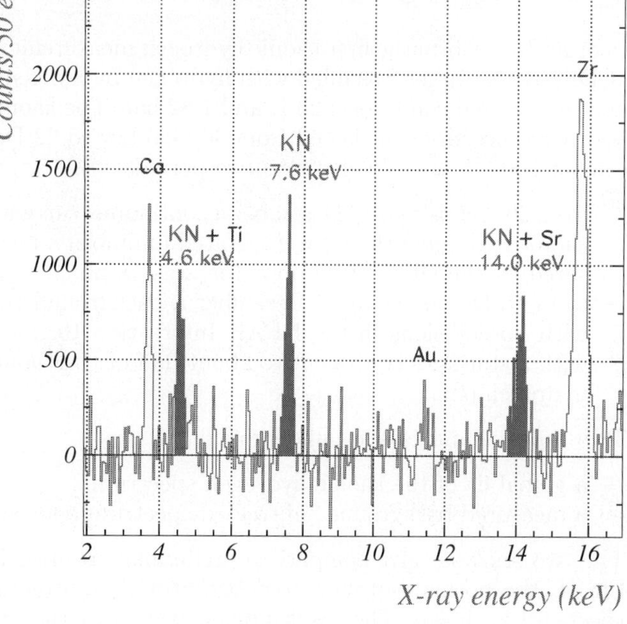

Fig. 3. Energy spectrum obtained for the run of October 2002 with kaonic nitrogen. Kaonic nitrogen transitions are clearly seen, as indicated in the figure

3 Kaonic nitrogen results

The measurement of kaonic nitrogen (whose transitions have yields 20 times higher than kaonic hydrogen and therefore a fast feedback can be obtained) performed by DEAR had the following primary objectives:

- to prove the feasibility of the DEAR technique to produce and detect exotic atoms using the K^- beam from the ϕ-decay at DAΦNE;
- to optimize the kaon stopping distribution inside the gaseous target.

These objectives were achieved in May 2001, when the first measurement of kaonic nitrogen $7 \to 6$ and $6 \to 5$ transitions, at 4.6 and 7.6 keV respectively, was performed. The results were published [6], suggesting a new method for a future precision measurement of the charged kaon mass.

In October 2002 the kaonic nitrogen spectrum was remeasured, just before the start of the kaonic hydrogen data taking.

The scientific aim of this run was the study of different degrader configurations to take into accout the effect of the boost in the ϕ-production.

A statistics corresponding to about 10.5 pb^{-1} integrated luminosity was collected. The overall X-ray energy spectrum, as measured by the 16-CCD 55, is shown in Fig. 2.

A refined analysis of the spectrum was performed [7]. This analysis allowed, for the first time,to disentangle a complex of three kaonic nitrogen transition:

1. the $7 \to 6$ transition, at 4.57 keV;
2. the $6 \to 5$ transition, at 7.59 keV;
3. the $5 \to 4$ transition, at 13.96 keV.

Table 1. Kaonic nitrogen results

Transition	Number of events	Yield of transition (%)
$7 \to 6$	2690 ± 650	33.7 ± 8.1
$6 \to 5$	5320 ± 395	55.5 ± 4.2
$5 \to 4$	1360 ± 330	66.4 ± 15.6

The continuous background subtracted spectrum is shown in Fig. 3.

The number of events for each transition is reported in Table 1.

Apart of the kaonic nitrogen transitions, in the spectrum are present:

- calcium line, at about 3.7 keV, due to the presence of calcium in the fiberglass reinforcement of the kapton target;
- silicon line - silicon being the material of the CCDs;
- aluminium line - from the top cover of the target;
- strontium line - from the ceramics-support of the CCDs;
- zirconium line - from the zirconium foil placed inside the setup with the purpose of energy scale calibration.

The transition yields, for the first time measured, were extracted by using Monte Carlo simulation of the setup and of the contributing physical processes. The preliminary results on the yields are reported in Table 1 and constitute important checks for cascade calculations in the field of exotic (kaonic) atom transitions.

The yields reported in Table 1 contain only the statistical errors; an accurate study on systematic errors is undergoing.

4 Kaonic hydrogen preliminary results

In order to perform the first kaonic hydrogen measurement at DEAR the target was filled with hydrogen in cryogenic and pressurized conditions: 23 K and 1.82 bar. The kaonic hydrogen measurement lasted from 30 October to 22 December 2002. It was divided into two periods:

– from 30 October to 16 December a continuous run with kaonic hydrogen; the total integrated luminosity, measured by the DEAR kaon monitor, was 58 pb^{-1};
– from 16 December to 22 December a background run with no collisions in the DEAR Interaction Region - with a statistics equivalent to about half of the kaonic hydrogen one.

Two types of analyses are undergoing:

– a global fit of the kaonic hydrogen spectrum;
– a measured background subtracted spectrum analysis.

The two analyses give compatible preliminary results. In Fig. 4, the normalized measured background subtracted spectrum is shown. The resulting spectrum was then fit with Voigtian functions (convolution between pure Breit-Wigner, for signal, and Gaussian, to take into account the experimental resolution). The K_α line of kaonic hydrogen, together with a bump corresponding to the $K-$complex are clearly evident. A peak corresponding to the $6 \rightarrow 5$ transition of kaonic carbon is as well present, with a statistical evidence of 2σ. The statistical evidence of the K-complex transition is (sum of individual contribution) more than 6σ. The very preliminary results on the shift and width for the kaonic hydrogen transitions are:

$$\epsilon = 6(150 \pm 45) \ eV \tag{7}$$
$$\Gamma = 6(250 \pm 90) \ eV \tag{8}$$

Further analyses and checks are undergoing.

5 Future plans

For the future, in order to achieve a percent level measurement of the shift and width for kaonic hydrogen and kaonic deuterium K_α lines an upgrade of the DEAR setup is in progress. One of the problems to cope with during the kaonic hydrogen measurement performed in 2002 was still the presence of a high background. The measurement was performed with a signal/background ratio of about 1/70. In order to further reduce the background, the use of a trigger system is a must. The CCD devices are non-triggerable, due to the fact that data read-out is slow (seconds). Consequently, the collaboration is planning to use new fast triggerable devices, namely large area Silicon Drift Detectors (SDD). The trigger is given by the entrance of the charged kaon in the target volume. The event can be identified and measured with high accuracy and low contamination by the use of a three-scintillator telescope, synchronized with the bunch frequency. The

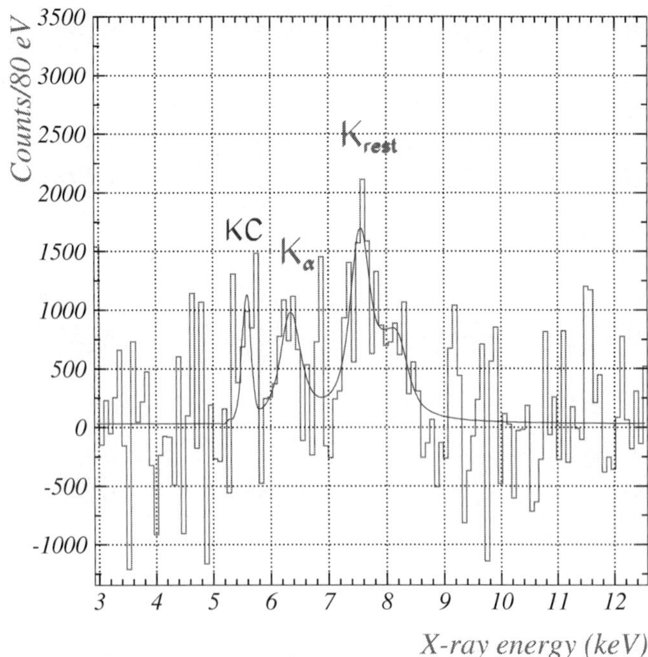

Fig. 4. Kaonic hydrogen measured background subtracted spectrum

timing window (1 μs) to be used in trigger, together with the excellent energy resolution (140 eV at iron position) of the SDD will allow a dramatic decrease of the background so to obtain a signal/background ratio of about 1/1 for kaonic hydrogen and 1/5 for kaonic deuterium.

Other exotic atoms (kaonic helium and other light atoms, as well as sigmonium atoms) and a high-precision measurement of the charged kaon mass will eventually become feasible.

Acknowledgements. The DAΦNE group in warmly acknowledged for the very good cooperation and team-work. Part of the work was supported by a "Transnational access to Research Infrastructure" (TARI) Contract No. HPRI-CT-1999-00088.

References

1. S. Bianco et al.: Rivista del Nuovo Cimento **22**, No. 11 (1999) 1
2. S. Deser et al.: Phys. Rev. **96**, 774 (1954); T.L. Truemann: Nucl. Phys. **26**, 57 (1961); A. Deloff: Phys. Rev. C **13**, 730 (1976)
3. C. Guaraldo: Proceedings of III International Workshop on Physics and Detectors for DAΦNE, November 16-19, Frascati, Italy, Frascati Physics Series Volume XVI (1999) 642
4. E. Reya: Rev. Mod Phys. **46**, 545 (1974); H. Pagels: Phys. Rep. **16**, 219 (1975)
5. R.L. Jaffe and C.L. Korpa: Comments Nucl. Part. Phys. **17**, 163 (1987)
6. G. Beer et al: Phys. Lett B **535**, 52 (2002)
7. T. Ishiwatari and P.D. Thesis: RIKEN (2003)

Eur Phys J A (2004) **19**, s01, 189–196
Digital Object Identifier (DOI) 10.1140/epjad/s2004-03-032-x

Quark matter in neutron stars

Marcello Baldo

INFN, and Dipartimento di Fisica dell' Universita' di Catania, via S. Sofia 64, 95123 Catania, Italy

Received: 7 Aug 2003 / Accepted: 14 Nov 2003 /
Published Online: 6 Feb 2004 – © Società Italiana di Fisica / Springer-Verlag 2004

Abstract. The density of nuclear matter in the interior of neutron stars can reach values, for the largest masses, which can be compatible with the onset of hadron deconfinement. For the study of this possibility the only viable method at present is the comparison between the available nucleon and quark Equations of State (EoS) at increasing baryon density. It is then possible to trace the transition to the deconfined phase or the appearence of a mixed phase. We present recent results on the structure of neutron stars based on this procedure. For the nucleon matter, the microscopic many-body theory of the Nuclear Equation of State is discussed in the framework of the Bethe-Brueckner-Goldstone method. The expansion is extended up to the three hole-line diagrams contribution. For the quark matter, different models are used to generate the quark EoS. Despite the maximum mass of neutron stars turns out to be only marginally sensitive to the considered quark EoS, it is found that the structure of neutron stars can drastically depend on the adopted model.

PACS. 21.65+f Nuclear matter – 97.60,JD Neutron Stars – 26.60+c Nuclear aspects of Neutron Stars – 24.10,Cn Many-body

1 Introduction

From a hystorical point of view, the nuclear saturation problem was the main motivation for the extensive studies, which have lasted for few decades, on infinite nuclear matter, since it is one of the fundamental problems in nuclear physics. However, infinite nuclear matter cannot be considered only an idealized system, since it is commonly believed that macroscopic portions of (asymmetric) nuclear matter form the interior bulk part of neutron stars (NS), usually associated with pulsars. The internal structure of NS is therefore directly linked to the Equation of State (EoS) of infinite nuclear matter. An accurate prediction of the EOS is highly demanded for most studies of neutron stars and related astrophysical applications. Of course, only indirect observations of NS structure are possible. However, the astrophysics of neutron stars is rapidly developing, in view of the observations coming from the new generation of artificial satellites, and one can expect that it will be possible in the near future to confront the theoretical predictions with more and more stringent phenomenological data.

At the hadronic level, several many-body theories and techniques have been developed and applied to the study of the nuclear Equation of State. The main difficulty in the many-body theory of nuclear matter is the treatment of the strong repulsive core, which dominates the short range behaviour of the nucleon-nucleon (NN) interaction. Simple perturbation theory cannot of course be applied, since the matrix elements of the interaction are too large. One way of overcoming this difficulty is to introduce the two-body scattering G-matrix, which has a much smoother behaviour even for large repulsive core. It is possible to rearrange the perturbation expansion in terms of the reaction G-matrix, in place of the original bare NN interaction, and this procedure is systematically exploited in the Bethe-Brueckner-Goldstone (BBG) expansion [1].

For quark matter, no calculation is available at the high baryon density which is encountered in neutron star interior. For the same reason, no theory, based on QCD, is available to describe the possible transition form hadron matter to the deconfined quark matter. At present, the only viable method to explore the possible appearence of the deconfined phase in tne core of neutron stars is the use of a well defined model for the quark-gluon plasma and the comparison of the corresponding EoS with the hadronic one to assess the possible phase transition (including eventually a mixed phase). In Sect. 2 we describe the hadronic EoS and its properties at increasing density. In Sect. 3 we confront this EoS with the quark EoS obtained within few models, and we extract the NS structure for each adopted quark matter model. It turns out that, while the maximum NS mass is only slightly dependent on the quark matter model, the NS structure is strongly dependent on it.

2 The BBG expansion and the nuclear EOS

The BBG expansion for the ground state energy at a given density, i.e. the EOS at zero temperature, can be ordered according to the number of independent hole-lines appearing in the diagrams representing the different terms of the expansion. This grouping of diagrams generates the so-called hole-line expansion [2]. The diagrams with a given number n of hole-lines are expected to describe the main contribution to the n-particle correlations in the system. At the two hole-line level of approximation the corresponding summation of diagrams produces the Brueckner-Hartree-Fock (BHF) approximation, which incorporates the two particle correlations. The BHF approximation includes the self-consistent procedure of determining the single particle auxiliary potential, which is an essential ingredient of the method. Once the auxiliary self-consistent potential is introduced, the expansion is implemented by introducing the set of diagrams which include "potential insertions". To be specific, the introduction of the auxiliary potential can be formally performed by splitting the hamiltonian in a modified way from the usual one

$$H = T + V = T + U + (V - U) \equiv H_0' + V' \qquad (1)$$

where T is the kinetic energy and V the nucleon-nucleon interaction. Then one consider $V' = V - U$ as the new interaction potential and H_0' as the new single particle hamiltonian. Then, the single particle energy $e(k)$ is given by

$$e(k) = \frac{\hbar^2 k^2}{2m} + U(k) \qquad (2)$$

while U must be chosen in such a way that the new interaction V' is, in some sense, "reduced" with respect to the original one V, so that the expansion in V' should be faster converging. The introduction of the auxiliary potential turns out to be essential, otherwise the hole-expansion would be badly diverging. The total energy E can then be written as

$$E = \sum_k e(k) + B \qquad (3)$$

where B is the interaction energy due to V'. The BHF sums the so called "ladder diagrams". The summation of these diagrams can be performed by solving the integral equation for the Brueckner G-matrix

$$\langle k_1 k_2 | G(\omega) | k_3 k_4 \rangle = \langle k_1 k_2 | v | k_3 k_4 \rangle +$$

$$+ \sum_{k_3' k_4'} \langle k_1 k_2 | v | k_3' k_4' \rangle \frac{\left(1 - \Theta_F(k_3')\right)\left(1 - \Theta_F(k_4')\right)}{\omega - e_{k_3'} - e_{k_4'}}. \qquad (4)$$

$$\langle k_3' k_4' | G(\omega) | k_3 k_4 \rangle$$

where $\Theta_F(k) = 1$ for $k < k_F$ and is zero otherwise, being k_F the Fermi momentum. The product $Q(k, k') = (1 - \Theta_F(k))(1 - \Theta_F(k'))$, appearing in the kernel of (4), enforces the scattered momenta to lie outside the Fermi sphere, and it is commonly referred as the "Pauli operator". This G-matrix can be viewed as the in-medium scattering matrix between two nucleons. The self-consistent single particle potential $U(k)$ is determined by the equation

$$U(k) = \sum_{k' < k_F} \langle k k' | G(e_{k_1} + e_{k_2}) | k k' \rangle_A \qquad (5)$$

with $|k k'\rangle_A = |k k'\rangle - |k' k\rangle$.

The first potential insertion diagram cancels out the potential part of the single particle energy of (2), in the expression for the total energy E. This is actually true for any definition of the auxiliary potential U. At the two hole-line level of approximation, one therefore gets

$$E = \sum_{k < k_F} \frac{\hbar^2 k^2}{2m} + \frac{1}{2} \sum_{k, k' < k_F} \langle k k' | G(e_k + e_{k'}) | k k' \rangle_A$$

$$\equiv \sum_{k < k_F} \frac{\hbar^2 k^2}{2m} + \frac{1}{2} \sum_{k < k_F} U(k) \qquad (6)$$

The result that only the unperturbed kinetic energy appears in the expression for E, and all the correlations are included in the potential energy part, holds true to all orders and it is a peculiarity of the BBG expansion. Of course, the modification of the momentum distribution, and therefore of the kinetic energy, is included in the interaction energy part, but it is treated on the same footing as the other correlation effects. This seems to present a noticeable advantage. In fact, the modification of the kinetic energy in itself is quite large and, of course, positive and should be therefore compensated by an extremely accurate calculations of the (negative) correlation energy. On the other hand, putting the two effects on the same footing, one can expect that strong cancellation occur order by order.

The BHF results indicate that already the two hole-line approximation is able to produce reasonable values for the saturation point. The remaining discrepancies can be summarized in the celebrated Coester band [3], the line along which the results for different "realistic" nucleon-nucleon forces appear to be approximately concentrated and which misses the phenomenological saturation point. According to the force used, either the saturation density is too high or the binding energy is too small. However, the results depend on the very definition of the single particle potential. The "standard" choice for $U(k)$ assumes that the potential is zero above the Fermi momentum k_F, while in the "continuous choice" [4] the definition of (5) is extended to momenta k larger than k_F, thus making U a continuous function through the Fermi surface. The final result of a hypothetically exact BBG calculation is independent of the auxiliary potential $U(k)$, but the rate of convergence can of course depend on the particular choice adopted. Therefore, the degree of dependence of the results on the choice of the potential can be considered an indication of the degree of convergence obtained at a given level of the BBG expansion. The results for both choices of $U(k)$ at the BHF level of approximation (Argonne v_{18} potential [5]) is reported in Fig. 1 in the case of symmetric nuclear matter (solid lines). As one can see, the saturation curves are different for the two prescriptions. It has to be noticed, however, that the apparent discrepancy of 4-5 MeV in the binding energies shown in Fig. 1, is about 10% of the calculated potential energy per particle, which is about -40 MeV around saturation. This is the degree of convergence obtained at the Brueckner level. In view of these results, obtained within the BHF approximations, it appears mandatory to consider the three hole-line diagrams. The value of their contribution can indeed

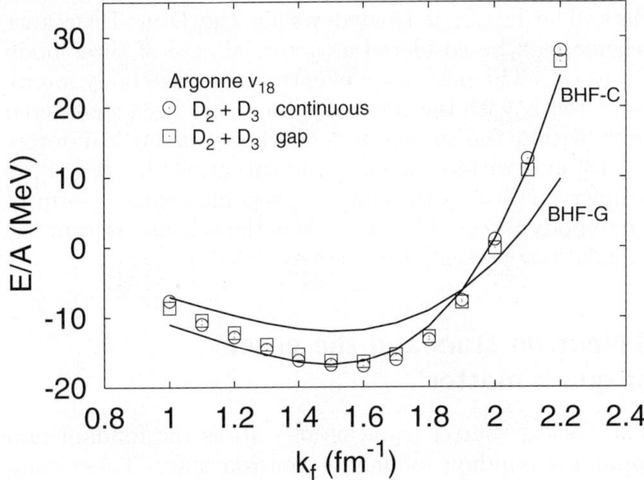

Fig. 1. Nuclear matter saturation curve for the Argonne v_{18} NN potential. The *solid lines* indicate the results at the Brueckner (*two hole-lines*) level for the standard (BHF-G) and the continuous choices (BHF-C) respectively. The results obtained adding the *three hole-line* contribution are given by the *open squares* (*standard choice*) and the *open circles* (*continuous choice*)

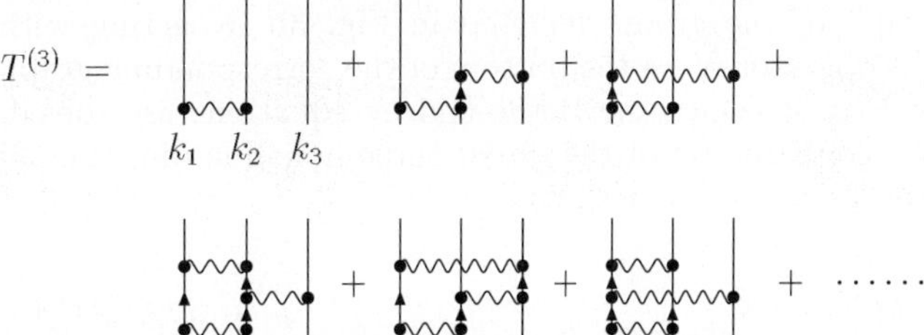

Fig. 2. The first few terms in the expansion of the Bethe-Fadeev integral equation

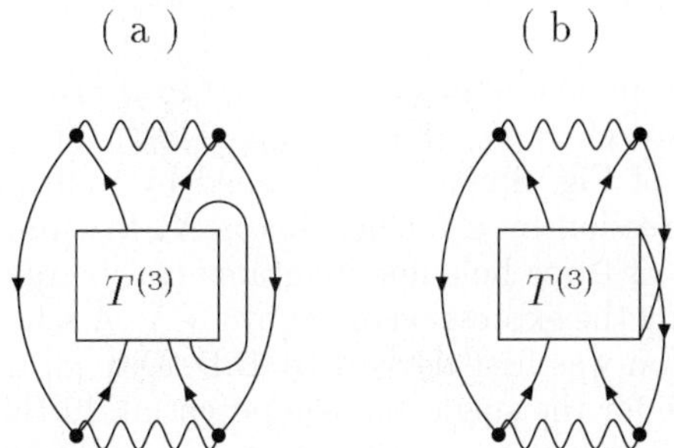

Fig. 3. Schematic representation of the direct **a** and exchange **b** three hole-line diagrams

provide a check of convergence and possibly an accurate EOS. According to the BBG expansion, this set of diagrams describes the irreducible three-nucleon correlations, i.e. the three-body correlations which cannot be reduced to a product of two-body correlations, already introduced at the BHF level. Since the two hole-line contribution has been summed up by introducing the G-matrix, the in-medium two-body scattering matrix, it is conceivable that the three hole-line diagrams can be summed up by introducing some similar generalization of the scattering matrix for three particles. The three-body scattering problem has a long history by itself, and has been given a formal solution by Fadeev [6]. For identical particles the original three integral Fadeev equations reduce to one, because of symmetry. The analogous equation and scattering matrix in the case of nuclear matter (or other many-body systems in general) has been introduced by Bethe [7]. The integral equation, the Bethe–Fadeev equation, reads schematically

$$\langle k|_1 k_2 k_3 | T^{(3)} | k'_1 k'_2 k'_3 \rangle = \langle k|_1 k_2 | G | k'_1 k'_2 | \delta \rangle_K (k_3 - k'_3) +$$

$$+ \langle k|_1 k_2 k_3 | G_{12} X \frac{Q_3}{e} T^{(3)} | k'_1 k'_2 k'_3 \rangle \ . \tag{7}$$

As one can see, the kernel contains the two-body scattering matrix G in place of the bare NN interaction, in line with the BBG scheme. The factor Q_3/e is the analogous of the similar factor appearing in the integral equation for the two-body scattering matrix G, see (4). Therefore, the projection operator Q_3 imposes that all the three particle states lie above the Fermi energy, and the denominator e is the appropriate energy denominator, namely the energy of the three-particle intermediate state minus the entry energy ω, in close analogy with the equation for the two-body scattering matrix G of (4). The real novelty

with respect to the two-body case is the operator X. This operator interchanges particle 3 with particle 1 and with particle 2, $X = P_{123} + P_{132}$, where P indicates the operation of cyclic permutation of its indices. It gives rise to the so-called "endemic factor" in the Fadeev equations, since it is an unavoidable complication intrinsic to the three-body problem in general. The reason for the appearance of the operator X in this context is that no two successive G matrices can be present in the same pair of particle lines, since the G matrix already sums up all the two-body ladder processes. In other words, the G matrices must alternate from one pair of particle lines to another, in all possible ways, as it is indeed apparent from the expansion by iteration of (7), which is represented in Fig. 2. Adding all terms with an arbitrary number of G-matrices, one gets a generalized ladder series for three-particles, analogous to the ladder series introduced for the two particles case in defining the G-matrix. Indeed, this is the basis for the integral (7). The introduction of the three-body scattering matrix $T^{(3)}$ allows to sum up the three hole-line diagrams, as schematically indicated in Fig. 3. Here the diagrams have been divided into two distinct groups, the directs ones (a) and the exchange ones (b).

Once the first interaction has occurred, the remaining part of the diagram describes the rescattering, in all possible way, of three particle-lines, since no further hole-line must be present in the diagram. This part of the diagram is indeed the three-body scattering matrix $T^{(3)}$, and the operator Q_3 in (7) assures, as already mentioned, that only particle lines are included. The set of diagrams indicated in part (b) can be obtained by the ones of part (a) by simply interchanging the final (or initial) point of one of the "undisturbed" hole-line with the final (or initial) point of the third hole-line. This means that one can obtain each graph of the group depicted in Fig. 3b by acting with the operator X on the bottom of the corresponding graph of Fig. 3a. Once the Bethe-Fadeev equations are solved, the contribution of the direct three hole-line diagrams of Fig. 3a can be written as

$$E_{3h}^{dir} = \frac{1}{2} \sum_{k_1,k_2,k_3 \leq k_F} \sum_{\{k'\},\{k''\} \geq k_F} \langle k|_1 k_2 | G | k_1' k_2' |_\rangle A \cdot$$

$$\cdot \frac{1}{e} \langle k|_1' k_2' k_3' | X T^{(3)} X | k_1'' k_2'' k_3'' |\rangle \frac{1}{e'} \langle k|_1'' k_2'' | G | k_1 k_2 |_\rangle A \quad , \tag{8}$$

In (8) the denominator $e = e_{k_1'} + e_{k_2'} - e_{k_1} - e_{k_2}$, and analogously $e' = e_{k_1''} + e_{k_2''} - e_{k_1} - e_{k_2}$. The exchange diagrams of Fig. 3b can be obtained by multiplying the same expression by a further factor X. In summary, the entire set of three hole-line diagrams can be obtained by multiplying the expression of (8) by $1+X$. A scheme of approximation was first devised by B.D. Day [8] within the gap choice for the single particle potential. In this scheme the first diagram in each one of the series (a) and (b) in Fig. 3, the so-called "bubble" and "ring" diagrams, are singled out from the whole set of three hole-line diagrams, while the remaining series of diagrams is summed up by solving the Bethe-Fadeev integral equation. This procedure turns out to be numerically convenient. We have checked [9] and extended these calculations to the continuous choice for the single particle potential. In the latter case a potential insertion diagram has to be added at the three hole-line level of approximation [1,9]. The final equation of State obtained by adding the three hole-line contribution is reported in Fig. 1, both for the gap choice (squares) and the continuous choice (stars) [10], again for the Argonne v$_{18}$ potential. Two conclusions can be drawn from these results. i) The two saturation curves in the gap and continuous choices, with the inclusion of the three hole-line diagrams, tend now to collapse in a single EOS, with some deviations only at the highest density. This is a strong indication that a high degree of convergence has been reached at this level of the expansion, according to the criterion discussed above. Notice that the saturation curves extend from a density which is about one half of saturation density to about five times saturation density, and, therefore, it appears unlikely that the agreement between the two choices can be considered as a fortuitous coincidence. ii) The Brueckner two hole-line EOS within the continuous choice turns out to be already close to the full EOS, since in this case the three hole-line contribution is quite small. In first approximation one can adopt the BHF results with the continuous choice as the nuclear matter EOS. Indeed, this is a further indication of convergence. The phenomenological saturation point for symmetric nuclear matter is, however, not reproduced, which confirms the finding in [8]. The binding energy per particle at the minimum of the saturation curve turns out to be close to the empirical value of about -16 MeV, but the corresponding density comes out about 20-30 % larger than the empirical one. The discrepancy can be accounted for by introducing three-body forces and relativistic ef-

fects. The latter, if treated within the Dirac-Brueckner scheme, can be cosidered as a special case of three-body forces [11]. The possibility of extracting thrre-body forces, consistently with the two-body forces, has been considered both within the meson-nucleon theory of nuclear forces [12,13] and within the chiral perturbation theory [14]. A simpler method is to adopt a phenomenological form of three-body forces [15], and this is the scheme used in the calculations presented in the next Section.

3 Neutron stars and the onset of quark matter

The nuclear matter equation of state is the fundamental input for building models of neutron stars. These compact objects, among the densest in the universe, are indeed characterized by values of the baryon density which ranges from the iron density at the surface up to eight-ten times normal nuclear matter density in the core. Therefore a detailed knowledge of the equation of state over a wide range of densities is required [16]. As we have seen in the previous section, the EOS of nucleon matter can be established up to densities relevant to neutron star studies. At increasing densities an additional complication arises. In fact, whereas at densities close to the saturation value the matter consists mainly of nucleons and leptons, at higher densities several species of particles may appear due to the fast rise of the nucleon chemical potentials. The Brueckner-Hartree-Fock scheme must be then generalized to include a possible fractions of hyperons, like Λ and Σ^-. An additional uncertainity comes from our limited knowledge of the nucleon-hyperon (NY) interaction, which we describe within the Nijmegen soft-core model [17]. No hyperon-hyperon interaction was taken into account, since no robust experimental data are available yet. For more details, the reader is referred to [18,19] and references therein. In Fig. 4 we show the chemical composition of β-stable and asymmetric nuclear matter containing hyperons (panel (a)) and the corresponding equation of state (panel (b)). The shown calculations have been performed using the Paris potential. We observe that hyperon formation starts at densities $\rho \simeq 2 - 3$ times normal nuclear matter density. The Σ^- baryon appears earlier than the Λ, in spite of its larger mass, because of the negative charge. The appearance of strange particles has two main consequences, i) an almost equal percentage of nucleons and hyperons are present at highest densities and ii) a strong deleptonization of matter, since it is energetically convenient to maintain charge neutrality through hyperon formation than β-decay. The equation of state is displayed in panel (b). The dotted line represents the case when only nucleons and leptons are present in stellar matter, whereas the solid line shows the case when hyperons are included as well. In the latter case the equation of state gets very soft, since the kinetic energy of the already present baryonic species is converted into masses of the new particles, thus lowering the total pressure.

The relation of NS mass with the matter EoS is apparent in the Tolman-Oppenheimer-Volkoff (TOV) [16],

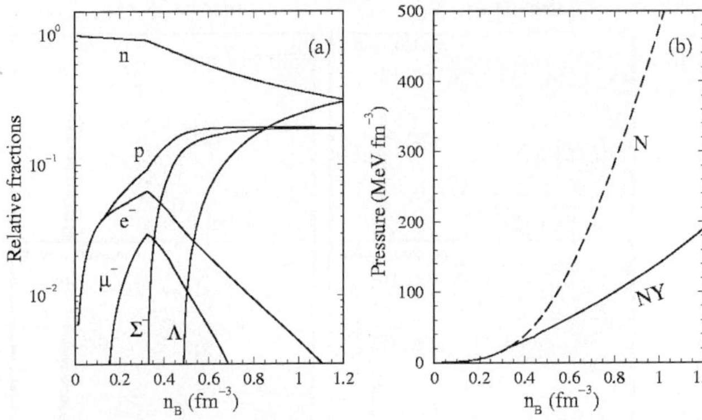

Fig. 4. In panel **a** we display the equilibrium composition of asymmetric and β-stable nuclear matter containing Σ^- and Λ hyperons. In panel **b** the *solid (dotted) line* represents the EoS obtained in the case when nucleons plus hyperons (*only nucleons*) are present

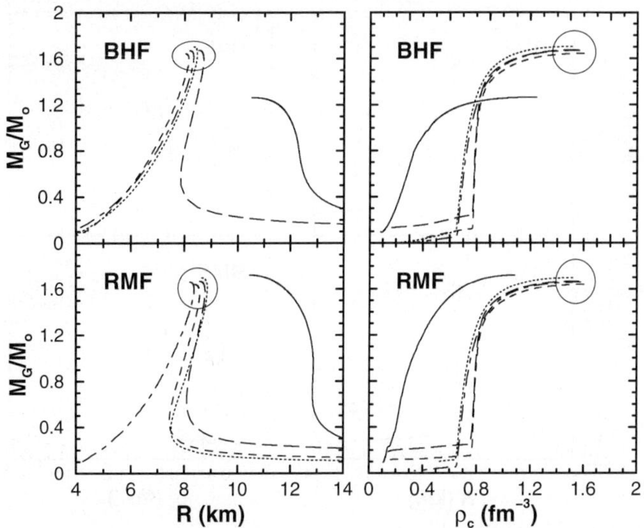

Fig. 5. The mass-radius (*left panel*) and the mass-central density (*right panel*) relations for $\epsilon_B = 0.8$ GeV fm^{-3}, see the text, and various parametrizations of the bag constant as a function of density. The *solid lines* indicate the resukts for the pure hadronic EoS discussed in the text

which determines the hydrostatic equilibrium configuration of the star (assumed to be spherically symmetric). The mass distribution is obtained by solving the TOV equations for the pressure $P(r)$ and for the mass $m(r)$ enclosed inside the sphere of radius r,

$$\frac{dP(r)}{dr} = -\frac{Gm(r)\epsilon(r)}{r^2}\frac{\left[1+P(r)/\epsilon(r)\right]\left[1+4\pi r^3 P(r)/m(r)\right]}{1-2Gm(r)/r} \tag{9}$$

$$\frac{dm(r)}{dr} = 4\pi r^2 \epsilon(r)$$

being G the gravitational constant and ϵ the energy density. The EoS establishes a relation between pressure and energy density, making this set of two coupled equations closed. Therfore, starting with a central mass density $\epsilon(r=0) \equiv \epsilon_c$, one integrates out until the pressure on the surface equals the one corresponding to the density of iron. This gives the stellar radius R and the corresponding gravitational mass M. It has to be noticed that the NS mass which is observable is the gravitational mass, given by

$$M_G \equiv m(R) = 4\pi \int_0^R dr \, r^2 \epsilon(r) \tag{10}$$

For the description of the NS crust, if present, one usually joins the hadronic equations of state with the ones by Negele and Vautherin [20] in the medium-density regime, and the ones by Feynman-Metropolis-Teller [21] and Baym-Pethick-Sutherland [22] for the outer crust.

However, the formation of a large fraction of strange matter at high density, and therefore in the interior of NS, can be only a hypothetical scenario, since another process can compete : the onset of a deconfined quark matter phase at large enough baryon density. This possibility has been considered recently by several authors, within specific models for the quark phase, and we will review briefly their findings.

The simplest quark matter model, widely used in model calculations, is the MIT bag model [23]. The model contains mainly two parameters, the bag constant B and

the strange quark mass m_s, besides the strong coupling constant α_S which describes possible perturbative corrections. A systematic study of NS structure, as the parameters are varied, has been presented in [24]. In this work, no strange matter is considered in the hadronic sector, whose EoS is described either within the relativistic mean field or within the non-relativistic many-body model of [25]. In the quark sector, the possibility of color superconductivity (CS) [26,27] is also considered (for a review on CS see [28].) Despite the presence of strong color superconductivity (gap $\Delta \sim 100$ MeV) can enhance appreciably the NS maximum mass, one of the main conclusions of this systematic study is that the NS mass cannot exceed values of about 1.6–1.7 solar mass. A larger value of the mass would require values of the bag constant so small to make symmetric nuclear matter unstable towards deconfinement even at saturation density. This results appears in agreement with our previous work [29, 30], where also the MIT bag model was used (without color superconductivity), in conjunction with the above discussed hadron EoS, which includes strangeness. Without the quark phase, our EoS would produce a maximum mass below the observational limit of 1.44 solar mass [31]. Mixed phase, along the general method of [32], was also included. The bag constant was allowed to vary with density, in order to tune the value of the baryon energy density ϵ_B at which deconfinement occurs in symmetric nuclear matter. If this value is restricted in a "reasonable interval" , i.e. $0.8 < \epsilon_B < 1.5$ GeV fm^{-3}, the resulting maximum mass depends only weakly on ϵ_B, and it does not exceed 1.7 solar mass. These findings are summarized in Figs. 5, 6, and 7

It has to be noticed that for the largest masses, as in [24] the NS is mainly composed of pure quark matter (hybrid star). This can be seen in Fig. 8, where QP indicates

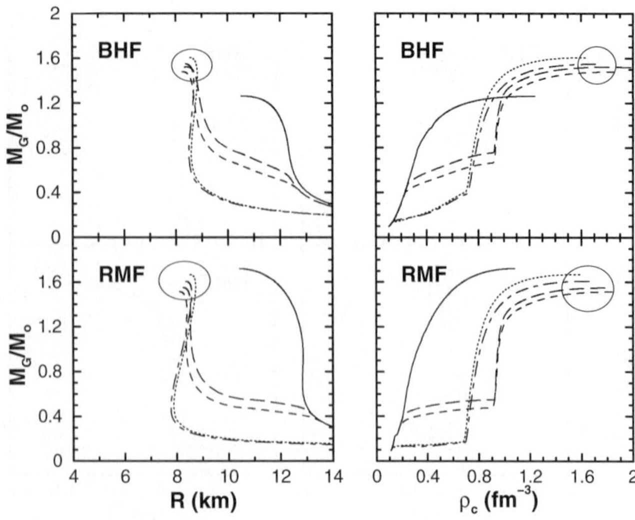

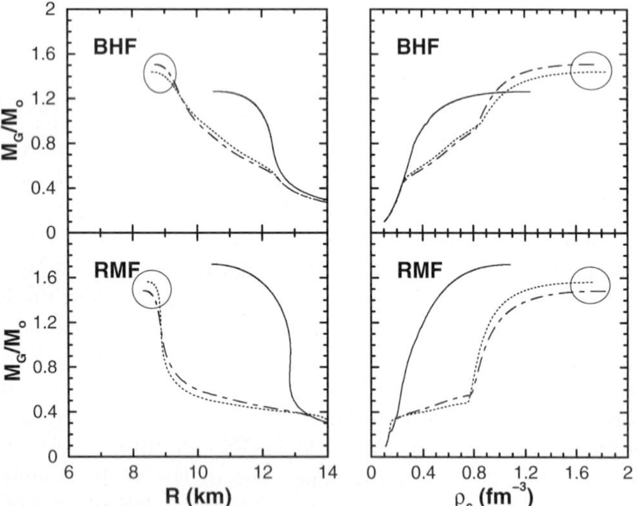

Fig. 6. The same as in Fig. 5, but for $\epsilon_B = 1.1$ GeV fm^{-3}, see the text

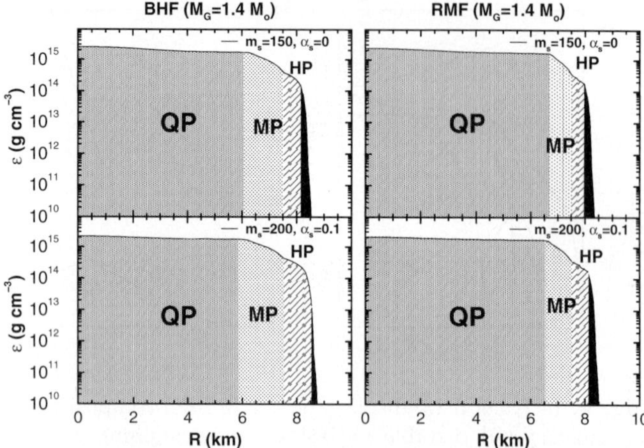

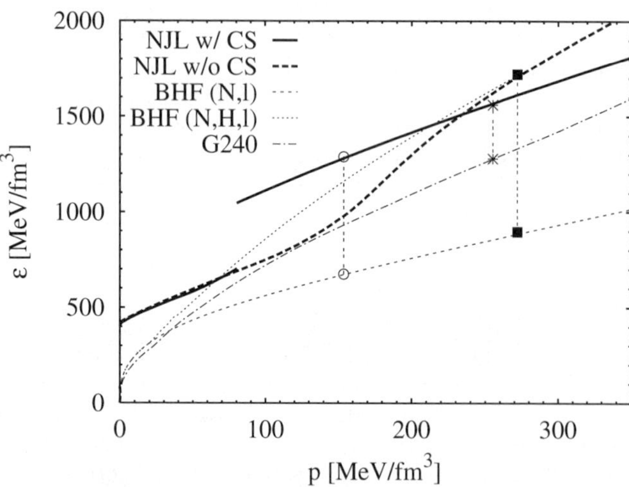

Fig. 8. Density profile of the different phase inside neutron stars when the MIT bag model is used for the quark phase. See the text for details

Fig. 7. The same as in Fig. 5, but for $\epsilon_B = 1.5$ GeV fm^{-3}, see the text

Fig. 9. Hadronic and quark matter equation of State. *Dashed line* indicates the hadron EoS without hyperons, the *dotted one* with hyperons. The *dashed-dotted line* indicates the EoS of the relativistic mean field with compressibility $K = 240$ MeV, see the text. Finally, the Eos for the NJL quark model with and without color superconductivity are indicated by the *bold solid and bold dashed lines* respectively. The points mark the values of the pressure where the corresponding hadron-quark phase transitions occur

pure quark phase, MP the mixed phase and HP the pure hadronic phase. Comparison is also made with a similar calculation where a relativistic mean field (RMF) EoS is used for the hadronic phase (including hyperons) [33,34]. According to [24], a strong color superconductivity can even render a pure quark mass possible.

A more microscopic model for the deconfined quark phase, which has some direct links with hadron phenomenology, is the Nambu-Jona Lasinio (NJL) model [35]. The parameters of the model are held fixed to the ones fitted to reproduce the masses of the meson octet. The NJL model generates dynamically the quark masses as a function of baryon density, through the presence of the quark condensate. The model has been generalized to include also color superconductivity [36]. The latter is also dynamically generated (gap equations), and therefore it does not introduce further parameters, since the value of the pairing gap is a well defined function of the baryon density.

Th resulting EoS for beta-stable neutron star matter is reported in Fig. 9, where for the hadron sector the EoS discussed above, see Fig. 4, is adopted.

The quark color superconductivity can actually appear in two different phase, the so-called two-flavour (2SC) phase, where only up and down quark partecipate, and the color-flavor locked (CFL) phase, where also the strange quark is included in a coherent combination with the other two quarks. The transition from the 2SC to the CFL phase occurs at a density (or pressure) where only the confined hadronic phase is present. The transition is responsible of the (almost) discontinuity of the quark EoS around $p \approx 100$ MeV/fm^3 in Fig. 9. Without color superconductivity, bold dashed line in Fig. 9, the EoS has, of course, a completely smooth behaviour. The different behaviour with and without CS of the quark EoS is due not only to the additional pairing energy but also to the different density dependence of the dynamically generated quark masses and the relative quark fractions, particularly for the strange quark. The phase transition from hadron to quark matter, that was assumed to be sharp for simplicity, is marked by the points and the corresponding vertical lines which join the different EoS. The transition from hadronic to superconducting quark matter occurs around $p \approx 160$ MeV/fm^3 (open circles), while, if the CS is excluded, the transition occur (full squares) at much higher pressure (and density).

If strange matter is allowed also in the hadonic phase, according to the EoS of Fig. 4, then no transition to quark matter is possible before the NS maximum mass is reached. As we have seen, the value of the maximum mass is in this case too low, which means that this particular EoS is not acceptable. This result does not exclude the possible appearence of hadronic matter with a strange component (hyperons), since the adopted EoS is strongly dependent on the particular hyperon-nucleon interaction used in the microscopic calculations. The interaction is not so well known, especially at density above saturation. Indeed, if one adopts the EoS of [37] (version with $K = 240$ MeV), derived within the relativistic mean field approximation, which also contains hyperon components, the transition to quark matter (with CS) is now possible, since this EoS is stiffer. The corresponding value of the transition pressure is marked by the stars in Fig. 9. This EoS differ from the previous one only in the region of density where hyperons are present, while it is quite similar in the pure nucleon sector. This is because the nucleon-hyperon effective interaction is less attractive, and therefore the hyperon content is smaller and the EoS is stiffer.

Once the EoS for NS matter is obtained, the NS density and composition profile can be obtained. The results for the total mass as a function of radius are summarized in Fig. 10. As already mentioned, the EoS of Fig. 4 which includes hyperons (dotted line) produces a too law maximum mass, with no possible transition to quark matter. If hyperon are not included, within the same scheme, and quark matter is described in the NJL model with no CS (bold dashed line), the maximum mass is substantially enhanced (about 1.8 solar mass). It has to be noticed that as soon as quark matter appears at increasing central density (or decreasing radius) just at the center, the NS becomes unstable toward collapse. Since we are considering a sharp transition between the two phases (no mixed phase), the transition produces the small kink at the maximum of the mass vs. radius curve which is apparent in Fig. 10. The inclusion of a mixed phase would smooth out this behaviour and only slightly modify the value of the maximum mass [38]. This result means also that the use of the NJL model for the quark phase actually does not allow any NS with a pure quark matter core. Possibly, only a small portion of mixed phase at the center would be possible.

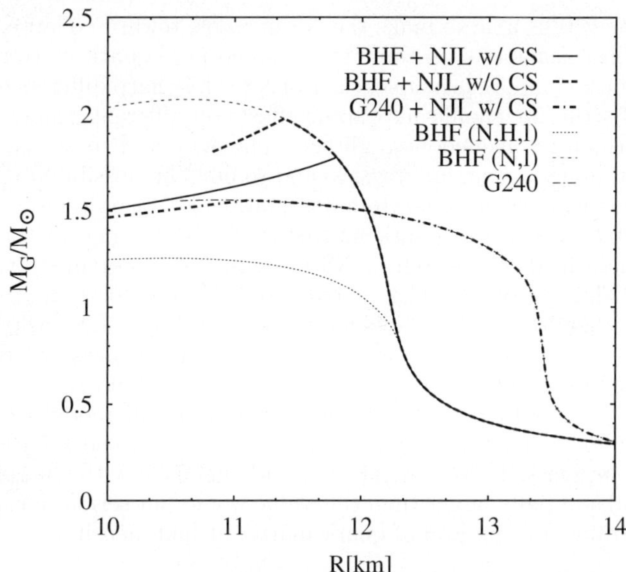

Fig. 10. Gravitational masses of comapct stars as a function of the radius for the different EOS shown in Fig. 9

This conclusion holds true also when CS is introduced in the quark phase (bold line). The effect of the CS is to reduce appreciably the maximum mass, which is in this case around 1.75 solar mass. Again, the presence of a mixed phase is expected to reduce slightly this value.

If the EoS of [37] is used, where also hyperons are included, keeping the same EoS for the quark (i.e. NJL with CS), the maximum mass further decreases to a value close to 1.55 solar mass, while the radius reduces substantially (by about 10%), making the object more compact. It can be then expected that a generic hadronic EoS with strangeness would produce a maximum mass lower than the value without strangeness, i.e. below about 1,75 solar mass. Indeed, an EoS with larger strangness content is expected to be softer, as we have seen in the discussion above.

4 Conclusions

We have discussed the results of NS structure calculations based, on one hand on a hadronic EoS which has been derived from microscopic many-body calculations, and on the other on simplified models for the possible quark phase in the core of the NS. The quark matter models include the MIT bag model and the Nambu-Jona Lasinio model, both with and without the possibility of color superconductivity. In all the calculations analysed, the maximum mass of NS never exceed values of 1.7 - 1.8 solar mass. Fot hybrid stars, i.e. stars which contain both hadronic and quark components, color superconductivity tends to reduce slightly the maximum mass. If CS is strong enough, even pure quark stars are possible for the MIT bag model. However, the structure of NS in the MIT and NJL models is drastically different. While the major fraction of NS is composed by quark matter for the MIT model, in the case of the NJL model just the onset of quark matter in the core is able to make the NS unstable toward collapse. This means that in the latter case no pure quark matter phase could, in principle, exist in NS. This sharp difference of structure should have some effects on different properties of NS, like cooling, glitches, magnetic and rotational properties, neutrino physics, and so on. The possibility of getting a clean signal from observational data on the possible presence of quark matter in NS is still one of the main filed of research in NS physics, but, unfortunately, is also one of the most controversial. More theoretical investigations and observational data will hopefully clarify in the future this fundamental issue in the physics of NS and in the physics of high density baryonic matter.

It has to be noticed, anyhow, that in all model explored so far the NS maximum mass never exceed 1.7-1.8 solar mass. Therefore, the observation of a NS with a mass substantially larger than this value would put serious constraints on the EoS of quark matter at high density.

Acknowledgements. The material presented in this contribution is the result of a fruitful collaboration, lasting for several years, with a number of people. Special thanks are due to Dr. I. Bombaci, Dr. G.F. Burgio, Prof. L.S. Ferreira, Dr. G. Giansiracusa, Prof. U. Lombardo, Dr. P.K. Sahu, Dr. H.-J. Schulze and Prof. H.Q. Song.

References

1. For a pedagogical introduction: see *Nuclear Methods and the Nuclear equation of State*, Edited by M. Baldo, World Scientific, Singapore, International Review of Nuclear Physics Vol. 9, 1999
2. B.D. Day: *Brueckner–Bethe Calculations of Nuclear Matter*, Proceedings of the School E. Fermi, Varenna 1981, Course LXXIX, ed. A. Molinari, (Editrice Compositori, Bologna, 1983), p. 1–72; Rev. Mod. Phys. **39**, 719 (1967)
3. F. Coester, S. Cohen, B.D. Day, and C.M. Vincent: Phys. Rev. C **1**, 769 (1970); B.D. Day: Comments Nucl. Part. Phys. **11**, 115 (1983)
4. J.P. Jeukenne, A. Lejeunne, and C. Mahaux: Phys. Rep. C **25**, 83 (1976); M. Baldo, I. Bombaci, L.S. Ferreira, G. Giansiracusa, and U. Lombardo: Phys. Rev. C **43**, 2605 (1991)
5. R.B. Wiringa, V.G.J. Stocks, and R. Schiavilla: Phys. Rev. C **51**, 38 (1995)
6. L.D. Fadeev: *Mathematical Aspects of the Three-Body Problem in Quantum Scattering Theory*, Davey, New York 1965
7. R. Rajaraman and H. Bethe: Rev. Mod. Phys. **39**, 745 (1967)
8. B.D. Day: Phys. Rev. C **24**, 1203 (1981); Phys. Rev. Lett. **47**, 226 (1981)
9. H.Q. Song, M. Baldo, G. Giansiracusa, and U. Lombardo: Phys. Rev. Lett. **81**, 1584 (1998)
10. M. Baldo, A. Fiasconaro, G. Giansiracusa, and U. Lombardo e H. Q. Song: Phys. Rev. C **65**, 017303 (2001)
11. G.E. Brown, W. Weise, G. Baym, and J. Speth: Comm. Nucl. Part. Phys. **17**, 39 (1987)
12. J. Fuyita and H. Miyazawa: Progr. in Theor. Phys. **17**, 360 (1957); C. Hadjuk, P.U. Sauer, and W. Streuve: Nucl. Phys. A **405**, 581 (1983)
13. P. Grangé, A. Lejeune, M. Martzolff, and J.-F. Mathiot: Phys. Rev C **40**, 1040 (1989)
14. B. Krippa, M.C. Birse, J.A. McGovern, and N.R. Walet: Phys. Rev C **67**, 031301 (2003)
15. J. Carlson, V.R. Pandharipande, and R.B. Wiringa: Nucl. Phys. A **401**, 59 (1983)
16. S.L. Shapiro and S.A. Teukolsky: *Black Holes, White Dwarfs and Neutron Stars*, (John Wiley & Sons, New York, 1983)
17. P. Maessen, T. Rijken, and J. de Swart: Phys. Rev. C **40**, 2226 (1989)
18. M. Baldo, I. Bombaci, and G.F. Burgio: Astron. and Astr. **328**, 274 (1997)
19. M. Baldo, G.F. Burgio, and H.-J. Schulze: Phys. Rev. **61C**, 055801 (2000)
20. J.W. Negele and D. Vautherin: Nucl. Phys. A **207**, 298 (1973)
21. R. Feynman, F. Metropolis, and E. Teller: Phys. Rev. **75**, 1561 (1949)
22. G. Baym, C. Pethick, and D. Sutherland: Astrophys. J. **170**, 299 (1971)
23. A. Chodos, R.L. Jaffe, K. Johnson, C.B. Thorn, and V.F. Weisskopf: Phys. Rev. D **9**, 3471 (1974)
24. M. and S. Reddy: Phys. Rev. **67C**, 074024 (2003)
25. A. Akmal and V.R. Pandharipande: Phys. Rev. C **56**, 2261 (1997)
26. M. Alford, K. Rajagopal, and F. Wilczek: Phys. Lett. B **422**, 247 (1998)
27. M. Alford, K. Rajagopal, and F. Wilczek: Nucl. Phys. B **537**, 443 (1999)
28. M. Alford: Ann. Rev. Nucl. Part. Sci. **51**, 131 (2001)
29. G.F. Burgio, M. Baldo, P.K. Sahu, A.B. Santra, and H.-J. Schulze: Phys. Lett. B **562**, 19 (2002)
30. M. Baldo and G.F. Burgio: *Microscopic Theory of the Nuclear equation of State and Neutron Star Structure*, in "Physics of Neutron Star Interiors", Eds. D. Blaschke, N. Glendenning, and A. Sedrakian, Lectures Notes in Physics, Springer, vol. 578 (2001), pp. 1-30
31. R.A. Hulse and J.H. Taylor: Astrophys. J. **195**, L51 (1975)
32. N.K. Glendenning: Phys. Rev. D **46**, 1274 (1992)
33. P.K. Sahu: Phys. Rev. C **62**, 045801 (2000)
34. P.K. Sahu and A. Ohnishi: Nucl. Phys. A **691**, 439 (2001)
35. T. Hatsuda and T. Kunihiro: Phys. Rep. **247**, 221 (1994)
36. M. Buballa and M. Oertel: Nucl. Phys. A **703**, 770 (2002)
37. N.K. Glendenning: Compact Stars (Springer, New York, 1996)
38. Burgio, M. Baldo, P.K. Sahu and H.-J. Schulze: Phys. Rev. C **66**, 025802 (2002)

Eur Phys J A (2004) **19**, s01, 197–201
Digital Object Identifier (DOI) 10.1140/epjad/s2004-03-033-9

EPJ A direct
electronic only

The Supernova–GRB connection

Color superconducting quark matter in compact stars

A. Drago[1], A. Lavagno[2], and G. Pagliara[1]

[1] Dipartimento di Fisica, Università di Ferrara and INFN, Sezione di Ferrara, 44100 Ferrara, Italy
[2] Dipartimento di Fisica, Politecnico di Torino and INFN, Sezione di Torino, 10129 Torino, Italy

Received: 1 Sep 2003 / Accepted: 14 Nov 2003 /
Published Online: 6 Feb 2004 – © Società Italiana di Fisica / Springer-Verlag 2004

Abstract. We study the effects of color superconductivity on the structure and formation of compact stars. We show that it is possible to satisfy most of recent observational boundaries on masses and radii if a diquark condensate forms in a hybrid or a quark star. Moreover, we find that a huge amount of energy, of the order of 10^{53} erg, can be released in the conversion from a (metastable) hadronic star into a (stable) hybrid or quark star, if the presence of a color superconducting phase is taken into account. Accordingly to the scenario proposed in Astrophys.J.586(2003)1250, the energy released in this conversion can power a Gamma Ray Burst. Possible experimental evidences, indicating a range of time delay between a Supernova explosion and a subsequent Gamma Ray Burst, are here discussed and interpreted.

PACS. 26.60.+c – 26.50.+x – 12.38.Mh – 97.60.Jd

1 Introduction

The new accumulating data from X-ray satellites provide important information on the structure and formation of compact stellar objects. Concerning the structure, the new data fix rather stringent constraints on the mass and the radius of a compact star. These data are at first sight difficult to interpret in a unique and self-consistent theoretical scenario, since some of the observations are indicating rather small radii and other observations are indicating large values for the mass of the star.

Concerning the formation scenario, crucial information are provided by the very recent observations of Gamma-Ray Bursts (GRB), indicating the possibility that some of the GRBs are associated with a previous Supernova (SN) explosion, with a delay between the first and the second explosion of the order of days or years [1,2]. A delay of order one week is also compatible with the results of the analysis of GRB030329 [3]. These observations could be explained associating the second explosion with the conversion of a (metastable) hadronic star (HS) into a more stable stellar object made at least in part of deconfined quark matter (QM). In the scenario proposed in [4] the HS can be metastable due to the presence of a non-vanishing surface tension at the interface separating hadronic matter (HM) from QM. The nucleation time (i.e. the time to form a critical-size drop of quark matter) can be extremely long if the mass of the star is small. Via mass accretion the nucleation time can be dramatically reduced and the star is finally converted into the stable configuration. The newly formed HyS or QS cools down emitting neutrinos and the subsequent neutrino-antineutrino annihilation can power a GRB.

In recent years, many theoretical works have investigated the possible formation of a diquark condensate in quark matter, at densities reachable in the core of a compact star [5,6,7]. The formation of this condensate can deeply modify the structure of the star [8,9,10].

We show that it is possible to satisfy the existing boundaries on mass and radius of a compact stellar object if a diquark condensate forms in a Hybrid Star (HyS) or a Quark Star (QS). Moreover, the formation of diquark condensate can significantly increase the energy released in the conversion from a purely HS into a more stable star containing deconfined QM.

2 Equation of state of beta-stable matter

To describe the high density EOS of matter we adopt standard models in the various density ranges. Concerning the hadronic phase we use the relativistic non-linear Glendenning-Moszkowski model (GM1-GM3) [11]. At very low density we have used the Negele-Vautherin [12] and the Baym-Pethick-Sutherland [13] EOS. For the quark matter phase we adopt a MIT-bag like model in which the formation of a diquark condensate is taken into account in a simple and effective way. To connect the two phases of our EOS, we impose Gibbs equilibrium conditions.

It is widely accepted that the Color-Flavor Locking phase (CFL) is the real ground state of QCD at asymp-

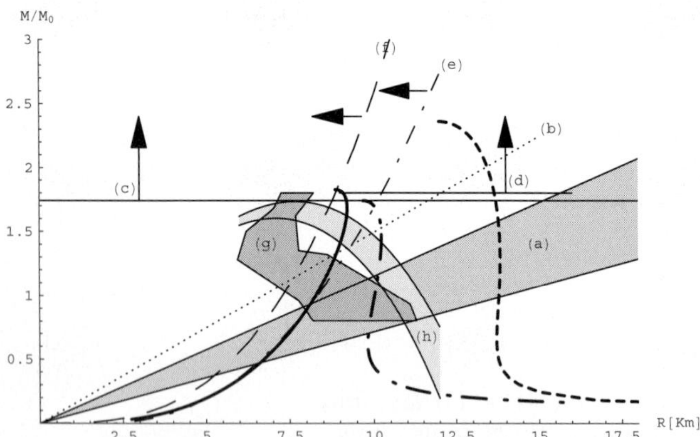

Fig. 1. Mass-radius plane with observational limits and a few representative theoretical curves: *thick solid line* indicates CFL quark stars, *thick dot-dashed line* CFL hybrid stars, *thick-dashed line* hadronic stars (*see text*). The observational limits come from: **a** Sanwal et al. 2002 [15], **b** Cottam et al. 2002 [16], **c** Quaintrell et al. 2003 [17], **d** Heinke et al. 2003 [18], **e,g** Dey et al. 1998 [19], **f** Li et al. 1999 [20], **h** Burwitz et al. 2002 [21]

toticly large densities. We are interested in the bulk properties of a compact star and we adopt the simple scheme proposed in [8,14] where the thermodynamic potential is given by the sum of two contributions. The first term corresponds to a "fictional" state of unpaired quark matter in which all quarks have a common Fermi momentum chosen to minimize the thermodynamic potential. The other term is the binding energy Δ of the diquark condensate expanded up to order $(\Delta/\mu)^2$. In [8] the gap is assumed to be constant, independent on the chemical potential μ. In the present calculation we consider a μ dependent gap resulting from the solution of the gap equation. The resulting EOS in our model reads therefore:

$$P = -\Omega_{CFL}(\mu) - B - \Omega^{electrons}(\mu_e) \tag{1}$$

$$E/V = \Omega_{CFL}(\mu) + \mu\rho + B + \Omega^{electrons}(\mu_e) + \mu_e\rho_e \tag{2}$$

where

$$\Omega_{CFL}(\mu) = \frac{6}{\pi^2}\int_0^\nu k^2(k-\mu)\,\mathrm{d}k$$
$$+ \frac{3}{\pi^2}\int_0^\nu k^2(\sqrt{k^2+m_s^2}-\mu)\,\mathrm{d}k - \frac{3\Delta^2\mu^2}{\pi^2} \tag{3}$$

with

$$\nu = 2\mu - \sqrt{\mu^2 + \frac{m_s^2}{3}}, \tag{4}$$

and the quark density ρ is calculated numerically by deriving the thermodynamic potential respect to μ.

3 Masses and radii of compact stellar objects

In Fig. 1 we have collected most of the analysis of data from X-ray satellites, concerning masses and radii of compact stellar objects [15,16,17,18,19,20,21]. Observing Fig. 1, we notice that the constraints coming from a few data

sets (labeled "e", "f"[1] "g" and maybe also constraint "h") indicate rather unambiguously the existence of very compact stellar objects, having a radius smaller than ~ 10 km. At the contrary, at least in one case ("a" in the figure), the analysis of the data suggests the existence of stellar objects having radii of the order of 12 km or larger, if their mass is of the order of 1.4 $M_\odot$. In this analysis one has also to take into account that it is difficult from an astrophysical viewpoint to generate compact stellar objects having a mass of the order of one solar mass or smaller. Therefore the most likely interpretation of constraint "a" is that the corresponding stellar object does not belong to the same class of objects which have a radius smaller than ~ 10 km. Concerning constraint "b", its interpretation is less clear, since it can be satisfied both with a very compact star or with a star having a larger radius. The apparent contradiction between the constraints "e", "f", "g" and the constraint "a" can be easily accommodated in our scheme, since it can be the signal of the existence of metastable purely hadronic stars which can collapse into a stable configuration when deconfined quark matter forms inside the star. In the next Section we will discuss the possible relation between this transition and at least some GRBs.

Finally, constraints ("c" and "d"[2]) do not provide stringent limits on the radius of the star, but they put strong constraints on the lower value of its mass. Constraints "c" and "d" are very important, since it is in general not easy to obtain solutions of the Tolman - Oppenheimer - Volkoff equation having both large masses and very small radii. As we will see, the existence of an energy gap associated with the diquark condensate helps in circumventing this difficulty, since the effect of the gap is to increase the maximum mass of QSs or of HySs having a huge content of pure quark matter.

In Fig. 1 we show a few theoretical M-R relations which correspond to the scenario we are proposing. More precisely, we show a thick-dashed line corresponding to HSs (GM1), a thick dot-dashed line corresponding to HySs (GM1, $B^{1/4} = 170$ MeV, Δ_2) and a thick solid line corresponding to QSs (GM1, $B^{1/4} = 170$ MeV, Δ_4). Both the HyS and the QS lines can satisfy essentially all the constrains derived from observations. The shapes of the gaps Δ_i are shown in Fig. 2.

Concerning the constraint "a", it is probably better satisfied by the HS line than by the HyS or QS lines, which would give stars having a mass smaller than $\sim 1.2M_\odot$. In conclusion, in our scheme most of the compact stars are either HySs or QSs having a mass in the range $1.2-1.8M_\odot$ and a radius $\sim 8.5-10$ km. Metastable HS can exist. As we will see in the next section their mass is probably smaller than $\sim 1.3M_\odot$.

---

[1] A very recent reanalysis of the data of the pulsar SAX J1808.4-3658, discussed in [20], seems to indicate slightly larger radii, of the order of 9-10 km for a star having a mass of 1.4-1.5 $M_\odot$ [22].

[2] If the observed X-ray emission is due to continuing accretion, then a smaller value for the mass is allowed, $M/M_\odot = 1.4$.

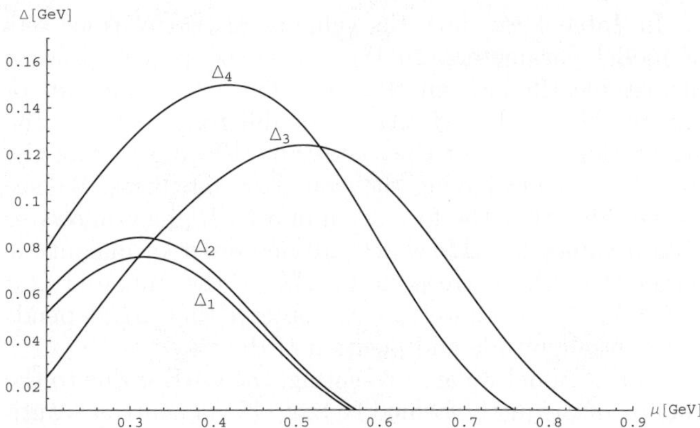

Fig. 2. Gap as function of the chemical potential, for four different parameter sets

4 Nucleation time and energy released

The existence of a possible relation between GRBs and SN explosions has been extensively discussed in the literature. It has not yet been clarified if the two explosions are always simultaneous or if, at least in a few cases, a time delay can exist, with the SN preceding the GRB. A very important information can be obtained from the analysis of the optical afterglow, since in two cases (GRB980425 / SN1998bw, GRB030329 / SN2003dh) the spectrum of a type Ic SN emerged once the spectrum of the GRB afterglow has been subtracted. It is therefore possible, at least in principle, to estimate the date of the SN explosion and to estimate the delay (if any) between the two explosions. Unfortunately, due to the uncertainties on the time dependence of the SN light curves, it is difficult to reduce the error in this estimate to a value smaller than a few days. The analysis of [3] suggests that the SN exploded within a few days of the GRB. More precisely, the difference between the light curves indicates that SN2003dh may have preceded its associated GRB by about 4-7 days. Actually the first indications of a possible delay between SN and GRB came from the analysis of the X-ray spectra of GRB990705 [1] and of GRB011211 [2]. In particular, in the case of GRB990705 the detection (for the first time) of Fe absorption lines indicated the existence of an Fe rich environment crossed by the GRB emission. The Fe abundance was assumed to be generated by a SN explosion preceding the GRB, with a time interval of order years. A similar analysis, performed in [2], suggested a time delay of order days for GRB011211.

The effect of the transition to deconfined QM on explosive processes like SNs and GRBs has been discussed by many authors. In particular, the possibility that the deconfinement transition takes place during the core-collapse of massive stars at the moment of the bounce, has been discussed e.g. in [23,24] and this mechanism can help the SN to explode. However, at the light of results like the ones presented in [25,26], it seems more plausible that the deconfinement takes place only when the proto-neutron star has deleptonized and cooled down to a temperature of a few MeV. In the model we are discussing, proto-neutron stars having a small enough mass can exist as metastable

HS if a non-vanishing surface tension is present at the interface between HM and QM. The process of quark deconfinement can be a powerful source for GRBs and it can also explain the delay between a SN explosion and the subsequent GRB observed in a few cases [1,2]. In the model proposed in [4] the central density of a pure HS increases, due to spin down or mass accretion, until its value approaches the deconfinement critical density. At this point a spherical virtual drop of QM can form. The potential energy for fluctuations of the drop radius R has the form [27]:

$$U(R) = \frac{4}{3}\pi R^3 n_q(\mu_q - \mu_h) + 4\pi\sigma R^2 + 8\pi\gamma R \qquad (5)$$

where n_q is the quark baryon density, μ_h and μ_q are the hadronic and quark chemical potentials, all computed at a fixed pressure P, and σ is the surface tension for the interface separating quarks from hadrons. Finally, the term containing γ is the so called curvature energy. For σ we use standard values from 10 to 40 MeV/fm^2 and we assume that it takes into account, in a effective way, also the curvature energy. The value of σ was estimated in [28] to be ~ 10 MeV/fm^2. Values for σ larger than ~ 30 MeV/fm^2 are probably not useful at the light of the result of [29, 30].

To compute the time needed to form a bubble of quarks having a radius larger than the critical one, we use the technique of quantum tunneling nucleation. We can assume that the temperature has no effect in our scheme: for values of $B^{1/4} \sim 160 - 180$ MeV the critical density ρ_1 separating pure HM from mixed phase is larger than $4\rho_0$ for $Z/A \sim 0.3$, i.e. for an isospin fraction typical of a newly formed and hot proto-neutron star [31]. This critical density typically exceeds the central density of hot and not too massive stars. Therefore the mixed phase can form only when the star has deleptonized and its temperature has dropped down to a few MeV [25,26]. When the temperature is so low, only quantum tunneling is a practicable mechanism.

The calculation proceed in the usual way: after the computation (in WKB approximation) of the ground state energy E_0 and of the oscillation frequency ν_0 of the virtual QM drop in the potential well $U(R)$, it is possible to calculate in a relativistic frame the probability of tunneling as [32]:

$$p_0 = \exp[-\frac{A(E_0)}{\hbar}] \qquad (6)$$

where

$$A(E) = 2\int_{R_-}^{R_+} dR\sqrt{[2M(R) + E - U(R)][U(R) - E]}. \qquad (7)$$

Here $R_\pm$ are the classical turning points and

$$M(R) = 4\pi\rho_h\left(1 - \frac{n_q}{n_h}\right)^2 R^3 , \qquad (8)$$

ρ_h being the hadronic energy density and n_h, n_q are the baryonic densities at a same and given pressure in the hadronic and quark phase, respectively. The nucleation time is then equal to

$$\tau = (\nu_0 p_0 N_c)^{-1}, \qquad (9)$$

where N_c is the number of centers of droplet formation in the star, and it is of the order of 10^{48} [32].

Let us recall once again the astrophysical scenario we have in mind. In a few cases a delay of the order of days or years between the SN explosion and the subsequent GRB have been postulated to explain the astrophysical data on the GRBs. In the scheme we are discussing, this delay is due to the formation of a metastable HS having a relatively small mass. The nucleation time, computed using (9), can be extremely long if the mass of the metastable star is small enough. Via mass accretion the nucleation time can be reduced from values of the order of the age of the universe down to a value of the order of days or years. We can therefore determine the critical mass M_{cr} of the metastable HS for which the nucleation time corresponds to a fixed small value (1 year in Table 1).

Table 1. Energy released ΔE (*measured in foe=10^{51} erg*) in the conversion to hybrid or quark star (*labeled with a* •), for various sets of model parameters, assuming the hadronic star mean life-time $\tau = 1$ yr (*see text*). M_{cr} is the gravitational mass of the hadronic star at which the transition takes place, for fixed values of the surface tension σ and of the mean life-time τ. BH indicates that the hadronic star collapses to a Black Hole. We indicate with a *dash* (–) situations in which the Gibbs construction does not provide a mechanically stable EOS

Hadronic Model	$B^{1/4}$ [MeV]	σ [MeV/fm^2]	$M_{cr}/M_\odot$	ΔE $\Delta = 0$	ΔE Δ_1	ΔE Δ_2	ΔE Δ_3	ΔE Δ_4
GM3	160	20	0.69	20	65•	69•	76•	148•
GM3	160	30	0.91	32	90•	95•	106•	196•
GM3	160	40	1.00	38	100•	105•	119•	216•
GM3	170	10	1.12	0	34	40	68	162•
GM3	170	20	1.26	4	44	50	86	185•
GM3	170	30	1.39	11	53	60	104	207•
GM3	170	40	1.49	BH	62	68	120	224•
GM3	180	10	1.55	BH	11	13	BH	–
GM3	180	20	1.61	BH	BH	22	BH	–
GM3	180	30	1.67	BH	BH	BH	BH	–
GM1	160	10	0.45	11	41•	44•	47•	96•
GM1	160	20	0.72	28	75•	79•	86•	160•
GM1	160	30	0.96	48	108•	114•	127•	220•
GM1	160	40	1.18	72	142•	148•	166•	276•
GM1	170	10	1.17	18	59	65	96	191•
GM1	170	20	1.33	33	79	85	124	226•
GM1	170	30	1.45	50	96	103	150	254•
GM1	170	40	1.60	BH	122	128	BH	290•
GM1	180	10	1.63	BH	BH	72	BH	–
GM1	180	20	1.72	BH	BH	BH	BH	–
GM1	180	30	1.79	BH	BH	BH	BH	–

In Table 1 we show the value of M_{cr} for various sets of model parameters. In the conversion process from a metastable HS into an HyS or a QS a huge amount of energy ΔE is released. ΔE is the difference between the gravitational mass of the metastable HS and that of the final HyS or QS having the same baryonic mass. We see in the table that the formation of a CFL phase allows to obtain values for ΔE which are one order of magnitude larger than the corresponding ΔE of the unpaired QM case ($\Delta = 0$). Moreover, we can observe that ΔE depends both on magnitude and position of the gap.

In the model we are presenting, the GRB is due to the cooling of the justly formed HyS or QS via neutrino - antineutrino emission. The subsequent neutrino-antineutrino annihilation generates the GRB. In our scenario the duration of the prompt emission of the GRB is therefore regulated by two mechanisms: 1) the time needed for the conversion of the HS into a HyS or QS, once a critical-size droplet is formed and 2) the cooling time of the justly formed HyS or QS. Concerning the time needed for the conversion into QM of at least a fraction of the star, the seminal work by [33] has been reconsidered by [34]. The conclusion of this latter work is that the stellar conversion is a very fast process, having a duration much shorter than 1s. On the other hand, the neutrino trapping time, which provides the cooling time of a compact object, is of the order of a few ten seconds [35], and it gives the typical duration of the GRB in our model.

5 Conclusions

We have studied the effect of color superconductivity on the EOS of quark matter and on the mass-radius relation for hybrid and quark stars. Comparing the theoretical curves with recent analysis of observational data, we find that color superconductivity is a crucial ingredient in order to satisfy all the constraints coming from observations. The most difficult problem posed by the astrophysical data is the indication of the existence of stars which are both very compact (R $\lesssim$ 9–10 km) and rather massive ($M \gtrsim 1.7 M_\odot$). We can satisfy these constraints either with hybrid or quark stars. In particular, concerning hybrid stars, the gap increases the maximum mass of the stable configuration, while keeping the corresponding radius $\lesssim$ 10 km.

The superconducting gap affects also deeply the energy released in the conversion from hadronic star into hybrid or quark star. In the scenario proposed in [4] the transition to deconfined quark matter takes place only when the star has deleptonized and cooled down. This is in agreement with the results of [25,26] and at variance with the scenario proposed in [23,24] where quark matter is produced at the moment of the bounce. This open the possibility to explain recent observations indicating a possible delay between a SN explosion and the subsequent Gamma Ray Burst [1,2,3], since we associate the second explosion with the transition from a metastable hadronic star to a stable star containing deconfined quark matter. The energy released, which will power the Gamma Ray

Burst through neutrino-antineutrino annihilation, is significantly increased by the effect of the superconducting gap and it can reach a value of the order of 10^{53} erg.

References

1. L. Amati et al.: Science **290**, 953 (2000)
2. J.N. Reeves et al.: Nature **414**, 512 (2002)
3. J. Hjorth et al.: Nature **423**, 847 (2003)
4. Z. Berezhiani, I. Bombaci, A. Drago, F. Frontera, and A. Lavagno: Astrophys. J. **586**, 1250 (2003)
5. M.G. Alford, K. Rajagopal, and F. Wilczek: Nucl. Phys. B **537**, 443 (1999)
6. M.G. Alford, J. Berges, and K. Rajagopal: Nucl. Phys. B **558**, 219 (1999)
7. T. Schafer and F. Wilczek: Phys. Rev. D **60**, 074014 (1999)
8. M.G. Alford and S. Reddy: Phys. Rev. D **67**, 074024 (2003)
9. M. Baldo, M. Buballa, F. Burgio, F. Neumann, M. Oertel, and H.J. Schulze: Phys. Lett. B **562**, 153 (2003)
10. D. Blaschke, H. Grigorian, D.N. Aguilera, S. Yasui, and H. Toki: AIP Conf. Proc. **660**, 209 (2003)
11. N.K. Glendenning and S.A. Moszkowski: Phys. Rev. Lett. **67**, 2414 (1991)
12. J. Negele and D. Vautherin: Nucl. Phys. A **207**, 298 (1973)
13. G. Baym, C. Pethick, and D. Sutherland: Astrophys. J. **170**, 299 (1971)
14. G. Lugones and J.E. Horvath: Phys. Rev. D **66**, 074017 (2002)
15. D. Sanwal, G.G. Pavlov, V.E. Zavlin, and M.A. Teter: Astrophys. J. **574**, L61 (2002)
16. J. Cottam, F. Paerels, and M. Mendez: Nature **420**, 51 (2002)
17. H. Quaintrell et al.: arXiv:astro-ph/0301243
18. C.O. Heinke, J.E. Grindlay, D.A. Lloyd, and P.D. Edmonds: Astrophys. J. **588**, 452 (2003)
19. M. Dey, I. Bombaci, J. Dey, S. Ray, and B.C. Samanta: Phys. Lett. B **438**, 123 (1998) [Addendum-ibid. B **447**, 352 (1999)]
20. X.D. Li, I. Bombaci, M. Dey, J. Dey, and E.P. van den Heuvel: Phys. Rev. Lett. **83**, 3776 (1999)
21. V. Burwitz, F. Haberl, R. Neuhaeuser, P. Predehl, J. Truemper, and V.E. Zavlin: Astron. Astrophys. **399**, 1109 (2003)
22. J. Poutanen and M. Gierlinski: arXiv:astro-ph/0303084
23. A. Drago and U. Tambini: J. Phys. G. **25**, 971 (1999)
24. D.K. Hong, S.D.H. Hsu, and F. Sannino: Phys. Lett. B **516**, 362 (2001)
25. O.G. Benvenuto and G. Lugones: Mon. Not. R. Astron. Soc. **304**, L25 (1999)
26. J.A. Pons, A.W. Steiner, M. Prakash, and J.M. Lattimer: Phys. Rev. Lett. **86**, 5223 (2001)
27. I.M. Lifshitz and Yu. Kagan: Zh. Eksp. Teor. Fiz. **62**, 385 (1972) [Sov. Phys. JETP, 35, 206]
28. M.S. Berger and R.L. Jaffe: Phys. Rev. C **35**, 213 (1987)
29. D.N. Voskresensky, M. Yasuhira, and T. Tatsumi: Nucl. Phys. A **723**, 291 (2003)
30. M.G. Alford, K. Rajagopal, S. Reddy, and F. Wilczek: Phys. Rev. D **64**, 074017 (2001)
31. M. Di Toro, A. Drago, V. Greco, and A. Lavagno: arXiv:nucl-th/0210052
32. K. Iida and K. Sato: Phys. Rev. C **58**, 2538 (1998)
33. A. Olinto: Phys. Lett. B **192**, 71 (1987)
34. J.E. Horvath and O.G. Benvenuto: Phys. Lett. B **213**, 516 (1988)
35. M. Prakash, I. Bombaci, M. Prakash, P.J. Ellis, J.M. Lattimer, and R. Knorren: Phys. Rept. **280**, 1 (1997)

Eur Phys J A (2004) **19**, s01, 203–205
Digital Object Identifier (DOI) 10.1140/epjad/s2004-03-034-8

EPJ A direct

electronic only

Ground state correlations in deep inelastic scattering and the Drell-Yan process

O. Linnyk, S. Leupold, and U. Mosel

Institut fuer Theoretische Physik, Universitaet Giessen, Heinrich-Buff-Ring, 16, D-35392 Giessen, Germany

Received: 1 Aug 2003 / Accepted: 14 Nov 2003 /
Published Online: 6 Feb 2004 – © Società Italiana di Fisica / Springer-Verlag 2004

Abstract. Such high energy processes as deep inelastic scattering (DIS) and Drell-Yan heavy lepton pair production are among the most important tools to probe the quark and gluon interaction. We calculated non-perturbative corrections to the LO cross section formulae for DIS and the Drell-Yan process. The interaction of partons was taken into account via dressing the incoming quark lines with spectral functions. We found that the effect of ground state correlation in DIS is large in the region of small Bjorken x and low momentum transfer Q^2. For a quark width of the order of 200 MeV, the cross section deviation reaches as much as 50% for $Q^2 = 10$ GeV^2. On the other hand, for the values of Q^2 well above the resonance region, e. g. 200 GeV^2, the effect of the initial quark off-shellness turned out to be small in DIS, but still substantial for the triple differential Drell-Yan cross section. Semi-inclusiveness of the latter process opens a possibility to extract from the experimental data important information on the shape of the quark spectral function in the nucleon. From the comparison to resent data on the Drell-Yan cross section from the NuSea collaboration, we obtained for the width of the quark spectral function the value about 200 MeV. The performed off-shell DIS and Drell-Yan cross section calculations allow for a better understanding of the quark and gluon interaction in the nucleon and, thus, the nucleon structure.

PACS. 13.60.Hb Total and inclusive cross sections (including deep-inelastic processes) – 13.85.Qk Inclusive production with identified leptons, photons, or other nonhadronic particles

1 Introduction

One of the major goals of present day research is to study the structure of the nucleon and other hadrons in terms of the fundamental quark-gluon dynamics. One would like to gain as good an understanding of hadron structure as our understanding of the compositeness of the nucleus in terms of nucleons and their interaction.

In high energy hadronic processes like deep inelastic scattering (DIS), the Drell-Yan process, jet production, etc. the soft and hard subprocesses can be disentangled. This allows one to apply the well established methods of perturbative QCD for the extraction of the information about the non-perturbative quark and gluon properties in a bound state from the experimental data. In this way, for instance, the distributions of partons, having different hadron light cone momentum fractions (parton distribution functions), are found [2].

The described method, based on factorization principle, is analogous to the plane wave impulse approximation (PWIA) for the quasi-elastic (e,e'p) scattering in nuclear physics.

In the theory of nuclei the importance of the effects beyond the PWIA, e.q. photon radiation, initial state in-teraction (ISI), and final state interaction (FSI) is well understood. Semi-exclusive processes offer an opportunity to study these effects. Cross section measurements, in which energy and momentum of the nucleon can be determined from the final state kinematics, can thus probe nuclear structure via the spectral function.

The neglected in PWIA initial and final state quark interaction effects on the observable hard scattering cross sections has not been studied yet. The primordial transverse momentum of the quarks in the nucleon has been considered in some works [1], but the effect of the parton virtuality on the observed inclusive and semi-exclusive processes has not been calculated so far.

As we show below, the off-shellness effects have the same order of magnitude as those of the intrinsic transverse momentum. Thus, the consistent treatment of the both off-shellness and non-collinearity is necessary.

By properly taking into account the ISI and off-shell kinematics of quarks, we succeeded to simultaneously describe the experimentally measured fully inclusive (for example, DIS) as well as semi-inclusive (triple-differential cross section of the Drell-Yan process) cross sections very well.

The applied technique and obtained results are presented in the following sections.

2 Method

The basic tool in the calculation of hard processes is the factorization into hard and soft physics:

$$d\sigma = f(Q^2, \xi) \otimes d\hat{\sigma}(\xi).$$

We additionally took into account the initial state interaction via dressing the incoming quark lines with a spectral function and used the generalized factorization:

$$d\sigma = f(Q^2, p_T, \xi) \otimes d\hat{\sigma}(\xi, m) \otimes Sp(m, \Gamma),$$

where: $d\hat{\sigma}(\xi, m)$ - off-shell partonic cross section, $Sp(m, \Gamma)$ - quark spectral function(s), $f(Q^2, p_T, \xi)$ - unintegrated quark distribution, ξ - hadron light cone momentum fraction carried by the struck parton, m - quark virtuality, Γ - quark width.

The hard part, i.e., partonic cross section is calculated using the rules of perturbative quantum chromodynamics (pQCD). We have calculated the pQCD cross section of electron scattering off a virtual quark and the cross section of the annihilation of an off-shell quark-antiquark pair into a pair of dileptons.

Both off-shell cross sections turn out to be gauge invariant. So, the modification of the vertex by the Ward identity was not necessary.

The full kinematics was taken into account as well. In case of DIS it reads:

$$\xi = \frac{x}{Q^2} \left(Q^2 - 2\mathbf{k}_\perp \cdot \mathbf{q}_\perp - m_i^2 \right),$$

Thus, the struck quark's virtuality and the hadron light cone momentum fraction, carried by this quark, are linearly connected. This is not the case in the Drell-Yan process. For the kinematics of Drell-Yan process in the hadron center of mass system, we obtained the following relations:

$$M_{DY}^2 = m_1^2 + m_2^2 + \xi_1\xi_2 P_1^+ P_2^- + \frac{\left(m_1^2 + \mathbf{p}_{1\perp}^2\right)\left(m_2^2 + \mathbf{p}_{2\perp}^2\right)}{\xi_1\xi_2 P_1^+ P_2^-}$$

$$x_F = \frac{1}{\sqrt{S}} \left(\xi_1 P_1^+ - \xi_2 P_2^- - \frac{\left(m_1^2 + \mathbf{p}_{1\perp}^2\right)}{\xi_1 P_1^+} + \frac{\left(m_2^2 + \mathbf{p}_{2\perp}^2\right)}{\xi_1 P_2^-} \right)$$

In the target rest frame, the connection between the observable mass (M_{DY}^2), Feynman variable (x_F), and transverse momentum (p_T) of the lepton pair on one hand side and the partonic variables on the other hand side is simpler. However, factorization in the usual form is not applicable in this system of [3].

The following parameterization is commonly used for the unintegrated parton distributions:

$$f(Q^2, p_T, \xi) = f(p_T) \cdot q(Q^2, \xi),$$

where $f(p_T)$ - a Gaussian, $q(Q^2, \xi)$ - conventional parton distribution functions. For the latter, we used the latest parameterizations by Glueck, Reya, Vogt [2].

In our calculations, a Breit-Wigner parameterizations for the quark spectral function was applied. The width was considered constant for a constant hard scale of the process (Q^2 for DIS and $M_{DY}^2 + p_T^2$ for Drell-Yan).

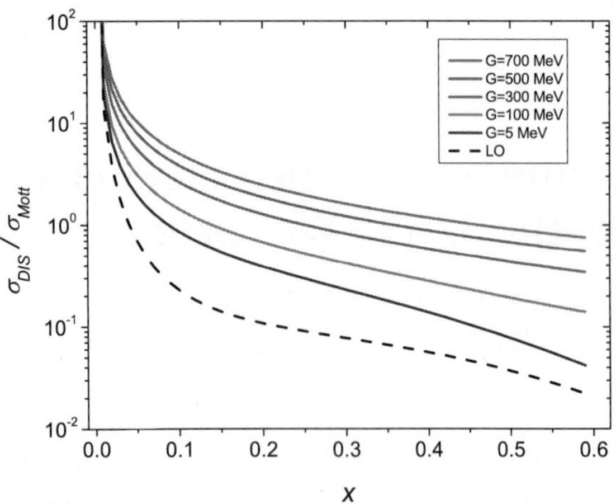

Fig. 1. DIS cross section for the range of quark spectral function widths from 5 MeV to 1 GeV and the parton model (*dashed line*)

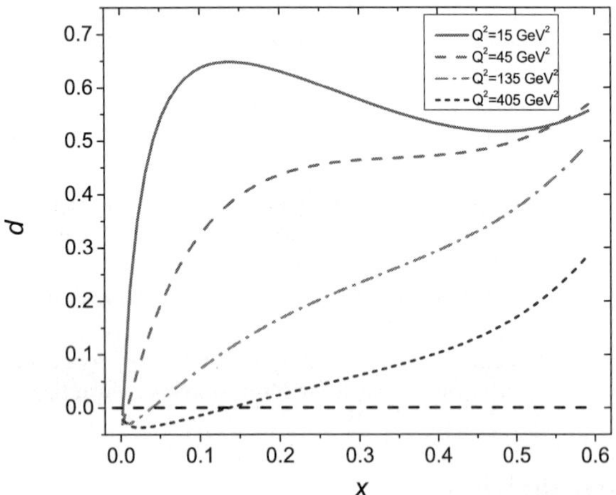

Fig. 2. Relative deviation of calculated cross section of DIS off the proton from parton model, $d = (\sigma_{ISI} - \sigma_{LO})/(\sigma_{ISI} + \sigma_{LO})$, for different Q^2

3 Results

The result of our calculations for DIS for a range of widths as compared to the parton model is shown in Fig. 1.

We found that the effect of the initial state interaction in DIS is large in the region of small Bjorken x and low momentum transfer Q^2. For the quark width 200 MeV, the cross section deviation reaches as much as 50% for $Q^2 = 10$ GeV^2.

On the other hand, for the values of Q^2 well above the resonance region, e. g. 200 GeV^2, the effect of the initial quark off-shellness accounts only to at most 10% of the LO cross section.

The found effect of the parton virtuality in DIS is Q^2-suppressed (Fig. 2). For the most of the experimentally investigated values of Q^2, the ambiguity in the parton distribution function parameterizations due to the renormalization scale uncertainty is of the same order as the ISI effect in DIS.

Thus, the value of the quark width in the nucleon cannot be extracted from the DIS data. This is the result

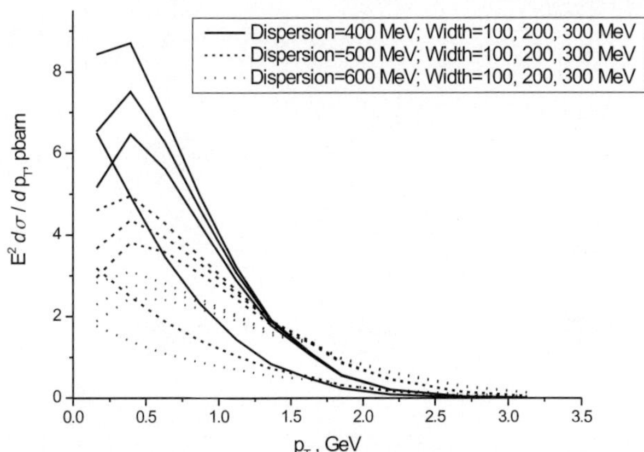

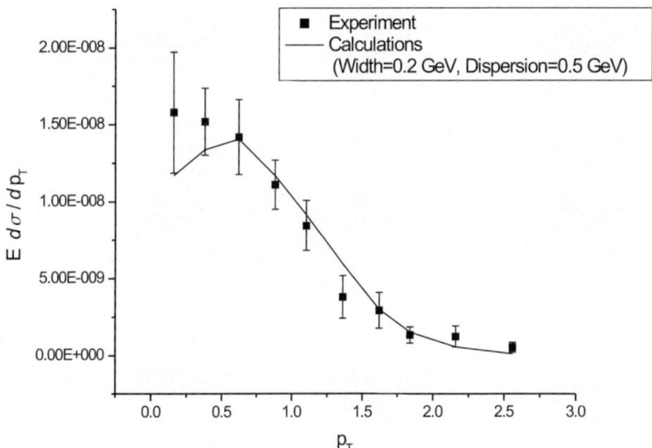

Fig. 4. Calculation result, compared to the data of the Fermilab experiment 866 for the continuum dimuon production in 800 GeV/c proton collision, $6.2 \leq M \leq 7.2$ GeV, $-0.05 \leq x_F \leq 0.2$. Only statistical errors shown

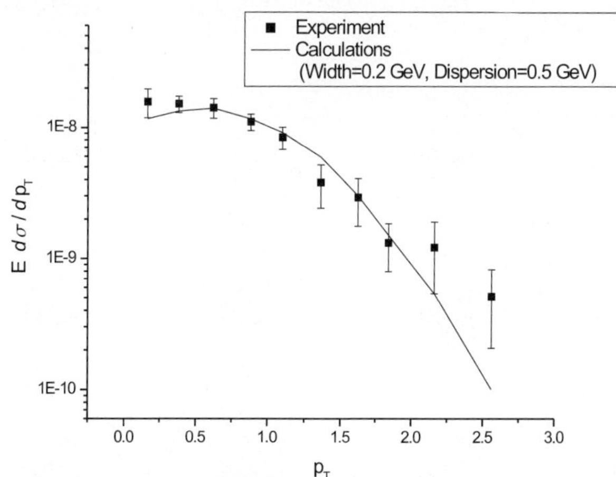

Fig. 3. Calculated distribution of the Drell-Yan lepton pair's transverse momentum for different values of the quark primordial transverse momentum dispersion and spectral function width

which was expected by an analogy to nuclear physics, because the DIS cross section is fully inclusive.

In contrast, for such a semi-exclusive observable as the transverse momentum distribution of the Drell-Yan lepton pair, we have found a substantial dependence on both the dispersion of the quark primordial transverse momentum and the spectral function width (Fig. 3). The LO QCD prediction for this distribution is the δ-function around 0, while the experimentally measured distribution is rather broad.

In the Figs. 5 and 4, an example of our description of the experimental data is presented. In this data set, the mass of the Drell-Yan pair is around 7 GeV and $0 < x_F < 0.2$. The optimal parameters for these values of M_{DY} and x_F are 500 MeV for the dispersion and 200 MeV for the width. The data are reproduced very well over the two orders of magnitude. The slight underestimation at $p_T > 2.5$ GeV is caused by the considerable contribution of the gluon Compton scattering process to the measurable cross section at these high p_T. Note that the convex shape of the distribution at small p_T cannot be described without the inclusion of the off-shell effects (Fig. 3).

Fig. 5. Calculation result, compared to the data of the Fermilab experiment 866 for the continuum dimuon production in 800 GeV/c proton collision, $6.2 \leq M \leq 7.2$ GeV, $-0.05 \leq x_F \leq 0.2$. Only statistical errors shown

In addition, by other models, neglecting off-shellness, the magnitude of the cross section is not correctly obtained and an additional overall K-factor is used. Our calculations yield not only the experimentally measured form of the cross section but also its amplitude without any K-factor.

4 Summary and outlook

We developed a formalism to study the quark and gluon structure of hadrons going further than the well known picture of collinear non-interacting partons and applied it to calculate cross sections of several high-energy processes. In this paper, the deep inelastic scattering and Drell-Yan pair production are considered.

We explicitly took into account the quark initial state interaction, missed in the standard perturbative consideration, by dressing the quark lines with spectral functions and using the method of generalized factorization.

There was discovered a substantial contribution of the quark off-shellness to the transverse momentum distribution of high-mass virtual photons produced in high-energy hadron-hadron collisions. The quark width in proton was estimated from comparison to resent data of the experiment E866 at Fermilab. For the mass of the Drell-Yan pair around 7 GeV and $0 < x_F < 0.2$, the quark width is 200 MeV. More details about the method and the results can be found in the long write-up by the same authors [3].

References

1. X.-N. Wang: Phys. Rev. C **61**, 064910 (2000); Y. Zhang, G. Fai, G. Papp, G. Barnafoldi, and P. Levai: Phys. Rev. C **65**, 034903 (2002)
2. M. Glueck, E. Reya, and A. Vogt: Eur. Phys. J. C **5**, 461 (1998)
3. O. Linnyk, S. Leupold, and U. Mosel: in preparation
4. J.C. Webb: PhD theses, hep-ex/0301031

Eur Phys J A (2004) **19**, s01, 207–212

Digital Object Identifier (DOI) 10.1140/epjad/s2004-03-035-7

EPJ A direct

electronic only

G^0 experiment status

G. Batigne, for the G^0 Collaboration

Laboratoire de Physique Subatomique et Cosmologie, 38028 Grenoble, France

Received: 18 Aug 2003 / Accepted: 14 Nov 2003 /
Published Online: 6 Feb 2004 – © Società Italiana di Fisica / Springer-Verlag 2004

Abstract. The G^0 project is a parity violation experiment dedicated to the measurement of the proton weak and axial form factors by means of elastic electron scattering. Combining these weak form factors with the known electromagnetic ones makes possible the extraction of the contribution of strange quarks to the charge and magnetization distributions in the nucleon. After introducing the physics case, this paper describes the G^0 apparatus and the measurements planned. An engineering run of this experiment took place recently ; first results are presented.

1 Introduction

The nucleon is naively viewed as a set of three valence quarks (uud for the proton and udd for the neutron) but its internal structure is far more complicated. In Quantum Chromodynamics (QCD), the quarks are bound by the strong interaction mediated by the exchange of gluons which can fluctuate into quark-antiquark pairs ($u\bar{u}$, $d\bar{d}$, $s\bar{s}$, ...). Thus the nucleon is a set of valence quarks surrounded by a "sea" of gluons and quark-antiquark pairs. At low energy, the strong interaction cannot be treated by the theory of perturbative QCD which implies that the structure of the nucleon cannot be easily described. Experiments are then required to constrain theoretical models and to gain a better understanding of the structure of hadrons.

Some experimental results have already given indications that the strange quarks contribute to the properties of the nucleon. The σ-term value in πN scattering suggests a contribution of strange quarks of about 130 MeV to the nucleon mass [1]. From eN Deep Inelastic Scattering data, strange quarks could contribute to the nucleon spin at the level of few percent ($\langle N|\bar{s}\gamma^\mu\gamma^5 s|N\rangle \simeq -10\%$ [2]). But those results are still questionable due to theoretical assumptions or unmeasured contributions.

The purpose of Parity Violating (PV) experiments in electron scattering is to extract the contribution of strange quarks to the charge and magnetization distributions of the nucleon. The formalism and the principle of these measurements are first presented. Then the G^0 experiment is described and first results from its engineering run are given.

2 Strangeness and parity violation

The extended structure of nucleons has been extensively studied by electron scattering. The virtual photon, exchanged during the reaction, probes the internal electromagnetic structure of the nucleons. The elastic cross section (eN) depends on Dirac and Pauli form factors, F_1^γ and F_2^γ respectively, which describe the electromagnetic structure of nucleons. These form factors depend only on the four-momentum carried by the virtual photon (Q^2). By varying the value of Q^2, one can tune the probed spatial dimension. One can also use the Sachs form factors, defined as follows :

$$G_E^\gamma = F_1^\gamma - \tau F_2^\gamma \qquad G_M^\gamma = F_1^\gamma + F_2^\gamma \qquad (1)$$

where $\tau = Q^2/(4M_N^2)$ with M_N the mass of the nucleon. In the Breit frame, G_E^γ and G_M^γ are respectively the Fourier Transforms of the spatial distributions of the charge and magnetization in the nucleon [3]. These form factors are related to the static properties of the nucleon (at $Q^2 = 0$) :

$$G_E^{\gamma,N}(0) = Q_N \qquad G_M^{\gamma,N}(0) = \mu_N \qquad (2)$$

where Q_N is the charge and μ_N the magnetic moment of the nucleon N. Another quantity of interest is the charge radius which is defined as follows :

$$\langle r^2 \rangle = -6 \left. \frac{dG_E}{dQ^2} \right|_{Q^2=0} \qquad (3)$$

As photons couple only to quarks in the nucleon, the form factors can be expressed as the sum of the contributions from the different quark flavors, $G_{E,M}^{q,N}$. The sea quarks are produced from the fluctuations of gluons in quark-antiquark pairs. The probability of this process is proportional to $1/M_q^2$ where q is the flavor of the produced quarks. At the energy of the current experiments, one can neglect the contribution of the heaviest quarks (c, b and

t) and express the electromagnetic Sachs form factors as follows :

$$G_{E,M}^{(\gamma,N)} = \sum_{q=u,d,s} e_q G_{E,M}^{q,N} \qquad (4)$$

The values of the electromagnetic form factors $G_{E,M}^{(\gamma,p)}$ and $G_{E,M}^{(\gamma,n)}$ have been or are presently measured for Q^2 values up to 1 $(GeV/c)^2$. These form factors provide four relations but there are twelve unknowns $(G_{E,M}^{q,N})$. To reduce this number to six, the hypothesis of isospin symmetry is made : $G_{E,M}^{u,p} = G_{E,M}^{d,n} = G_{E,M}^{u}$, $G_{E,M}^{d,p} = G_{E,M}^{u,n} = G_{E,M}^{d}$ and $G_{E,M}^{s,p} = G_{E,M}^{s,n} = G_{E,M}^{s}$. The measurement of the contribution of the strange quarks $(G_{E,M}^s)$ requires then two additional independent combinations of $G_{E,M}^{q,N}$ to solve the system.

Similarly to the electromagnetic interaction, one can also define Sachs form factors for the weak interaction, $G_{E,M}^{(Z,N)}$ (Z boson exchange in eN scattering). As the Z boson also couples only to quarks in the nucleon, the weak form factors of the proton can be written as :

$$G_{E,M}^{(Z,p)} = \sum_{q=u,d,s} C_V^q G_{E,M}^q \qquad (5)$$

where $C_V^q = 2T_3 - 4Q_q \sin^2 \theta_W$ is the weak vector charge of quarks of flavor q [5]. As the weak interaction violates parity symmetry, there is also an axial form factor G_A^e. The two weak form factors of the proton represent then two new independent combinations. The strange contributions to charge and magnetization can then be extracted using the following formula [4]:

$$G_{E,M}^s = (1 - 4\sin^2 \theta_W)G_{E,M}^{(\gamma,p)} - G_{E,M}^{(\gamma,n)} - G_{E,M}^{(Z,p)} \qquad (6)$$

Thus the determination of strange quark contributions requires the measurement of weak form factors of the proton, as, contrary to electromagnetic form factors, these weak form factors are still mostly unknown.

The elastic scattering (eN) cross section depends in fact on both electromagnetic and weak form factors. But for values of Q^2 around 1 $(GeV/c)^2$, the weak element matrix, M_Z, is 10^{-5} smaller than the electromagnetic one associated to the exchange of a virtual photon, M_γ. The experimental systematic errors related to normalization prevent then the weak form factors to be obtained from unpolarized cross section measurements. Contrary to the electromagnetic interaction, the parity symmetry is violated in the weak interaction. Consequently the cross section of elastic scattering of longitudinally polarized electrons on nucleons ($\vec{e}N$) depends on the helicity of the incident electrons (i.e. $\sigma_+ \neq \sigma_-$, where $\pm$ refers to the two possible helicity states). Thus the weak part of the scattering can be extracted by a parity violation asymmetry measurement :

$$A_{PV} = \frac{\sigma_+ - \sigma_-}{\sigma_+ + \sigma_-} \qquad (7)$$

Because of the smallness of the weak interaction, the asymmetry is of the order of 10^{-5} but this time many normalization terms (i.e. luminosity, acceptance, ...) cancel in

Table 1. Values in part per million (ppm) of $A_{s=0}$, η, χ and ξ for the kinematical condition of the G^0 experiment and for Q^2=0.5 $(GeV/c)^2$

	θ_e	$A_{s=0}$ (ppm)	η (ppm)	χ (ppm)	ξ (ppm)
A_F	13°	-16.7	<u>60.8</u>	25.0	1.4
A_B	110°	-29	18.0	<u>40.3</u>	8.5
A_D	110°	-39.8	14.7	9.0	<u>10.1</u>

the ratio. This parity violation asymmetry is a function of electromagnetic, weak and axial form factors [5,6]:

$$A_{PV}(Q^2,\theta_e) = -\frac{G_F Q^2}{4\sqrt{2}\pi\alpha} \frac{f_E G_E^{(Z,N)} + f_M G_M^{(Z,N)} + f_A G_A^e}{A_D} \qquad (8)$$

where

$$f_E = \epsilon G_E^{(\gamma,p)}, \; f_E = \tau G_M^{(\gamma,p)}, \; f_A = \epsilon' G_M^{(\gamma,p)}$$
$$A_D = \epsilon \left(G_E^{(\gamma,p)}\right)^2 + \tau \left(G_M^{(\gamma,p)}\right)^2 \qquad (9)$$

The kinematical factors ϵ, τ and ϵ' are function of Q^2 and θ_e, the electron scattering angle in the laboratory frame. Formula 8 can also be expressed as a function of G_E^s, G_M^s and G_A^e :

$$A_{PV} = A_{s=0} + \eta G_E^s + \chi G_M^s + \xi G_A^e \qquad (10)$$

where $A_{s=0}$, the value of the asymmetry with no strange quarks contribution, is derived from the Standard Model predictions and the measured electromagnetic form factors. The terms η, χ and ξ depend on kinematics of the electron scattering and electromagnetic form factors. Formula 10 shows that the asymmetry is sensitive to a linear combination of G_E^s, G_M^s and G_A^e. Thus for a given value of Q^2, a set of three measurements must be performed to separate them. The first two measurements correspond to electron scattering on proton in the forward (A_F) and backward (A_B) directions. The axial form factor, G_A^e, can be calculated from theoretical predictions. However due to difficulties of estimating its value and Q^2 dependence, it appears worth also measuring it. Thus a third measurement (A_D) is performed on deuterons at backward angles in quasi-elastic scattering. The different values of $A_{s=0}$, η, χ and ξ are reported in Table 1 for Q^2=0.5 $(GeV/c)^2$ and for the electron scattering angles used in the G^0 experiment. From the values quoted in this table, one can see that the measurement at forward angle (A_F) is the most sensitive to G_E^s. The backward angle measurement on protons (A_B) is more sensitive to G_M^s and a measurement on deuterons (A_D) gives a way to extract the axial form factor.

Two parity violation experiments have already published results. The first one, SAMPLE at MIT-Bates, measured backward angle scattering ($130° < \theta_e < 170°$) asymmetries on LH_2 at $Q^2 = 0.1$ $(GeV/c)^2$ [7]. A second measurement was performed at the same kinematics but on deuterium [8]. The combination of the two results gives the following values for G_M^s and the isovector part of G_A^e :

$$G_M^s = 0.14 \pm 0.29 \pm 0.31 \qquad (11)$$
$$G_A^e(T = 1) = 0.22 \pm 0.45 \pm 0.39 \qquad (12)$$

where the first error corresponds to statistics and the second to systematics (errors on beam properties, electromagnetic form factors, etc.). The second experiment, HAPPEX at Jefferson Laboratory, measured the asymmetry on LH_2 at Q^2=0.48 $(\mathrm{GeV}/c)^2$ and forward scattering angle ($\theta_e = 12.5°$) [9,10]. This experiment extracted a linear combination of G_E^s and G_M^s :

$$G_E^s + 0.392 G_M^s = 0.025 \pm 0.020 \pm 0.014 \qquad (13)$$

The effect of strange quarks is small but the individual electric and magnetization contributions of strange quarks cannot be separated in a single measurement ; also a limited range of Q^2 is presently covered. An experimental effort is thus still underway. The PVA4 experiment [11] at Mainz will separate G_E^s and G_M^s at Q^2=0.225 $(\mathrm{GeV}/c)^2$. The HAPPEX2 experiment on the proton and on ^{4}He, at JLab, will be able to make the separation of G_E^s and G_M^s at the same Q^2 as the SAMPLE experiment.

This contribution details the G^0 experiment which is now starting to take data. The aim of this project is to separate G_E^s, G_M^s and G_A^e for three values of Q^2 : 0.3, 0.5 and 0.8 $(\mathrm{GeV}/c)^2$ [12]. The expected precision on G_E^s, G_M^s and G_A^e and the wide range of Q^2 covered by this experiment can be seen in Fig. 1 to be quite unique.

Theoretical models have been used to predict the contributions of strange quarks [13]. Models are based on chiral perturbation theory [14], Vector Dominance Meson approaches [15] and lattice QCD techniques [16] and provide values of G_E^s and G_M^s. Some of these predictions are reported in Fig. 1, compared with projected experimental results. One can see that the models give different strange quark contributions. So experimental results are required to constrain these theoretical models. About the isovector axial form factor, its value at $Q^2 = 0$ $(\mathrm{GeV}/c)^2$ was calculated by Zhu *et al* [17] and its Q^2 dependence by Maekawa *et al* [18]. Their results were combined and are reported in Fig. 1. The only published experimental result is given by the SAMPLE collaboration and has a large error. In these figures are also reported expected errors for HAPPEX2 and G^0 experiments. So this set of new experiments will strongly constrain theoretical models.

3 G^0 experiment

The values of asymmetries to be measured by parity violation experiments are of the order of 10^{-5}. However one requires an absolute precision of $10^{-6} - 10^{-7}$, and thus all false asymmetry sources must be controlled to high precision. There are thus general requirements that must be fulfilled for a violation parity experiment. This is in particular the case for the beam properties. First the polarization has to be known at the few percent level because

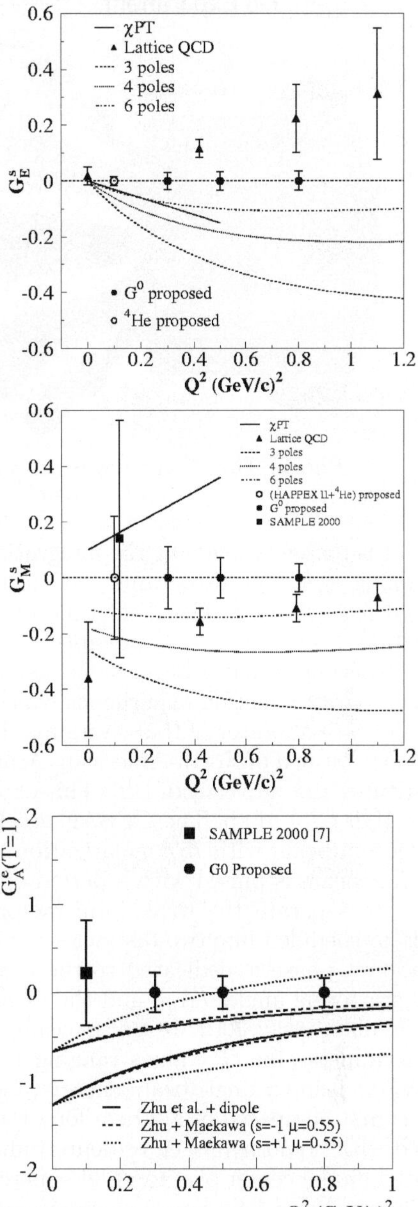

Fig. 1. Predictions of some theoretical models on G_E^s, G_M^s and $G_A^e(T = 1)$. The expected errors of approved experiment at JLab and the results of SAMPLE are also shown (figures taken from [19], references therein)

the measured asymmetry is proportional to the beam polarization. The helicity reversal must happen with a high frequency (30-60 Hz typically) to minimize the effects of slow drift, especially on intensity. Feedback systems on intensity, energy and position are used to keep the beam properties as close as possible in the two helicity states and thus to limit corrections to be applied in the asymmetry calculation. To achieve a statistical precision of 10^{-7}, one has to record 10^{14} events, requiring a large luminosity. The parity violation experiments use long (between 20 and 40 cm) cryogenic targets and high beam intensity. A full data acquisition event by event is also excluded. The

G0 Experiment

Fig. 2. The G^0 apparatus

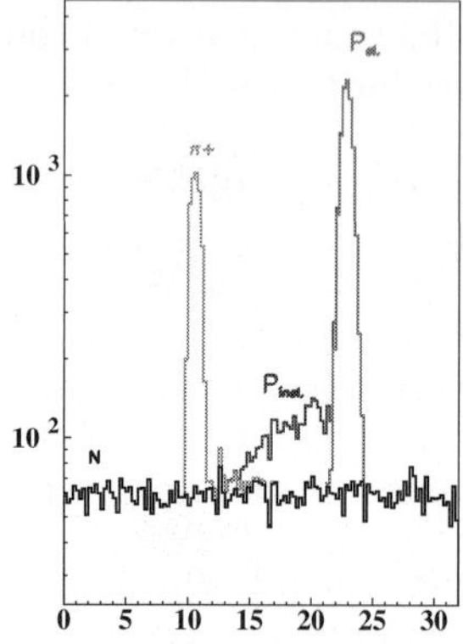

Fig. 3. Example of a simulated ToF spectrum. The vertical axis represents relative counting rates and the horizontal axis ToF is expressed in ns

measurement is therefore made by the integration of a signal or summing events in histograms. Also fast detectors (typically 10 ns) are used. The dead time of electronics and acquisition must be as low as possible.

The G^0 experiment that is being described in this contribution is a parity violation experiment which is taking place at Jefferson Laboratory (JLab, Virginia, USA). The collaboration is formed of North-American, Armenian and French institutes and universities [12]. This experiment is installed in Hall C of JLab. The CEBAF accelerator delivers us electron beam with 75% polarization and 40 μA intensity. This beam is pulsed with a period of 32 ns.

In order to separate the weak form factors, the G^0 experiment is scheduled into two distinct sets of measurements. The first phase is dedicated to the measurement at forward scattering angles (A_F) and the final data taking will take place early 2004. It will measure the parity violation asymmetry for Q^2 values ranging between 0.1 and 1 $(\text{GeV/c})^2$ using a single beam energy (3 GeV). The G^0 set-up is first composed of a 20 cm long target which can be filled with liquid hydrogen or deuterium. This target is located at the center of a toroidal superconducting magnet. The produced field bends charged particles on a focal plane where are located a set of detectors (FPD[1]) formed of scintillator pairs, optically coupled to phototubes. Eight sectors of detection (octants) are used, each formed of 16 FPD detectors. The set-up is shown in Fig. 2. The shape and position of each detector are designed in such a way that it covers a given range in Q^2. The scattering angle of electrons is varying between 6° and 22°. As the detection of electrons is very difficult at these angles, the recoil protons instead are detected ($48° < \theta_p < 78°$). The background formed by neutrals (γ and neutrons) is mostly removed by a set of internal collimators preventing direct view of the target. The background formed by charged particles (π^+ and inelastically scattered protons) is rejected by a Time of Flight (ToF) measurement. For each FPD detector, a ToF spectrum is constructed. Figure 3 shows an example of a simulated ToF spectrum. The beam time structure is chosen so that all particles

produced by a beam burst every 32 ns are detected before the arrival of the next beam burst. The elastically scattered protons can be selected by cuts on ToF. Custom electronics were designed and allow to discriminate photomultiplier signals, make the coincidence between the two layers of scintillator for each detector[2] and generate the ToF spectra. These histograms are read during each helicity reversal which occurs at a frequency of 30 Hz. This procedure eliminates deadtime related to the data acquisition.

The second phase of the G^0 experimental program will be dedicated to back angle measurements (both on LH_2 and LD_2) and should take place in 2005-2006. In contrast with the first phase, electrons are detected. Their scattering angle is around 110° and the value of Q^2 varies slowly in the acceptance of the collimators. Therefore the measurements will be performed at three beam energies (420, 580 and 800 MeV) and values of Q^2 equal to 0.3, 0.5 and 0.8 $(\text{GeV/c})^2$ will be accessible. To detect particles scattered at backward angles, the overall apparatus must be turned around. The detected electrons are ultrarelativistic so a ToF measurement cannot be used to reject the background[3]. Thanks to the magnetic field of the spectrometer, elastically scattered electrons have different trajectories compared to inelastically scattered electrons and π^-. So eight hodoscopes composed of nine additional scintillator detectors, called CEDs[4], will be installed at the exit of the magnet cryostat and associated to each octant of FPD. A coincidence matrix between CED and FPD detectors allows one to separate elastic and inelastic events as shown in Fig. 4. The asymmetry will be calculated from the counting rates corresponding to the matrix locations associated to elastically scattered electrons.

Pions are however also detected in some of the matrix locations associated to elastically scattered electrons. To discriminate and reject these pions, which rates would become very high with a deuterium target, an aerogel

---

[1] Focal Plan Detector

[2] This allows additional rejection of neutrals.

[3] In this part of the experiment the charged background is π^- and inelastically scattered electrons.

[4] Cryostat Exit Detectors

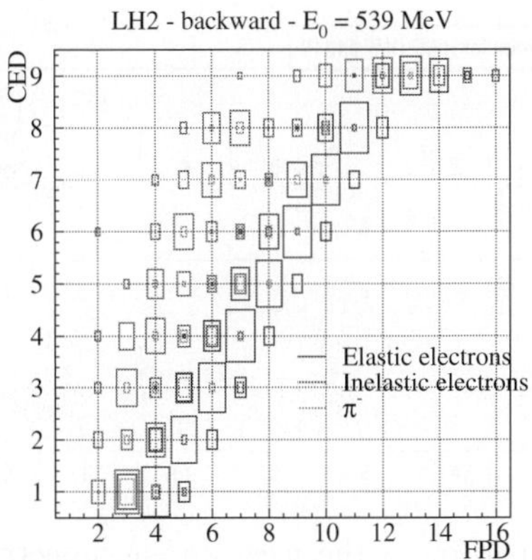

Fig. 4. Typical simulated coincidence matrix between CED and FPD detectors

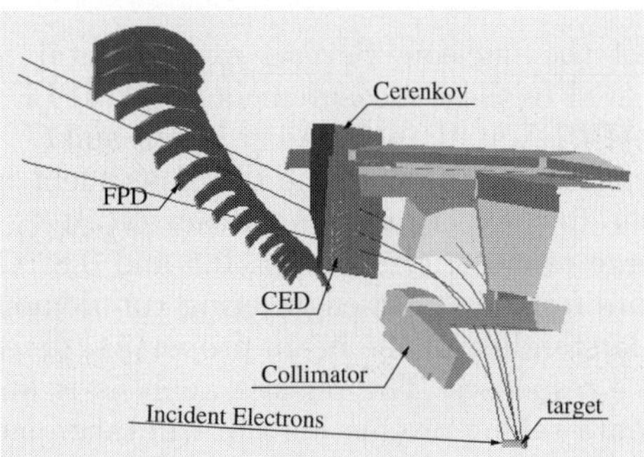

Fig. 5. Sketch of one octant of detectors in the back angle configuration. The collimators are used to select the kinematical range but also to shield detectors from the line of sight of the target

Čerenkov detector will be installed close to the CED hodoscope (see Fig. 5). The refractive index of the aerogel is chosen so that only electrons produce light when passing through the Čerenkov detector. The Čerenkov signal is consequently used to validate the encoding of an event. This procedure allows the rejection of typically 95% of the pions. The Čerenkov detectors are under construction and test. New custom electronics is presently designed and will form the coincidence matrix and process the Čerenkov signal.

4 Engineering run

The first engineering run of the G^0 experiment took place from October 2002 to January 2003 in the forward angle configuration (A_F). During this period, not only the apparatus was tested but also many special features required of the CEBAF beam. First the accelerator succeeded to deliver the G^0 beam with use of a new laser system. The standard CEBAF beam is pulsed with a 2 ns period with an intensity up to 100 μA. The challenge was to build a 40 μA beam with a 32 ns period which means to have 6 times more electrons than usual per burst. Various feedback systems were also tested. The goal is to achieve, for the 700 h of data taking scheduled for the production run,

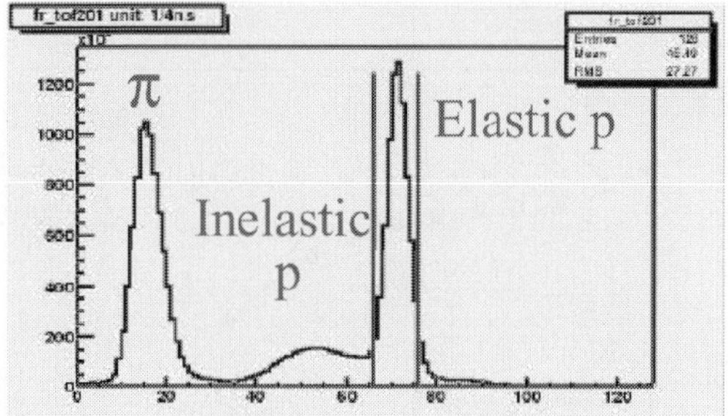

Fig. 6. Example of a measured ToF spectrum. The time binning is 0.250 ns

an overall charge asymmetry less than $\Delta I/I < 1$ ppm and differences in position less than $\Delta X < 20$ nm. The amount of useful data for asymmetry calculation recorded during this engineering run represents only 1/16 of the final statistics. For these data, the charge asymmetry was less than 5 *ppm* and the position asymmetries less than 50 nm which meet the requirements for these recorded data. The target system was also tested ; in particular we checked that it can stand the beam intensity without boiling. A key point for the experiment concerned the operation of the spectrometer. The nominal value of current was achieved in its superconducting mode. The magnetic field is directly related to the Q^2 range measured by the experiment $(0.1 < Q^2 < 1 \,(\mathrm{GeV/c})^2)$. The first part of this engineering revealed that the hall background was higher than expected. It was found that the main source of background were neutrals generated at the beam pipe. So the shielding was properly improved. Concerning the detectors, the high voltages for photomultipliers (PMT) were set at values allowing high detection efficiency and the PMTs were able to stand rates at nominal beam current. The gains were matched and their stability was satisfactory over time and different beam conditions. Discriminator thresholds were adjusted in order to remove the noise and low energy particle background while keeping 100% efficiency for elastic proton detection. All electronic channels worked fine and ToF spectra were found stable in time. An example of ToF spectrum is given in Fig. 6 (the vertical lines in this figure represent typical ToF cuts to select elastic protons). Some studies are still in progress, in particular regarding the deadtime corrections.

The results of the engineering run also show that the rates of inelastic events were higher than expected from the simulations. The main source of these additional events was assigned to reactions occuring in the aluminium target cell windows. So the thickness of these windows will be reduced for the next engineering run scheduled this fall.

A check of agreement between simulation, and the understanding of the set-up, and measurement has been made using the ToF difference between pions and elastic protons. Figure 7 shows the results of this study for each detector. The observed agreement is quite good (around 100 ps) and required the inclusion of detailed geometry and event generators in the simulation. This ToF difference is also important as it allows to determine the mean

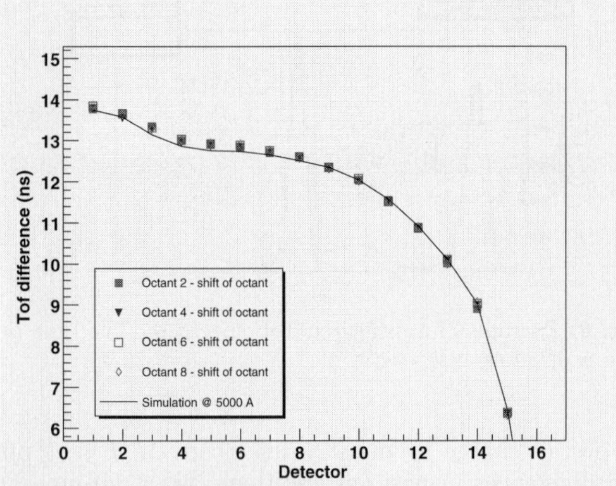

Fig. 7. Comparison of the ToF difference between elastic proton and π^+ extracted from the simulation (*black line*) and measurement (*points*)

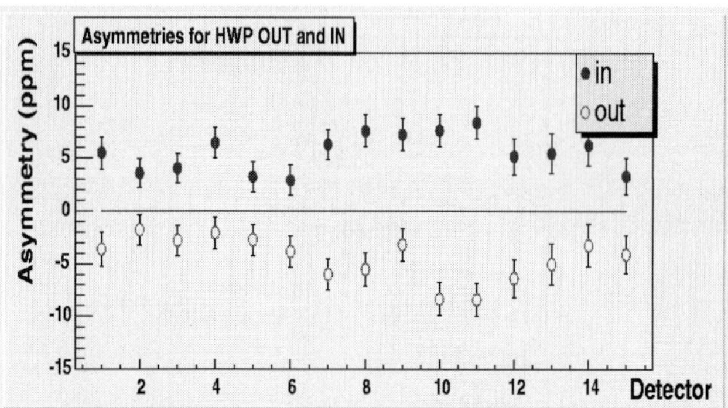

Fig. 8. Comparison of the measured asymmetries with and without the Insertable Half Wave Plate (IHWP)

value of Q^2 associated with each FPD detector. Indeed pions are ultrarelativistic with almost fixed ToF contrary to protons for which the kinetic energy is proportional to Q^2. In order to keep the systematic error introduced by the uncertainty on Q^2 below 5% of the statistical error, the value of Q^2 must be known with a precision of 1%. This is achieved with the observed precision of 100 ps on ΔToF.

The parity violation asymmetries have been calculated for each detector. The statistical error associated with this engineering run is about four times larger than the one expected in production run. A Half Wave Plate (IHWP) can be inserted at the injector to reverse the electron helicity compared to the helicity signal sent to the electronics. If there is no false asymmetry generated from the beam or the electronics the results with and without this IHWP must be exactly the same but with opposite sign. Figure 8 shows that it is the case within errors (only statistical ones are displayed). The analysis is still underway. The asymmetry of inelastic events has to be subtracted from the measured asymmetry in order to extract the elastic proton asymmetry. Another study performed at present is related to the electronics response to the charge asymmetry. The goal is to achieve a systematic error at the order of 1% of the charge asymmetry. The main contribution to this systematic error is the deadtime which is proportional to the counting rates. The understanding of the beam properties and feedback systems is also in progress. Improvements and changes will be tested during the next engineering run this fall.

5 Conclusions

The purpose of parity violation measurements in (eN) elastic scattering is to determine the contribution of strange quarks to the charge and magnetization distributions of the nucleon. Several experimental programs are dedicated to these measurements : SAMPLE at MIT-Bates, HAPPEX at JLab, PVA4 at Mainz and G^0 at JLab. The special contribution of the G^0 experiment is that it will be the first experiment to extract G_E^s, G_M^s and G_A^e over a large range of Q^2 (= 0.3, 0.5 and 0.8 $(Gev/c)^2$). The results from the first engineering run demonstrate a good understanding of the beam properties, detector and electronics responses. The physics analysis is under way and the data taking production run will take place at the beginning of the next year.

References

1. J. Glaser, H. Leutwyler, and M. Saino: Phys. Lett. B **253**, 252 (1991)
2. K. Abe et al: Phys. Rev. D **58**, 112003 (1998)
3. F.J. Ernst, R.G. Sachs, and K.C. Wali: Phys. Rev. **119** 1105 (1960)
4. D.H. Beck and R.D. McKeown: Ann. Rev. Nucl. **51** 189 (2001)
5. M. Musolf et al.: Phys. Rep. **239**, 1 (1994)
6. P.A. Souder and K.S. Kumar: Prog. Part. Nucl. Phys. **45**, S333 (2000)
7. R. Hasty et al.: Science **290**, 2117 (2000)
8. D. Spayde et al.: Phys. Rev. Lett. **84**, 1106 (2000)
9. K.A. Aniol et al.: Phys. Rev. Lett. **82**, 1096 (1999)
10. K.A. Aniol et al.: Phys. Lett. B **509**, 211 (2001)
11. F.E. Maas et al. (PVA4 Collaboration): Eur. Phys. J. A **17**, 339 (2003)
12. G^0 web page: http://www.npl.uiuc.edu/exp/G0/G0Main.html
13. D.H. Beck and B.R. Holstein: Int. Jour. Mod. Phys. E **10**, 1 (2001)
14. T.R. Hemmert, B. Kubis, and U.-G. Meißner: Phys. Rev. C **60**, 045501 (1999)
15. H. Forkel: Phys. Rev. C **56**, 510 (1996)
16. S.J. Dong, K.F. Liu, and A.G. Williams: Phys. Rev. D **58**, 074504 (1998)
17. S. Zhu et al.: Phys. Rev. D **62**, 033008 (2000)
18. C.M Maekawa, J.S. Veiga, and U. Van Kolck: Phys. Lett. B **488**, 167 (2000)
19. R. Tieulent: PhD Thesis, Grenoble University, ISN 02-27

Eur Phys J A (2004) **19**, s01, 213–221

Digital Object Identifier (DOI) 10.1140/epjad/s2004-03-036-6

EPJ A direct

electronic only

The study of high density matter at RHIC

Survey of experimental results from RHIC

Thomas S. Ullrich

Brookhaven National Laboratory, Upton New York 11973-5000, USA

Received: 30 Jul / Accepted: 14 Nov 2003 /
Published Online: 6 Feb 2004 – © Società Italiana di Fisica / Springer-Verlag 2004

Abstract. QCD predicts a phase transition between hadronic matter and a Quark Gluon Plasma at high energy density. The Relativistic Heavy Ion Collider (RHIC) at Brookhaven National Laboratory is a new facility dedicated to the experimental study of matter under extreme conditions. Already the first round of experimental results at RHIC indicated that the conditions to create a new state of matter are indeed reached in the collisions of heavy nuclei. Studies of particle spectra and their correlations at low transverse momenta provide evidence of strong pressure gradients in the highly interacting dense medium and hint that we observe a system in thermal equilibrium. Recent runs with high statistics allow us to explore the regime of hard-scattering processes where the suppression of hadrons at large transverse momentum, and quenching of di-jets are observed thus providing further evidence for extreme high density matter created in collisions at RHIC.

PACS. 25.75.-q Relativistic heavy-ion collisions

1 Introduction

At high temperature, hadronic matter dissolves into a soup of its constituent quarks and gluons. For an asymptotically free field theory such as QCD, the state of matter at high energy density is simple: long range (low momentum) interactions are screened, and short range (high momentum) interactions are weak, leading to an ideal gas equation of state in the high energy density limit [1]. At temperature $T \gg \Lambda_{\mathrm{QCD}}$ matter is a gas of deconfined, weakly interacting quarks and gluons (the "Quark-Gluon Plasma", or QGP), whereas at $T \ll \Lambda_{\mathrm{QCD}}$ quarks and gluons are confined and matter consists of strongly interacting hadrons.

The QCD phase diagram has a complex structure [2]. At low temperature and low baryon density the phase is hadronic (confined phase) and chiral symmetry is broken. Color-superconducting and other phases may exist at high baryon density and low temperature [3], whereas at high temperature the quarks and gluons are deconfined and chiral symmetry is restored. The early universe descended from high T at extremely small μ_B. Neutron star cores have high μ_B and very low T.

First-principles calculations of finite temperature QCD can only be carried out numerically on the lattice [4]. These calculation show that the energy density ϵ exhibits a sharp rise in the vicinity of the critical temperature T_C, indicating a rapid change in the number of underlying degrees of freedom. However, the ideal gas Stefan-Boltzmann limit ϵ_{SB} has not yet been achieved at $T \sim 4T_C$. Putting in physical values, $T_C \sim 175$ MeV, resulting in critical energy density $\epsilon_C = (6 \pm 2)T_C^4 \sim 0.7$ GeV/fm^3. This value should be kept in mind for comparison to conditions achieved in laboratory experiments.

The order of the deconfinement phase transition can be determined in some limiting cases [4]. It is first order for pure gauge and for three light quarks, second order for two light and one heavy quark. For physical quark masses the order of the transition, or indeed whether it is a smooth cross over, has not been determined. The extension of lattice calculations to $\mu_B > 0$ is a long-standing problem, but there has been significant recent progress in determining the phase boundary and equation of state for finite μ_B [6].

In the early universe, the confinement transition (QGP $\rightarrow$ hadrons) at very low μ_B occurred about 10 μs after the big bang [7]. A strongly first order phase transition may have generated primordial black holes, strange quark nuggets, or local baryon asymmetries affecting primordial nucleosynthesis, though no relics of this transition have been observed thus far. In the cores of neutron stars, a QGP phase at low temperature and high μ_B may generate observable millisecond pulsar phenomena [8].

It is natural to ask whether the deconfinement and chiral symmetry restoration transitions can be studied in accelerator-based experiments. We require a system having temperature of order the pion mass in equilibrium over a large volume. The best possibility to accomplish this is

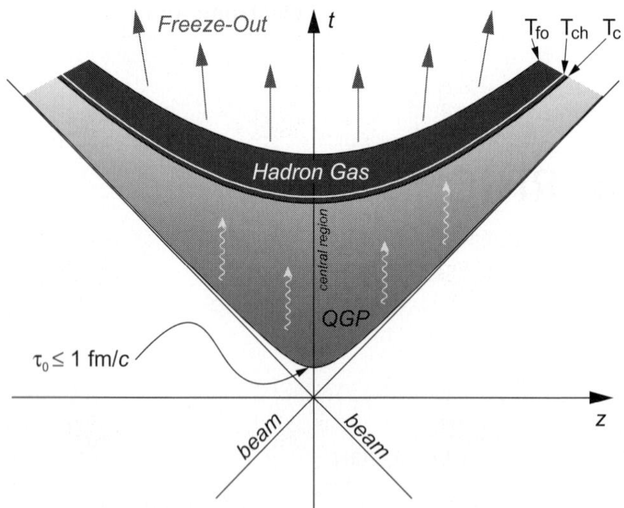

Fig. 1. The space-time picture of relativistic heavy-ion collisions, proceeding through the following stages: formation, deconfined state (QGP), hadron gas (T_C), chemical freezeout (T_{ch}), thermal freezeout (T_{fo}), and free streaming

through the collision of heavy nuclei at the highest possible energy.

However, nuclear collisions at high energy are highly dynamic. Even if a hot, equilibrated system is created early in the collision, it will immediately expand and cool. If a QGP is formed it will hadronize after a brief period, and signatures from the deconfined phase may be masked by those from the hot, interacting hadron gas. The central issue is to find and measure those experimental observables that are sensitive to the state of matter early in the collision [9].

Broadly speaking, several epochs in the evolution of high energy nuclear collisions can be sketched as depicted in Fig. 1. Immediately following the collision is a brief formation time $\tau < 1\mathrm{fm}/c$, during which large momentum transfer (short-distance) processes occur. It is also during this period that the highest energy density is achieved and the QGP may result. After an expansion time of perhaps a few fm/c the plasma hadronizes into a dense, interacting hadron gas. Further expansion and cooling causes the temperature to fall to the point at which inelastic collisions among hadrons cease (*chemical freezeout*) and the relative populations of the various long-lived hadron species are established. Elastic collisions continue until *kinetic freezeout*, after which point the hadrons fly effectively undisturbed to the detectors. Thus, the relative populations of stable hadrons reflect the conditions at chemical freezeout, whereas their momentum spectra reflect the conditions at kinetic freezeout. More penetrating probes (dileptons, direct photons, jets, heavy quarks,...) may carry information on the conditions around the time of their formation earlier in the collision.

Collisions of heavy nuclei at ultra-relativistic energies have been studied in fixed target experiments at the Brookhaven AGS and CERN/SPS for the past 15 years. In 1999, the Relativistic Heavy Ion Collider (RHIC)

at Brookhaven National Laboratory was commissioned, bringing into operation the first facility largely dedicated to the study of matter at high energy density.

I will first describe the RHIC machine and experiments and then discuss some of the main results from its heavy ion physics program. I will briefly touch on "soft physics" observables ($p_T \lesssim 2$ GeV/c) and summarize what has been learned from them. The second part of the review concentrates on the major new development at RHIC: jet production and indications of partonic interactions with dense matter, which may directly probe the energy density achieved early in the collision.

2 The relativistic heavy ion collider and experiments

RHIC consists of two concentric superconducting rings, 3.8 km in length. It has enormous flexibility in beam masses and energies, with capability to collide gold ions from $\sqrt{s_{\mathrm{NN}}} = 20$ to 200 GeV per nucleon pair, protons up to $\sqrt{s} = 500$ GeV, and asymmetric systems, most importantly protons and deuterons with heavy nuclei. The top center-of-mass energy for heavy nuclei is a factor of 10 larger than for the fixed target experiments at the CERN/SPS, extending significantly the statistical and transverse momentum reach of many observables and opening up new channels. RHIC is also the first polarized proton collider, creating new opportunities to study the spin content of the proton, in particular the contribution of the gluon at low x_{Bj}. More details of the machine and its first year performance can be found in [10].

Design luminosity for Au+Au at 200 GeV is $\mathcal{L} = 2 \cdot 10^{26}$ cm^{-2}sec^{-1}, giving an interaction rate of about 2 kHz. While this luminosity appears to be tiny relative to other modern colliders, recall that the rate for hard processes such as jet production in nuclear collisions scales as $\sim A^2$ (A=atomic mass), so that hard process rates in Au+Au at design luminosity are the same as at a proton collider with $\mathcal{L} \approx 10^{31}$ cm^{-2}sec^{-1}. The design luminosity for p+p collisions at 500 GeV is $2 \cdot 10^{32}$ cm^{-2}sec^{-1}, with an interaction rate of about 8 MHz. Design polarization for p+p is 70%.

RHIC has six intersection regions, of which four are currently instrumented with experiments [11] (see Fig. 2). PHENIX consists of an axial field magnet and four independent spectrometers: two at midrapidity containing tracking, ring imaging Cerenkov counters, time of flight, and electromagnetic calorimetry, which are optimized for precision lepton, photon and hadron measurements, and two forward muon arms. STAR has conventional collider detector geometry, with a large solenoidal magnet, Time Projection Chamber for tracking, large coverage EM calorimetry, and an inner silicon-based tracker. STAR is designed for hadron, jet, lepton and photon measurements over large acceptance, as well as studies of eventwise fluctuations in high multiplicity nuclear collisions. BRAHMS consists of two small acceptance spectrometers for inclusive identified hadron measurements over wide

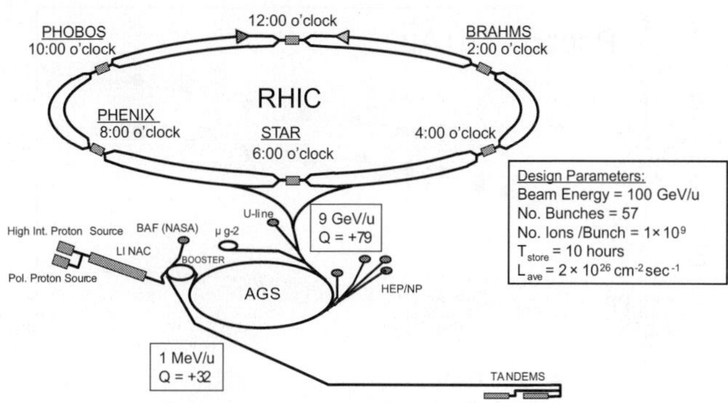

Fig. 2. The Relativistic Heavy Ion Collider (RHIC) accelerator complex at Brookhaven National Laboratory

phase space. PHOBOS has very wide phase space coverage for charged particles using silicon detectors, and a mid-rapidity spectrometer based on a dipole magnet, silicon tracking and Time of Flight. STAR and PHENIX each have about 450 collaborators, whereas PHOBOS and BRAHMS each have fewer than 100.

RHIC had a brief commissioning run in 1999. The data reported here are from a Au+Au run at $\sqrt{s_{NN}}$ =130 GeV in 2000 and Au+Au and polarized p+p (∼15% polarization) runs at $\sqrt{s_{NN}}$ = 200 GeV in 2001-2. The integrated luminosity is about 80 μb^{-1} for Au+Au and 1 pb^{-1} for p+p. A comprehensive view of the physics program of RHIC and what has been achieved thus far can be found in the proceedings of the recent Quark Matter conferences [12,13].

Nuclei are extended objects, and high energy nuclear collisions can be characterized experimentally as head-on ("central") or glancing ("peripheral"). The RHIC experiments all have different methods of determining the geometry, or centrality, of the collisions. In general, all measure some final state variable (or combination of these variables) that can be related to fractions of the total measured hadronic cross-section. Typical final state variables are the charged particle multiplicity at central or forward rapidities and the forward-going "spectator" neutrons that are measured in zero-degree calorimeters situated downstream of the interaction region. These centrality bins are then mapped to variables that allow for direct comparison between experiments. The primary variables that are used for comparison are *(i)* the number of participating nucleons in the collision N_{part} and *(ii)* the number of independent binary N+N collisions N_{bin}, which are determined via the application of the Glauber model phenomenology. The basic concept of the Glauber model is to treat a Au+Au collision as a superposition of many independent nucleon-nucleon (N+N) collisions. Thus, the only parameters that the model depends on are the nuclear density profile (Woods-Saxon) and the non-diffractive inelastic N+N cross-section. The former is well measured in e+Au scattering experiments and the latter is well established by many previous experiments. With this, N_{bin} and N_{part} can be calculated as a function of impact parameter. Once the distributions $d\sigma/dN_{part}$ and $d\sigma/dN_{bin}$ are deter-

mined, these histograms are binned according to fractions of the total cross-section. This then determines the mean values of N_{bin} and N_{part} for each centrality class.

Phenomenologically it has been found that total particle production scales roughly as N_{part}, whereas the rate of hard processes will scale as N_{bin} in the absence of nuclear effects. Such scaling rules can be used to uncover the new physics present in nuclear collisions. For instance, the violation of N_{bin} scaling at high-p_T indicates significant effects of the nuclear medium on high-p_T processes.

3 General characteristics of RHIC collisions

3.1 Global observables

One of the earliest probes suggested for QGP formation involves a study of the global parameters of the events, e.g. the energy deposition, multiplicity, and the average transverse momentum of the emitted particles, as a function of center-of-mass energy $\sqrt{s_{NN}}$, mass number A, and centrality of the collision. For example, by studying the multiplicity of the produced particles one might estimate theoretically the entropy produced in the collision. Sudden changes in behavior with varying centrality or A would be indicators of a phase transition. So far, however, no such anomalous changes have been observed, at either the AGS, SPS, or RHIC. All results on global observables so far indicate a rather smooth evolution in centrality and $\sqrt{s_{NN}}$. This, of course, does not necessarily imply the absence of a phase transition, but might be rather an indication of either the insensitivity of these observables to the early phase of the collision and/or might suggest a second order phase transition (or a cross-over). With the commencement of the RHIC program the question of multiparticle production in nuclear collisions became more complex due to the poorly understood role of perturbative QCD (hard processes).

Figure 3 from the PHOBOS Collaboration [14] shows charged particle multiplicity distributions $dN_{ch}/d\eta$ vs η over the full RHIC phase space, for Au+Au collisions at all three collision energies studied so far. The longitudinal phase space growth with increasing energy is apparent, as well as the increase in multiplicity for more central collisions. For the most central collisions at $\sqrt{s_{NN}}$ = 130 GeV, 4200 charged particles are produced in the full phase space.

The distributions exhibit a central plateau near $\eta \sim 0$, indicating that the system has approximate longitudinal boost invariance. More detailed considerations of identified particle spectra show that boost invariance holds over a rather smaller region $\Delta y \sim 1$ [15].

The interpretation of the scaling of the multiplicity at mid-rapidity as a function of N_{part} appears still ambiguous. A simple model by Kharzeev and Nardi (KN) explains the dependence in a two-component approach differentiating between soft processes scaling with N_{part} and hard processes scaling with N_{bin} [16]. When fit to the data as shown in Fig. 4 by the PHOBOS collaboration the model allows one to extract the fraction of particles

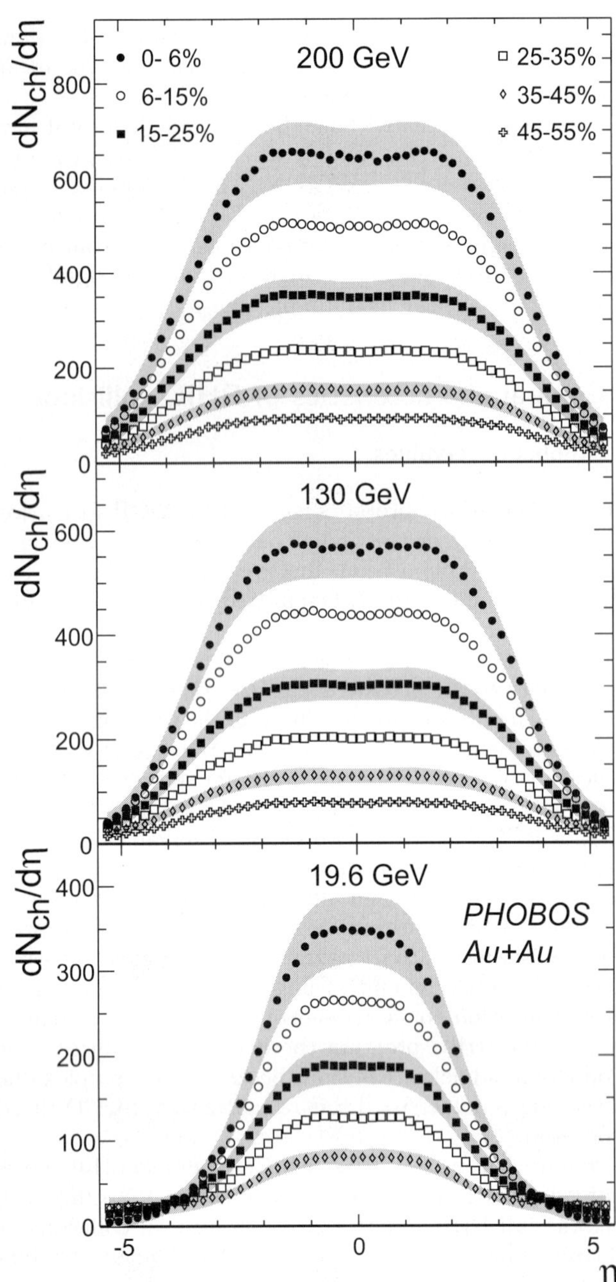

Fig. 3. $dN_{ch}/d\eta$ for various collision centralities over the full RHIC phase space [14]

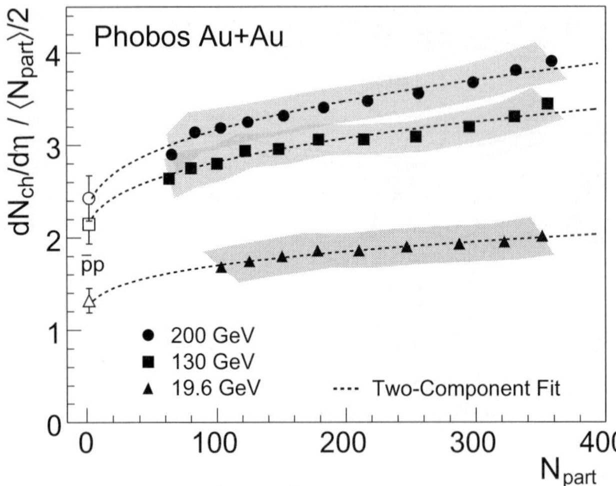

Fig. 4. Charged particles per participant pair as a function of number of participants in $\sqrt{s_{NN}} = 19.6, 130$ and 200 GeV Au+Au collisions measured by PHOBOS. The curves are two-component fits described in the text and in [16]

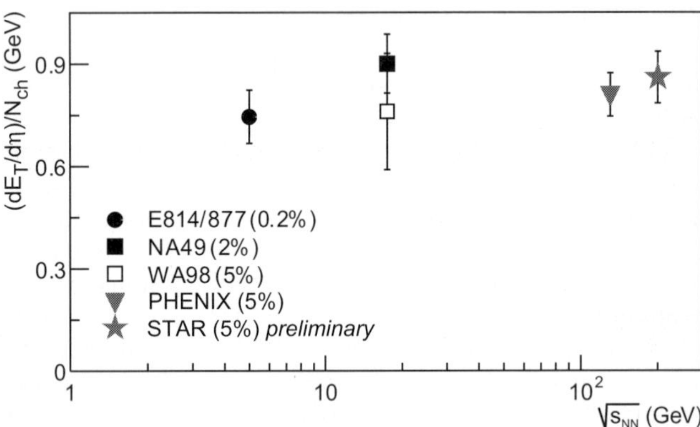

Fig. 5. $dE_T/d\eta|_{\eta=0}/dN_{ch}/d\eta|_{\eta=0}$ versus $\sqrt{s_{NN}}$ for 5% most central events at AGS, SPS, and RHIC ([20,21])

produced from hard processes and one obtains values of 36% for $\sqrt{s_{NN}} = 130$ GeV, and 45% for 200 GeV, respectively [17]. A second class of calculations is based on parton saturation [18,19]; since the parton densities in the initial stage of the collision can be related to the density in the final state a parametrized dependence of the saturation scale Q_s on $\sqrt{s}$ and impact parameter allows one to predict $dN_{ch}/d\eta$. However, the predictions from these and related models have been found to be almost indistinguishable when applied to RHIC data, especially because of the large experimental uncertainties in the calculation of N_{part} for very peripheral collisions, the region where

the differences between the various models become more apparent. An exception are models based on final state saturation who significantly overpredict the yield at low N_{part}. The systematic uncertainties in N_{part} for peripheral events can only be reduced when data from collisions of light ions, A< 100, become available.

Another important observable for characterizing the global properties of bulk matter is the transverse energy E_T. This was studied in detail by the PHENIX collaboration for $\sqrt{s_{NN}} = 130$ and PHENIX and STAR for 200 GeV [20,21] in the mid-rapidity region. They find that $dE_T/d\eta$ and $dN_{ch}/d\eta$ increase with N_{part} in a very similar fashion resulting in an almost constant ratio $\langle E_T \rangle / \langle N_{ch} \rangle \sim 0.9$ GeV. This holds for $\sqrt{s_{NN}} = 130$ *and* 200 GeV. Even more surprising is the fact that studies from Au+Au collisions at $\sqrt{s_{NN}} = 4.8$ and Pb+Pb collisions at 17.2 GeV yield very similar values, suggesting that the increased energy put into the system results solely in an increased particle production leaving the average energy per particle almost constant (see Fig. 5).

Bjorken studied boost invariant hydrodynamics [22] and derived a useful pocket formula for the energy density achieved in the central region:

$$\epsilon_{\text{Bj}} = \frac{1}{\pi R_A{}^2 \tau} \frac{dE_T}{dy}, \tag{1}$$

where E_T is the transverse energy, R_A is the nuclear radius, and τ is the formation time, typically taken as ~ 1 fm/c, after which the hydrodynamic description is valid. From the measured $dE_T/d\eta$ for the 2% most central Au+Au collisions at 200 GeV PHENIX estimated the Bjorken energy density to be $\epsilon_{\text{Bj}} \approx 5.5$ GeV/fm^3, assuming a conservative formation time of $\tau = 1$ fm/c. Similar studies at SPS in Pb+Pb collisions at $\sqrt{s_{\text{NN}}} = 17.2$ GeV give $\epsilon_{\text{Bj}} \approx 3.2$ GeV/fm^3 [23]. These values represent of course only a lower limit for the initial energy density since the longitudinal expansion of the system reduces the transverse energy considerably. Recent lattice results on QCD thermodynamics estimate the critical energy density to be $\epsilon \approx 0.70 \pm 0.35$ GeV/fm^3 [24], a value significantly surpassed already at SPS.

3.2 Thermodynamical behavior

One of the most important issues in the physics of heavy ion collisions is the question if, and if so at what stage, the produced system thermalizes and to what extent a thermal description is appropriate for the evolving system. The use of thermodynamic concepts to describe multi-particle production has a long history beginning with Hagedorn in the early 1960's [25]. The concept of a *temperature* applies, strictly speaking, only to systems in at least local thermal equilibrium. The measured hadron spectra contain two pieces of information: *(i)* their yields and ratios providing the chemical composition of the fireball at the chemical freeze-out point and *(ii)* their transverse momentum spectra that provide information about thermalization of the momentum distributions and collective flow. It is obvious that the observed single particle spectra do not reflect earlier conditions, *i.e.* the hot and dense deconfined phase, where chemical and thermal equilibrium may have been established, since rescattering erases most traces from the dense phase. Only those effects which are accumulative during the expansion, such as flow, remain.

As argued above, the relative population of the stable hadrons is fixed at chemical freezeout. Equilibrium at that point should be evident in the measured population ratios.The assumption of a locally thermalized source in chemical equilibrium can be tested by using statistical thermal models to describe the ratios of various emitted particles. This yields a baryon chemical potential μ_B, a strangeness saturation factor γ_s, and the temperature T_{ch} at chemical freeze-out. Because of the absence of any dynamic assumptions many details can never be fully absorbed by these models. Discrepancies between model and data up to 30% should be considered inside the systematic uncertainty of the thermal model approach [26]. So far these models are remarkably successful in describing

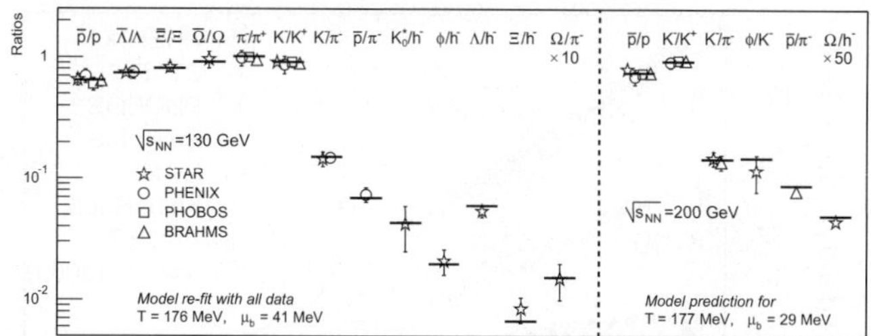

Fig. 6. Left panel: comparison between RHIC experimental particle ratios for $\sqrt{s_{\text{NN}}} = 130$ GeV and statistical model calculations with $T_{\text{ch}} = 176$ MeV and $\mu_B = 41$ MeV (from [29]). Right panel: comparison between RHIC ratios at $\sqrt{s_{\text{NN}}} = 200$ GeV and prediction discussed in [29]

particle ratios at SPS [27,28] and now also at RHIC [29, 30]. This observation, together with the large collective flow (radial and elliptic) measured at RHIC, is generally considered a strong hint that chemical equilibrium is indeed reached. The wide reaching implications of thermal statistical models and the models themselves are still the subject of intense debate.

While the net-baryon chemical potential at chemical freeze-out is essentially determined by the baryon to antibaryon ratios ($\bar{p}/p$, $\bar{\Lambda}/\Lambda$ etc.), the non-identical particle ratios are the "thermometer" of the thermal statistical models. Figure 6 shows the comparison of measured particle ratios with *(i)* a statistical model fit for 130 GeV and *(ii)* model-related predictions for 200 GeV Au+Au collisions [29]. The predictions match well with the current results and indicate no significant change in T_{ch} but a drop in μ_B from ~ 41 MeV at $\sqrt{s_{\text{NN}}} = 130$ GeV to 29 MeV at 200 GeV. The chemical freeze-out temperature is naturally limited by the confinement phase-transition temperature assumed to be around 175 MeV, although T_{ch} is actually not constrained in thermal model fits.

Transverse momentum spectra of identified particles reflect the system at kinetic freeze-out and allow us to extract information from the latest stage of the evolution when the system was still thermally coupled and governed by elastic interactions among its constituents. The measured inverse slope parameter is determined by two components: the actual temperature at the freeze-out and the transverse flow component, *i.e.*, collective radial expansion of the system. A thermalized system without any flow would yield plain exponential spectra in transverse mass $m_T = (p_T{}^2 + m^2)^{1/2}$, while the flow component results in a bowed shape; the higher the mass the more bowed the spectra.

The average inverse slope T of the spectra can, in simple terms, be approximated as $T = T_{\text{fo}} + m\langle\beta_T\rangle^2$ where β_T is the transverse flow velocity and T_{fo} the temperature of the system at kinetic freeze-out. This ansatz, however, has the disadvantage that attempts to extract T_{fo} and β_T are strongly dependent on the range in which the slopes were determined. Of even greater concern is the assumption of a fixed flow velocity which oversimplifies the problem considerably. To overcome these problems full hydrodynamical model calculation are necessary that, at RHIC

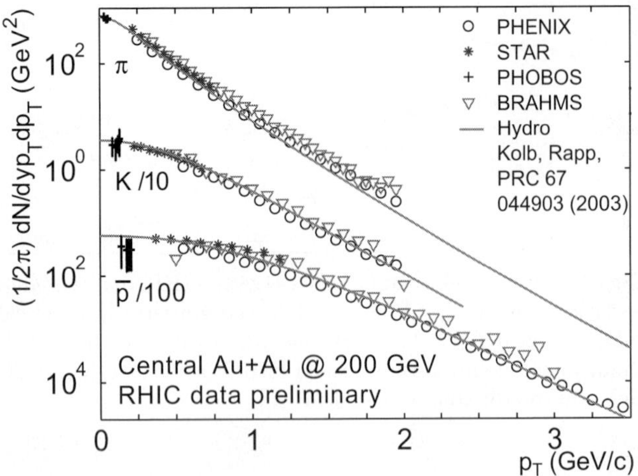

Fig. 7. Compilation of preliminary transverse momentum spectra of π^-, K^-, and $\overline{p}$ for 200 GeV central Au+Au collisions from all RHIC experiments. The curves are calculation from a hydrodynamical model calculation [33]

energies, describe the observed spectra very well [33] (see Fig. 7). To overcome the complexity of adjusting the initial energy density and the equation of state in a full hydrodynamical model calculation, many studies now use the so-called 'blastwave' parametrization [34]. The parametrization contains the essential hydrodynamic features but allows to extract flow velocity and freeze-out temperature by a simple fit to the spectra. Results are similar to those obtained from hydrodynamic models. While the so obtained freeze-out parameters at CERN/SPS energies yield $T_{\mathrm{fo}} = 122 - 127$ MeV and $\langle \beta_T \rangle = 0.48 \pm 0.01$ [32], fits to RHIC data give a slightly smaller freeze-out temperature $T_{\mathrm{fo}} \sim 110$ MeV but a higher flow value $\langle \beta_T \rangle = 0.55 - 0.6$ due to the higher pressure in the system [31].

3.3 Summary of soft physics

I cannot hope to address here all the soft physics observables that have been studied at RHIC. To conclude this section I will summarize a few of the main results, more detailed discussions can be found in ([13,15]).

- Low net-baryon density: antibaryon/baryon ratios at midrapidity are ∼0.6-1.0. The system is close to net-baryon free, similar to the early universe, but not precisely so. Finite baryon number is transported $\Delta y \sim$ 5.5 rapidity units from the beam.
- There are strong indications that hadronic chemical equilibrium has been achieved at a temperature T_{ch} near the lattice critical temperature T_C.
- Hydrodynamic calculations describe well the transverse momentum spectra and collective flow.. The mass dependence of these observables, which is a sensitive test of the hydrodynamic picture, is described in detail.
- The energy density achieved early in the collision is estimated to be ~ 5.5 GeV/fm^3, exceeding the critical

energy density derived from lattice QCD calculations by more than a factor of 5.
- Identical two-particle correlations (Hanbury Brown-Twiss correlations) are sensitive to the space-time evolution of the source. The extracted source radii and duration of freezeout show no significant increase relative to lower energy nuclear collisions. Such correlations measure only a piece of a dynamically expanding source and these results may indicate a very explosive expansion resulting from high early pressure.

4 Hard scattering at RHIC

The increase in $\sqrt{s_{\mathrm{NN}}}$ for nuclear collisions at RHIC relative to fixed target experiments opens up new channels to probe the dense medium generated in the collision. Jets with $E_T \sim 40$ GeV and higher are produced in sufficient numbers to provide robust observables. The measurement of jets in nuclear collisions poses a special problem, however: while the presence of a hard scattering in a nuclear collision can be detected (though in a biased way) via high p_T leading hadrons, the huge soft multiplicities contaminate any finite jet cone, spoiling the jet energy measurement. At sufficiently high E_T this effect may be minor (e.g. nuclear collisions at the LHC), but for the jet E_T currently accessible at RHIC it is fatal. We therefore restrict our considerations to leading particles and their correlations. We show below that hadrons with $p_T > 4$ GeV/c are produced dominantly from jet fragmentation, even in the most central Au+Au collisions.

Twenty years ago Bjorken [35] proposed that hard scattered partons in nuclear collisions could provide a sensitive probe of the surrounding medium. The energy loss dE/dx due to elastic scattering of the partons in a Quark Gluon Plasma depends on the temperature as T^2_{plasma} and results in a suppression of the observed rate of jets or their leading hadrons at fixed p_T. Later work showed that dE/dx from elastic scattering is negligible but that radiative energy loss in dense matter could be considerable [36, 37,38]. The energy loss is directly sensitive to the gluon density of the medium, ρ_{glue}. While not a direct signature of deconfinement, measurement of ρ_{glue} that is substantially larger than in cold nuclear matter [39] is incompatible with the presence of a hadronic medium, thus large energy loss serves as an indirect signature of deconfinement.

There are currently several sets of measurements which address the question of partonic energy loss in dense matter, of which two I will discuss in turn: suppression of inclusive spectra and correlations of high-p_T hadron pairs.

4.1 High-p_T: Suppression of inclusive spectra

In p+p collisions at RHIC energies, hadrons with $p_T \sim$ 4 GeV/c typically carry 75% of the energy of their parent jet, leading to the possibility that partonic energy loss in Au+Au collisions is reflected in the suppression of leading hadrons [38]. Hadron suppression is measured via the

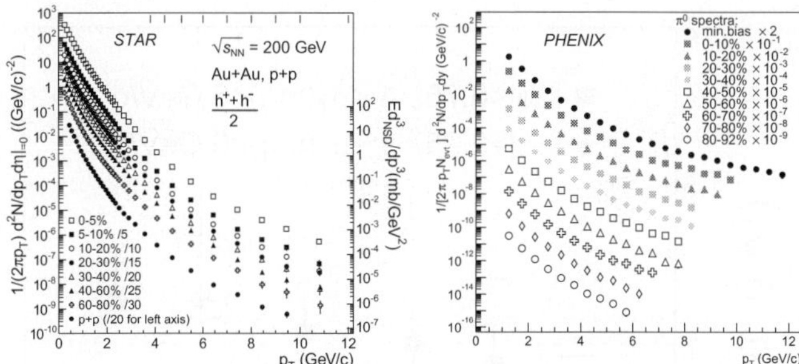

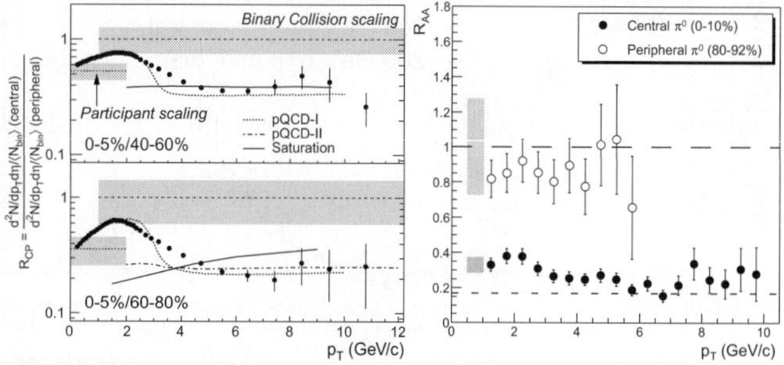

Fig. 8. Inclusive charged hadron invariant distributions from STAR (*left*) and for neutral pions from PHENIX (*right*) for various event centralities [40, 41]

nuclear modification factor:

$$R_{AA}(p_T) = \frac{d\sigma_{AA}/dydp_T{}^2}{\langle N_{\mathrm{bin}}\rangle d\sigma_{\mathrm{pp}}/dydp_T{}^2} \qquad (2)$$

where $\sigma_{\mathrm{pp}}/dyd^2p_T$ is the inclusive cross section measured in elementary nucleon-nucleon collisions and $\langle N_{\mathrm{bin}}\rangle$ accounts for the geometric scaling from the elementary to nuclear collision. $R_{AA}(p_T)$ is normalized to be unity if Au+Au collisions are an incoherent superposition of p+p collisions. In addition to partonic energy loss, $R_{AA}(p_T)$ may be altered by nuclear effects such as gluon shadowing, which will reduce $R_{AA}(p_T)$, and the Cronin effect (multiple soft scattering in the initial state), which will increase it. These effects must be disentangled using measurements in simpler systems (in particular p+Au). Similar to $R_{AA}(p_T)$ is $R_{CP}(p_T)$ which describes the $\langle N_{\mathrm{bin}}\rangle$-normalized ratio of central and peripheral Au+Au spectra. The practical advantage of $R_{CP}(p_T)$ is that it extends to higher p_T than $R_{AA}(p_T)$, with significantly smaller uncertainties.

First indication of a strong suppression of high-p_T hadrons in nuclear collisions was reported from the 130 GeV data by both PHENIX and STAR [42]. The 200 GeV data have much higher statistics and both collaborations have now pushed this study well into the perturbative regime. Figure 8 shows on the left the inclusive charged hadron spectra as a function of centrality for p+p and Au+Au collisions at 200 GeV as measured by STAR [40], extending to $p_T = 12$ GeV. Shown on the right is the π^0 spectra from PHENIX [41]. Figure 9 shows $R_{CP}(p_T)$ and $R_{AA}(p_T)$ for these data. Suppression factors of 4-5 are observed in central collisions for both π^0s and charged hadrons, with weak, if any, dependence on p_T above $p_T \sim 5$ GeV/c. At asymptotically high jet E_T the relative energy loss should be negligible and $R_{AA}(p_T)$ and $R_{CP}(p_T)$ should return to unity. No evidence of this limiting high energy behavior is seen.

$R_{CP}(p_T)$ for $p_T < 6$ GeV/c is similar to measurements at $\sqrt{s_{\mathrm{NN}}} = 130$ GeV [42], but is now seen to be approximately constant for $5 < p_T < 12$ GeV/c. While different suppression factors have been reported for meson and baryon production in central Au+Au collisions for $2 < p_T < 4$ GeV/c [43], recent measurements indicate that all hadron species attain a similar suppression at the on-

Fig. 9. Nuclear modification factor for Au+Au at $\sqrt{s_{\mathrm{NN}}} = 200$ GeV. Left: $R_{CP}(p_T)$ for charged hadrons for two peripheral centrality bins. Error bars include both statistical and systematic uncertainties [40]. Right: $R_{AA}(p_T)$ for π^0 in central and peripheral collisions. The error bars include all point-to-point experimental errors [41]. In both plots the *shaded bands* represent uncertainties in the normalization that can move all points up or down together

set of the constant suppression region observed here [44]. It is consistent with $\langle N_{\mathrm{part}}\rangle$ scaling at $p_T \sim 4$ GeV/c as reported in [45], but is significantly below $\langle N_{\mathrm{part}}\rangle$ scaling at higher p_T.

The p_T-dependence of the suppression in Fig. 9 is well reproduced for $p_T > 5$ GeV/c by various perturbative QCD calculations [46, 47]. All pQCD models incorporate nuclear shadowing of initial-state parton densities, the Cronin effect [48], and partonic energy loss containing energy loss. Other popular models such as the saturation picture [49] fails for the most peripheral bin. The magnitude of suppression is fit to the central collision data in the pQCD models but is predicted in the saturation calculation. Attenuation of initial jet formation due to multiple nucleon interactions [50] generates an increase in partonic $R_{AA}(p_T)$ for central collisions of a factor ~ 2 in $5 < E_T < 12$ GeV. Though the model does not incorporate fragmentation, a similar p_T-dependence would be expected for high-p_T hadrons, in contrast to observations. Suppression in the final state due to in-medium scattering of fragmentation hadrons also results in a rising $R_{AA}(p_T)$ with increasing p_T due to the dependence of hadron formation time on the total jet energy [51], though detailed comparison of this model to data requires further theoretical development.

4.2 High-p_T: Two-particle correlations

While jet fragmentation is reasonably expected to dominate hadron production at sufficiently high-p_T, it remains an open question whether the current measurements have achieved that limit. A full jet reconstruction in Au+Au events is not possible due to the complex underlying event, but intra-jet correlations amongst high-p_T hadrons are still visible and one can use these to map the transition in p_T from soft to hard physics. The rate of back-to-back di-jets may be especially sensitive to partonic energy loss effects: if one escapes from the surface, its partner has enhanced probability to plow through the bulk matter.

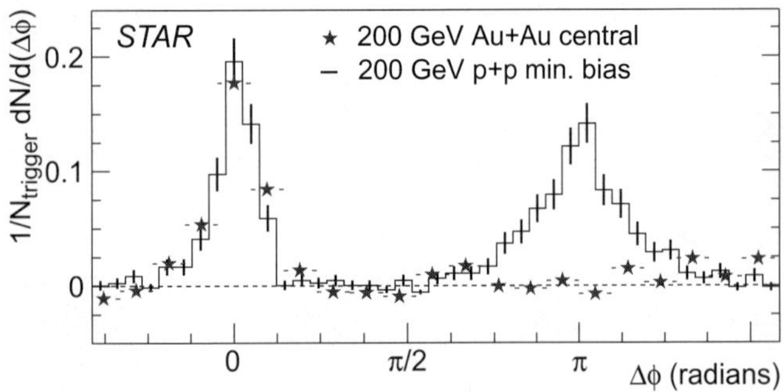

Fig. 10. Background subtracted azimuthal correlations of high-p_T hadron pairs in Au+Au compared to p+p [52]

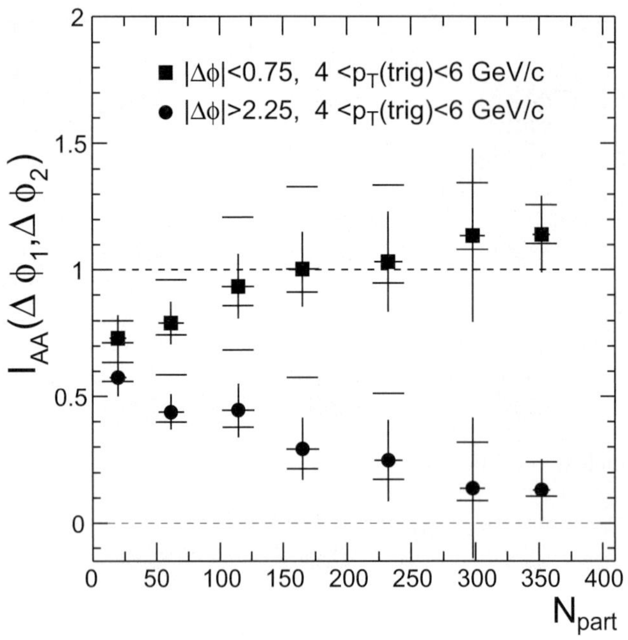

Fig. 11. I_{AA} vs centrality (*large N_{part}=central collisions*) for near-angle (*upper*) and back-to-back (*lower*) angular correlations [52]. *Left:* trigger $p_T > 4$ GeV/c. *Right:* trigger $p_T > 3$ GeV/c

High-p_T hadron pairs with large azimuthal angle difference (back-to-back pairs) are the ideal tool to search for the effect of partonic energy loss on the di-jet (di-hadron) production rate [52]. Figure 10 shows the background subtracted azimuthal angular distribution between pairs for the most central Au+Au collisions at 200 GeV and, as a reference, for p+p collisions at the same energy. The background contribution of elliptic flow, *i.e.*, the azimuthal anisotropy in the particle emission in non-central events, and other non-jet effects are accounted for by means described in [52]. At all centralities, the near-angle peak is similar to the one observed in p+p, as is the large angle peak for peripheral collisions. Thus, back-to-back hadron pair production in peripheral nuclear collisions is well described by an incoherent superposition of 'background' sources (elliptic flow) and jet-like correlations measured in p+p. However, the large angle peak in the most central collisions is absent as depicted in Fig. 10. We are therefore led to a striking observation: back-to-back jet production is strongly suppressed in the most central Au+Au collisions.

The full centrality dependence of this effect can be quantified by plotting I_{AA}, the ratio of the integrated correlation peaks in Au+Au over those in pp, as a function of event centrality. Figure 11 shows I_{AA} as a function of N_{part} for trigger particles with $4 < p_T < 6$ GeV/c. The near-angle peak strength ($|\Delta\phi| < 0.75$) is near or above unity for all centralities and both thresholds: there is no suppression in the near-angle correlation. In contrast, the back-to-back correlation strength ($|\Delta\phi| > 2.25$) in Au+Au relative to p+p decreases smoothly from peripheral to central collisions. For the most central collisions the strength is consistent with zero, which would correspond to a complete suppression of the away-side hadron correlation rate above the threshold.

Various nuclear effects may contribute to the features seen in Fig. 11. Initial state multiple scattering might generate both the increase of the near-angle and suppression of the back-to-back correlation strength for more central collisions. The suppression in the most peripheral bin for Au+Au relative to p+p for both the near-angle and back-to-back peaks could be due to nuclear shadowing, or possibly the interplay between multiple scattering and absorption in matter. These issues will be clarified once the analysis of the data from the d+Au run at RHIC is completed.

4.3 Summary of high-p_T physics

This section reveals the following facts about high-p_T hadron production in Au+Au collisions at RHIC energies:

- Inclusive hadron production is suppressed in central collisions by a factor 4-5, with no strong p_T dependence within $5 < p_T < 12$ GeV/c.
- Near-angle two particle correlations show clear jet-like correlations for $p_T > 4$ GeV/c.
- Back-to-back two particle correlations show a striking suppression in central collisions.

All of these phenomena suggest a picture in which the system generated in nuclear collisions at RHIC is largely opaque to high energy partons and only those jets produced on the periphery of the reaction zone and heading outwards survive and are observed. This scenario naturally generates the suppression of the inclusive spectra as well as the strong suppression of back-to-back pairs. However, this picture as presented is only qualitative, and it remains to be shown whether a surface emission model can simultaneously describe all observed phenomena.

5 Conclusions and outlook

RHIC has had three successful data-taking periods. It has met its design luminosity goal for Au+Au, and has operated as the world's first polarized proton collider. The runs in 2003 allow the study of asymmetric (d+Au) collisions for the first time in a collider. A long Au+Au run at top energy will occur in 2004, possibly with an energy scan at lower integrated luminosity.

An extensive array of measurements in the soft physics sector indicate the production of an equilibrated system at high temperature and pressure which blows apart rapidly. In the high-p_T sector, striking signals have been observed that suggest strong partonic energy loss resulting from a system at high energy density at early time, though there remain important puzzles and open questions.

I have not discussed the study of heavy quark production, which is in its infancy at RHIC. Charmonium suppression, a promising signature of deconfinement [9]. PHENIX has measured the inclusive electron spectrum, due primarily to charm [53], and recently reported the first measurement of J/Ψ at RHIC [54]. This physics will be a main focus of future runs.

References

1. J.C. Collins and M.J. Perry: Phys. Rev. Lett. **34**, 1353 (1975)
2. See for example K. Rajagopal these proceedings
3. K. Rajagopal and F. Wilczek: hep-ph/0011333
4. K. Kanaya: hep-ph/0209116
5. F. Karsch: Nucl. Phys. A **698**, 199c (2002)
6. Z. Fodor: hep-lat/0209191
7. D. Boyanovsky: hep-ph/0102120
8. N.K. Glendenning and F. Weber: astro-ph/0003426
9. J.W. Harris and B. Müller: Annu. Rev. Nucl. Part. Sci. B **46**, 71 (1966)
10. T. Roser: Nucl. Phys. A **698**, 23c (2002)
11. The Relativistic Heavy Ion Collider and Experiments: Nucl. Inst. Meth. A **499**, Issues 2-3 (2003)
12. Proceedings of Quark Matter 2001, Nucl. Phys. A **698**, (2002)
13. Proceedings of Quark Matter 2002: Nucl. Phys. A **715** (2003)
14. B.B. Back et al.: nucl-ex/0210015
15. T.S. Ullrich: Nucl. Phys. A **715**, 399c (2003)
16. D. Kharzeev and M. Nardi: Phys. Lett. B **507**, 121 (2001)
17. B.B. Back et al.: Phys. Rev. C **65**, 061901 (2002)
18. K.J. Eskola, K. Kajantie, and K. Tuominen: Phys. Lett. B **497**, 29 (2001)
19. D. Kharzeev and E. Levin: Phys. Lett. B **523**, 79 (2001)
20. A. Bazilevski: Nucl. Phys. A **715**, 486c (2003)
21. STAR collaboration: private communication
22. J.D. Bjorken: Phys. Rev. D **27**, 140 (1983)
23. T. Alber et al.: Phys. Rev. Lett. **75**, 3814 (1995)
24. F. Karsch: Nucl. Phys. A **698**, 199 (2002)
25. R. Hagedorn: Suppl. A. Nuovo Cimento Vol III, No.2 150 (1965)
26. U. Heinz: Nucl. Phys. A **661**, 140 (1999)
27. P. Braun-Munzinger, I. Heppe, and J. Stachel: Phys. Lett. B **465**, 15 (1999)
28. F. Becattini et al.: Phys. Rev. C **64**, 024901 (2001)
29. P. Braun-Munzinger et al.: Phys. Lett. B **518**, 41-46 (2001)
30. F. Becattini: J. Phys. G **28**, 1553 (2002)
31. G. Van Buren et al.: Nucl. Phys. A **715**, 129c (2003)
32. M. van Leeuwen et al.: Nucl. Phys. A **715**, 161c (2003)
33. P.F. Kolb and R. Rapp: Phys. Rev. C **67**, 044903 (2003)
34. E. Schnedermann, J. Sollfrank, and U. Heinz: Phys. Rev. C **48**, 2462 (1993)
35. J.D. Bjorken: FERMILAB-Pub-82/59-THY
36. M. Gyulassy and M. Plümer: Phys. Lett. B **432**, 121 (1990); R. Baier et al.: Phys. Lett. B **345**, 277 (1995)
37. R. Baier, D. Schiff, and B.G. Zakharov: Annu. Rev. Nucl. Part. Sci. B **50**, 37 (2000)
38. X.N. Wang and M. Gyulassy: Phys. Rev. Lett. **68**, 1480 (1992); X.N. Wang: Phys. Rev. C **58**, 2321 (1998)
39. E. Wang and X.N. Wang: Phys. Rev. Lett. **89**, 162301 (2002); F. Arleo: Phys. Lett. B **532**, 231 (2002)
40. J. Adams et al.: nucl-ex/0305015
41. S.S. Adler et al.: nucl-ex/0304022
42. K. Adcox et al.: Phys. Rev. Lett. **88**, 022301 (2002); C. Adler et al.: Phys. Rev. Lett. **89**, 202301 (2002)
43. J. Velkovska: Proceedings of Strange Quark Matter 2003, J. Phys. G. to be published
44. P. Sorenson et al.: nucl-ex/0305008
45. B.B. Back et al.: nucl-ex/0302015
46. X.N. Wang: nucl-th/0305010; private communication. Calculations use model parameters $\mu_0 = 2.0$ GeV and $\epsilon_0 = 2.04$ GeV/fm
47. I. Vitev and M. Gyulassy: Phys. Rev. Lett. **89**, 252301 (2002)
48. D. Antreasyan et al.: Phys. Rev. D **19**, 764 (1979); P.B. Straub et al.: Phys. Rev. Lett. **68**, 452 (1992)
49. D. Kharzeev, E. Levin, and L. McLerran: Phys. Lett. B **561**, 93 (2003); D. Kharzeev: private communication
50. R. Lietava, J. Pisut, N. Pisutova, and B. Tomasik: Eur. Phys. J. C **28**, 119 (2003)
51. K. Gallmeister, C. Greiner, and Z. Xu: Phys. Rev. C **67**, 044905 (2003)
52. C. Adler et al.: Phys. Rev. Lett. **90**, 082302 (2003)
53. K. Adcox et al.: Phys. Rev. Lett. **88**, 192302 (2002)
54. J. Nagle et al.: nucl-ex/0209015

Eur Phys J A (2004) **19**, s01, 223–228
Digital Object Identifier (DOI) 10.1140/epjad/s2004-03-037-5

EPJ A direct
electronic only

ALICE

E. Vercellin, for the ALICE collaboration

Dipartimento di Fisica Sperimentale dell'Universitá di Torino and INFN Torino, Italy

Received: 30 Oct 2003 / Accepted: 14 Nov 2003 /
Published Online: 6 Feb 2004 – © Società Italiana di Fisica / Springer-Verlag 2004

Abstract. ALICE, the dedicated heavy-ion experiment at LHC, will be presented in this paper. The physics goals of the experiment will be briefly discussed, with emphasys on those items which are new or more relevant with respect to RHIC and SPS energies. The general features and the performance of the ALICE detector will be outlined, together with the planned data taking program.

PACS. 25.75.-q – 25.75.Nq

1 Introduction

Ultra-relativistic heavy-ion experiments are performed to produce and study the Quark-Gluon Plasma (QGP). According to lattice-QCD calculations, this new state of matter is expected to exist at high temperatures and/or high baryon densities. This prediction has triggered an extensive experimental activity, which has been carried out at different accelerators, able to deliver higher and higher bombarding energies. The extremely interesting results obtained by the SPS heavy-ion experiments [1], corroborated by the first RHIC data [2], represent a strong experimental motivation to set up a heavy-ion programme at the forthcoming CERN Large Hadron Collider (LHC). In this frame, the leading role will be played by ALICE [3],[4],[5], the LHC experiment fully devoted to (and specifically designed for) the study of heavy-ion collisions. While at previous accelerator facilities (AGS, SPS and RHIC) different Physics observables relevant for QGP studies were addressed by different experiments, ALICE is conceived as a general–purpose detector and will address most of the phenomena related to the QGP formation at LHC energies. If, on one hand, this all-in-one approach makes the experimental task very challenging, on the other it is also clear that the LHC energy regime will offer unprecedented conditions for the formation and the study of the Quark-Gluon Plasma.

The paper is organized as follows. The main Physics issues which are new or more relevant at LHC energies with respect to the SPS and RHIC will be outlined in Sect. 2. A description of the ALICE detector will be given in Sect. 3 and its performance will be outlined in Sect. 4. Finally, the data taking scenario will be overviewed in Sect. 5.

2 Heavy-ion physics at LHC

The advent of the LHC will bring heavy-ion Physics into a completely new energy region, previously accessible only in the interactions of the highest energy cosmic rays. In fact, the LHC will deliver Pb-ions at a center of mass energy of 5.5 TeV per nucleon pair, which represents a jump of more than one order (two orders) of magnitude with respect to the RHIC (SPS) energy. The higher colliding energy will lead to ideal conditions for the formation and the study of the QGP. First of all, the higher energy will improve by large factors all parameters (such as energy density, size and lifetime of the system) relevant to QGP formation. Secondly, the initial temperature will largely exceed the calculated critical temperature for QGP formation, therefore allowing the study of the QGP in its asymptotic ideal gas form. Finally, the net baryon density in the central region will essentially vanish: this will make the experimental conditions close to the ones of latice QCD calculations as well as to those of the early universe.

Another relevant point is that the increase of the energy opens the possibility to exploit a wider set of observables as compared to the previous accelerators, leading to a better and more comprehensive understanding of the properties of the system. In this respect, an outstanding example is represented by the hard probes, which are sensitive to the nature of the medium in its early stages. In fact, since high-p_t partons are expected to lose a significant fraction of their energy when crossing a deconfined medium [6], a suppression of the hadronic activity at high transverse momenta (the so-called jet quenching) should be observed in case of QGP formation [7]. This powerful QGP signature can be extensively studied with high statistics at LHC since, at this energy, the cross section will be dominated by semi-hard and hard processes: about 30 (3×10^{-3}) partons with transverse energy larger than 10 GeV (100 GeV) are in fact expected per central Pb-Pb collision [8].

The LHC heavy-ion program will allow to probe a novel range of Bjorken-x values, down to about 10^{-5} [8].

This region looks extremely interesting, since at small x the gluon density is expected to be close to saturation (gluon shadowing), leading to modifications of the particle production rates.

The traditional study of heavy quarkonia production and suppression will be improved with respect to previous accelerators as well. At LHC energies, in fact, the whole spectrum of heavy quarkonia (J/ψ and Υ families) will be measured and, for the first time, energy densities high enough to melt the Υ (1S) could be reached. Together with heavy quarkonia, also open heavy flavours will be abundantly produced (about 100 $c\text{-}\bar{c}$ and few $b\text{-}\bar{b}$ pairs per central Pb-Pb collision), making a detailed study of D and B meson production feasible [9].

The temperature of the system should be high enough to allow the detection of thermal photons and dileptons.

This kind of measurement would provide direct information on the temperature of the system.

The very high multiplicity (of the order of 2500 to 6000 charged particles per unit of rapidity, according to extrapolations based on RHIC data) will allow the measurement of several observables on a event-by-event basis. The single event analysis of multiplicity, particle composition and spectra and HBT parameter of the system will be used to identify non-statistical fluctuation related to critical phenomena.

3 Overview of the ALICE detector

ALICE is conceived as a general-purpose detector, in which the most part of hadrons, leptons and photons produced in the interaction can be measured and identified. The design of the detector was driven by two main considerations. The first one is the need to cope with the high multiplicities expected in Pb-Pb collisions: ALICE is designed to operate at multiplicities up to 8000 charged particles per unit of rapidity, i.e. a number which represents a reasonable safety margin with respect to the current expectations, previously quoted in Sect. 2. The second one is that, because of ion losses due to electromagnetic processes, the maximum luminosity with Pb-beams will be limited to 10^{27} cm^{-2}s^{-1}. Given this luminosity and an inelastic cross section of 8 b, the maximum event rate for Pb-Pb collisions will $\sim$ 8000 minimum bias collisions per second. Such a low interaction rate, coupled to the high multiplicity, has led to the choice of slow but high granularity detectors, like the time projection chamber (TPC) and the silicon drift detectors (SDD).

The layout of the ALICE experiment is shown in Fig. 1. The main component of the ALICE detector is the central barrel, where hadrons, photons and electrons are measured in the central rapidity region (-0.9$\leq \eta \leq$0.9). It consists of a complex system relying on high-granularity, yet relatively slow drift detectors, on a weak solenoidal magnetic field (B$\leq$ 0.5 T) and on detectors devoted to particle identification. The detection of muons is performed by a dedicated forward spectrometer (2.5$\leq \eta \leq$4), based on a large warm dipole equipped with tracking and trigger chambers. The set-up is completed by a set of small detectors, located at large rapidities, devoted to event characterization. The main features of these three components are briefly summarized in the following.

3.1 The central barrel

The primary purpose of the central detectors is to provide safe and robust track finding. The main tracking detectors are the Time Projection Chamber (TPC) and the Inner Tracking System (ITS). As it will be discussed below and in Sect. 4.1, the tracking performance is further improved when the information from these two detectors is combined with the one from the Transition Radiation Detector.

The ALICE TPC [10] (88 m^3 in volume) is the largest TPC ever built. Together with track finding and momentum measurement, it is design to provide particle identification via dE/dx. It has a cylindrical shape, with an inner radius of 90 cm (given by the maximum acceptable hit density of 0.1 cm^{-2}) and an outer radius of 250 cm. The latter is determined by the track length needed to achieve a dE/dx resolution better than 10%. To optimize the double-track resolution, the detector is operated with a 90/10 Ne/CO$_2$ gas mixture. The total number of channels is 570,000.

The role of the Inner Tracking System [11] is to provide secondary vertex reconstruction for hyperon and charmed meson decays, tracking and identification of low-p_t particles and to improve the momentum resolution for high momentum particles crossing the TPC. It consists of six cylindrical layers, located at radii ranging from 4 cm to 44 cm. Because of the particle density and to achieve an impact parameter resolution better than 100 μm, silicon pixel detectors have been chosen for the two innermost layers and silicon drift for the following two. The two outer layers consist of double-sided micro-strip silicon detectors. Four layers will have analog read-out to perform particle identification via dE/dx measurement in the $1/\beta^2$ region, in such a way to allow the use of the ITS as a standalone spectrometer for low-p_t particles.

Particle identification in the ITS and in the TPC is performed via dE/dx measurements and is therefore restricted to relatively low transverse momenta. PID is extended to higher transverse momenta thanks to dedicated detectors, specially developed for this purpose. These are the Time Of Flight detector (TOF), the Transition Radiation Detector (TRD) and the ring imaging Cherenkov detector for High-Momentum PID (HMPID).

The TOF system [12] is based on Multigap Resistive Plate Chambers (MRCP): a new-concept detector with excellent time resolution, much better than 100 ps. The multigap RPC consists of a stack of resistive plates which are kept few hundred microns apart each other in such a way to obtain a series of gas gaps. The voltage between the two external plates is fixed, while the intermediate plates take the correct voltage due to electrostatics. The high resistivity of the plates makes them "transparent" for the avalanche signals, so that the signal induced on the

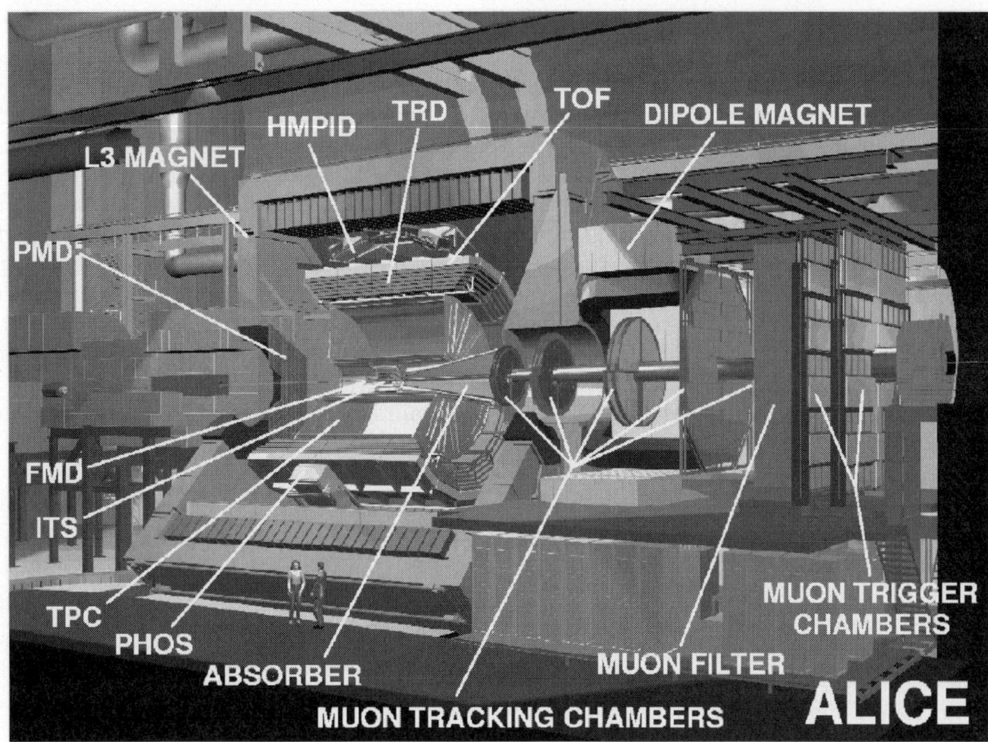

Fig. 1. ALICE layout

external plates (which is read-out by means of strips or pads) corresponds to the "analog sum" of the avalanches in all the gaps. The ALICE MRPC design consists of a double stack of 2×5 gaps and the plate material is low-cost glass. Several tests have shown that resolutions of the order of 50-60 ps are achieved by this detector. The ALICE TOF system will consist of a MRPC cylindrical layer with radius equal to 3.7 m and a total surface of 140 m^2; the total number of channels is 160,000.

The identification of particles with higher momentum will be provided (in a restricted area) by the HMPID [13], a proximity-focus RICH detector placed at a distance of about 4.5 m from the beam axis. The HMPID consists of seven modules, each 1.5×1.5 m^2, for a total of over 160,000 readout channels.

The Transition Radiation Detector (TRD) [14] will play the main role in providing electron identification. It consists of six layers of radiator followed by Time Expansion Chambers filled with $Xenon/CO_2$. This detector provides electron identification for $p_t \geq 1$ GeV/c (the e/π rejection power is about 100 at $p_t \geq 3$ GeV/c) and electron trigger for $p_t \geq 3$ GeV/c. The TRD is operated in conjuction with the ITS and the TPC for precise momentum measurements. In this way it is possible to achieve the invariant mass resolutions (similar to those of the forward muon spectrometer, see Sect. 3.2) needed for the identification of heavy vector mesons in the e^+e^- decay channel. There is indeed another synergy between the TRD and the main tracking detectors (ITS and TPC). Although the TRD chief goal is electron identification, its excellent localizing properties (about 500 μm in rϕ) can be exploited to increase the tracking lever arm, hence obtain-

ing a significant improvement in momentum resolution, in particular at high momenta. In addition, thanks to its fast tracking capabilities, it can be used to trigger on high p_t electrons and hadrons. The latter option is essential for selecting jet leading partiles. The detector covers the whole ALICE central barrel; its radial position is about 3 m, the total surface is 800 m^2 and the total number of electronics channels is about one million.

Prompt photons, π^0's and η's are measured in the PHOS [15], a single-arm, high-resolution electromagnetic calorimeter. The accuracy of the single inclusive photon spectra will be determined by the systematic errors on photon-reconstruction efficiency and by the knowledge of the decay background. An acceptable systematic error can be obtained only at low channel occupancy and therefore requires a calorimeter with small Molière radius, R_M, at a large distance (≈ 5 m) from the vertex. The acceptance has been defined such as to keep the statistical errors below the expected systematic ones. The PHOS is located 5 m vertically beneath the interaction region and is built from $PbWO_4$, a crystal with small Molière radius and high light output.

3.2 The muon spectrometer

The complete spectrum of heavy quark vector mesons (as well as the invariant mass continuum) will be measured in the $\mu^+\mu^-$ channel by the ALICE Forward Muon Spectrometer [16]. The choice of the forward geometry allows the detection of low p_t J/ψ's. In fact, muon identification is only feasible for muon momenta above 4 GeV/c

due to the amount of material (absorber) required to reduce the flux of hadrons. Hence, detection of low p_t charmonia is possible only at small angles, where the muons are Lorentz-boosted. The spectrometer is designed to detect muons in the angular interval between 2 and 9 degrees, which corresponds to the pseudorapidity interval $2.5 \leq \eta \leq 4.0$. The spectrometer consists of a front absorber ($\sim 10\ \lambda_{INT}$), to absorb hadrons and photons from the interaction vertex, a large dipole magnet (nominal field 0.7 T, field integral of 3 T·m) a high-granularity tracking system (10 detection planes) and a trigger system (4 detection planes) placed behind a passive muon filter wall ($\sim 10\ \lambda_{INT}$). The tracking and trigger detectors are protected from particles and secondaries produced at large rapidity by a high-density shield placed around the beam pipe all along the spectrometer. The tracking system is based on low-thickness (about $0.03X_0$) cathode pad chambers. These are arranged in five stations (each one made of two chambers): two of them are placed before, one inside and two after the dipole. To keep the occupancy at the 5% level, a high segmentation of the readout pads is needed, leading to a total number of channels of about one million. Thanks to the high space resolution (better than 100 μm in the bending plane) and to the bending power of the magnet, an invariant mass resolution as good as 70 MeV/c^2 (100 MeV/c^2) is achieved in J/ψ (Υ) region, allowing to resolve all the heavy quark vector mesons. The aim of the trigger system is to select events containing a pair of high-p_t muons of opposite sign emitted in heavy quarkonia decays. The p_t selection is made by two trigger stations, each one consisting of two planes of Resistive Plate Chambers (RPCs) operated in streamer mode and equipped with a new-concept dual-threshold front-end chip which allows to reach time resolutions of the order of 1 ns. The trigger electronics (based on programmable circuits working in pipeline) performs the comparison of the coordinates measured in the first and second station and selects the muon p_t in a time of about 700 ns.

3.3 Large rapidity detectors

The ALICE setup is completed by several small detector systems (ZDC, PMD, FMD, T0 and V0) located at small angle (i.e. at large rapidities, both forward and backward). The purpose of these detectors is to measure global event characteristics and to provide trigger signals which enable the first level of event selection.

The impact parameter is measured by a set of Zero Degree Calorimeters (ZDCs) [17], placed at about 116 m from the Interaction Point. At this distance from IP, spectator protons are spatially separated from neutrons by the magnetic elements of the LHC beam line and therefore are detected in two different calorimeters. These are quartz-fiber calorimeters: the shower generated by incident particles in a dense absorber (passive material) produces Cerenkov light when crossing the quartz fibers (active material) interspersed in the absorber.

The Photon Multiplicity Detector (PMD) [18] measures the ratio of photons to charged particles and is also used for determining the reaction plane. It consists of a few m^2 pre-shower gaseous detector which covers the pseudorapidity region $-3.5 \geq \eta \geq -2.5$.

The Forward Multiplicity detector (FMD) measures the pseudorapidity distribution of charged particles over a large pseudorapidity interval ($-5.1 \geq \eta \geq -1.7$ and $1.7 \geq \eta \geq 3.4$) almost complementary to the one covered by the ITS. It consists of a mosaic of silicon pads detector arranged in 5 discs surrounding the beam pipe.

The task of the T0 Cerenkov counter array is to determine the event time with a precisione better than 50 ns, while the one of the V0 scintillator array is to provide a fast measurement of the event multiplicity (to be used in the main interaction trigger) and to locate the interaction vertex.

4 ALICE performance

The performance of the ALICE detector is discussed in this section, with emphasys on the relevant sectors of tracking, particle identification, lepton (both electrons and muons) and jet measurements.

4.1 Tracking

A detailed simulation of the ALICE tracking system has been carried out taking into account the details of the detectors involved. Vertex finding is the first step of the tracking procedure: it is performed by means of the first two layers of the ITS (silicon pixels). This method allows to identify the interaction vertex with a precision of 5 μm in z (i.e. along the beam axis) and of 15 μm in the transverse plane for Pb-Pb collisions (averaged over the centrality). At design multiplicity the tracking efficiency of the TPC is better than 90%, almost independent of p_t down to about 100 MeV/c. At lower multiplicities, the efficiency increases, reaching a value of about 97% below 4000 charged particles per unit of rapidity. When the ALICE tracking detectors (i.e. ITS and TPC) are used in conjunction with the TRD, a momentum resolution better than 1.5% is reached for momenta between 0.2 and 2 GeV/c and it still remains of the order of 12% at 100 GeV/c. Such a good resolution at very high momenta is relevant for jet physics, allowing a detailed study of the fragmentation functions. If the relevance of a good tracking performance at high momenta has been widely discussed in this paper, it is also important to underline here the capabilities of the ALICE tracking system at low momenta, below 100 MeV/c. In this region, the tracking is accomplished by the ITS used as a standalone spectrometer. This feature is relevant to reconstruct the low-p_t particles emitted in hyperon decays. Another relevant role played by the ITS is the identification of the secondary vertex. In this respect, the crucial parameter is the impact parameter resolution: it improves when increasing p_t, reaching a value of 60 μm at 1 GeV/c. This value is adequate for the detection of short-lived B and D mesons ($c\tau$ of the order of 100 to 300 μm).

4.2 Particle identification

Protons, pions and kaons are identified over the whole acceptance of the ALICE barrel by combining the $\mathrm{d}E/\mathrm{d}x$ measurements in the ITS and in the TPC with the precise measurement of the time-of-flight performed by the TOF system. This allows π/K (K/p) discrimination from ~ 100 MeV/c up to ~ 2.5 GeV/c (~ 4 GeV/c), leading to a number of identified particles, for each Pb-Pb event, high enough to perform event-by-event studies. In addition, hadron identification is improved by the HMPID up to higher momenta (K/π to 3 GeV/c and p/K to 5 GeV/c) on a limited area corresponding to about 15% of the acceptance of the ALICE barrel. Thanks also to the good momentum and angular resolution, the decay $\phi \to KK$ can be measured with great precision, with an invariant mass resolution of 2 MeV/c. In addition, the excellent vertexing capability of the ALICE tracking system allow the detection of hyperons (including the rare Ω) and of the hadronic decays of charmed mesons. Besides its relevance from the point of view of Physics (already underline in Sect. 2), the latter represents an oustanding example of the combined power of the PID, tracking and vertexing capabilities of the ALICE detector system. The result is a measurement of neutral D mesons via $K\pi$ decay with a significance of about 37, which allows a direct study of the p_t dependence of charmed meson production down to transverse momenta of the order of 1 GeV/c (significance ~ 12).

4.3 Leptons

Both electrons and muons are measured in ALICE: the former in the central barrel (electron identification is provided by the TRD) and the latter in the forward muon spectrometer. This will allow the detection of the whole spectrum of heavy quark vector mesons (charmonium and bottomonium states) both in the $\mathrm{e}^+\mathrm{e}^-$ and $\mu^+\mu^-$ decay channels.

Although the global performance and the expected statistics (few thousands Υ's and few hundred thousands J/ψ's detected in one month of Pb-Pb data taking), are similar, the two experimental methods are indeed complementary under different points of view: first of all, the rapidity windows covered (midrapidity in the central barrel, forward region in the muon spectrometer). A second complementarity is represented by the different transverse momentum regions in which the detection of charmonium states is performed. Such a difference is due to the different p_t cuts on single leptons applied at the trigger level (3 GeV/c for the TRD, 1 GeV/c for the muon arm). The small value of the p_t cut on muons allow the detection of charmonium states down to $p_t \sim 0$ in the muon channel, while the higher p_t cut on electrons prevents charmonium detection below $p_t \sim 5$ GeV/c in the electron one (we note explicitly that, due to their high mass, bottomonium states are detected down to zero transverse momentum both in the TRD and in the muon arm). Furthermore, while vertexing is not performed in the forward rapidity region where the muons are detected, the vertexing capabilities of the ITS can be coupled to the TRD. In this way, in the electron channel it will be possible to distinguish between primary and secondary J/ψ. While the identification of the former is relevant for QGP studies, the identification of the latter allows a direct measurement of the B meson production cross section. In addition, vertexing can be also used to perform a direct measurement of the D and B meson production via their semileptonic decay, while for the muon channel the yields of these particles can be inferred only by the analysis of inclusive spectra. Finally, $c\bar{c}$ and $b\bar{b}$ cross sections can be measured in ALICE via electron-muon coincidence - the only leptonic channel that gives direct access to correlated $c\bar{c}$ and $b\bar{b}$ pairs. For this measurement, the TRD and the muon spectrometer are operated in conjunction, the electron being identified in the central barrel and the muon in the forward spectrometer. This methods allows the study of open charm and beauty in the rapidity window $1\leq y \leq 3$, therefore bridging the acceptances of the central part and of the muon spectrometer. By combining all the different methods outlined above ALICE can measure charm and beauty production in the interval $-1\leq y \leq 4$.

4.4 Jets

Jets are measured in ALICE by reconstructing charged particles in a cone around a "seed" particle of high p_T. The TRD triggering capability provides the possiblity to record $\approx 10^6$ events per month of Pb-Pb data-taking with jet energy above 100 GeV (10^4 above 200 GeV). Therefore ALICE will collect sufficient statistics to exploit its excellent tracking and PID capability for the study of detailed jet fragmentation functions, which is a very sensitive tool for the study of jet quenching. Using the photon measured in the PHOS as a tag, ALICE can study photon-jet back-to-back pairs. In this way, the average jet energy is defined by the photon measurement, and the fast parton energy loss is directly accessible. The range of jet energies accessible with such a measurement would be greatly enhanced by the proposed large acceptance electromagnetic calorimeter. If available, the EMCal would also significantly improve the measurement of the jet energy, by complementing the measurement of the charged particles performed by the tracking system, and provide an improved jet trigger.

5 Perspectives

The LHC is at present scheduled to start operation in 2007. The acceleration of nuclear beams is a part of the initial program and a pilot run with heavy ion is foreseen in the first LHC year. The LHC will operate according to the scheme already adopted at the SPS, with 4-6 weeks (corresponding to about 10^6 s of useful beam time) dedicated to heavy ions each year. The experimental program with heavy ions at present foresees Pb-beams for two or three years, for one year a run with proton-nucleus collisions and finally a run with lighter ions (probably Ar-Ar) to vary the energy density. The plan for the the following years will be decided on the basis of the results of the initial program. It is important to underline that ALICE also plans to partecipate to the standard p-p runs. In fact, proton-proton data, together with p-A, represent the baseline necessary for understanding the data collected with heavy-ions. As already mentioned, the maximum luminosity with Pb-beams will be limited to 10^{27} cm^{-2}s^{-1}, a value which in any case allows to carry out the Pb-Pb part of the ALICE program. The p-p run (in parallel with the other LHC experiments) will be carried out at full beam intensity but at reduced luminosity, below $3 \cdot 10^{30}$ cm^{-2}s^{-1}.

In order to be ready for data taking at the start of the beam, the ALICE experiment has entered the construction phase: mass production already started or will start before the end of year 2003 for the most part of the ALICE subsystems.

References

1. J. Stachel: these proceedings
2. T. Ullrich: these proceedings and M. Leitch, these proceedings
3. ALICE Technical Proposal: CERN/LHCC/95-71 (1995)
4. The Forward Muon Spectrometer: ALICE TP Addendum, CERN/LHCC/96-32
5. A Transition Radiation Detector for Electron Identification within the ALICE Central Detector, ALICE TP Addendum: CERN/LHCC/99-13
6. R. Baier et al.: Phys. Lett. B **345**, 277 (1995)
7. X.N. Wang, M. Gyulassy, and M. Plumer: Phys. Rev. D **51**, 3436 (1995)
8. B. Alessandro et al.: ALICE-INT 2002-25
9. A. Dainese et al.: ALICE-INT 2002-05
10. ALICE Collaboration: Time Projection Chamber Technical Design Report, CERN/LHCC 2000-01, ALICE TDR 7
11. ALICE Collaboration: Inner Tracking System Technical Design Report, CERN/LHCC 99-12, ALICE TDR 4
12. ALICE Collaboration: Time of Flight Technical Design Report, CERN/LHCC 2000-12, ALICE TDR 8 and CERN/LHCC 2002-16 Addendum to ALICE TDR 8
13. ALICE Collaboration: HMPID Technical Design Report, CERN/LHCC 98-19, ALICE TDR 1
14. ALICE Collaboration: Transition Radiation Detector Technical Design Report, CERN/LHCC 2001-021, ALICE TDR 9
15. ALICE Collaboration: Photon Spectrometer Technical Design Report, CERN/LHCC 99-04, ALICE TDR 2
16. ALICE Collaboration: Dimuon Forward Spectrometer Technical Design Report, CERN/LHCC 99-22, ALICE TDR 5 and CERN/LHCC 2000-46 Addendum to ALICE TDR 5
17. ALICE Collaboration: Zero Degree Calorimeter Technical Design Report, CERN/LHCC 99-05, ALICE TDR 3
18. ALICE Collaboration: Photon Multiplicity Detector Technical Design Report, CERN/LHCC 99-32, ALICE TDR 6

Eur Phys J A (2004) **19**, s01, 229–237
Digital Object Identifier (DOI) 10.1140/epjad/s2004-03-038-4

EPJ A direct

electronic only

$b\bar{b}b\bar{b}$ production in proton-proton and proton-nucleus collisions at the CERN LHC

A. Del Fabbro and D. Treleani

Dipartimento di Fisica Teorica dell'Università di Trieste and INFN, Sezione di Trieste, Strada Costiera 11, Miramare-Grignano, I-34014 Trieste, Italy

Received: 27 Jul / Accepted: 14 Nov 2003 /
Published Online: 6 Feb 2004 – © Società Italiana di Fisica / Springer-Verlag 2004

Abstract. Given the large parton luminosities, sizable rate of events, with several pairs of b-quarks produced contemporarily by multiple parton interactions, are foreseen in hadronic and nuclear collisions at very high energies. We compare the different contributions to $b\bar{b}b\bar{b}$ production, due to single and double parton scatterings, in pp and in pA collisions at the CERN LHC within the acceptance of the ALICE and of the LHCb detectors.

PACS. 14.20.Dh – 13.40.Gp – 21.10.Ft

1 Introduction

The production mechanism of heavy quarks in hadronic collisions is non trivial and also the simplest observable quantity, the integrated inclusive cross section, cannot be easily reproduced in perturbation theory [1]. Comparisons with the experimental data of the D0 Collaboration [2] at TEVATRON indeed have shown that the NLO pQCD calculations in α_S [3] underestimate the cross section by a factor $\sim 2, 3$.

A complementary approach to heavy quarks production, which keeps explicitly into account that transverse momenta and virtualities of the interacting partons become increasingly important in the kinematical regime of $s \gg m_b^2 \sim \hat{s} \gg \Lambda^2$, and includes terms at every order in α_S in the calculation of the cross section, is the k_t-factorization, where the interaction is factorized into un-integrated structure functions and off shell matrix elements [4,5,6]. Phenomenologically the k_t-factorization is not inconsistent with HERA and TEVATRON data and allow one to reproduce both the value of the integrated inclusive cross section and various differential distributions, including the correlation in the azimuthal angle between the produced b quarks, where different approaches are less successfully compared with experiment [7].

Interestingly, although the value of the integrated inclusive cross section cannot be obtained trivially, one may find several cases where the overall effect of higher order corrections amounts to a simple rescaling of the lowest order parton model result. Several distributions, derived either using the k_t-factorization approach or by working out the cross section at the NLO pQCD, are in fact rather similar (apart form normalization) to those obtained with a simplest lowest order calculation [8]. In a few cases, the

whole effect of higher order corrections is hence (approximately) reduced to a single numerical value, the K factor:

$$ K = \frac{\sigma(b\bar{b})}{\sigma_{LO}(b\bar{b})} . \tag{1} $$

where $\sigma(b\bar{b})$ is the actual inclusive cross section and $\sigma_{LO}(b\bar{b})$ the result of the lowest order calculation in pQCD.

Although the expected inclusive cross section of b production is hence still pretty uncertain at LHC energies, all estimates point in the direction of rather large values because of the high parton luminosity [9]. The fairly large flux of partons make it also plausible to expect a sizable rate of events, where two or more $b\bar{b}$ pairs are produced contemporarily by different partonic collisions in a single pp interaction [10]. One of the reasons of interest in the production of multiple pairs of b quarks is that such a process allows to generate and study a completely new sector of hadon spectroscopy, namely double heavy barions and double b tetraquarks [11,12]. We think that it's hence interesting to estimate the production rates of multiple $b\bar{b}$ pairs at the LHC, comparing the two possible production mechanisms, multiparton interactions and the more conventional single parton collision process [13].

We limit our considerations to the simplest cases, where the whole effect of higher order corrections is taken into account by the overall normalization factor. Given the lack of information on higher order corrections in the $2 \to 4$ processes, we assume that the K factors of the $gg \to b\bar{b}b\bar{b}$ and of the $gg \to b\bar{b}$ processes are equal. The $gg \to b\bar{b}$ process is worked out in the k_t-factorization approach, fixing the input parameters by comparing with the TEVATRON data. The cross section is then extrapolated at LHC energies, identifying a few differential distributions where the effect of higher order corrections reduces to a simple rescaling of the lowest order result. The value of the K-factor derived in this way is used to renormalize the double $(gg \to b\bar{b})^2$ and the single $gg \to b\bar{b}b\bar{b}$ parton scattering cross sections, which are evaluated by working out all Feynman diagrams at order α_S^4.

2 $b\bar{b}$ cross section at TEVATRON and LHC and K-factor

In the k_t-factorization approach the $b\bar{b}$ production cross section is expressed as [4,5]

$$\sigma(pp \to b\bar{b}) = \int \frac{d^2 q_{t1}}{\pi} \frac{d^2 q_{t2}}{\pi} \, dx_1 dx_2 \, f(x_1, q_{t1}, \mu) \times f(x_2, q_{t2}, \mu) \, \hat{\sigma}(x_1, q_{t1}; x_2, q_{t2}; \mu) \quad (2)$$

where $f(x, q_t, \mu)$ is the unintegrated structure function, representing the probability to find a parton with momentum fraction x, transverse momentum q_t at the factorization scale μ, while $\hat{\sigma}$ is the off-shell partonic cross section of the process $g^* g^* \to Q\bar{Q}$.

To evaluate the inclusive cross section we use two different prescriptions for constructing the k_t-distributions from the usual integrated parton densities. The first is based on the conventional DGLAP evolutions equations [14], with virtual corrections re-summed in the survival probability factor $T_a(k_t^2, \mu^2)$ [15]. For the second prescription we follow [16], where the un-integrated structure functions are obtained from the leading order BFKL equation and are expressed as the convolution of the collinear gluon densities $G(x, \mu^2)$ with the universal function $\mathcal{G}(x, k_t^2, \mu^2)$:

$$\mathcal{F}(x, k_t^2, \mu^2) = \int_x^1 d\xi \, \mathcal{G}(\xi, k_t^2, \mu^2) \, G\left(\frac{\xi}{x}, \mu^2\right) \quad (3)$$

The weight factors $\mathcal{G}(\xi, k_t^2, \mu^2)$ are known in double-logarithmic approximation and depend on the quantity $\bar{\alpha}_s = 3\alpha_s/\pi$, which in the BFKL formalism is a fixed quantity, related to the pomeron intercept $\alpha(0) = 1 + \Delta$, with $\Delta = 4\bar{\alpha}_s \log 2$. Following [17] we take $\Delta = 0.35$.

To obtain the functions $\mathcal{F}(x, k_t^2, \mu^2)$ we use the parton distributions set GRV94 [18] with factorization scale $\mu_F^2 = \hat{s}$, while the cross section at the lowest order in pQCD is evaluated with the MRS99 parton distributions [19], with factorization and renormalization scale equal to the transverse mass of the b-quark. Comparing the total cross sections, we obtain $K \sim 5.5$.

In Fig. 1 the integrated cross section of $b\bar{b}$ production is plotted as a function of the minimum value of the transverse momentum p_t^{min} of the b-quarkat both at TEVATRON($\sqrt{s} = 1.8$TeV) and at LHC($\sqrt{s} = 14$TeV). The dotted curves represent the cross section derived using the unintegrated gluon structure function, according with the BFKL prescription of (4), whereas the dashed lines are evaluated by using the prescription in (3). The continuous lines represent the result of the lowest order calculation multiplied by the K factor. At TEVATRON

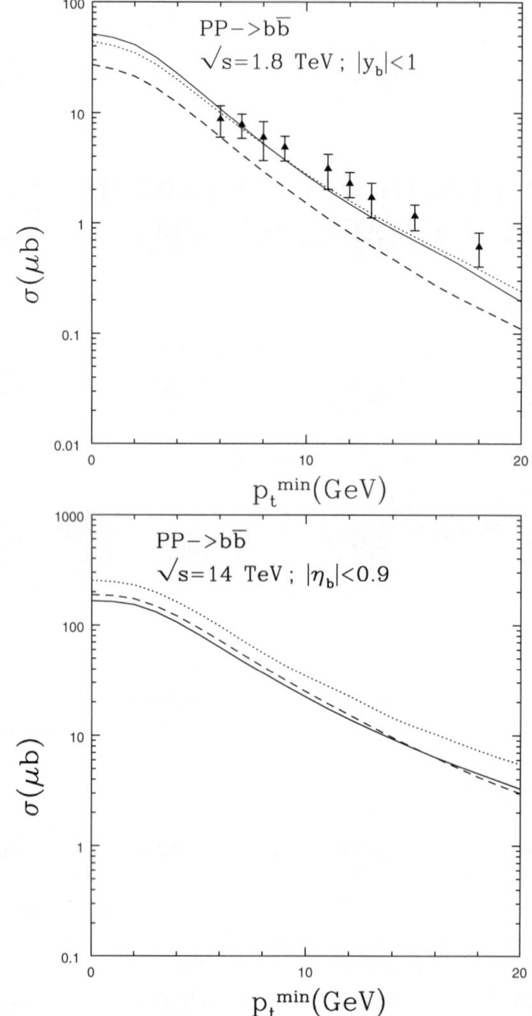

Fig. 1. $p\bar{p} \to b\bar{b}$ production cross section as a function of p_t^{min} at $\sqrt{s} = 1.8 TeV$, with the b-quark within the rapidity range $|y_b| < 1$, experimental data from [2], and at $\sqrt{s} = 14$ TeV with the b-quark within the pseudo-rapidity range $|\eta| < 0.9$

energy the b-quark distributions are within the rapidity interval $|y| < 1$ and are compared with the D0 experimental data [2]. The same distributions, extrapolated at LHC energy, are plotted as a function of p_t^{min} within the pseudorapidity interval $|\eta| < 0.9$, corresponding to the acceptance of the ALICE detector.

In Fig. 2 the rapidity(y) and pseudorapidity(η) distributions are shown, normalized to one and within $|\eta| < 0.9$. The continuous histograms are the result of the lowest order calculation, whereas the dashed histograms represent the distributions evaluated with the k_t-factorization approach.

As one may see in the cases considered the whole effect of higher orders reduces to a simple rescaling.

3 $b\bar{b}b\bar{b}$ cross section

We will compare the two competing mechanisms of $b\bar{b}b\bar{b}$ production, the leading order QCD $2 \to 4$ [20] and the

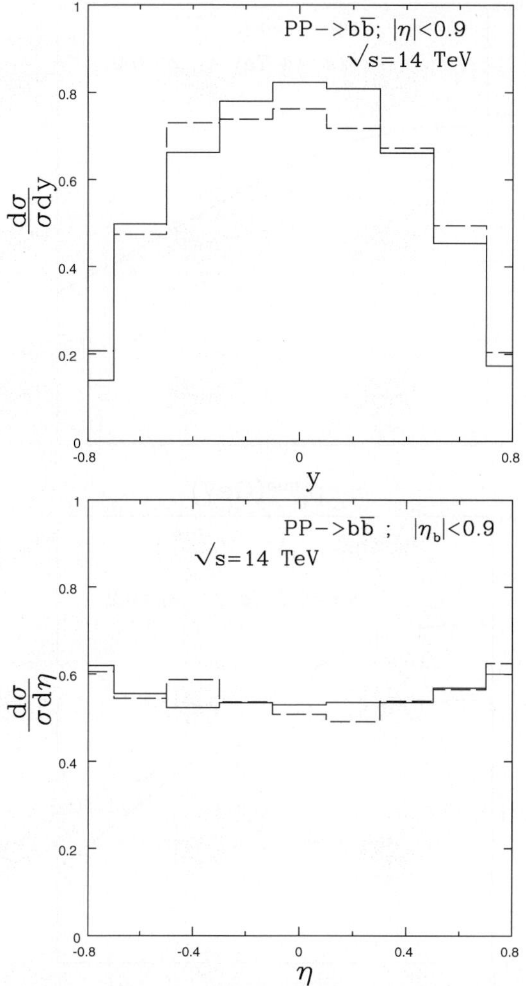

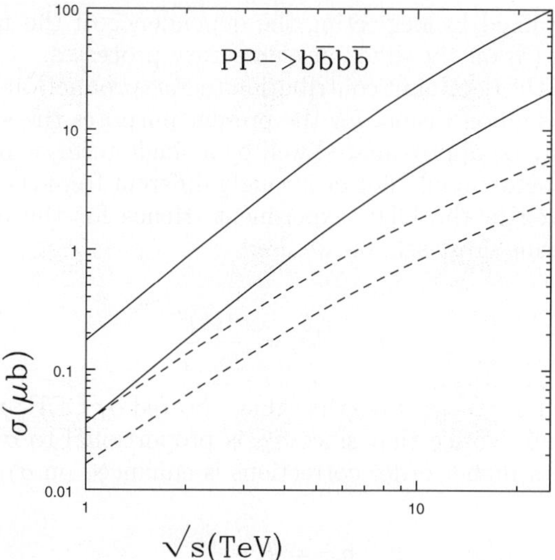

Fig. 3. $b\bar{b}b\bar{b}$ total cross section as a function of centre of mass energy. *Lower curves $K = 2.5$, higher curves $K = 5.5$*

Fig. 2. Normalized rapidity (y) and pseudorapidity (η) distributions for $b\bar{b}$ production at ALICE with the k_t factorization approach (*dashed histograms*) and at the lowest order in pQCD multiplied by the K-factor (*continuous histograms*)

double parton scattering $(2 \rightarrow 2)^2$ [21], in proton-proton collisions in the kinematical range of the ALICE and of the LHCb detectors, namely at center-of-mass energies of 5.5 and 14 TeV, within the pseudorapidity regions $|\eta| < 0.9$ and $1.8 < \eta < 4.9$, down to very low transverse momenta.

For the leading order in α_s single scattering subprocesses, one needs to evaluate 14 Feynman diagrams for each flavor in the initial state, for quarks initiated process, and 76 diagrams for gluon fusion. We generate the matrix elements of the partonic amplitudes with MadGraph [22] and HELAS [23]. As a value of the bottom quark mass we take $m_b = 4.6$ GeV and we use the MRS99 parton distributions [19]. The resulting cross section is then multiplied by the K factor, obtained as described in the previous section, while the multi-dimensional integrations are performed by VEGAS [24].

To evaluate the double parton scattering cross section the input two-body parton distribution functions $\Gamma(x_1, x_2, s)$ is needed [10]. Here $x_{1,2}$ are the fractional momenta of the two partons belonging to the same hadron

and s their distance in transverse space. We make the usual simplifying assumption of neglecting all parton correlations in fractional momenta, so we factorize the two-body parton distribution as

$$\Gamma(x_1, x_2, s) = G(x_1)G(x_2)F(s) \tag{4}$$

where $G(x)$ are the usual one-body parton distributions and $F(s)$ is a function normalized to 1 and representing the parton pair density in transverse space. The double scattering cross section is hence written as [25]

$$\sigma_D(b\bar{b}; b\bar{b}) = \frac{1}{2} \sum_{ij} \Theta^{ij} \sigma_i(b\bar{b})\sigma_j(b\bar{b}) \tag{5}$$

where the indices i, j label the different cases where each $b\bar{b}$ pair is originated either by a $q\bar{q}$ annihilation, discriminating the cases of sea and valence, or by two gluons and $\sigma_i(b\bar{b})$ represents the inclusive cross sections for $b\bar{b}$ production in a hadronic collision, the index i labelling a definite parton process. The weight factors Θ^{ij} have dimension an inverse cross section and result from integrating the product of the two-body parton distributions in transverse space, while the factor $1/2$ is a consequence of the symmetry of the expression for exchanging i and j. The dependence of Θ^{ij} on the indices i, j accounts for the possibility, for different pairs of partons in the hadron, to be characterized by different values of their relative average transverse distance [25,21].

When looking at the experimental analysis of double parton scatterings [26,27,28] one finds the cross section expressed as:

$$\sigma_D = \frac{m}{2} \frac{\sigma_S(A)\sigma_S(B)}{\sigma_{eff}} \tag{6}$$

where $m = 1$ if the two parton processes A and B are identical, while $m = 2$ if they are different and σ_S is the single scattering inclusive cross section. The relation may be obtained by neglecting the dependence of the factors Θ^{ij} in (5) on the different elementary processes.

As the dominant contribution to $b\bar{b}b\bar{b}$ production at the LHC is gluon fusion, for the present purposes the sum in (5) may be approximated well by a single term, while the scale factor should not be strongly different from the value observed in the CDF experiment. Hence for the double scattering cross section we use:

$$\sigma_D(b\bar{b}b\bar{b}) = \frac{\sigma(b\bar{b})^2}{2\sigma_{eff}} \, . \tag{7}$$

where for σ_{eff} we take the value reported by CDF, $\sigma_{eff} = 14.5$ mb. Notice that since σ_D is proportional to σ_S^2, the effect of higher order corrections is enhanced on σ_D:

$$\sigma_S = K\,\sigma_S^{LO}$$
$$\sigma_D = K^2\,\sigma_D^{LO} \tag{8}$$

where $\sigma_{S,D}^{LO}$ refers to the lowest order expressions of the cross section.

4 Results for $b\bar{b}b\bar{b}$ production in pp collisions

In Fig. 3 the cross section for $b\bar{b}b\bar{b}$ production is shown as a function of the c.m. energy. The continuous curves refer to the double parton scattering contribution, while the dotted curves to single scattering. In both cases the lower curve is evaluated with $K = 2.5$, while the higher curve with $K = 5.5$, which are the typical estimates of the NLO-pQCD and the result of our calculation within the k_t-factorization approach.

In Fig. 4 the two contributions to the integrated cross section, are plotted as a function of p_t^{min}, the minimum value of the transverse momenta of the b quarks (which are required to be all inside the pseudorapidity interval $|\eta| < 0.9$), at 14 and 5.5 TeV. The continuous curves refer to the double parton scattering contribution, while the dotted curves to single scattering. The double parton cross section decreases faster with p_t^{min} than the single parton cross section, the two contributions are of the same order at $p_t^{min} = 8-10$ GeV. Pseudorapidity, and rapidity distributions at 14 TeV are plotted in Fig. 5, where continuous and dashed histograms have the same meaning as in the previous cases.

In Fig. 6 we plot the pseudorapidity (η) distributions at $\sqrt{s} = 14$ TeV, requiring both b-quarks to be in the pseudorapidity interval $1.8 < \eta < 4.9$, corresponds to the acceptance of the LHCb detectort. In the same figure we also compare the single and double parton scattering contributions, integrated within the rapidity acceptance of the LHCb, as a function of p_t^{min}.

The overall indication is that double parton scatterings dominate the $b\bar{b}b\bar{b}$ integrated cross section by a large factor, both in the central rapidity region and at the larger rapidity values of the LHCb experiment. In both cases the contribution of the single parton scattering term becomes important only after applying cuts to the transverse momenta of the order of 8-10 GeV.

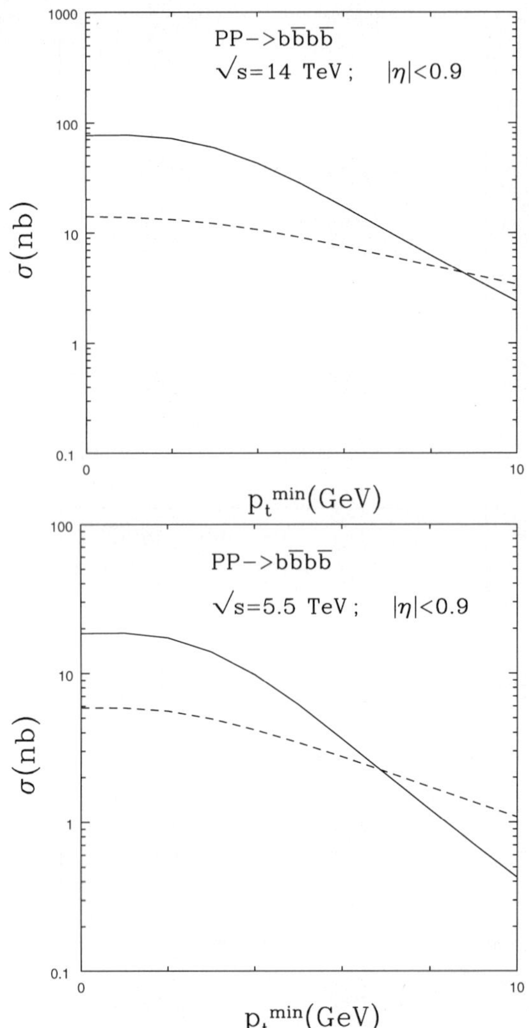

Fig. 4. $b\bar{b}b\bar{b}$ production cross section at $\sqrt{s} = 14$ TeV and at $\sqrt{s} = 5.5$ TeV as a function of p_t^{min} with all the four b-quarks in the pseudo-rapidity interval $|\eta| < 0.9$

5 $b\bar{b}b\bar{b}$ production in pA collisions

All production rates are significantly enhanced in proton-nucleus collisions, which may offer considerable advantages for studying multiparton interactions [29]. Quite in general [30,31] the expression of the double parton scattering cross section to produce two $b\bar{b}$ pairs is given by

$$\sigma_A^D(b\bar{b}, b\bar{b}) = \frac{1}{2}\sum_{ij} \int \Gamma_p(x_i, x_j; s_{ij})\hat{\sigma}(x_i, x_i')\hat{\sigma}(x_j, x_j')$$
$$\times \; \Gamma_A(x_j', x_j'; s_{ij})dx_i dx_i' dx_j dx_j' d^2 s_{ij}, \tag{9}$$

where the index A refers to the target nucleus, the indices i, j to the different kinds of partons that annihilate to produce a $b\bar{b}$ pair and the factor $1/2$ is a consequence of the symmetry of the expression for exchanging i and j.

The most suitable conditions are those where the nuclear distributions are additive in the nucleon parton distributions. In such a case one may express the nuclear

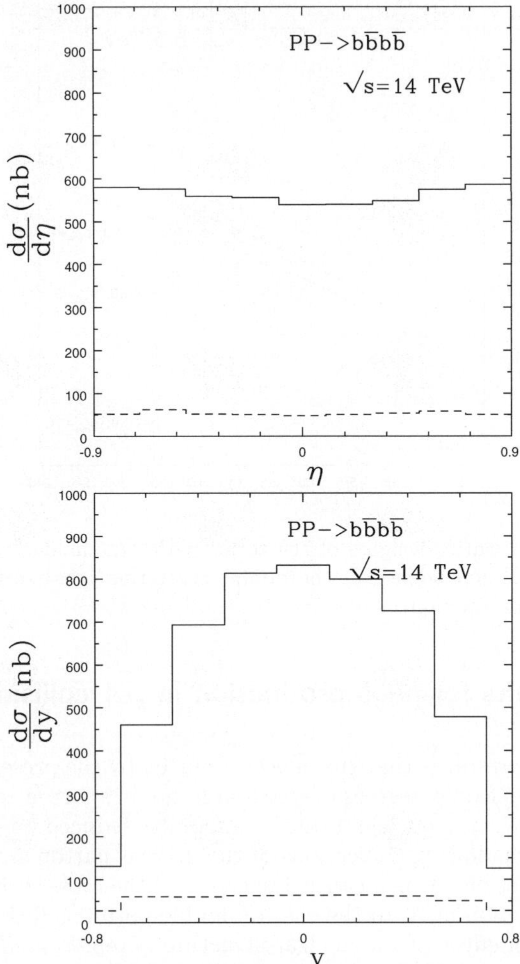

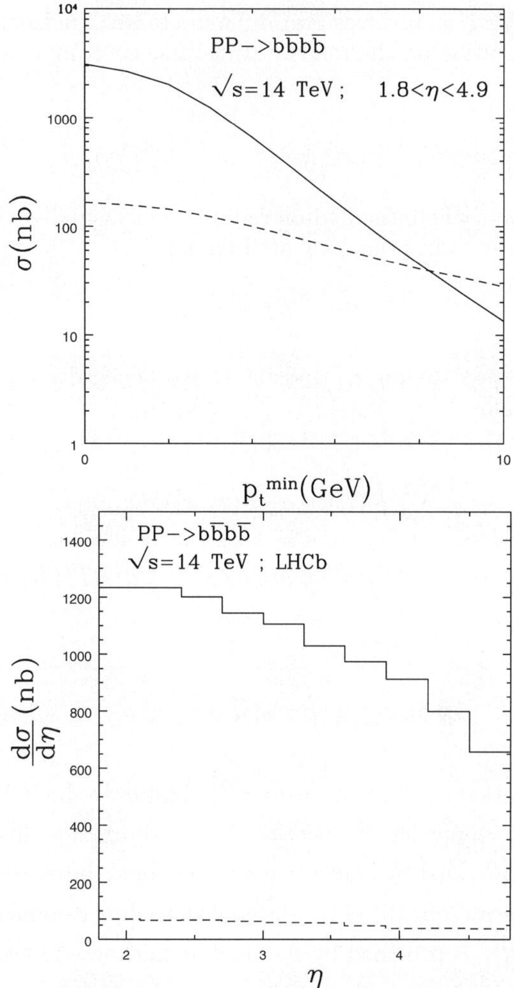

Fig. 5. $b\bar{b}b\bar{b}$ production with the two equal sign b-quarks in the pseudo-rapidity interval $|\eta_b| < 0.9$. η-distributions and y_b-distributions at $\sqrt{s} = 14$ TeV. The *continuous histograms* refer to the contribution of double parton scatterings while the *dashed histograms* to single parton scattering

Fig. 6. $b\bar{b}b\bar{b}$ production with the two equal sign b-quarks in the pseudo-rapidity interval $1.8 < \eta < 4.9$ at $\sqrt{s} = 14$ TeV. Production cross section as a function of p_t^{min} and η. The *continuous line and histogram* refer to the contribution of double parton scatterings while the *dashed line and histogram* to single parton scattering

parton pair density, $\Gamma_A(x'_j, x'_j; s_{ij})$, as the sum of two well defined contributions, where the two partons are originated by either one or by two different parent nucleons:

$$\Gamma_A(x'_i, x'_j; s_{ij}) = \Gamma_A(x'_i, x'_j; s_{ij})\Big|_1 + \Gamma_A(x'_i, x'_j; s_{ij})\Big|_2 \quad (10)$$

and correspondingly $\sigma_A^D = \sigma_A^D|_1 + \sigma_A^D|_2$. The two terms $\Gamma_A|_{1,2}$ are easily expressed in terms of the nuclear nucleon's density by introducing the transverse parton coordinates $B \pm \frac{s_{ij}}{2}$, where B is the impact parameter of the hadron-nucleus collision. One may write

$$\Gamma_A(x'_i, x'_j; s_{ij})\Big|_{1,2} = \int d^2B\, \gamma_A\Big(x'_i, x'_j; B + \frac{s_{ij}}{2}, B - \frac{s_{ij}}{2}\Big)\Big|_{1,2} \quad (11)$$

where $\gamma_A|_{1,2}$ are given by

$$\gamma_A\Big(x'_i, x'_j; B + \frac{s_{ij}}{2}, B - \frac{s_{ij}}{2}\Big)\Big|_1 = \Gamma_N(x'_i, x'_j; s_{ij}) T(B)$$

$$\gamma_A\Big(x'_i, x'_j; B + \frac{s_{ij}}{2}, B - \frac{s_{ij}}{2}\Big)\Big|_2 = G_N(x'_i) G_N(x'_j)$$
$$\times\ T\Big(B + \frac{s_{ij}}{2}\Big) T\Big(B - \frac{s_{ij}}{2}\Big) \quad (12)$$

with $T(B)$ is the nuclear thickness function, normalized to the atomic mass number A and G_N nuclear parton distributions divided by the atomic mass number.

The first term in (10) gives a simple rescaling of the double parton distribution of a isolated nucleon:

$$\Gamma_A(x'_i, x'_j; s_{ij})\Big|_1 = \Gamma_N(x'_i, x'_j; s_{ij}) \int d^2B\, T(B) \quad (13)$$

The resulting contribution to the cross section is the same as in a nucleon-nucleon interaction, enhanced by the atomic mass number factor A:

$$\sigma_A^D\Big|_1 = A\sigma_N^D \quad (14)$$

The $\sigma^D_A|_2$ term involves two different target nucleons in the integration on the relative transverse coordinate s_{ij}:

$$\int ds_{ij}\Gamma_p(x_i,x_j;s_{ij})T\left(B+\frac{s_{ij}}{2}\right)T\left(B-\frac{s_{ij}}{2}\right) \quad (15)$$

In the limit where the hadron radius is much smaller than the nuclear radius one may approximate

$$T\left(B\pm\frac{s_{ij}}{2}\right)\simeq T(B) \quad (16)$$

The integrations on s_{ij} and on B are hence decoupled. One obtains:

$$\sigma^D_A\Big|_2 = \frac{1}{2}\sum_{ij}\int G_p(x_i,x_j)\hat{\sigma}(x_i,x_i')\hat{\sigma}(x_j,x_j')$$
$$\times\; G_N(x_i')G_N(x_j')dx_idx_i'dx_jdx_j'\int d^2BT^2(B), (17)$$

where

$$G_p(x_i,x_j) = \int d^2s_{ij}\Gamma_p(x_i,x_j;s_{ij}) \quad (18)$$

Remarkably the two terms $\sigma^D_A\Big|_1$ and $\sigma^D_A\Big|_2$ have very different properties. In fact the correct dimensionality of $\sigma^D_A\Big|_1$ is provided by transverse scale factors related to the *nucleon* scale, cfr. (5),(14). The analogous dimensional factor in $\sigma^D_A\Big|_2$ is provided by the *nuclear* thickness function, which is at the second power, being two the target nucleons involved in the interaction.

As pointed out in [29], while on general grounds σ^D_A depends both on the longitudinal and transverse parton correlations, the $\sigma^D_A\Big|_2$ term depends solely on the longitudinal momentum fractions x_i,x_j so that, when the $\sigma^D_A\Big|_2$ term is isolated, one has the capability of measuring the longitudinal and, a fortiori, also the transverse parton correlations of the hadron structure in a model independent way.

The additivity of the nuclear structure functions may not be a bad approximation for a sizable part of the kinematical regime of bottom quarks production at the LHC. In the case of a central calorimeter with the acceptance of the ALICE detector ($|\eta| < 0.9$), the average value of momentum fraction of the initial state partons, in a $pp \to b\bar{b}b\bar{b}$ process, is $\langle x\rangle \approx 6\times 10^{-3}$. By introducing a cut in the transverse momenta of the b quarks of 5 GeV one obtains $\langle x\rangle \approx 10^{-2}$, while a cut of 20 GeV in p_t, within the same pseudorapidity range, gives $\langle x\rangle \approx 2\times 10^{-2}$. If considering a more forward detector, as LHCb ($1.8 < \eta < 4.9$), the average value of momentum fraction is $\langle x\rangle \approx 5\times 10^{-2}$. Deviations from additivity at low x are less than 10% for $x \geq 2\times 10^{-2}$ [32] and, although increasing with the atomic mass number, non additive corrections are at most a 20% effect, on the considered kinematical regime.

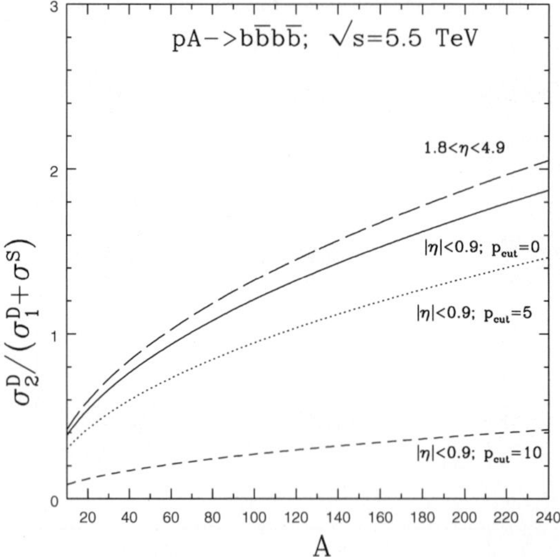

Fig. 7. Relative weights of the terms with "anomalous" and "usual" A-dependence in the double scattering cross section for $b\bar{b}b\bar{b}$ production

6 Results for $b\bar{b}b\bar{b}$ production in pA collisions

A major result is that the effects, induced by the presence of the nucleonic degrees of freedom in double parton scatterings with a nuclear target, cannot be reduced to the simple shadowing corrections of the nuclear parton structure functions, which cause a decrease of the cross section as a function of A. In the case of double parton collisions, the main effect of the nuclear structure is represented by the presence of the $\sigma^D_A|_2$ term in the cross section, which scales with a different power of A as compared to the single scattering contribution, producing a additive correction to the cross section.

The 'anomalous' dependence of the double parton scattering cross section, as a function of the atomic mass number, is emphasized in Fig. 7, where the ratio $\sigma^D_2/(\sigma^D_1+\sigma^S)$ is plotted as a function of A. The ratio represents the contribution to the cross section of the processes where two different nuclear target nucleons are involved in the interaction, scaled to the contribution where only a single target nucleon is involved. The dependence on the atomic mass number of the latter terms is the same of all hard processes usually considered, where nuclear effects may be wholly absorbed in the shadowing corrections to the nuclear structure functions. The contribution to the cross section of the σ^D_2 term is, on the contrary, "anomalous", involving two different target nucleons in the interaction. The ratio above hence represents the relative weights of the "anomalous" to the "usual" contributions to the double parton scattering cross section on a nuclear target. The plots in Fig. 7 refer to the cases $1.8 \leq \eta \leq 4.9$ and $|\eta| \leq .9$, with different cuts on the transverse momenta of the produced b-quarks ($p_{cut} = 0, 5, 10$ GeV/c).

The different contributions to the cross section due to interactions with a single or with two different target nucleons, are shown in Fig. 8, in the case of a central calorimeter. The upper figure shows the cross section as

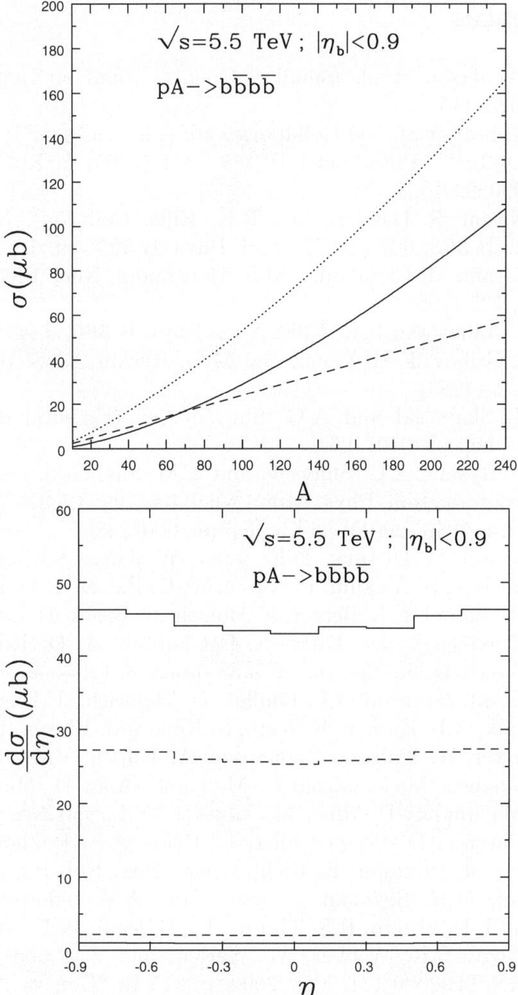

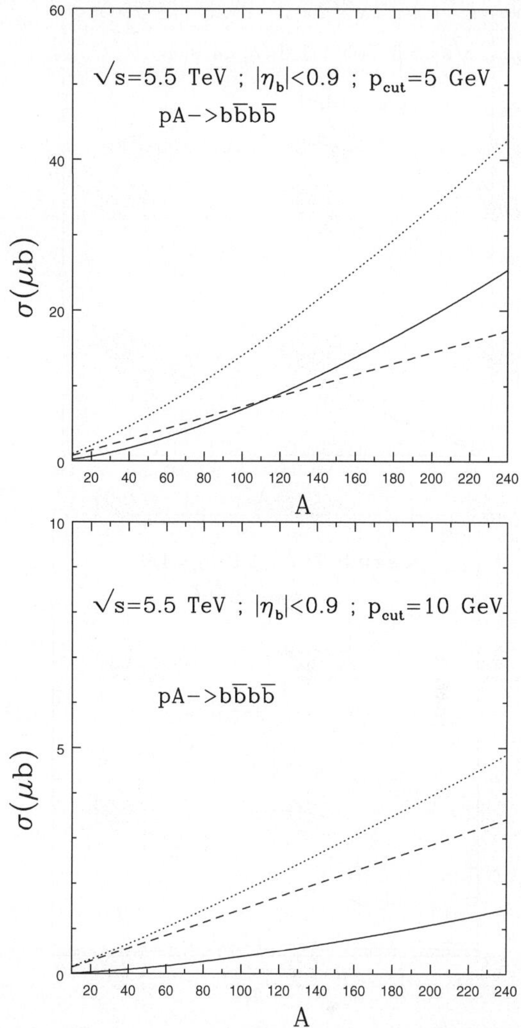

Fig. 8. Different contributions to the cross section for $b\bar{b}b\bar{b}$ production, in a central calorimeter. *Upper figure:* cross section as a function of A of the one-nucleon (*dashed line*) and of the two nucleons (*continuous line*) terms. The *dotted line* is the sum of the two terms. *Lower figure:* differential pseudorapidity distributions of a b-quark produced in an event with two $b\bar{b}$ pairs. One-nucleon (*dashed histograms*) and two-nucleon contributions (*continuous histograms*) in the case of a heavy (*higher histograms*) and of a light nucleus (*lower histograms*)

Fig. 9. Different contributions to the cross section for $b\bar{b}b\bar{b}$ production in a central calorimeter as a function of A, after applying a cut of 5 GeV (*upper figure*) and of 10 GeV (*lower figure*) in the transverse momenta of each produced b-quark: one-nucleon (*dashed line*), two nucleons (*continuous line*) processes and total (*dotted line*)

a function of the atomic mass number of the one-nucleon (dashed line) and of the two nucleons (continuous line) contributions to the cross section. The dotted line is the sum of the two terms. The lower figure shows the two contributions to the oseudorapidity distribution of a b-quark produced in an event with two $b\bar{b}$ pairs: one-nucleon (dashed histograms) and two-nucleon contributions (continuous histograms) in the case of a heavy (higher histograms) and of a light nucleus (lower histograms).

The A-dependence of the two different contributions, as a function of A, are shown in Fig. 9 in the case of a central calorimeter, after applying a cut of 5 GeV/c (upper figure) and of 10 GeV/c (lower figure) in the transverse momenta of each produced b-quark. Dashed, continuous and dotted lines have the same meaning as in Fig. 8.

The case of a forward calorimeter, $1.8 \leq \eta \leq 4.9$, is shown in Fig. 10, where the dashed, continuous and dotted lines have the same meaning as before.

Summarizing the large size of the cross section of $b\bar{b}b\bar{b}$ production in hadron-nucleus collisions at the LHC (the values are of the order of one hundreds of μb) suggests that the production of multiple pairs of b-quarks is fairly typical at high energies, hence representing a convenient channel to study multiple parton interactions. A rather spectacular feature is the "anomalous" dependence on A. The effects induced by the presence of the nucleonic degrees of freedom in the nuclear structure are in fact not limited to the usual shadowing corrections to the nuclear structure functions, which cause a limited *decrease* (not larger than 20%, in the kinematical regime considered here) of the cross section for a hard interaction in hadron-nucleus collisions. When considering double parton scatterings, all

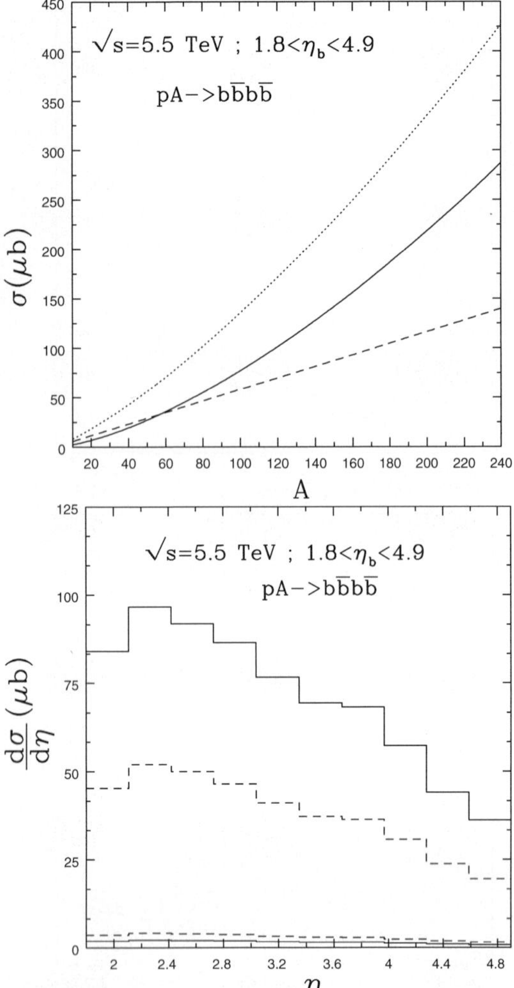

Fig. 10. Different contributions to the cross section for $b\bar{b}b\bar{b}$ production, in a forward calorimeter. Upper figure: cross section as a function of A of the one-nucleon (*dashed line*) and of the two nucleons (*continuous line*) contributions. The dotted line is the sum of the two terms. Lower figure: differential rapidity distributions of a b-quark produced in an event with two $b\bar{b}$ pairs. One-nucleon (*dashed histograms*) and two-nucleon contributions (*continuous histograms*) in the case of a heavy (*higher histograms*) and of a light nucleus (*lower histograms*)

nuclear effects are exhausted in the shadowing corrections only in the $\sigma_A^D|_1$ term. The dominant effect of the nuclear structure is on the contrary due to the presence of the $\sigma_A^D|_2$ term in the cross section, which scales with a different power of A as compared to single scattering term, giving rise to a sizably larger correction, with opposite sign as compared to shadowing correction, namely to an *increase* of the cross section which becomes sizably larger than 100% in the case of a heavy nucleus.

Acknowledgements. This work was partially supported by the Italian Ministry of University and of Scientific and Technological Researches (MIUR) by the Grant COFIN2001.

References

1. B. Anderson et al. [Small x Collaboration]: arXiv:hep-ph/0204115
2. B. Abbott et al. [D0 Collaboration]: collisions at s**(1/2) = 1.8-TeV," Phys. Lett. B **487**, 264 (2000) [arXiv:hep-ex/9905024]
3. P. Nason, S. Dawson, and R.K. Ellis: Collisions, Nucl. Phys. B **303**, 607 (1988); Nucl. Phys. B **327**, 49 (1989)
4. S. Catani, M. Ciafaloni, and F. Hautmann: Nucl. Phys. B **366**, 135 (1991)
5. J.C. Collins and R.K. Ellis: Nucl. Phys. B **360**, 3 (1991)
6. L.V. Gribov, E.M. Levin, and M.G. Ryskin: Phys. Rept. **100**, 1 (1983)
7. Y.M. Shabelski and A.G. Shuvaev: experimental data, arXiv:hep-ph/0107106
8. M.G. Ryskin, A.G. Shuvaev, and Y.M. Shabelski: beauty hadroproduction, Phys. Atom. Nucl. **64**, 1995 (2001) [Yad. Fiz. **64**, 2080 (2001)] [arXiv:hep-ph/0007238]
9. S. Catani, M. Dittmar, D.E. Soper, W. James Stirling, S. Tapprogge, S. Alekhin, P. Aurenche, C. Balazs, R.D. Ball, G. Battistoni, E.L. Berger, T. Binoth, R. Brock, D. Casey, G. Corcella, V. Del Duca, A. Del Fabbro, A. De Roeck, C. Ewerz, D. de Florian, M. Fontannaz, S. Frixione, W.T. Giele, M. Grazzini, J.P. Guillet, G. Heinrich, J. Huston, J. Kalk, A.L. Kataev, K. Kato, S. Keller, M. Klasen, D.A. Kosower, A. Kulesza, Z. Kunszt, A. Kupco, V.A. Ilyin, L. Magnea, Michelangelo L. Mangano, Alan D. Martin, K. Mazumdar, P. Mine, M. Moretti, W.L. van Neerven, G. Parente, D. Perret-Gallix, E. Pilon, A.E. Pukhov, I. Puljak, J. Pumplin, E. Richter-Was, R.G. Roberts, G.P. Salam, M.H. Seymour, N. Skachkov, A.V. Sidorov, H. Stenzel, D. Stump, R.S. Thorne, D. Treleani, W.K. Tung, A. Vogt, B.R. Webber, M. Werlen, and S. Zmouchko: CERN-TH-2000-131, May 2000. p. 115 In *Geneva 1999, Standard model physics (and more) at the LHC* 1-115 arXiv:hep-ph/0005025
10. P.V. Landshoff and J.C. Polkinghorne: Phys. Rev. D **18**, 3344 (1978); Fujio Takagi: Phys. Rev. Lett. **43**, 1296 (1979); C. Goebel, F. Halzen, and D.M. Scott: Phys. Rev. D **22**, 2789 (1980); N. Paver and D. Treleani: Nuovo Cimento A **70**, 215 (1982); B. Humpert: Phys. Lett. B **131**, 461 (1983); M. Mekhfi: Phys. Rev. D **32**, 2371 (1985), ibid. D **32**, 2380 (1985); B. Humpert and R. Odorico: Phys. Lett. **154B**, 211 (1985); T. Sjostrand and M. Van Zijl: Phys. Rev. D **36**, 2019 (1987); F. Halzen, P. Hoyer, and W.J. Stirling: Phys. Lett. **188B**, 375 (1987); M. Mangano: Z. Phys. C **42**, 331 (1989); R.M. Godbole, Sourendu Gupta, and J. Lindfors: Z. Phys. C **47**, 69 (1990)
11. M. Rosina, D. Janc, D. Treleani, and A. Del Fabbro: AIP Conf. Proc. **660**, 377 (2003) [arXiv:hep-ph/0301136]
12. D. Janc, M. Rosina, D. Treleani, and A. Del Fabbro: Few Body Syst. Suppl. **14**, 25 (2003) [arXiv:hep-ph/0301115]
13. A. Del Fabbro and D. Treleani: Phys. Rev. D **66**, 074012 (2002) [arXiv:hep-ph/0207311]
14. M.A. Kimber, A.D. Martin, and M.G. Ryskin: Eur. Phys. J. C **12**, 655 (2000) [arXiv:hep-ph/9911379]
15. G. Marchesini and B.R. Webber: Radiation, Nucl. Phys. B **310**, 461 (1988)
16. J. Blumlein: Report No. DESY 95-121, hep-ph/9506403
17. S.P. Baranov and N.P. Zotov: Hera, Phys. Lett. B **458**, 389 (1999)
18. M. Gluck, E. Reya, and A. Vogt: Z. Phys. C **67**, 433 (1995)

19. A.D. Martin, R.G. Roberts, W.J. Stirling, and R.S. Thorne: Eur. Phys. J. C **14**, 133 (2000)

20. V.D. Barger, A.L. Stange, and R.J. Phillips: Phys. Rev. D **44**, 1987 (1991)

21. A. Del Fabbro and D. Treleani: Phys. Rev. D **63**, 057901 (2001) [arXiv:hep-ph/0005273]

22. T. Stelzer and W.F. Long: Comp. Phys. Comm. **81**, 357 (1994)

23. E. Murayama, I. Watanabe, and K. Hagiwara, HELAS: HELicity Amplitude Subroutines for Feynman Diagram Evaluations, KEK report 91-11, January 1992

24. G.P. Lepage: J. Comput. Phys. **27**, 192 (1978)

25. G. Calucci and D. Treleani: Phys. Rev. D **60**, 054023 (1999)

26. T. Akesson et al. [Axial Field Spectrometer Collaboration]: Z. Phys. C **34**, 163 (1987)

27. F. Abe et al. [CDF Collaboration]: 1.8-TeV, Phys. Rev. Lett. **79**, 584 (1997)

28. F. Abe et al. [CDF Collaboration]: Phys. Rev. D **56**, 3811 (1997)

29. M. Strikman and D. Treleani: Phys. Rev. Lett. **88**, 031801 (2002) [arXiv:hep-ph/0111468]

30. N. Paver and D. Treleani: Nuovo Cim. A **70**, 215 (1982)

31. M. Braun and D. Treleani: Eur. Phys. J. C **18**, 511 (2001) [arXiv:hep-ph/0005078]

32. P. Amaudruz et al. [New Muon Collaboration]: Nucl. Phys. B **441**, 3 (1995) [arXiv:hep-ph/9503291]

Laboratory physics programs and new facilities

Eur Phys J A (2004) **19**, s01, 239–241
Digital Object Identifier (DOI) 10.1140/epjad/s2004-03-039-3

EPJ A direct
electronic only

New results from LEPS at SPring-8

Takashi Nakano, for the LEPS collaboration

RCNP, Osaka University, Ibaraki, Osaka 567-0047, Japan

Received: 31 Jul / Accepted: 14 Nov 2003 /
Published Online: 6 Feb 2004 – © Società Italiana di Fisica / Springer-Verlag 2004

Abstract. The photon beam at SPring-8 is produced by backward-Compton scattering of laser photons from 8 GeV electrons. The maximum energy of the photon beam is 2.4 GeV which is above the $s\bar{s}$ production threshold. We report the status of the new facility and the prospect of hadron physics study with this high quality beam. Results from the first physics run on the subjects of ϕ photo-production and a $S = +1$ baryon are presented.

PACS. 13.60.Le Meson production – 13.60.Rj Baryon production

1 Laser-electron photon beam at SPring-8 (LEPS)

The Spring-8 facility is the most powerful third-generation synchrotron radiation facility in the world. The energy of the electrons in the storage ring is 8 GeV and the beam current is 100 mA. The laser-electron photon (LEP) beam at the SPring-8 is generated by Backward-Compton scattering of laser photons with the 8-GeV electrons (Fig. 1). The maximum energy of the beam is currently 2.4 GeV for a 351-nm (3.5 eV) Ar laser, which is well above the threshold for $s\bar{s}$ productions. The polarization of the LEP beam is about 95 % at the the maximum energy when laser lights are linearly polarized. The polarization drops as the photon energy decreases and crosses zero at the half of the maximum energy. An energy of laser photons can be changed so that the polarization remains reasonably high in the energy region of interest.

The LEP energy is determined by measuring the energy of a recoil electron with a tagging counter which measures the deviation of the recoil electron from the 8-GeV electron beam orbit. Since the tagging counter does not cover the region very near the beam orbit, a photon with an energy below 1.5 GeV cannot be tagged. The energy resolution of the tagged photon is 15 MeV (σ) mainly due to the energy spread of the electron beam and an uncertainty of a photon-electron interaction point in the 7.8-m straight section.

The operation of the laser-electron photon beam at SPring-8 started in July, 1999. The intensity of the beam is about 2.5×10^6 photons/sec for a 5 W laser-output, and a typical tagger rate is 8×10^5/sec.

2 LEPS detector

Figure 2 shows a schematic drawing of the LEPS detector. For the tracking of the charged particles, a silicon-strip vertex detector (SSD) and 3 drift chambers are used. The SSD consists of single-sided silicon-strip detectors (vertical and horizontal planes) with the strip pitch of 120 μm. The first drift chamber located before a 0.7-T magnet consists of 6 wire planes (3 vertical planes, 2 planes at $+45°$, and 1 plane at $-45°$), and the other two drift chambers after the magnet consist of 5 planes (2 vertical planes, 2 planes at $+30°$, and 1 plane at $-30°$). A time-of-flight (TOF) scintillator array is positioned 3 m behind the dipole magnet.

Electron-positron pairs produced before the target are rejected online by a plastic counter, and the pairs produced at the target are rejected by an aerogel Čerenkov counter (AC) with the index of 1.03. The electron and positorons which escape the online trigger rejections are blocked by lead bars which were set horizontally along the median plane inside the magnet gap. Pions with a momentum higher than ~ 0.6 GeV/c are vetoed online by the aerogel Čerenkov counter (AC). A 0.5-cm thick plastic scintillator (SC) located 9.5 cm downstream from the 5-cm thick liquid-hydrogen (LH$_2$) target ensures at least one charged particle produced in the LH$_2$ target. The events from the SC is turned out to be very useful to study events generated from neutrons in carbon nuclei at the SC

The angular coverage of the spectrometer is about ± 0.4 rad and ± 0.2 rad in the horizontal and vertical directions, respectively. The momentum resolution (σ) for 1-GeV/c particles is 6 MeV/c. The timing resolution (σ) of the TOF is 150 psec for a typical flight length of 4 m from the target to the TOF. The momentum-dependent mass resolution is about 30 MeV/c^2 for a 1-GeV/c kaon.

3 First physics run and results

The physics run with a 5-cm long liquid H$_2$ target wad carried out during December, 2000 to June, 2001. The

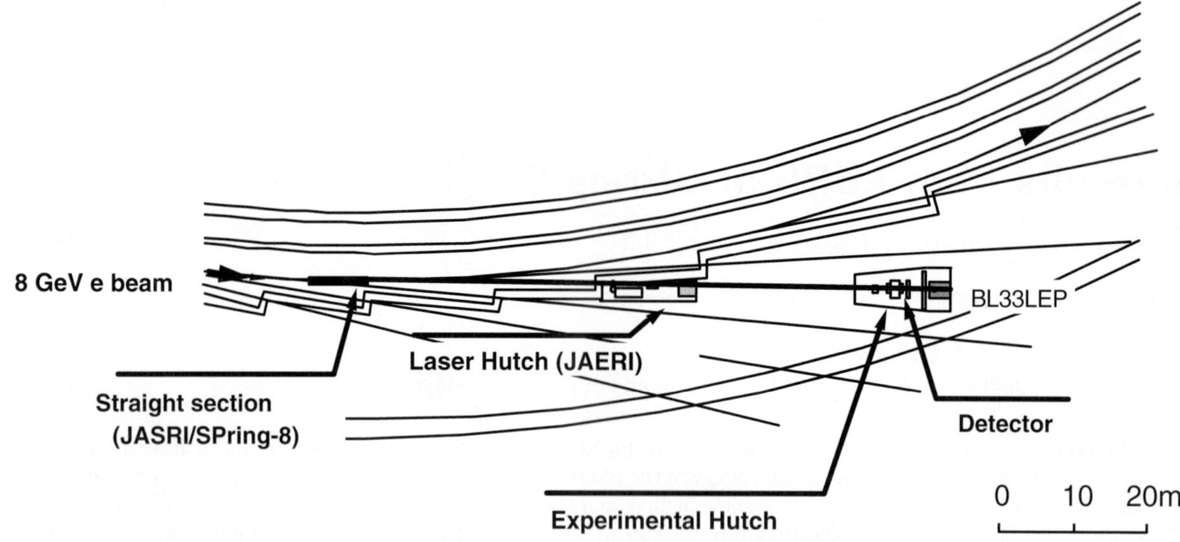

Fig. 1. The LEPS beamline (BL33LEP)

trigger required a tagging counter hit, no charged particle before the target, charged particles after the target, no signal in the aerogel Čerenkov counter, at least one hit on the TOF wall. A typical trigger rate was about 20 counts per second.

3.1 ϕ photo-production

A ϕ meson is almost pure $s\bar{s}$ state. Therefore, diffractive photo-production of a ϕ meson off a proton in a wide energy range is well described as a pomeron-exchange (multi gluon-exchange) process [1,2,3]. However, at low energies other contributions arising from meson (π, η)-exchange [3], a scaler $(0^{++}$ glueball)-exchange [4], and $s\bar{s}$ knock-out [5] are possible. These contributions fall off rapidly as the incident γ-ray energy increases, and can be studied only in the low energy region near the production threshold. The experimental separation of these contributions are difficult if one measure only differential cross-sections because they have similar photon-energy and momentum transfer dependences. Linearly polarized photons are an ideal probe to decompose these contributions. For natural-parity exchange such as pomeron and 0^{++} glueball exchanges, the decay plane of K^+K^- is concentrated in the direction of the photon polarization vector. For unnatural-parity exchange processes like π and η exchange processes, it is perpendicular to the polarization vector.

Pichowsky and Lee [3] predicts that the meson exchange processes dominate in the low photon energy region around 2.3 GeV. However, a preliminary analysis of the KK decay asymmetry in the forward angles showed the natural-parity exchange contributions were still dominant in the region. [6].

3.2 Observation of a $S = +1$ baryon resonance

We searched for baryon resonances with strangeness quantum number S=+1 in the K^- missing mass spectrum for the $\gamma + n \rightarrow K^+ + K^- + n$ reaction [7]. The search was motivated in part by a recent paper by Diakonov, Petrov and Polyakov [8] where masses and widths of an anti-decuplet baryons were predicted from the chiral soliton model. The lightest member of the anti-decuplet is the Θ^+ which is an exotic 5-quark state with a quark configuration of $uudd\bar{s}$ that subsequently decays into a K^+ and a neutron. The model predicts the mass of the Θ^+ to be ~ 1530 MeV/c^2 with a narrow width of ≤ 15 MeV/c^2.

For the present analysis, we selected K^+K^- pair events produced in the SC, which accounted for about half of the K^+K^--pair events. The missing mass $MM_{\gamma K^+K^-}$ of the N(γ, K^+K^-)X reaction was calculated by assuming that the target nucleon (proton or neutron) has the mean nucleon mass of 0.9389 GeV/c^2 (M_N) and zero momentum. Subsequently, events with $0.90 < MM_{\gamma K^+K^-} < 0.98$ GeV/c^2 were selected. The main physics background events due to the photo-production of the ϕ meson were eliminated by removing the events with the invariant K^+K^- mass from 1.00 GeV/c^2 to 1.04 GeV/c^2.

In order to eliminate photo-nuclear reactions of $\gamma p \rightarrow K^+K^-p$ on protons in ^{12}C and ^{1}H at the SC, the recoiled protons were detected by the SSD. The direction and momentum of the nucleon in the final state was calculated from the K^+ and K^- momenta. And we rejected such events in which the recoiled nucleon was out of the SSD acceptance or the recoiled proton hit was found in the SSD. A total of 109 events satisfied all the selection criteria ("signal sample").

In case of reactions on nucleons in nuclei, the Fermi motion has to be taken into account to obtain appropriate missing-mass spectra. The missing mass corrected for the Fermi motion, $MM^c_{\gamma K\pm}$, is deduced as

$$MM^c_{\gamma K\pm} = MM_{\gamma K\pm} - MM_{\gamma K^+K^-} + M_N. \qquad (1)$$

The validity of the correction was checked with the $\gamma n \rightarrow K^+\Sigma^- \rightarrow K^+\pi^-n$ sequential process, where the K^+ and π^- were detected.

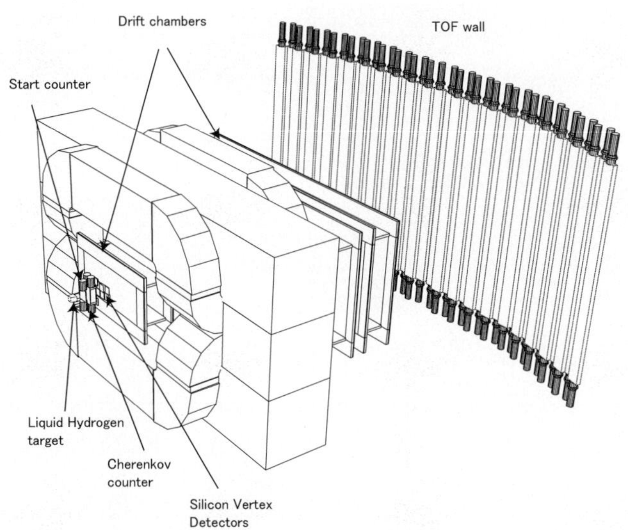

Fig. 2. The LEPS detector setup

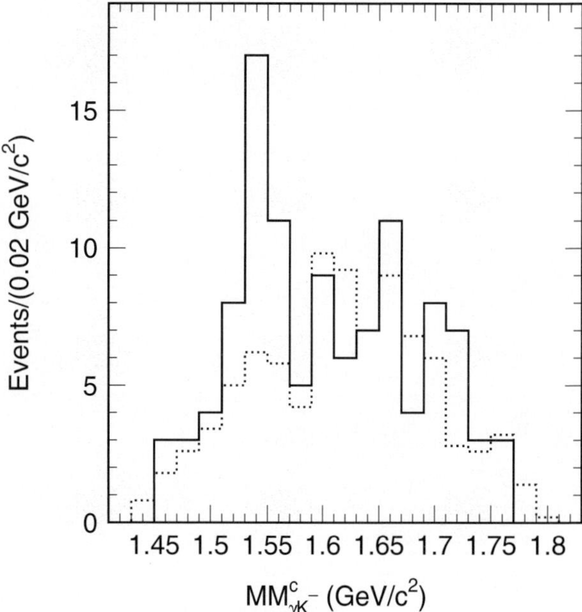

Fig. 3. The $MM^c_{\gamma K^-}$ spectrum for the signal sample (*solid histogram*) and for events from the LH$_2$ (*dotted histogram*) normalized by a fit in the region above 1.59 GeV/c^2

The corrected K^+ missing-mass distribution for the events that satisfy all the selection conditions is compared with that for the events for which a coincident proton hit was detected in the SSD. In the latter case, a clear peak due to the $\gamma + p \rightarrow K^+\Lambda(1520) \rightarrow K^+K^-p$ reaction is observed while the $\Lambda(1520)$ peak does not exist in the signal sample. This indicates that the signal sample is dominated by events produced by reactions on neutrons. Fig. 3 shows the corrected K^- missing mass distribution of the signal sample. A prominent peak at 1.54 GeV/c^2 is found. The broad background centered at ~ 1.6 GeV/c^2 is most likely due to non-resonant K^+K^- production and the background shape in the region above 1.59 GeV/c^2 has been fitted by a distribution of events from the LH$_2$. The estimated number of the events above the background level is 19.0 ± 2.8, which corresponds to a Gaussian significance of 4.6 σ.

After subtracting the background from the signal sample, the spectrum in the region of $1.47 \leq MM^c_{\gamma K^-} < 1.61$ GeV/c^2 was compared with Monte Carlo simulations assuming a Breit-Wigner function for a resonance distribution. The best fit to the spectrum gives the mass of the resonance to be 1.54 ± 0.01 GeV/c^2. And the upper limit for the width was determined to be 25 MeV/c^2 with a 90 % C.L.. This narrow peak strongly indicates the existence of an $S = +1$ resonance which may be attributed to the exotic 5-quark baryon proposed as the Θ^+.

Acknowledgements. The authors gratefully acknowledge the dedicated efforts of the staff of the SPring-8 for providing a good quality beam. This research was supported in part by the Ministry of Education, Science, Sports and Culture of Japan, by the National Science Council of Republic of China (Taiwan), and by KOSEF of Republic of Korea.

References

1. T.H. Bauer et al.: Rev. Mod. Phys. **50**, 261 (1978)
2. A. Donnachie and P.V. Landshoff: Nucl. Phys. B **267**, 690 (1986)
3. M.A. Pichowsky and T.-S.H. Lee: Phys. Rev. D **56**, 1644 (1997)
4. T. Nakano and H. Toki: in *Proc. of Intern. Workshop on Exciting Physics with New Accelerator Facilities*, SPring–8, Hyogo, 1997, World Scientific Publishing Co. Pte. Ltd., p.48 (1998)
5. A.I. Titov, Y. Oh, and S.N. Yang: Phys. Rev. Lett. **79**, 1634 (1997); A.I. Titov, Y. Oh, and S.N. Yang: Phys. Rev. C **58**, 2429 (1998)
6. W.C. Chang and T. Mibe: Nucl. Phys. A **721**, 739c (2003)
7. T. Nakano et al. (LEPS collaboration): Phys. Rev. Lett. **91**, 012002 (2003)
8. D. Diakonov, V. Petrov, and M. Polyakov: Z. Phys. A **359**, 305 (1997)

Eur Phys J A (2004) **19**, s01, 243–247
Digital Object Identifier (DOI) 10.1140/epjad/s2004-03-040-x

EPJ A direct

electronic only

Research perspectives at Jefferson Lab: 12 GeV and beyond

Kees de Jager

Jefferson Laboratory, Newport News, VA 23606, USA

Received: 30 Sep 2003 / Accepted: 14 Nov 2003 /
Published Online: 6 Feb 2004 – © Società Italiana di Fisica / Springer-Verlag 2004

Abstract. The plans for upgrading the CEBAF accelerator at Jefferson Lab to 12 GeV are presented. The research program supporting that upgrade is illustrated with a few selected examples. The instrumentation under design to carry out that research program is discussed. Finally, a conceptual design of a future upgrade which combines an electron-ion collider facility at a luminosity of up to 10^{35} cm^{-2}s^{-1} and a CM energy of up to 65 GeV with a 25 GeV fixed-target facility.

PACS. PACS-key 13.60.-r – 25.30.-c – 29.17.+w

1 Introduction

The design parameters of the Continuous Electron Beam Accelerator Facility (CEBAF) at the Thomas Jefferson National Accelerator Facility (JLab) were defined nearly two decades ago. In that period our understanding of the behaviour of strongly interacting matter has evolved significantly, providing important new classes of experimental questions which can be optimally adressed by a CEBAF-type accelerator at higher energy. The original design of the facility, coupled to developments in superconducting RF technology, makes it feasible to triple the initial design value of CEBAF's beam energy to 12 GeV in a cost-effective manner (at about 15% of the cost of the initial facility).

The research program with the 12 GeV upgrade will provide breakthroughs in two key areas: (1) gluonic excitations of mesons and the origin of quark confinement and (2) the unified description of the quark-gluon content of the nucleon. The upgrade will provide important advances in two additional areas: (1) a unified description of high-density cold nuclear matter and (2) measurements which test the Standard Model. A detailed overview of the upgrade research program is given in the recent pre-Conceptual Design Report [1].

2 Research program

The dynamics of quarks and gluons leading to the phenomenon of confinement are among the outstanding unsolved problems in physics. Theory and computation in lattice QCD have led to the concept of quarks interacting through the development of flux tubes. These flux tubes contain the gluonic degrees of freedom, which can be excited to generate "hybrid" states. Whereas a few experimental indications for the gluonic excitation of mesonic

states have been reported, the 12 GeV upgrade would be the unique facility for a systematic search of such states. It would provide a hitherto unavailable combination of kinematical reach and statistics.

Recent theoretical advances [2] have established, within the framework of QCD, the impact of Generalized Parton Distributions (GPD), which are accessible through deeply virtual and exclusive reactions. These GPDs contain the correlations between quark states of different momenta and thus reach beyond the standard parton distributions. They provide, among other properties, a simultaneous determination of the longitudinal momenta of quarks and of their transverse position within the nucleon: for the first time it will become possible to build a three-dimensional picture of the nucleons from experimental data. It also promises to determine the contribution of the quark orbital angular momentum to the nucleon spin. Exclusive reactions, such as deeply virtual Compton scattering and meson production, will be measured with unprecedented precision and kinematical range with the 12 GeV upgrade.

With the 12 GeV upgrade Jefferson Lab will be unique in its ability to attain in nuclear reactions very high values of Q^2 and x_B along with an excellent resolution. This ability represents a unique tool for studying high-density configurations in the nuclear ground state. Such studies will determine the limits of hadron-based nuclear physics from a comparison to results from ab-initio nucleon-based calculations. Short-range nucleon-nucleon correlations can be studied with great sensitivity in quasi-elastic electro-disintegration of light nuclei, in deep-inelastic electron-nucleus scattering at x_B-values greater than 1, in semi-inclusive measurements of a nucleon from the target fragmentation region and in the elastic form factors of few-nucleon systems. In reactions at high momentum transfer the color-singlet nature of strongly interacting matter will be emphasized. Examples of such experiments are photodisintegration of the deuteron, studies aimed at identifying point-like configurations (color transparency) in nuclei and nuclear hadronization processes.

Precision parity-violating electron scattering experiments with the 12 GeV upgrade have the sensitivity to search for deviations from the Standard Model that could signal the presence of new gauge bosons, of leptoquarks or of particles predicted by supersymmetry. Of clear importance will be a new measurement of the weak charge of the electron. A Møller scattering experiment similar to the SLAC E158 experiment could achieve half the uncertainty achieved in E158. An 11 GeV measurement of parity-violating DIS could easily attain the precision required to determine whether the NuTeV result is a true indication of new physics. The present program of measuring the $\pi^\circ \to \gamma\gamma$ decay width via the Primakoff effect can be extended at 11 GeV to include the η and $\eta\prime$ mesons. Such measurements will significantly enhance our knowledge of the u and d quark mass ratio.

3 Accelerator

At present CEBAF accelerates electrons to 6 GeV by recirculating the beam four times through two superconducting linacs, each producing an energy gain of 600 MeV per pass. Both linac tunnels provide sufficient space to install five additional newly designed cryomodules. The new cryo-modules will each provide over 100 MV (compared to the 28 MV from the existing ones), by increasing the gradient to 20 MV/m and the number of cavity cells from five to seven. This will result in a maximum energy gain per pass of 2.2 GeV, providing a maximum beam energy to Halls A, B and C of 11 GeV. The new Hall D will be provided with the desired maximum energy of 12 GeV by adding a tenth arc and recirculating the beam a fifth time through one linac. A total of 90 μA of CW beam can be provided at the maximum beam energy. Further modifications required are changing the dipoles in the arcs from C-type to H-type magnets, replacing a large number of power supplies and doubling the central helium liquifier capacity to 10 kW. An overview of the upgrade of the accelerator is shown in Fig. 1.

4 Hall upgrades

4.1 Hall A

The present base instrumentation in Hall A has been used with great success for experiments which require high luminosity and high resolution in momentum and/or angle of at least one of the reaction products. The central elements are the two High Resolution Spectrometers (HRS). Both devices provide a momentum resolution of better than 2×10^{-4} and an angular resolution of better than 1 mrad with a design maximum central momentum of 4 GeV/c.

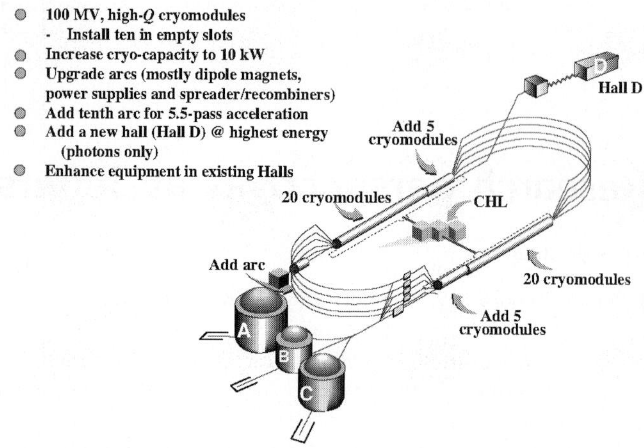

Fig. 1. Overview of the accelerator upgrade to 12 GeV

Two instrumentation upgrades are proposed to allow an optimal study of the intended experiments: a magnetic spectrometer, dubbed MAD (Medium Acceptance Detector), and an electro-magnetic calorimeter. The proposed MAD device is a magnetic spectrometer built from two combined-function (quadrupole and dipole) superconducting magnets. The design provides a maximum central momentum of 7.5 GeV/c at a total bend angle of 20°. It is expected that even higher momenta can be accommodated for specific experiments by decreasing the deflection of the second magnet. MAD has a momentum resolution of better than 10^{-3} and an angular resolution of better than 1 mrad. A pointing accuracy of better than 0.5 mrad required for an accurate L/T separation can be obtained. The angular acceptance is 28 msr at angles larger than 35° and decreases linearly to 6 msr at an angle of 12°. An angle as small as 5° can be reached by using one of the existing septum magnets. This design has resulted in the characteristics shown in Table 1.

Table 1. The design characteristics of the MAD spectrometer shown along with the HRS performance

Parameter	MAD design	HRS performance
Central momentum range	0.4 - 7.5 GeV/c	0.2 - 4.3 GeV/c
Scattering angle range	5° - 130°	6° - 150°
Momentum acceptance	±15%	±5%
Momentum resolution	0.1%	0.02%
Angular acceptance	28 msr ($\geq$35°)	6 msr (standard)
	4-6 msr (5° − 12°)	12 msr (forward)
Angular resolution (hor)	1 mrad	0.5 mrad
Angular resolution (ver)	1 mrad	1 mrad
Target length acc. (90°)	50 cm	10 cm
Vertex resolution	0.5 cm	0.1 cm
Maximum DAQ rate	20 kHz	5 kHz
e/h Discrimination	0.5×10^5 at 98%	1×10^5 at 98%
π/K Discrimination	1000 at 95%	1000 at 95%

The basic detector package for the MAD spectrometer, covering the full momentum and angular acceptance, includes: fast high-resolution tracking chambers, a hydrogen-gas Čerenkov counter, trigger scintillator counters and a lead-glass hadron rejector. For the detection of hadrons a variable-pressure gas Čerenkov counter, two diffusely reflecting aerogel counters, a Ring Imaging Čerenkov counter and a Focal Plane Polarimeter will also be available. The data acquisition system is designed with a new generation of pipeline digitizing front-end electronics in order to handle event rates up to 20 kHz.

In combination with the MAD spectrometer, a 100 msr lead-glass calorimeter is available for studies of nucleon form factors and of Real Compton Scattering. A large acceptance, high granularity calorimeter with a 1296 element array of PbF_2 crystals is proposed to optimally study Generalized Parton Distributions through Deep Virtual Compton Scattering. It would also benefit other experiments, such as photo-production of neutral mesons at large transverse momenta.

With the cross section dropping rapidly with increasing Q^2, a large acceptance spectrometer that can operate at a high luminosity is crucial for precision measurements. In combination with the existing HRS or calorimeter in Hall A, MAD will open up a window to a rich research program, including measurements of:

- Polarized and unpolarized quark distributions up to large x, using polarized proton and ^3He targets. Values of x larger than 0.8 can be probed once spin duality has been established;
- Semi-inclusive $\pi^\pm$ and $K^\pm$ electroproduction in order to separate the spin and flavor components of the sea quark distributions up to $x \approx 0.65$;
- Nucleon form factors to significantly larger Q^2-values, over 5 $(\text{GeV/c})^2$ for G_E^n and over 14 $(\text{GeV/c})^2$ for G_E^p;
- Real Compton Scattering up to $s \approx 20$ $(\text{GeV/c})^2$ and $|t| \approx 15$ $(\text{GeV/c})^2$ and Deep Virtual Compton Scattering up to $Q^2 \approx 7$ $(\text{GeV/c})^2$ and $x \approx 0.5$;
- Deuteron photo-disintegration up to photon energies of 7 GeV over a large range of CM angles, with recoil polarization up to a photon energy of 4 GeV;
- Few-body elastic form factors to Q^2-values of over 10 $(\text{GeV/c})^2$;
- Parity violation in deep inelastic scattering with very high accuracy to probe possible extensions of the Standard Model,
- Threshold ψ photoproduction to study the color Van der Waals-type component of the NN force.

4.2 Hall B

The CEBAF Large Acceptance Spectrometer (CLAS) in Hall B is used for experiments that require the detection of several, loosely correlated particles in the hadronic final state at a limited luminosity. CLAS is a magnetic toroidal multi-gap spectrometer. Its magnetic field is generated by six superconducting coils. The detection system consists of drift chambers to determine the track of a charged particle, gas Čerenkov counters for particle identification, scintillation counters for the trigger and for measuring time-of-flight and electromagnetic calorimeters to detect showering particles like electrons and photons. CLAS presently operates at a luminosity of 10^{34} cm^{-2}s^{-1}.

The CLAS upgrade CLAS++ incorporates two major improvements: (1) increasing the luminosity by an order of magnitude to account for the lower cross section values, (2) providing a more complete detection of the hadronic final state. The use of major components (torus magnet, scintillators, Čerenkov counters and EM calorimeters) will be retained. The tracking chambers will be replaced and a new central detector added. Table 2 summarizes the expected performance of CLAS++.

Table 2. CLAS++: acceptance and resolution

	Forward detector	Central detector
Angular coverage		
Tracks (inbending)	$8° - 37°$	$40° - 135°$
Tracks (outbending)	$5° - 37°$	$40° - 135°$
Photons	$3° - 37°$	$40° - 135°$
Track resolution		
$\delta p/p$	$0.003 + 0.001p$	$\delta p_T/p_T = 0.02$
$\delta\theta$ [mrad]	1	8
$\delta\phi$ [mrad]	2 - 5	2
Photon detection		
Energy range [MeV]	> 150	> 60
$\delta\theta$ [mrad]	4 at 1 GeV	15 at 1 GeV
Neutron detection		
η_{eff}	$0.5\ (p > 1.5\ \text{GeV/c}$	NA
Particle ID		
Electron/pion	$> 1000\ (p < 4.8\ \text{GeV/c})$	NA
	$> 100\ (p > 4.8\ \text{GeV/c})$	NA
π^+/π^-	full range	$< 0.65\ \text{GeV/c}$
K/π	full range	$< 0.65\ \text{GeV/c}$
$K^+/p, K^-/\bar{p}$	$< 4.5\ \text{GeV/c}$	$< 0.90\ \text{GeV/c}$
$\pi^° \to \gamma\gamma$	full range	full range
$\eta \to \gamma\gamma$	full range	full range

With these modifications and additions to the existing CLAS components, CLAS++ will be able to carry out a core program for the study of internal nucleon dynamics and hadronization processes as listed here:

- Quark-gluon dynamics and nucleon tomography through measurement of deeply virtual Compton scattering and deeply virtual meson production, both with unpolarized and polarized hydrogen and deuterium targets;
- Polarized and unpolarized valence quark distributions at high x_B, using polarized hydrogen and deuterium targets, and by employing a novel technique of neutron tagging. Values of x_B up to 0.85 can be accessed in deep inelastic processes. A broad program of semi-inclusive measurements will allow quark-flavor tagging and give access to transverse quark structure functions;
- The magnetic structure of the neutron will be probed through magnetic form factor measurements up to $Q^2 \approx 14$ $(\text{GeV/c})^2$, and the C2/M1 ratio for the $N \to \Delta(1232)$ up to $Q^2 \approx 12$ $(\text{GeV/c})^2$. Higher mass resonance transitions can be studied in multiple meson decays at high Q^2 as well;
- Space-time characteristics of quark hadornization and color transparency can be studied in nuclei in highly sensitive processes;

– Meson spectroscopy on $^{3,4}He$ with a small-angle quasi-real photon tagger allows to eliminate baryonic background. Heavy baryon spectroscopy (*e.g.* Ξ) can be studied on hydrogen targets.

4.3 Hall C

The Hall C facility has generally been used for experiments which require high luminosity at moderate resolution. The core spectrometers are the High Momentum Spectrometer (HMS) and the Short Orbit Spectrometer (SOS). These two devices have been used flexibly as either electron or hadron arms, at times in coincidence with each other, at times in coincidence with a third experiment-specific arm. The HMS has a maximum momentum of 7.6 GeV/c, the SOS a value limited to only 1.7 GeV/c.

For the upgrade the SOS spectrometer will be replaced by the so-called Super High Momentum Spectrometer (SHMS). The SHMS design consists of two superconducting quadrupoles and one combined-function magnet. It will have a maximum momentum of 11 GeV/c, a minimum scattering angle of 5.5°, a momentum acceptance (resolution) of ±20 (0.2)% and an angular acceptance (resolution) of 2 msr (2 mrad). The basic configuration of the detector stack would consist of DC tracking chambers, trigger scintillator hodoscopes and a lead glass calorimeter. The overall specifications of the Hall C spectrometer set-up for 11 GeV running are summarized in Table 3.

Table 3. Summary of the design specifications for the SHMS and the HMS performance

Parameter	SHMS Specification	HMS Performance
Central momentum range	2.5 - 11 GeV/c	0.4 - 7.3 GeV/c
Scattering angle range	5.5° - 25°	10.5° - 90°
Momentum acceptance	-15 to +25%	±10%
Momentum resolution	<0.2%	0.1%
Angular acceptance	2 msr (SSA tune)	6.7 msr
	4 msr (LSA tune)	
Angular resolution (hor)	2-4 mrad	0.8 mrad
Angular resolution (ver)	1-1 mrad	1 mrad
Target length acc. (90°)	50 cm	10 cm
Vertex resolution	0.2-0.6 cm	0.3 cm
Maximum DAQ rate	10 kHz	2 kHz
e/h Discrimination	10^3 at 98%	10^3 at 98%
π/K Discrimination	100 at 95%	100 at 95%

With the new equipment Hall C will be able to deliver, amongst others:

– A charged pion from factor measurement up to $Q^2 \approx$ 6 $(GeV/c)^2$;
– Deep exclusive pion and kaon electroproduction up to $Q^2 \approx 10$ $(GeV/c)^2$, including precise longitudinal-transverse separations and spin-dependent measurements;
– Nucleon elastic and transition form factors to $Q^2 \approx 18$ $(GeV/c)^2$;
– Real Compton Scattering up to $s \approx 20$ $(GeV/c)^2$;
– Complete separation of the F_L, F_T, g_1 and g_2 inclusive structure functions of the proton (in the valence quark region) up to $Q^2 \approx 10$ $(GeV/c)^2$;

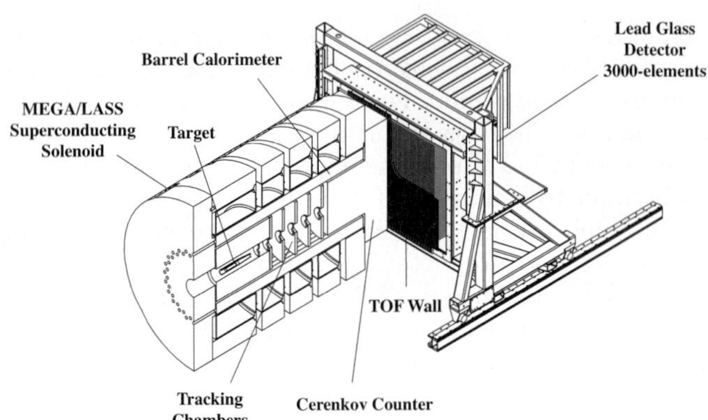

Fig. 2. Schematic view of the detector in Hall D

– Precision measurements of the Q^2-dependence of nuclear effects in both inclusive structure functiuons and (deep) exclusive scattering, crossing the charm threshold;
– A parity violation deep inelastic scattering experiment with unprecedented precision, to search for extensions of the Standard Model.

4.4 Hall D

The GlueX experiment in Hall D will be focused on a definitive measurement of the spectrum of exotic hybrid mesons, which are expected in a mass range from 1 to 2.5 GeV/c^2. Lattice QCD calculations have convincingly illustrated [3] the linear quark-quark potential necessary for confinement. However, very little is still known about the direct excitation of the flux tube. The observation of such direct manifestations of gluonic degrees of freedom will provide understanding of confinement [4]. The quantum numbers of the flux tube, added to those of a $q\bar{q}$ meson, can produce exotic hybrids with unique J^{PC} quantum numbers. These excitations can be probed far more effectively with photons than with π^- or K-mesons, because the quark spins are aligned in the virtual vector-meson component of the photon. For a full partial-wave analysis of such excitations linearly polarized photons are a requisite.

The optimum photon energy for production of exotic hybrids in its expected mass range is between 8 and 9 GeV. Linearly polarized photons in this energy range are optimally produced by coherent bremsstrahlung with 12 GeV electrons. The Hall D detector provides a nearly hermetic acceptance for both charged and neutral particles and includes several particle identification systems. Figure 2 is a schematic representation of the proposed detector. Momentum analysis of charged particles is achieved with a superconducting solenoid and tracking chambers. The final planned photon flux is 10^8 photons/s. At this flux the experiment will acumulate in one year of running a factor of 100 more meson data than are presently available even from pion production. The primary characteristics of the detector are given in Table 4. Its hermetic design makes it an ideal tool to:

- Determine the masses and quantum numbers of mesons in the mass range of 1.5 to 3.5 GeV/c^2;
- Study properties of hybrid mesons produced at rates as low as a few percent of that of normal mesons;
- Map out the poorly known spectra of $s\bar{s}$ mesons.

Table 4. Summary of the GlueX detector's characteristics

	Range
Charged particles	
Coverage	$1 \leq \theta \leq 170°$
Momentum resolution	$\sigma_p/p \approx 1 - 2\%$
Position resolution	$\sigma \approx 150 - 200\mu m$
dE/dx measurements	$20 \leq \theta \leq 140°$
Vertex detector	$\sigma \approx 500\mu m$
Time-of-flight scintillators	$\sigma_t \approx 50$ ps
Čerenkov for π/K separation	$\theta \leq 14°$
Barrel time resolution	$\sigma_t \approx 250$ ps
Photon detection	
Energy measurements	$1 \leq \theta \leq 120°$
Veto capability	$\theta \geq 120°$
Lead glass energy resolution	$\sigma_E/E \approx 2 + 5\%/\sqrt{E}$
Barrel energy resolution	$\sigma_E/E \approx 4.4\%/\sqrt{E}$
Barrel position resolution	$\sigma_z \approx 1$ cm
DAQ/ trigger	
Level 1	200 kHz
Event rate	15 kHz to tape
Data rate	100 MB/s

5 Electron-ion collider

Conceptual design studies [5] have indicated that the JLab facility can be further upgraded to provide a high-luminosity Electron Light-Ion Collider (ELIC) in the 20-65 GeV center-of-mass energy range. Such a facility would uniquely address the following research goals: (1) to complete our understanding of how quarks and gluons provide the binding and the spin of the nucleon, (2) to understand how quarks and gluons evolve into hadrons via the dynamics of confinement and (3) to refine our understanding of how the nuclear binding arises from QCD. A schematic lay-out of the ELIC collider is shown in Fig. 3. Longitudinally polarized electrons are injected into the CEBAF accelerator and accelerated to 5-7 GeV in a single pass (after all of the remaining 20 MV cryo-modules have been replaced by new 100 MV modules). The electrons are then injected into a circulator ring, in which they collide with the ions. After appr. 100 revolutions they are extracted and transported back for deceleration and energy recovery. ^1H, ^2H and ^3He ions are injected longitudinally polarized and accelerated in a conventional RF linac to a maximum

energy of 200 MeV. After pre-boosting to 2-3 GeV the ions are stacked and accelerated in a "Fig. 8" ring to a maximum of 150 GeV. The "Fig. 8" ring provides flexible spin manipulation and preservation. This design results in a maximum CM energy of appr. 65 GeV. A luminosity of 5×10^{34} cm^{-2}s^{-1} appears feasible through the use of electron cooling and crab crossings. Ions of up to mass 40, albeit unpolarized, will also be available. This design will also provide a 25 GeV beam in fixed-target mode.

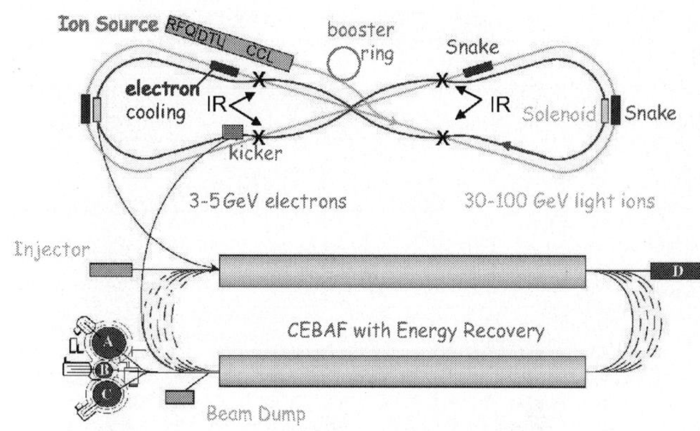

Fig. 3. Schematic view of ELIC

Acknowledgements. The author gratefully acknowledges detailed discussions with L. Merminga and Y. Derbenev on the electron-ion collider. This work was supported by DOE contract DE-AC05-84ER40150 under which the Southeastern Universities Research Association (SURA) operates the Thomas Jefferson National Accelerator Facility.

References

1. Pre-Conceptual Design Report for the Science and Experimental Equipment for the 12 GeV Upgrade of CEBAF, 2003, L.S. Cardman et al., editors: http://www.jlab.org/div_dept/physics_division/pCDR_public
2. X. Ji: Phys. Rev. Lett. **78**, 610 (1997); A. Radyushkin: Phys. Lett. B **380**, 417 (1996)
3. G.S. Bali et al.: Proceedings of Int. Conf. on Quark Confinement and the Hadron spectrum, World Scientific, p. 225 (1995)
4. N. Isgur, R. Kokoski, and J. Paton: Phys. Rev. Lett. **54**, 869 (1985); S. Godfrey and J. Napolitano: Rev. Mod. Phys. **71**, 1411 (1999)
5. L. Merminga, D.R. Douglas, and G.A. Krafft: Annu. Rev. Nucl. Part. Sci. **53**, 387 (2003)

Eur Phys J A (2004) **19**, s01, 249–256
Digital Object Identifier (DOI) 10.1140/epjad/s2004-03-041-9

EPJ A direct
electronic only

Selected recent HERMES results

Klaus Rith, on behalf of the HERMES collaboration

Physikalisches Institut, Universität Erlangen-Nürnberg, D-91058 Erlangen

Received: 19 Aug 2003 / Accepted: 14 Nov 2003 /
Published Online: 6 Feb 2004 – © Società Italiana di Fisica / Springer-Verlag 2004

Abstract. The first phase of HERMES data taking covers the years 1995 to 2000. Data were taken with longitudinally polarised ^{3}He (1995), hydrogen (1996-1997) and deuterium (1998-2000) targets, but also with several other unpolarised targets (H_2, D_2, ^{3}He, ^{4}He, N_2, Ne and Kr). Some selected results from this data taking period are discussed: the spin-dependent structure function g_1 from deuterium, which has been measured with high accuracy down to $x = 0.0021$; quark and antiquark helicity distributions from semi-inclusive deep-inelastic scattering on hydrogen and deuterium, which indicate that the antiquark distributions are consistent with zero and with each other and do not favour a substantial negative polarisation of the strange quark sea; and the first measurement of nuclear attenuation of identified fast charged and neutral pions, charged kaons, protons and antiprotons produced from nitrogen and krypton targets.

PACS. 13.60.-r – 13.87.Fh – 13.88.+e – 14.20.Dh – 14.40.-n – 14.65.-q

1 Introduction

HERMES is one of the four experiments at the HERA electron proton collider at DESY. It uses the high current longitudinally polarised electron or positron beams of HERA with a beam energy E of about 27.6 GeV together with polarised and unpolarised gas targets internal to the storage ring. Scattered electrons and particles produced in the deep-inelastic electron-nucleon interactions are detected and identified by an open-geometry forward spectrometer with large momentum and solid angle acceptance [1]. HERMES is based on two special techniques: longitudinal electron polarisation in a high energy storage ring [2], which is achieved by a system of spin rotator magnets [3], and a storage cell target [4] where the nuclear-polarised atoms from a high intensity polarised source [5, 6] are present as pure atomic species without dilution from unpolarised target material.

The primary scientific goal of HERMES was and still is the detailed investigation of the spin structure of the nucleon. But the physics reach of the experiment extends well beyond this specific aspect of hadronic physics and the experiment explores many details of hadron structure, hadron production and hadronic interactions with electromagnetic probes at cen-tre-of-mass energies $\sqrt{s}$ of around 7 GeV.

The first phase of HERMES data taking covers the years 1995 to 2000. Data were taken with longitudinally polarised ^{3}He (1995), hydrogen (1996-1997) and deuterium (1998-2000) targets, but also with several other unpolarised targets (H_2, D_2, ^{3}He, ^{4}He, N_2, Ne and Kr). These data yielded a lot of high precision results on in-clusive and semi-inclusive polarised deep-inelastic scattering and have confirmed and extended our understanding of the contribution of quark spins to the spin of the nucleon. The results include: the measurement of the spin-dependent structure function g_1 for ^3He, H and D, the determination of the generalised Drell-Hearn-Gerasimov integral for proton, neutron and deuteron, the flavour decomposition of the quark helicity distributions from semi-inclusive polarised deep-inelastic scattering, a first indication of a positively polarised gluon distribution from a double-spin asymmetry in the production of pairs of hadrons with high transverse momenta, the observation of a double-spin asymmetry in the cross section for exclusive electro-production of vector mesons, an experimental indication for effects of transversity in the nucleon, azimuthal single-spin asymmetries in hard exclusive electro-production of real photons and pions which can be related to generalised parton distributions, the investigation of longitudinal spin-transfer in electro-production of Λ hyperons in the current fragmentation region and the observation of a positive transverse polarisation of Λ particles produced in quasi-real photo-production. A comprehensive review of these results on spin-asymmetries in deep-inelastic electron-nucleon scattering can be found in [7]. Additional results include: the determination of a flavour asymmetry in the light quark sea from charged pion production from unpolarised hydrogen and deuterium targets [8], multiplicity distributions in pion electro-production [9], cross sections and decay angular distributions for hard exclusive electro-production of vector mesons [10,11], the first observation of a coherence length effect on exclusive coherent and incoherent electro-production of ρ^0 mesons from nuclear targets and their Q^2 dependence [12,13], and the measurement of the attenuation of fast hadrons produced from nuclear targets [14,15]. Many more results from ongoing analyses- based on data collected in this first phase of data taking will be published in the near future. In this paper only a few aspects are discussed: the spin-dependent structure function g_1 from deuterium, quark and antiquark helicity distributions from semi-inclusive deep-inelastic scattering on D and H, and the attenuation of fast hadrons produced from nuclear targets.

The second phase of HERMES data taking will cover the years until the end of 2006. The physics program, which is summarised in the contribution by W.D. Nowak to this conference [16], will concentrate on measurements with a transversely polarised hydrogen target for a first determination of the so-called *transversity* distribution and measurements with high density unpolarised hydrogen targets in conjunction with a new recoil detector around the storage cell target to study in detail hard exclusive processes.

2 The spin-dependent deuteron structure function $g_1^d(x, Q^2)$

In the years 1998 to 2000 data were taken at HERMES with the longitudinally polarised deuterium target. Altogether about 9 million DIS events were collected, corresponding to an integrated luminosity of about 80 pb^{-1}. The average beam polarisation was $\langle |P^{\mathrm{B}}| \rangle = 0.53$ and the average target polarisation $\langle |P^{\mathrm{T}}| \rangle = 0.845$. The kinematic range of the data is $0.0021 < x < 0.85$, $0.1 < y < 0.91$, $Q^2 > 0.1$ GeV^2 and $W^2 > 3.24$ GeV^2. Here $x = Q^2/2M\nu$ is the Bjorken variable, $-Q^2$ is the squared four-momentum transfer, M is the nucleon mass, ν is the energy of the virtual photon, $y = \nu/E$, and W is the invariant mass of the photon-nucleon system.

The preliminary HERMES result for the structure function ratio $g_1^{\mathrm{d}}/F_1^{\mathrm{d}}$ for the data from the year 2000 is shown in Fig. 1 together with the corresponding data from SMC [17], E143 [18] and E155 [19]. Here

$$g_1^d(x, Q^2) = \frac{1}{2}[g_1^p + g_1^n](x, Q^2) = \frac{1}{4}\sum_q e_q^2\, \Delta q(x, Q^2) \quad (1)$$

$$= \frac{5}{36}[\Delta u + \Delta\bar{u} + \Delta d + \Delta\bar{d} + \Delta s + \Delta\bar{s}](x, Q^2)$$

$$- \frac{1}{12}[\Delta s + \Delta\bar{s}](x, Q^2)$$

is the spin-dependent structure function and F_1^d the corresponding spin-independent structure function. $\Delta q(x, Q^2)$ is the quark helicity distribution for flavour q and e_q is the quark charge in units of the elementary charge. For each data point in x the mean Q^2 is different for each experiment as can be seen from the lower panel of the figure. The seven data points from SMC at even lower values of x down to $x \cong 10^{-4}$ are not shown here. No corrections for smearing effects, which will increase somewhat the error bars and influence the central values mainly at very

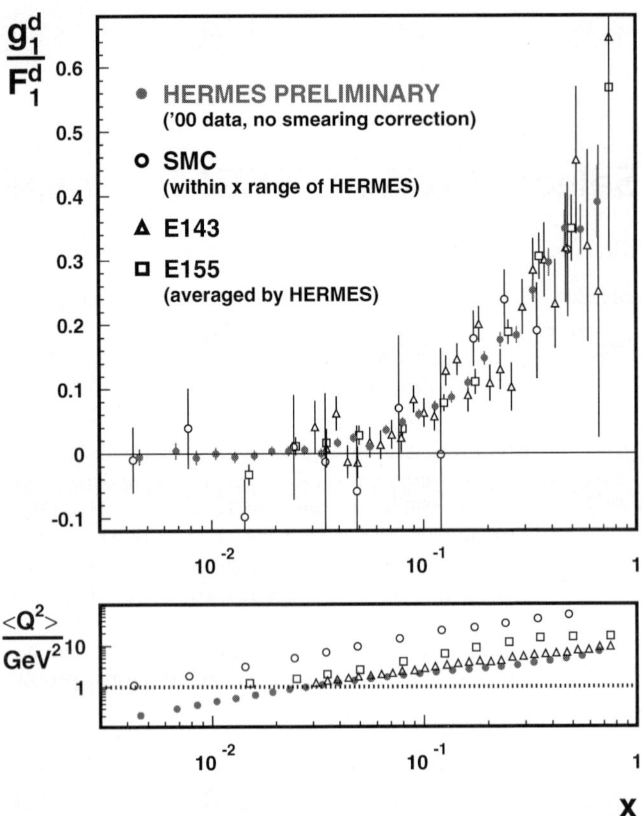

Fig. 1. Preliminary HERMES results for the structure function asymmetry $g_1^{\mathrm{d}}/F_1^{\mathrm{d}}$ as a function of x together with the corresponding data from E143, E155 and SMC in the x range covered by the HERMES data

low x, have been applied yet to these preliminary HERMES data. They have substantially smaller error bars and much smaller point to point fluctuations than those from the other experiments, and determine the x dependence of this ratio now very well. This is especially evident in the range of very small x, but also at higher x where the central values of various previous data sets scattered substantially. At x below about 0.03 the asymmetry is compatible with zero, with a slight tendency to negative values. One can conclude from this behaviour that at these small x values $\Delta d(x)$ and $\Delta u(x)$ have about the same magnitude, but opposite sign. If the asymmetry really gets negative then one can speculate that either the magnitude of the down-quark helicity distribution must get bigger than the up-quark helicity distribution or that the net sea-quark helicity distribution must be negative. At large x the rise of the asymmetry with x is much less steep than for the proton data [17,18,20,21,22], the limit at $x \to 1$ possibly being substantially smaller than unity.

The preliminary HERMES result from the 2000 data for $g_1^{\mathrm{d}}(x)$ at the measured values of Q^2 is shown as a function of x together with the corresponding data from SMC in Fig. 2. It is evident that with these new data also the x dependence of g_1^{d} is very well determined in the range $x > 2\cdot10^{-3}$. Below x values of about $3\cdot10^{-2}$ it is compatible with zero with a slight tendency to negative values.

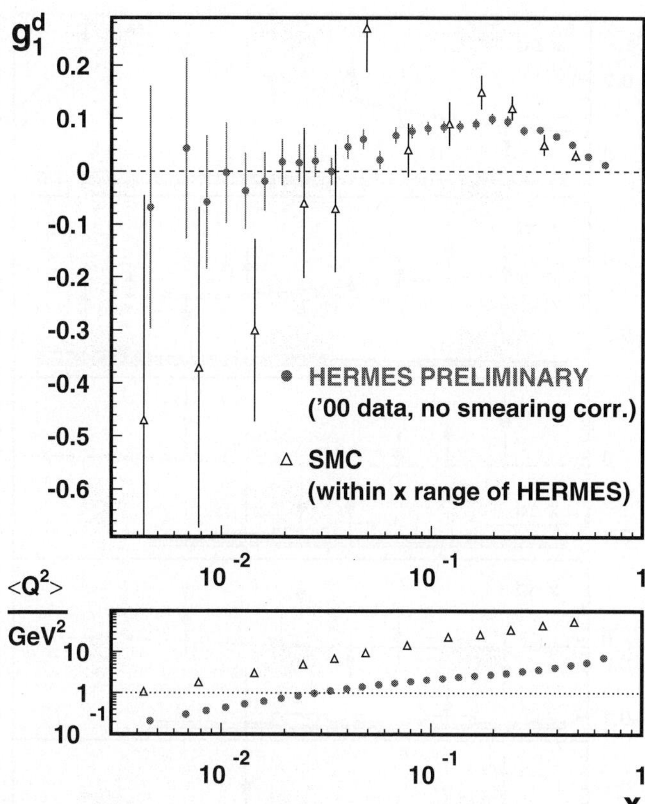

Fig. 2. Preliminary HERMES results for the structure function g_1^d together with the SMC results at the measured values of Q^2 as a function of x in the x range covered by the HERMES data

3 Quark helicity distributions from semi-inclusive DIS

3.1 Semi-inclusive asymmetries

The up-quark helicity distribution, $\Delta u(x)$, is rather well constrained by the inclusive polarised proton and deuteron data as well in shape as in magnitude, while NLO-QCD fits, like e.g. [23, 24, 25, 26] give a much larger possible range for $\Delta d(x)$. The sea-quark helicity distributions and especially $\Delta s(x)$ are essentially undetermined from such fits and very different solutions are possible for the gluon helicity distribution $\Delta g(x)$. A detailed understanding of the spin structure of the nucleon requires, therefore, a direct determination of the x-dependence and of the moments of the helicity distributions of quarks and antiquarks of the various flavours, and of the gluon. This can in principle be achieved in semi-inclusive spin-dependent deep-inelastic scattering when in addition to the scattered lepton also a leading hadron is detected. Since the charge and the identity of high energy forward hadrons are correlated to the flavour of the struck quark, this type of reaction can be used to study the flavour dependence of the quark helicity distribution. Precise double-spin asymmetries in the cross sections for leptoproduction of various types of hadrons allow a separation of the individual contributions of each quark flavour to the nucleon spin.

In leading order (LO) QCD, the semi-inclusive cross section for the production of a particular hadron h takes the factorised form

$$\sigma^h(x, z) \propto \sum_q e_q^2 \, q(x) \, D_q^h(z) , \qquad (2)$$

where the fragmentation function $D_q^h(z)$, which parameterises all our ignorance of the non-perturbative fragmentation process, is the probability density that a struck quark of flavour q fragments into a hadron h with energy E_h, or fractional virtual photon energy $z = E_h/\nu$, respectively. For simplicity the weak logarithmic dependence on Q^2 is suppressed here. The double-spin asymmetry A_1^h in the above cross section is then given by

$$A_1^h(x) = \frac{\sum_q e_q^2 \Delta q(x) D_q^h(z,)}{\sum_{q'} e_{q'}^2 q'(x) D_{q'}^h(z)} \cdot \frac{1 + R(x)}{(1 + \gamma^2)}, \qquad (3)$$

where the factor involving $\gamma^2 = Q^2/\nu^2$ and R, the ratio of longitudinal to transverse virtual photon cross sections, accounts for the longitudinal component that is included in experimentally determined parameterisations of $q(x)$ but not in $\Delta q(x)$. It is assumed that the spin-dependent and spin-independent fragmentation functions are the same at least for pseudo-scalar mesons like pions and kaons.

3.2 Quark polarisations and purities

Integrating (2) over the range $0.2 \leq z \leq 0.8$, where one can assume that the hadrons originated from the struck quark and not from the target remnant and where also the contribution from exclusive processes is small, (3) becomes

$$A_1^h(x_i) = \sum_q P_q^h(x_i) \cdot \frac{\Delta q(x_i)}{q(x_i)} \cdot \frac{1 + R(x)}{(1 + \gamma^2)} \quad , \qquad (4)$$

where $\Delta q(x_i)/q(x_i)$ are the quark polarisations and $P_q^h(x_i)$ are the spin-independent purities:

$$P_q^h(x_i) \equiv \frac{e_q^2 q\left(x_i, Q_i^2\right) \mathcal{D}_q^h\left(x_i, z_i, Q_i^2\right)}{\sum_{q'} e_{q'}^2 q'\left(x_i, Q_i^2\right) \mathcal{D}_{q'}^h\left(x_i, z_i, Q_i^2\right)} \quad . \qquad (5)$$

Here $\mathcal{D}_q^h\left(x_i, z_i, Q_i^2\right)$ are *effective* fragmentation functions in a bin i with mean values of the kinematical quantities x_i, z_i, and Q_i^2 derived by averaging over all events in the bin, which take into account also the acceptance function of the spectrometer. Note that these purities only depend on one single parameter x_i in this definition, as the mean values Q_i^2 and z_i are fixed within one x-bin i.

The purities may be interpreted as the probability that, in a given bin of x, the photon struck a quark of type q in the nucleon when a hadron of type h is detected in the experiment. Obviously, the purities for each hadron species add up to one in each bin.

The purities provide a simple way to separate the quark polarisations from other quantities, which are related to unpolarised quark distributions and fragmentation functions. Their input parameters are known with good precision from a large number of DIS experiments on unpolarised targets, augmented by data on fragmentation functions from e^+e^- annihilation experiments.

For the determination of the various quark polarisations one combines inclusive asymmetries and semi-inclusive asymmetries for positive and negative hadrons and for identified hadrons of different species from various targets and obtains a system of linear equations which can be written for each bin i in a matrix form as:

$$\mathbf{A}\left(x_i\right) = \mathbf{P}\left(x_i\right)\mathbf{Q}\left(x_i\right) . \tag{6}$$

Here the vector $\mathbf{A}$ contains as elements the measured inclusive and semi-inclusive hadron asymmetries for the various targets, the vector $\mathbf{Q}$ contains the quark and antiquark polarisations for each flavour and $\mathbf{P}$ is the purity matrix with elements P_q^h. This system of asymmetries is typically over-determined, as the number of measured asymmetries is larger than the number of various quark polarisations to be extracted.

The polarisations for each bin x_i are then obtained by solving (6) for the vector $\mathbf{Q}$ by χ^2-minimisation, accounting for the correlations between x-bins and between the various asymmetries.

3.3 Quark and antiquark helicity distributions

The high statistics deuterium data allowed a precise determination of semi-inclusive double-spin asymmetries for identified positive and negative pions and kaons. Together with the semi-inclusive asymmetries for positive and negative hadrons from the proton target [27] and the inclusive proton and deuteron asymmetries eight different asymmetries could be used as input for the asymmetry vector $\mathbf{A}(x) = (A_{1p}, A_{1p}^{\pi+}, A_{1p}^{\pi-}, A_{1d}, A_{1d}^{\pi+}, A_{1d}^{\pi-}, A_{1d}^{K+}, A_{1d}^{K-})$. The large data sample allowed a five parameter fit using the following polarisation vector:

$$\mathbf{Q}(x) = \left(\frac{\Delta u}{u}, \frac{\Delta d}{d}, \frac{\Delta \bar{u}}{\bar{u}}, \frac{\Delta \bar{d}}{\bar{d}}, \frac{\Delta s}{s}, \frac{\Delta \bar{s}}{\bar{s}} \equiv 0 \pm \frac{1}{\sqrt{3}}\right) . \tag{7}$$

The quark and antiquark helicity distributions $\Delta q(x)$ are determined as the product of the extracted polarisations and the corresponding spin-independent parton distribution functions [29] at the mean $\langle Q^2 \rangle = 2.5 \; GeV^2$ of the present work. The results [28] are shown in Fig. 3 together with two LO-QCD fits [25, 26] to previously published inclusive data.

In these fits the first moments of the valence helicity distribution Δu_v and Δd_v have been constrained using the F and D decay constants experimentally determined from the weak decays of the members of the spin-$\frac{1}{2}$ baryon octet. Explicitly flavour symmetry among the three sea-quark helicity distributions has been assumed in order to reduce the number of free fit parameters. Together with

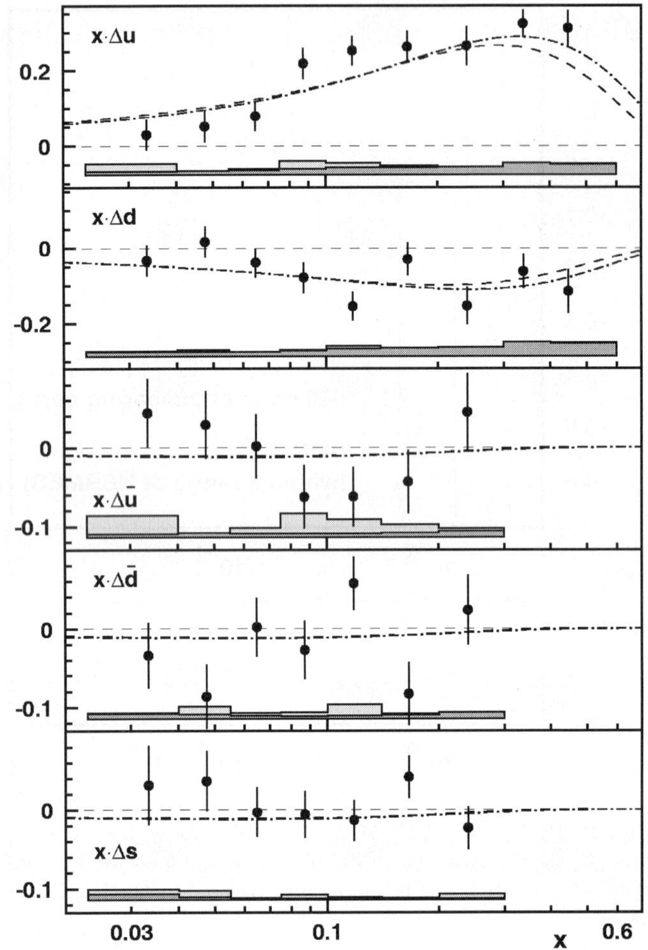

Fig. 3. Quark helicity distributions at $\langle Q^2 \rangle = 2.5 \; GeV^2$ as a function of x compared to two LO QCD fits to previously published inclusive data shown as dashed [25] and dot-dashed [26] curves

the above constraint this leads to a small and slightly negative sea-quark helicity distribution. The helicity distributions $\Delta u(x)$ and $\Delta d(x)$ extracted from the HERMES data are consistent with previous (semi)-inclusive results and with the LO-QCD fits. $\Delta u(x)$ is positive and $\Delta d(x)$ is negative over the measured range of x. Over most of the x-range the ratio of the magnitude of the two distributions is around 0.4, which is consistent with the negative value of $g_1^n(x)$ but substantially larger than the ratio $\frac{1}{4}$ predicted for the moments in the naive quark parton model. The sea-quark helicity distributions $\Delta \bar{u}(x)$, $\Delta \bar{d}(x)$ and $\Delta s(x)$, extracted separately from these data for the first time, are consistent with zero and with each other. The statistical error bars are, however, still substantially larger than the magnitude of the distribution obtained from the NLO-QCD fits. Within errors there is no indication for a substantial negative polarisation of the strange sea that was deduced from the analysis of only inclusive data assuming SU(3) symmetry, but given the size of the errors it is difficult to conclude whether there really is a disagreement or not.

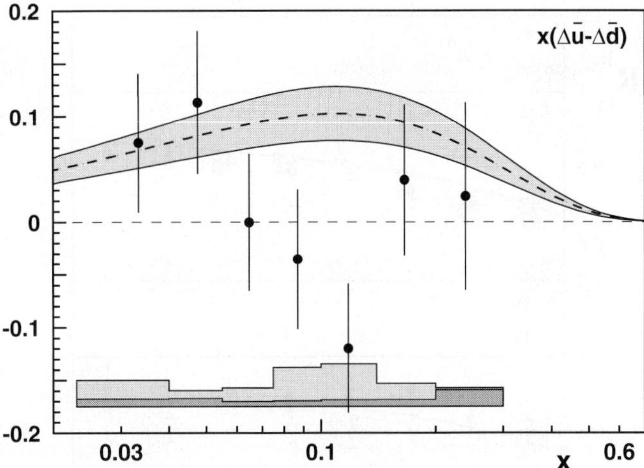

Fig. 4. The light quark sea flavour asymmetry $\Delta\bar{u} - \Delta\bar{d}$ in the helicity distributions, at $\langle Q^2 \rangle = 2.5 \ GeV^2$, compared to a theoretical prediction [30] (*dashed curve with theoretical uncertainty band*)

Figure 4 shows the difference of the light quark sea helicity distribution $\Delta\bar{u}(x) - \Delta\bar{d}(x)$ together with a theoretical prediction of this quantity based on the chiral quark soliton model [30]. There is no evidence in the data for a large flavour asymmetry, and the data appear to be below the theoretical prediction. Given the substantial point to point fluctuations there is, however, no real discrepancy between this theoretical prediction and the experimental data.

4 Hadronisation in nuclei

In QCD hadrons are colourless objects and quarks with nonzero colour quantum number can therefore not exist as free particles. This phenomenon of confinement leads to the process of hadronisation in high energy reactions like deep-inelastic scattering: as well the struck quark as the target remnant have to fragment into hadrons. This fragmentation process is described by the fragmentation function $D_q^h(z)$, introduced above, which is a measure for the probability that a quark of flavour q produces a hadron of type h carrying a fraction z of the energy of the struck quark in the target rest frame.

Experimental information on the hadronisation process can be obtained by embedding the formation process in an atomic nucleus. The nuclear environment may thereby influence the hadronisation process for instance by multiple interactions of the quark with the surrounding medium and induced gluon radiation resulting in energy loss of the quark and thereby a reduction of the hadron yield at high z. If the hadron is formed inside the nucleus, the hadron can interact via the relevant hadronic interactions, causing a further reduction of the hadron yield. This reduction depends on the distance traversed by the quark before the hadron is formed which is proportional to the so called formation time and the quark energy ν. At high ν the hadron will be formed outside the nucleus and the

reduction of yield will be small. By studying the properties of leading hadrons emerging from nuclei, information on the characteristic time-distance scales of hadronisation can be derived. The understanding of quark propagation in the nuclear medium is crucial for the interpretation of ultra-relativistic heavy ion collisions, as well as high energy proton-nucleus and lepton-nucleus interactions [31].

Theoretical descriptions of the hadronisation process in the nuclear medium include phenomenological string models, final state interactions of the hadrons with the surrounding medium or in-medium modifications of the quark fragmentation functions due to different origins like parton energy loss, higher twist contributions or gluon bremsstrahlung. For a collection of corresponding references see [14,15], some aspects of the different approaches have been discussed at this conference [32].

The HERMES experiment is ideally suited for such studies: the energy domain ranges from about 2 to 23 GeV and the corresponding length scales involved in the hadronisation process are of the same order as the nuclear size, the nuclear sizes of a few fm can be varied by using several atomic species ranging from hydrogen to krypton, and the gaseous targets are so dilute that very little secondary interactions of the produced hadrons with the target material occur. The dual-radiator RICH detector [33] allows to identify pions, kaons, protons and antiprotons over most of the kinematic range covered by the spectrometer and to study the nuclear effects separately for these different hadron species.

The experimental results for semi-inclusive deep-inelas-tic scattering on nuclei are usually presented in terms of the hadron multiplicity ratio R_M^h, which is defined as the number of hadrons of type h produced per deep-inelastic scattering event for a nuclear target of mass A to that from a deuterium target (D). This ratio depends on ν, Q^2, z and p_t^2, where p_t is the hadron momentum component transverse to the virtual photon direction.

The multiplicity ratio is defined as:

$$R_M^h(z, \nu, p_t^2, Q^2) = \frac{\left. \frac{N_h(z, \nu, p_t^2, Q^2)}{N_e(\nu, Q^2)} \right|_A}{\left. \frac{N_h(z, \nu, p_t^2, Q^2)}{N_e(\nu, Q^2)} \right|_D}, \qquad (8)$$

where N_h is the yield of semi-inclusive hadrons in a given (z, ν, p_t^2, Q^2)-bin, and N_e the yield of inclusive deep-inelastic scattering leptons in the same (ν, Q^2)-bin. In the following the ratio R_M^h is evaluated as a function of ν, z and p_t^2, while integrating over all other kinematic variables.

In this contribution results are presented on the hadron multiplicities on krypton ^{84}Kr and nitrogen ^{14}N relative to deuterium for beam energies of 12 and 27.6 GeV. These are the first measurements of the multiplicity ratio for identified charged and neutral pions, charged kaons, protons and antiprotons. Additionally the previously published [14] nitrogen data for charged hadrons and identified pions are reevaluated, now covering a wider kinematic range. The 27.6 GeV krypton data were obtained during dedicated runs in 1999, when high density krypton gas was injected into the storage cell and target areal densities up to 1.4×10^{16} nucleons per cm^2 were obtained. This made it

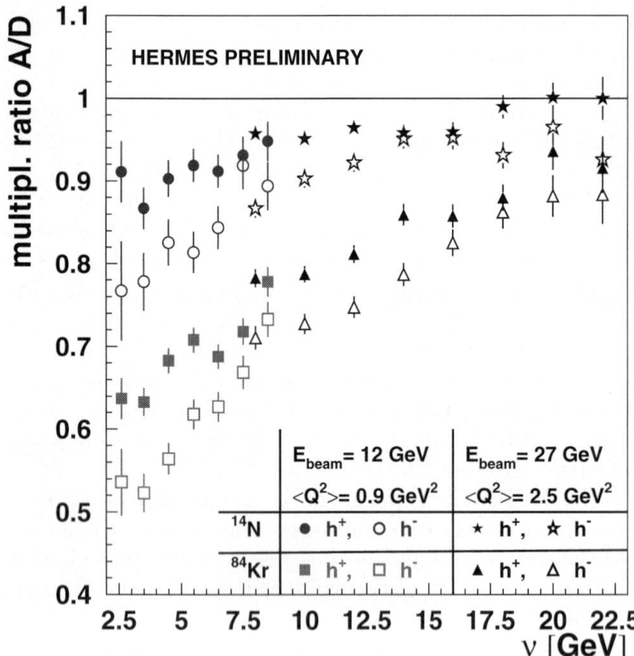

Fig. 5. Hadron attenuation for nitrogen and krypton as a function of ν for the two beam energies 12 GeV and 27.6 GeV

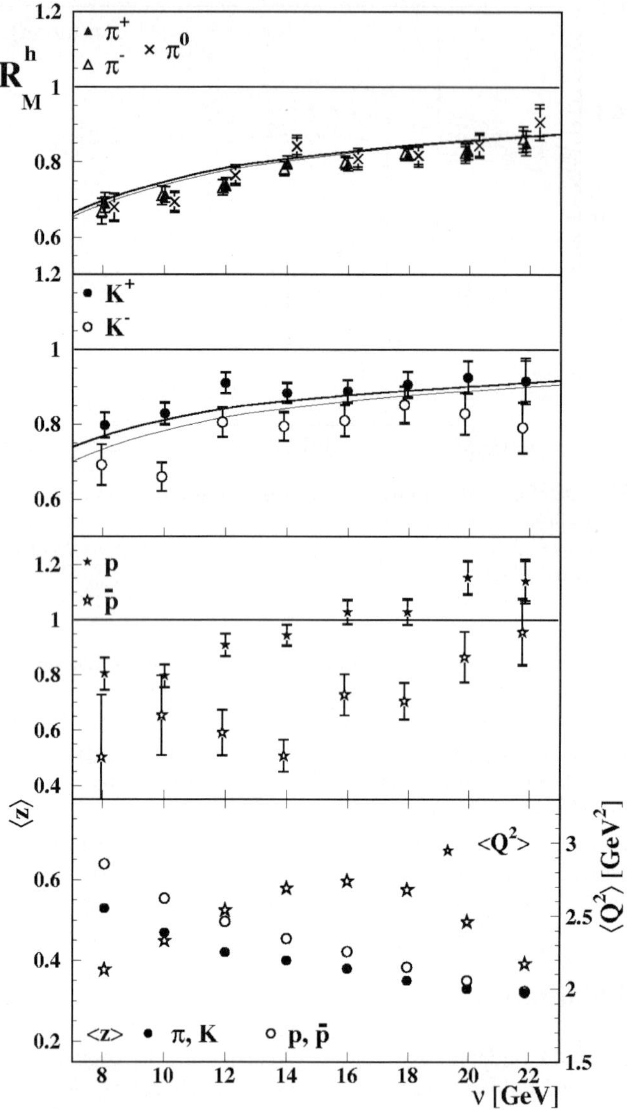

Fig. 6. Hadron attenuation in krypton as a function of ν for a beam energy of 27.6 GeV

possible to accumulate the krypton statistics within a few days. The polarised deuterium data were collected over a period of one year (1999) with an about hundred times less dense polarised target. The yields from deuterium were averaged over the two spin orientations. The 12 GeV data for deuterium, nitrogen and krypton have been obtained in fall 2000, when HERA has operated for a short period at this energy.

In Fig. 5 preliminary results for the multiplicity ratio as a function of ν are shown for both beam energies and for nitrogen and krypton [34].

Following features can be observed: the attenuation effect is substantially stronger for krypton than for nitrogen; for negative hadrons R^h_M decreases continuously with decreasing ν down to values of about 0.5 (0.75) at ν around 0.2 for krypton (nitrogen); at high values of ν the ratio approaches unity, in agreement with earlier EMC data [35], where at ν values around 50 GeV the ratio was compatible with unity for a ^{12}C target and about 0.97 for Cu, indicating that the nuclear effects become very small when the hadron is formed far outside the nucleus and also that effects due to hadron interactions are the dominant source for the attenuation; and there is a significant difference between the ratio for positive and negative hadrons, in agreement with the previously published HERMES data for nitrogen [14].

The ν dependence of the multiplicity ratio for identified pions, kaons, protons and antiprotons for $z > 0.2$ obtained at a beam energy of 27.6 GeV is shown in Fig. 6. The inner (outer) error bars represent the statistical (total) uncertainties, the thick (thin) solid curves represent the calculations of [36] for positive (negative)

charge states. In the bottom panel the average z and Q^2 values are displayed for each bin in ν.

The data show the following features: the attenuation ratio increases with increasing ν for all hadron species; R^h_M is identical for positive, negative and neutral pions and also for negative kaons within the total experimental uncertainties, while the attenuation effect for positive kaons is significantly smaller; an even larger difference is observed between protons and antiprotons and between these baryons and mesons. The charge dependence shown in Fig. 5 must therefore be due to positive kaons and protons. These differences can be interpreted in terms of different formation times of baryons and mesons [37], in terms of different hadron-nucleon interaction cross sections or a mixing of quark and gluon fragmentation functions and their modification in nuclei [38].

In Fig. 7 the multiplicity ratios for identified hadrons are presented as a function of z for $\nu > 7 GeV$. In the

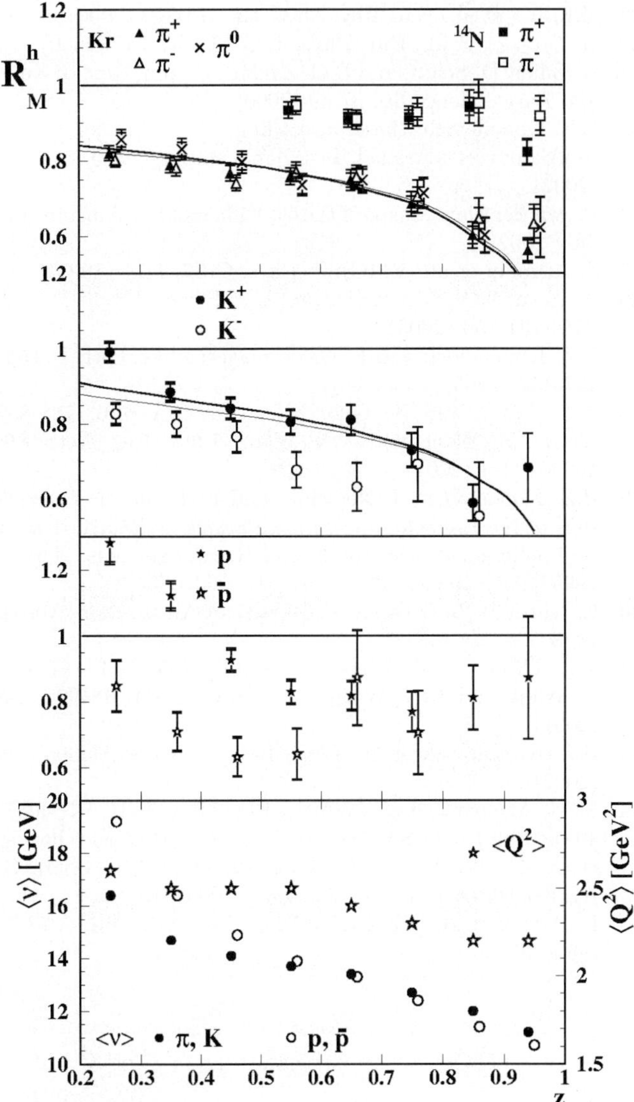

Fig. 7. Hadron attenuation as a function of z for 27.6 GeV

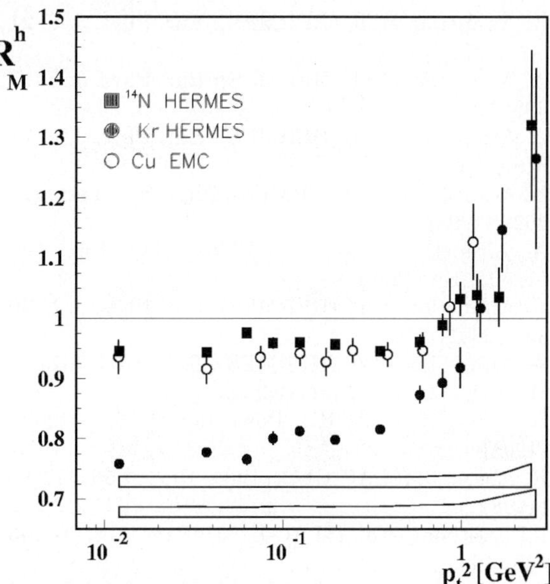

Fig. 8. Hadron attenuation in charged lepton-nucleon scattering as a function of p_t^2

$z > 0.2$, together with data from EMC [35] for Cu in the range $10 < \nu < 80$ GeV.

The data for $p_t^2 < 0.7$ GeV^2 show the attenuation discussed above, while at larger values of p_t^2 one observes an enhancement. Such a behaviour can be explained by multiple scattering of the quarks and hadrons propagating in the nuclear medium which broadens the transverse momentum distribution. Such an effect, but larger in magnitude, has been previously observed in hadron-nucleus, and nucleus-nucleus reactions and is known as the Cronin effect [41]. Recent theoretical calculations [42,43] predict this enhancement to occur at a p_t scale of about $1-2$ GeV. These HERMES data may help to interpret the new relativistic heavy-ion results from SPS [44] and RHIC [45], which show a weaker p_t enhancement than expected from the proton-nucleus data.

Acknowledgements. This work was supported by the German Bundesministerium für Bildung und Forschung, BMBF, grant numbers 06 ER 928I and 06 ER 125I.

upper panel the ratio for identified charged pions from the reanalysed nitrogen data is also shown for comparison.

At least for the mesons the attenuation ratio decreases substantially with increasing z. An indication of such a behaviour has already been observed in the previously published nitrogen data [14]. It is much stronger and clearly visible in the krypton data. The data are in good agreement with the model calculations of [36], but also in fair agreement with those of the gluon-bremsstrahlung model [39].

The data for identified charged pions for $z > 0.5$ were used to estimate the mass number dependence of the attenuation [40]. They are closer to the $A^{2/3}$-dependence predicted in [38] than to the $A^{1/3}$-dependence expected from models based on nuclear absorption effects only.

In Fig. 8 the multiplicity ratio R_M^h for charged hadrons from ^{14}N and Kr is shown as a function of of the squared transverse hadron momentum, p_t^2, for $\nu > 7$ GeV and

References

1. K. Ackerstaff et al. (HERMES): Nucl. Instr. and Meth. A **417**, 230-265 (1998)
2. D.P. Barber et al.: Phys. Lett. B **343**, 436-443 (1995)
3. J. Buon and K. Steffen: Nucl. Instr. and Meth. A **245**, 248-261 (1986)
4. C. Baumgarten et al.: Nucl. Instrum. and Meth. A **496**, 277-285 (2003)
5. A. Nass et al.: Nucl. Inst. and Meth. A **505**, 633-644 (2003)
6. C. Baumgarten et al.: Nucl. Instrum. and Meth. A **482**, 606-618 (2002)
7. K. Rith: Prog. Part. Nucl. Phys. **49**, 245-324 (2002)
8. K. Ackerstaff et al. (HERMES): Phys. Rev. Lett. **81**, 5519-5523 (1998)

9. A. Airapetian et al. (HERMES): Eur. Phys. J. C **21**, 599-606 (2001)
10. A. Airapetian et al. (HERMES): Eur. Phys. J. C **17**, 389-398 (2000)
11. A. Airapetian et al. (HERMES): Eur. Phys. J. C **18**, 303-316 (2000)
12. K. Ackerstaff et al. (HERMES): Phys. Rev. Lett. **82**, 3025-3029 (1998)
13. A. Airapetian et al. (HERMES): Phys. Rev. Lett. **90**, 052501, 1-6 (1998)
14. A. Airapetian et al. (HERMES): Eur. Phys. J. C **20**, 479-486 (2001)
15. A. Airapetian et al. (HERMES): hep-ex/0307023
16. W.D. Nowak: these proceedings
17. B. Adeva et al. (SMC): Phys. Rev. D **58**, 112001, 1-17 (1998)
18. K. Abe et al. (SLAC-E143): Phys. Rev. D **58**, 112003, 1-54 (1998)
19. P.L. Anthony et al. (SLAC-E155): Phys. Lett. B **463**, 339-345 (1999)
20. A. Airapetian et al. (HERMES): Phys. Lett. B **442**, 484-492 (1998)
21. B. Adeva et al. (SMC): Phys. Rev. D **60**, 072004, 1-9 (1999); Phys. Rev. D **62**, 079902, 1-2(E) (2000)
22. P.L. Anthony et al. (SLAC-E155): Phys. Lett. B **493**, 19-29 (2000)
23. B. Adeva et al. (SMC): Phys. Rev. D **58**, 112002, 1-15 (1998)
24. Y. Goto et al.: Phys. Rev. D **62**, 034017, 1-18 (2000)
25. M. Glück et al.: Phys. Rev. D **63**, 094005, 1-12 (2001)
26. J. Blümlein and H. Böttcher: Nucl. Phys. B **636**, 225-263 (2002)
27. K. Ackerstaff et al. (HERMES): Phys. Lett. B **464**, 123-134 (1999)
28. A. Airapetian et al. (HERMES): hep-ex/0307064
29. H.L. Lai et al.: Eur. Phys. J. C **12**, 375-392 (2000)
30. B. Dressler et al.: Eur. Phys. J. C **14**, 147-157 (2000)
31. R. Baier, D. Schiff, and B.G. Zakharov: Ann. Rev. of Nucl. and Part. Science **50**, 37-69 (2000)
32. B.Z. Kopeliovich: these proceedings
33. N. Akopov et al.: Nucl. Instr. and Meth. A **479**, 511-530 (2002)
34. P. van der Nat: Master's thesis, University of Amsterdam, July 2002
35. J. Ashman et al. (EMC): Z. Phys. C **52**, 1-11 (1991)
36. A. Accardi, V. Muccifora, and H.J. Pirner: Nucl. Phys. A **720**, 131-156 (2003)
37. B.Z. Kopeliovich and F. Niedermayer: Phys. Lett. B **151**, 437-438 (1985)
38. X.N. Wang and X. Guo: Nucl. Phys. A **696**, 788-832 (2001); E. Wang and X.N. Wang: Phys. Rev. Lett. **89**, 162301, 1-4 (2002)
39. B.Z. Kopeliovich, J. Nemchik, and E. Predazzi: *Proceedings of the workshop on Future Physics at HERA, Vol. 2*, G. Ingelman, A. De Roeck, and R. Klanner, eds. (DESY 1995/1996) 1038-1042 and nucl-th/9607036
40. E. Garutti: Ph.D thesis, University of Amsterdam, March 2003
41. J.W. Cronin et al.: Phys. Rev. D **11**, 3105-3123 (1975)
42. E. Wang and X.N. Wang: Phys. Rev. C **64**, 034901, 1-8 (2001)
43. B.Z. Kopeliovich et al.: Phys. Rev. Lett. **88**, 232303, 1-4 (2002)
44. M.M. Aggarwal et al. (WA98): Phys. Rev. Lett. **81**, 4087-4091 (1998); and **84**, 578-579(E) (2000); H. Appelshauser et al. (NA49): Phys. Rev. Lett. **82**, 2471-2475 (1999); G. Agakishiev et al. (CERES): hep-ex/0003012
45. K. Adcox et al. (PHENIX): Phys. Rev. Lett. **88**, 022301, 1-6 (2002)

Eur Phys J A (2004) **19**, s01, 257–260
Digital Object Identifier (DOI) 10.1140/epjad/s2004-03-042-8

EPJ A direct
electronic only

Mesonproduction and experimental studies of the excitation spectrum of the nucleon

Berthold H. Schoch

Physikalisches Institut, Universität Bonn, Germany

Received: 31 Jul 2003 / Accepted: 14 Nov 2003 /
Published Online: 6 Feb 2004 – © Società Italiana di Fisica / Springer-Verlag 2004

Abstract. Key issues of the structure of the nucleon can be addressed by experimental studies of Meson-production. Depending on the selected channels and kinematics, different aspects can be investigated. The measurements of charged channels at low t constitute a large part of the cross section and exhibit a special sensitivity of the meson cloud. These contributions can be strongly suppressed by using a polarized beam and target. Instead, the excitation of resonances are emphasized in this case. The choice of neutral channels in the final state shows also dominant sensitivities for resonance excitation. The sequential decay of resonances can be studied systematically for the first time.

1 Introduction

Several key questions in low energy QCD can be addressed by studying the complete spectrum of excited states of the nucleon . This includes the determination of the quantum numbers and decay branching ratios of these states as well as the extraction of the photocouplings of the states. The identification of the relevant degrees of freedom is one of the most important issues to settle. From the knowledge of the low energy part of the excitation spectrum it can be inferred that the nucleon is built from the degrees of freedom of spin -1/2 fermions confined to a valence qqq system. Further questions wait for answers: Are there, in addition, excitations of the gluonic and sea quark degrees of freedom? Can as many states be found as results from calculations based on symmetric quark models suggest or are degrees of freedom frozen out? The quarks considered in the qqq system are quasiparticles with masses of a third of the nucleon mass. Do the properties of these quasiparticles change when excited to the highest energies? Does the relatively large energy splitting of the parity doublets change to get degenerate at higher excitation energies? An answer to these questions leads to an understanding how in the framework of QCD baryons are formed and provides the basis for a description of the origin of the forces between nucleons. This motivation and the recent progress concerning advanced equipment revives programs of studies of the excitation spectrum of the nucleon in several laboratories. With the study of reactions like $\gamma + N \Longrightarrow N + n \cdot mesons$ a comprehensive experimental program can be carried out. Using polarized beams and targets as well as large acceptance detectors, the relevant observables can be extracted. Illustrated by data taken recently at ELSA , a general picture of photo - meson production will be presented.

2 Total photon cross sections

2.1 The total photon absorption cross section

The total photon absorption cross [1] section on the proton (see Fig. 1) shows two remarkable features: The pronounced resonance structure at low photon energies ($E_\gamma \leq 1.5 GeV$) and the seemingly structureless shape of the cross section at higher energies ($E_\gamma \geq 1.5 GeV$). The magnetic and electric dipole excitations of the nucleon are seen in the first and second peak, respectively. The structureless part is well described by an extrapolation of a Regge fit for photon energies with $6 GeV \leq E_\gamma \leq 200 GeV$ [2]. The successful extrapolation into the resonance region suggests that the same absorption mechanism as in the high energy regime is responsible for this "background" contributions, as this part of the cross section is called in the resonance region. The detailed study of this "background" contribution is interesting by itself, because it constitutes a mayor part of the total absorption cross section which determines via dispersion theory such static properties of the nucleon like the electric and magnetic polarizabilities. These again, are test cases for the validity of extended calculations based on the Chiral Perturbation Theory, the low energy realization of QCD.

2.2 Total meson production cross sections

Figure 2 shows, together with the total photon absorption cross section, the total cross sections of the reactions $\gamma + p \Longrightarrow p + \pi^+ + \pi^-$, $\gamma + p \Longrightarrow p + \pi^+ + \pi^- + \pi^0$, $\gamma + p \Longrightarrow p + \rho^0$, $\gamma + p \Longrightarrow p + \omega$ and $\gamma + p \Longrightarrow p + K^+ + K^-$ as measured with the SAPHIR detector at ELSA [3,4].

These cross sections represent a large portion of the total absorption cross section for $E_\gamma \geq 1 GeV$. A first step to

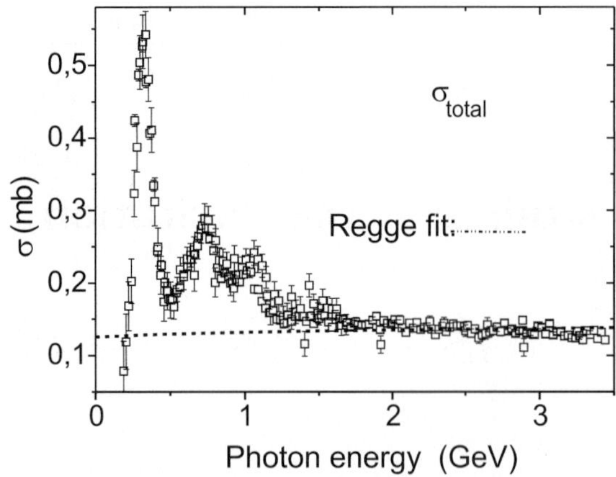

Fig. 1. The total absorption cross section of the proton

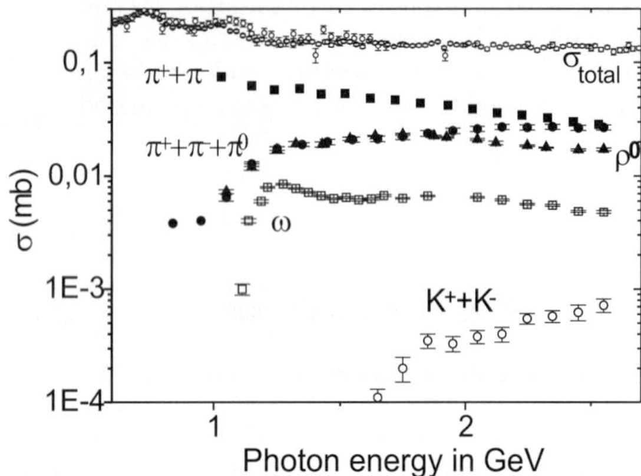

Fig. 2. Total meson production cross sections

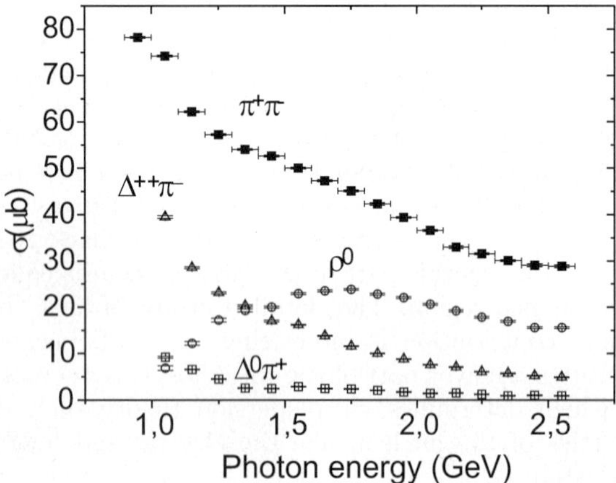

Fig. 3. Decomposition of the two pion cross section

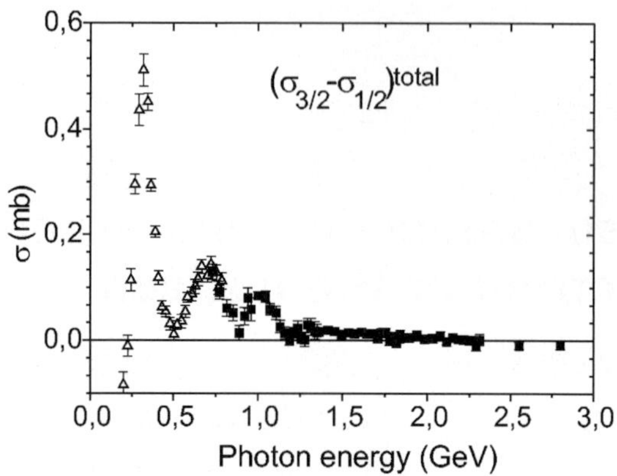

Fig. 4. Total cross section difference

2.3 Total photon absorption cross sections for a polarized beam and target

With the development and implementation of polarized targets and beams for $\gamma-$ induced reactions new classes of experiments become possible. The spin structure in the resonance region can be investigated in a more direct way by using longitudinally polarized $\gamma-$ beams in combination with longitudinal polarized targets. A mayor experimental achievement has been the development of polarized targets [5] which are suitable to be used in the whole angular acceptance range of 4π- detectors. Linearly polarized $\gamma-$beams, prepared by using coherent bremsstrahlung, are suitable to investigate transitions due to convection currents and allow to separate parity even and parity odd transitions. For the first time the total absorption cross section for circular polarized photons has been determined by performing double polarization experiments. By measuring the difference of the spin projected cross sections $\sigma_{3/2}- \sigma_{1/2}$ the spin response of the nucleon excitation spectrum up to $E_\gamma = 3GeV$ has been extracted. Figure 4 shows the total photon absorption cross section difference $\sigma_{3/2}- \sigma_{1/2}$ in the photon energy range $.2GeV \leq E_\gamma \leq 3GeV$ as has been measured in a double polarization experiment at MAMI ($0.2GeV \leq E_\gamma \leq 0.8GeV$) [6] and ELSA ($0.75GeV \leq E_\gamma \leq 3GeV$) [7,8].

At a first glance by looking at the cross section the very pronounced peaks in the resonance region become apparent. The missing of the above mentioned "background" terms constitutes the second remarkable feature. This signifies the suppression, especially of the diffractive or peripheral contributions to the cross section, by performing double polarization experiments. With these data the GDH - sum rule has been checked, a fundamental relation between the total absorption cross section for circular polarized photons on longitudinal polarized protons and the anomalous magnetic moment of the proton. Extensive experimental programs are on the horizon to use double polarization experiments to disentangle and identify via partial wave analyses the different production amplitudes. More detailed information concerning the reaction mechanism yields the t- dependence of the different reaction channels. A few typical examples in selected γ- energy ranges will be presented.

get insight into the reaction mechanism provides the decomposition of the $\gamma+p \Longrightarrow p+\pi^{+}+\pi^{-}$ cross section into the different reaction channels $\gamma + p \Longrightarrow p + \rho^{0}$, $\gamma + p \Longrightarrow \Delta^{++}+\pi^{-}$ and $\gamma+p \Longrightarrow \Delta^{0}+\pi^{+}$ which is shown in Fig. 3. The dominance of the reactions $\gamma+p \Longrightarrow \Delta^{++} + \pi^{-}$ and $\gamma + p \Longrightarrow p + \rho^{0}$ indicates diffractive processes. Is there a general way to "switch off" the dominance of the diffractive processes? The results of recent measurements point into the direction of using polarized beams and targets.

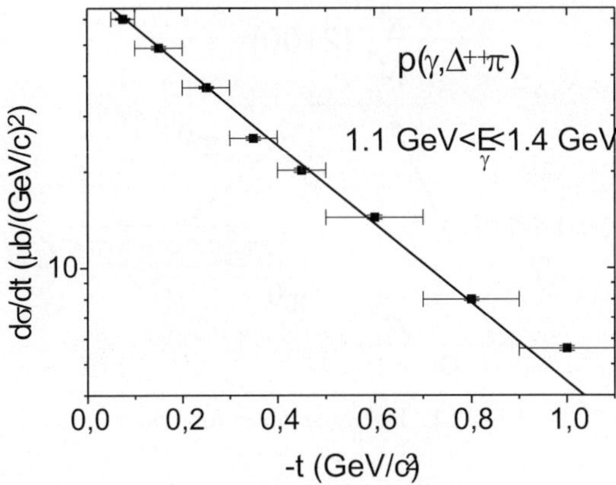

Fig. 5. t-dependence:$\gamma + p \Longrightarrow \Delta^{++} + \pi^-$.

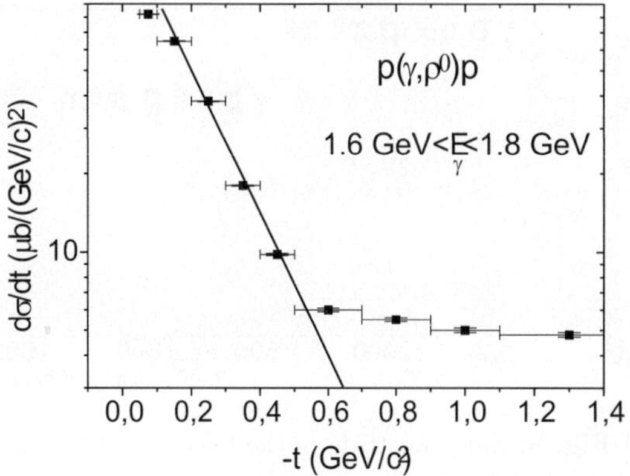

Fig. 7. t-dependence:$\gamma + p \Longrightarrow p + \rho^0$

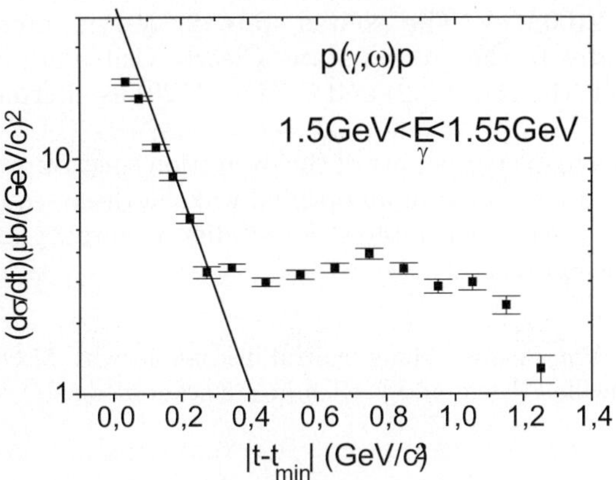

Fig. 6. t-dependence:$\gamma + p \Longrightarrow p + \omega$

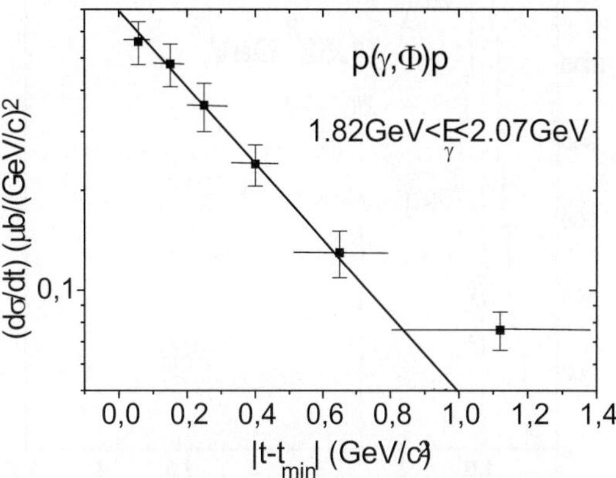

Fig. 8. t-dependence:$\gamma + p \to p + \Phi$

3 Differential cross sections

Figure 5 shows the t- dependence of the cross section for the reaction $\gamma + p \Longrightarrow \Delta^{++} + \pi^-$.

The large cross section at small t and its linear decrease over the whole range as a function of t on a logarithmic scale signifies the dominance of a diffractive process. For the reaction $\gamma + p \Longrightarrow p + \omega$, shown in Fig. 5, a different behavior of the cross section can be seen: In the low t- range again the diffractive process dominates but in addition, at larger t, the cross section flattens out. This behavior of the differential cross section indicates contributions of certain partial waves due to resonances.

A similar picture as with the $\gamma + p \Longrightarrow p + \omega$ reaction can be seen in the $\gamma + p \Longrightarrow p + \rho^0$ reaction, Fig. 7.

Both reactions hold, therefore, the promise to be very useful channels to find so far unidentified resonances. Remains to be considered (Fig. 8) the last vector meson with the quantum numbers of the photon, the Φ−meson. The Φ- meson production is dominated by the diffractive process as expected due to the composition of the Φ−meson by strange quarks. Two directions become visible to use these reactions for investigations of the excitation spectrum of the nucleon. The first one, to perform a partial wave analysis with the identification of the quantum numbers of resonance contributions, is under way. The second one, will be the observation of ex-

cited states of the nucleon by using the inelasticity of the diffractive process. As an example [9] the reactions $\gamma + p \Longrightarrow Roper + \rho^0 \Longrightarrow n + \pi^+ + \pi^+ + \pi^-$ and $\gamma + p \Longrightarrow Roper + \omega \Longrightarrow n + \pi^+ + \pi^+ + \pi^- + \pi^0$ are considered. Especially interesting will be a systematic study of scalar excitations as proposed in [9] and a careful examination of the isospin sector due to the "availability" of virtual isovector -and isoscalar vector mesons in the photon beam. Photon beams with photon energies $E_\gamma \geq 3.2 GeV$ are mandatory in order to cover the whole excitation spectrum of the nucleon via these unique inelastic vector meson scattering reactions.

4 The observation of multi photon states in the final state

A decomposition of the total γ−absorption cross section into its final states exhibits a strong dominance of charged mesonic states as can be seen e.g. in Fig. 2. Because of large "background" contributions no obvious resonance structures are seen in the total cross sections. However, by choosing reactions like $\gamma + p \Longrightarrow p + \eta$, $\gamma + p \Longrightarrow p + \eta'$, $\gamma + p \Longrightarrow p + \pi^0 + \pi^0$, $\gamma + p \Longrightarrow p + \eta + \pi^0$ resonances are expected to stand out of the background very clearly as the example of experimental studies on the $S_{11}(1535)$ have demonstrated,e.g. [10]. At ELSA an experimental program has been started to investigate, in a first round of

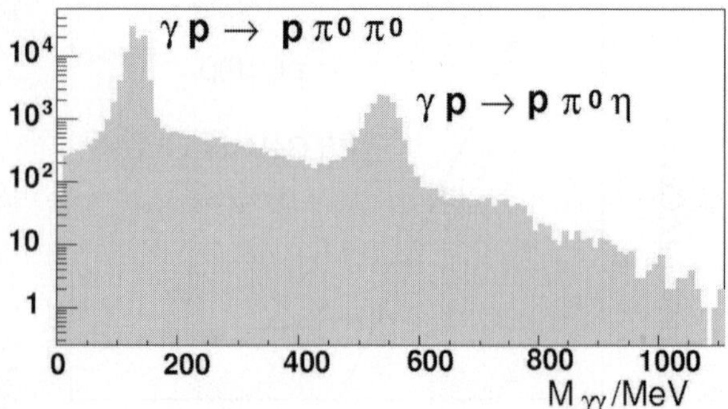

Fig. 9. Sum spectrum of the barrel: 4 photons

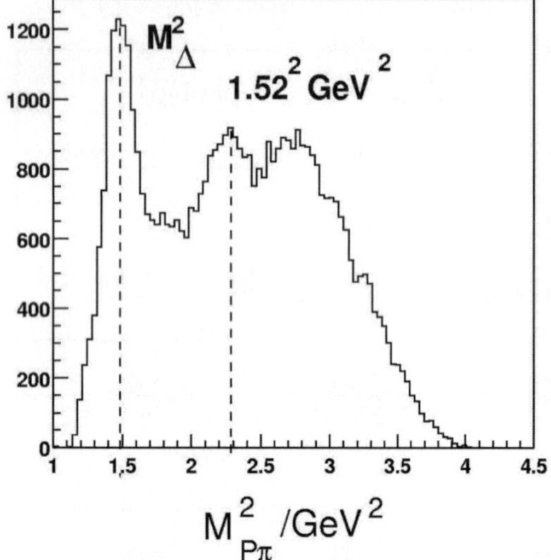

Fig. 10. The reconstructed mass spectrum of the reaction: $\gamma + p \Longrightarrow p + \pi^0 + \pi^0$,

experiments, the multi photon decay channels. The high segmented (1380 CsJ- crystals) CRYSTAL BARREL [11] detector serves as the main instrument in an experimental set-up with a photon tagger, a forward time of flight wall and fiber detectors inside the barrel. Figure 9 shows a spectrum of 4 photons in the final state.

By identifying the hit pattern and adding up the collected energy, the expected mesonic final states stand up very clearly. These many photon final states can be measured almost background free. Besides transitions from excited states leading to the ground state, cascade transitions can be observed for the first time in a systematic

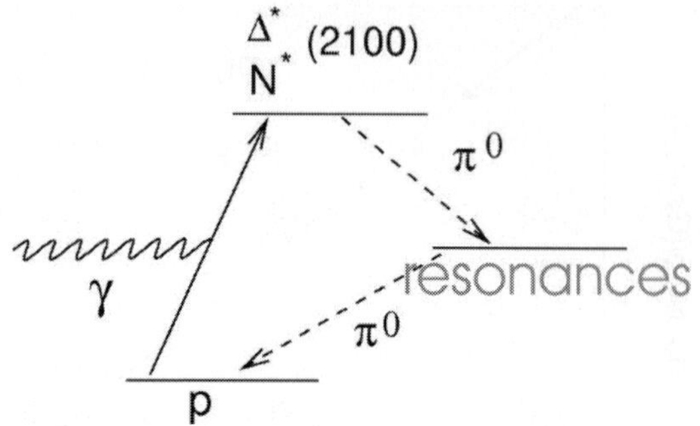

Fig. 11. Decay scheme for two π^0

way. Figure 10 shows a cascade decay, as indicated in Fig. 11, for the two pion channel, photo excited up to $\sqrt{s} = 2100 MeV$. The excited state decays via intermediate states to the ground state. Clearly visible are in this example the $\Delta_{33}(1232)$ and the $D_{13}(1520)$ as intermediate states.

The high energy part of the excitation spectrum, which can be investigated in an optimal way via these sequential decays, are of great interest for studies of chiral symmetry restoration [12].

Acknowledgements. Many fruitful discussions with M.Ostrick, W.Schwille,F.Klein and E.Klempt are acknowledged.

References

1. K. Hagiwara et al.: Phys. Rev. D **66**, 010001 (2002)
2. A. Donnachie and P.V. Landshoff: Particle World **2**, 7 (1991)
3. J. Barth: PhD-Thesis, Physikalisches Institut, University Bonn, (2002)
4. Ch. Wu and W. Schwille: private communication
5. C. Bradke et al.: Nucl. Instr. Meth. A **436**, 430 (1999)
6. J. Ahrens et al.: Phys. Rev. Lett. **87**, 022003 (2001)
7. T. Speckner: PhD-Thesis, Physikalisches Institut, University Erlangen, (2002) to be published
8. G. Zeitler: PhD-Thesis, Physikalisches Institut, University Erlangen, (2002) to be published
9. M. Soyeur, Nucl. Phys. A **671**, 532 (2000)
10. B. Schoch: Progr. Part. Nucl. Phys. **34**, 43 (1995)
11. E. Aker et al.: Nucl. Instr. Meth. A **321**, 69 (1992)
12. E. Klempt: Phys. Rev. C **66**, 052002 (2002)

Eur Phys J A (2004) **19**, s01, 261–265
Digital Object Identifier (DOI) 10.1140/epjad/s2004-03-043-7

EPJ A direct
electronic only

HERMES run II

Wolf-Dieter Nowak, for the HERMES collaboration

DESY, D-15738 Zeuthen, Platanenallee 6

Received: 29 Sep 2003 / Accepted: 14 Nov 2003 /
Published Online: 6 Feb 2004 – © Società Italiana di Fisica / Springer-Verlag 2004

Abstract. The two main physics objectives for running the HERMES experiment in the years 2003-2006 are the study of the transverse spin structure of the nucleon and measurements of cross section asymmetries in exclusive reactions in order to obtain information on Generalized Parton Distributions.

1 Introduction

Spin-dependent ('polarized') parton distribution functions (PDFs) attracted the attention of experimentalists and theorists in the late eighties when it was found in Deep Inelastic Scattering (DIS) experiments that – in contrast to the expectation – only a very small fraction (about 0.1) of the *longitudinal* spin of the nucleon is made up by the spins of the quarks [1,2]. The experiment HERMES at HERA [3] very recently published [4] the first five-fold flavor separation analysis in leading order QCD from which a somewhat larger fraction (about 0.4) was derived.

The spin-dependent PDFs describe the imbalance between the distributions $q^\uparrow(x)$ and $q^\downarrow(x)$ ($q^\rightarrow(x)$ and $q^\leftarrow(x)$) in which quark spins are aligned parallel and anti-parallel to that of the transversely (longitudinally) polarized parent nucleon: for the quark species q the transversity distribution is given by $h_1^q(x) \equiv \delta q(x) = q^\uparrow(x) - q^\downarrow(x)$ and the helicity distribution by $g_1^q(x) \equiv \Delta q(x) = q^\rightarrow(x) - q^\leftarrow(x)$. The sum of either pair constitutes the spin-independent ('unpolarized') distribution $f_1^q(x) \equiv q(x)$. Here x is the fraction of the nucleon's momentum carried by the struck parton, in a system where the nucleon has infinite momentum. Note that the additional (weak) dependence of the PDFs on Q^2, the four-momentum transfered by the virtual photon in DIS, is omitted for simplicity. The 1st moment $\delta\Sigma = \sum_q \int_0^1 dx(h_1^q(x) - \bar h_1^q(x))$ describes the tensor charge of the nucleon while its axial charge is given by $\Delta\Sigma = \sum_q \int_0^1 dx(g_1^q(x) + \bar g_1^q(x))$.

The *transverse* spin structure of the nucleon has not yet been studied in any detail. It is the last unknown piece in the complete leading-twist spin structure of the nucleon and comprises new information on the dynamics of quarks inside hadrons, thereby allowing the verification of the validity of two yet untested QCD predictions:
i) the tensor charge is larger than the axial charge,
ii) the transversity distribution has a weaker Q^2 evolution than the helicity distribution.

Generalized Parton Distributions (GPDs) constitute a unified theoretical description of inclusive and (hard) ex-clusive processes. The GPDs H^q and $\tilde H^q$ reduce to the ordinary PDFs $f_1^q(x)$ and $g_1^q(x)$, respectively, for vanishing momentum transfer t at the nucleon vertex, while the GPDs E^q and $\tilde E^q$ are not accessible in DIS. The GPDs bear qualitatively new information on the structure of the nucleon, especially on parton-parton correlations, and thus provide a more complete picture of the nucleon as compared to ordinary PDFs.

GPDs can in principle be accessed through measurements of hard exclusive processes. Their t-dependence provides information on the distribution of partons *transverse* to the nucleon's direction of motion. This may eventually lead to the development of a 3-dimensional picture of the partonic structure of the nucleon [5,6,7]. A measurement of the 2nd moment of the sum of the unpolarized GPDs H^q and E^q, in the limit of vanishing t, may eventually allow the determination of the *total angular momentum* carried by quarks in the nucleon [8].

At HERMES, the determination of the u-quark transversity distribution $h_1^u(x) \equiv \delta u(x)$ is the main goal of the measurements planned until 2004, facilitated by the installation of a transverse target magnet surrounding an internal polarized hydrogen target. Projections will be discussed in chapter 2. For a possible measurement of $Im\,H$, considered to be the ultimate goal of HERMES running in the years 2005-06, projections will be discussed in chapter 3. More details on transversity and GPDs, as seen from an experimentalist's point of view, can be found in [9].

2 Transverse spin structure of the nucleon

Recently the decades-old theoretical debate was revived on how to correctly interpret data from measurements of single-spin azimuthal asymmetries for pion production in semi-inclusive deep inelastic scattering (SIDIS). The most recent status, including explanation of variables, (some) theoretical predictions and a full list of references is given in, e.g., [10]. The observed non-zero cross section asymmetries can be explained by two different mechanisms,

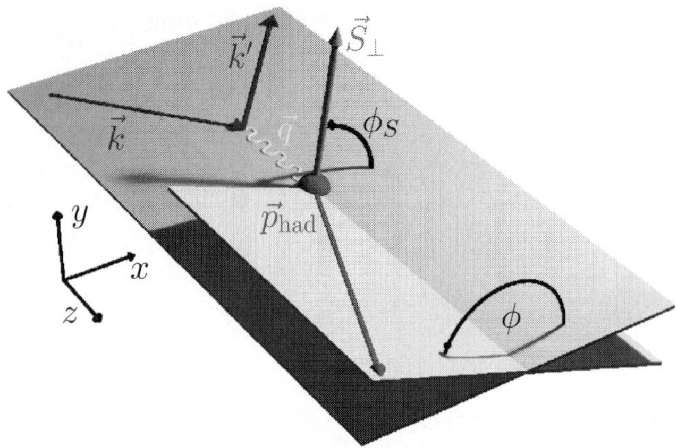

Fig. 1. Definition of azimuthal angles for transversity measurements

named after their inventors 'Sivers mechanism' [11] and 'Collins mechanism' [12]. The former explains the asymmetries observed in the azimuthal distribution of the outgoing quark, i.e. of the outgoing pion after the fragmentation, by the existence of a quark distribution function $(f_{1T}^\perp)$ that depends on the quark's transverse momentum $k_\perp$. The latter mechanism assumes, as the observed cross section asymmetries are chirally even, that the two involved chirally-odd functions are non-zero. In SIDIS these are the transversity $h_1^q(x)$ and the 'Collins fragmentation function' $H_1^\perp$. It appears that the relative weight of both mechanisms can eventually only be determined by experimental data.

Experimentally, a differentiation between both mechanisms requires data obtained from semi-inclusive scattering of unpolarized (U) leptons on a transversely (T) polarized target: $eN^\uparrow \longrightarrow e'\pi X$. The azimuthal dependence of the asymmetry in the measured count rates $N^\uparrow, N^\downarrow$ reads

$$A(\phi) = \frac{1}{\langle P \rangle} \cdot \frac{N^\uparrow(\phi) - N^\downarrow(\phi)}{N^\uparrow(\phi) + N^\downarrow(\phi)},$$

where $\langle P \rangle$ stands for the average target polarization while the arrows' notation was explained above. The azimuthal angles are defined in Fig. 2, where $k(k')$ is the 4-vector of incoming (outgoing) lepton, p_{had} the outgoing pion momentum vector, and $S_\perp$ denotes the target spin vector. For the Sivers and Collins mechanism the weighted cross section asymmetries, or the $\sin\phi$-moments of the cross section, are respectively given by

$$A_{UT}^{\sin(\phi-\phi_S)} \sim \sum_q e_q^2 f_{1T}^{\perp(1),q}(x_B) D_1^q(z),$$

$$A_{UT}^{\sin(\phi+\phi_S)} \sim \sum_q e_q^2 h_1^q(x_B) H_1^{\perp(1),q}(z).$$

Here e_q denotes the quark charge, the superscript (1) indicates that the corresponding function was integrated over $k_\perp$, x_B is Bjorken's scaling variable, and D_1^q is the spin-independent fragmentation function with z being the fraction of the virtual photon's energy carried by the outgoing pion.

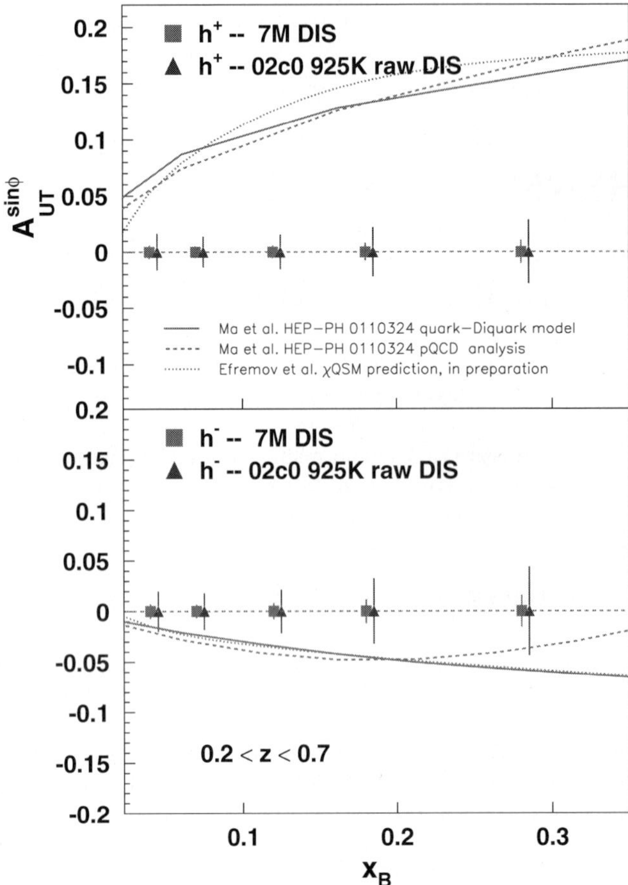

Fig. 2. Projected accuracy of HERMES measurements of the single transverse target-spin asymmetry

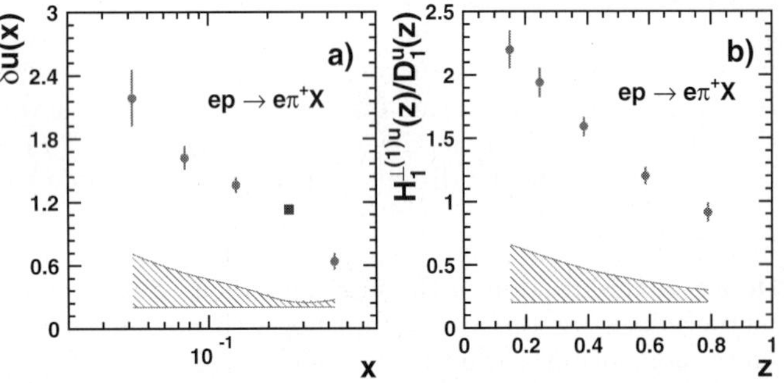

Fig. 3. Projected accuracy of HERMES measurements of transversity and the ratio of polarized to unpolarized fragmentation functions for the u-quark. The *square* denotes the normalization point (*see text*)

Data taking at HERMES in 2002/2003, using a transversely polarized hydrogen target, resulted in 725k events after data quality cuts corresponding to 925k raw DIS events. The analysis is in progress; the expected statistical accuracy in the asymmetry $A_{UT}^{\sin\phi}$ is shown in Fig. 2 for the present data set and the projected final data set of 7M events, in comparison to several model predictions for the Collins mechanism. Distinguishing between contemporary model predictions will at best be possible with the maximum expected statistics, the more important issue of this first measurement of $A_{UT}^{\sin\phi}$ being its overall size and shape. For the final data set a simultaneous extraction of the u-quark transversity distribution and the ratio of polarized to unpolarized fragmentation functions has been simulated [13], as can be seen from Fig. 3. Note that here the Sivers mechanism was not considered and the basically undetermined overall normalization was taken from the expected similarity of transversity and helicity distribution at low Q^2 and medium x.

3 Towards measuring GPDs in DVCS

The – theoretically – simplest hard exclusive process in electroproduction is Deeply Virtual Compton Scattering (DVCS): $\gamma^* p \to \gamma p$. In the Bjorken limit the dominating pQCD subprocess is described by the 'hand-bag diagram' shown in the left panel of Fig. 4. The Bethe-Heitler (BH) process that leads to an identical final state is shown in the right panel of the same figure.

At HERMES kinematics the DVCS process cannot be measured directly as it is dominated by the BH process. Interference between both, however, opens access to *both* the real and imaginary parts of (certain combinations of) DVCS amplitudes. They can be accessed by measuring the azimuthal dependences of certain cross section asymmetries, the most prominent ones being the lepton charge asymmetry A_C (unpolarized beam, unpolarized target):

$$A_C \sim d\sigma(e^+ p) - d\sigma(e^- p) \sim \cos(\phi_\gamma) \times \mathrm{Re}\,\mathcal{I},$$

and the lepton helicity asymmetry A_{LU} (L: longitudinally polarized beam, U: unpolarized target):

$$A_{LU} \sim d\sigma(\overrightarrow{e^+} p) - d\sigma(\overleftarrow{e^+} p) \sim \sin(\phi_\gamma) \times \mathrm{Im}\,\mathcal{I}.$$

Here $\mathcal{I}$ represents the interference term in the DVCS/BH cross section. The definition of ϕ_γ is explained in Fig. 5. The GPDs are buried in generally complicated integrals describing the DVCS amplitudes. Practically, a direct determination of GPDs is impossible. Models of GPDs must be used to construct 'predictions' that have to be compared to corresponding experimental results, thus launching an iterative procedure whose precision is very hard to assess at present. Similarly as in the case of transversity measurements, a measurement of the azimuthal (ϕ_γ) dependence of the (above given) cross section asymmetries is the tool of choice.

Presently, the HERMES collaboration is pursuing the construction of a Recoil Detector [14] that is expected to be commissioned in 2004-05. It will surround an unpolarized gas target with a silicon tracker that resides inside the beam pipe. The next layers are a scintillating fibre detector, a photon detector and a 1 Tesla superconducting solenoid magnet. The Recoil detector, designed for measuring mainly recoil protons leaving the target under large polar angles (between 0.1 and 1.35 rad) will serve two main purposes:

i) the t-resolution at low values of $-t$, where the kinematic dependence of GPDs is especially interesting, will be improved dramatically, by more than one order of magnitude (from 0.2 to below 0.02 GeV/c^2);

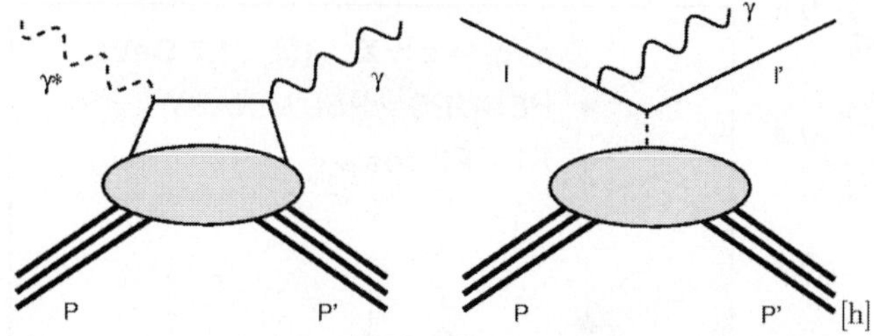

Fig. 4. *Left*: Deeply Virtual Compton Scattering. *Right*: Bethe-Heitler process

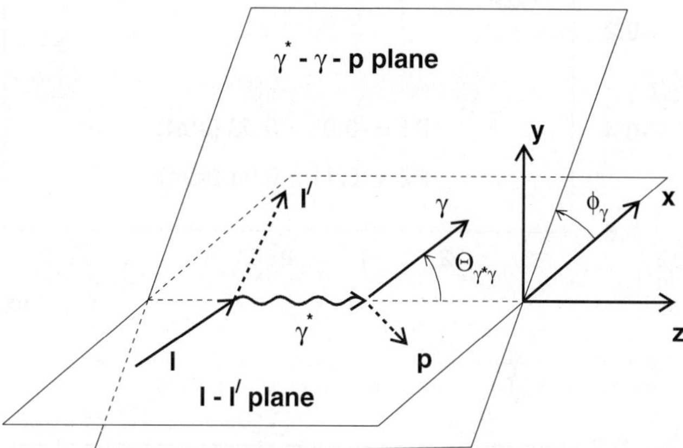

Fig. 5. Definition of the azimuthal angle ϕ_γ in DVCS

ii) the detection of recoil protons will allow the reduction of the undetected fraction of events that have an excited nucleon state instead of a proton ('associated DVCS') down to below 1%.

Preliminary HERMES results on the azimuthal dependence of the lepton charge asymmetry in Fig. 6 can be compared with projections in Fig. 7 based on an integrated luminosity of 2 fb^{-1} obtained on a hydrogen target [15]. The latter are displayed for $x_B > 0.2$ as the asymmetry exhibits a quite different behavior at low x_B. As can be seen, the projected data for HERMES Run II can clearly distinguish between GPD predictions. For details on the different GPD models used we refer to [15]. Note that all projections given here and in the following are based on a sample of purely exclusive events.

Preliminary HERMES results on the missing mass dependence of the lepton helicity asymmetry are shown in Fig. 8. Projections on its t-dependence, obtained as explained above, are displayed separately for low and high values of x_B in Fig. 9 [16]. Quite detailed measurements of the t-dependence can be expected in both regions of x_B.

On the basis of the same expected statistics, an attempt has been made to obtain projections for measuring $Im\,H$ [17]. Working in the region of low $-t$ ($-t < 0.15$ (GeV/c)^2) it has been found that the relative contribution of the GPD H largely dominates and the lepton helicity asymmetry $A_{LU}^{\sin\phi}$ mainly depends on $Im\,H$. For two different GPD parameterizations (A) and (B) the dependence on the skewedness variable ξ is shown in Fig. 10. As $\xi = x_B/(2 - x_B)$, this variable resembles for small val-

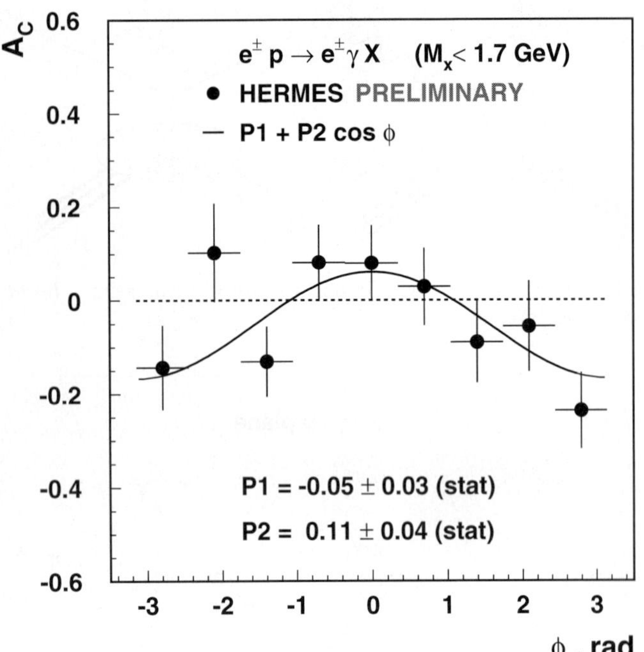

Fig. 6. Preliminary HERMES results on the azimuthal dependence of the lepton charge asymmetry in DVCS

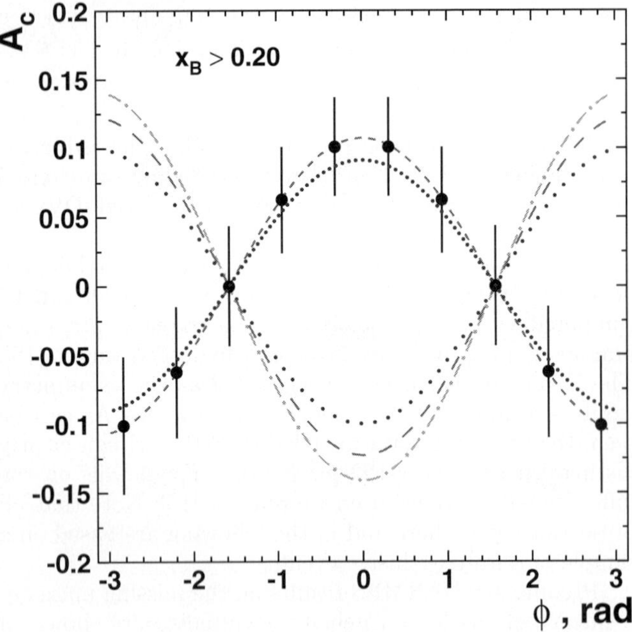

Fig. 7. Projections for the azimuthal dependence of the lepton charge asymmetry in DVCS from 2005-06 running (*see text*)

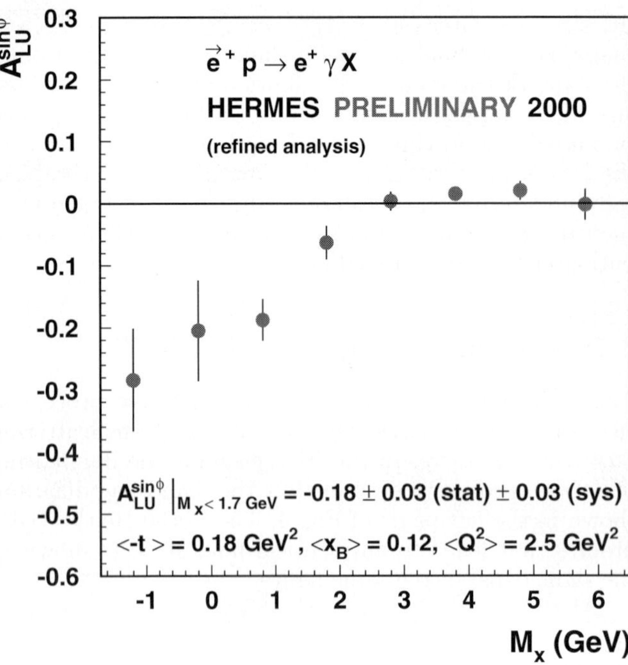

Fig. 8. Preliminary HERMES results on the missing mass dependence of the lepton helicity asymmetry in DVCS

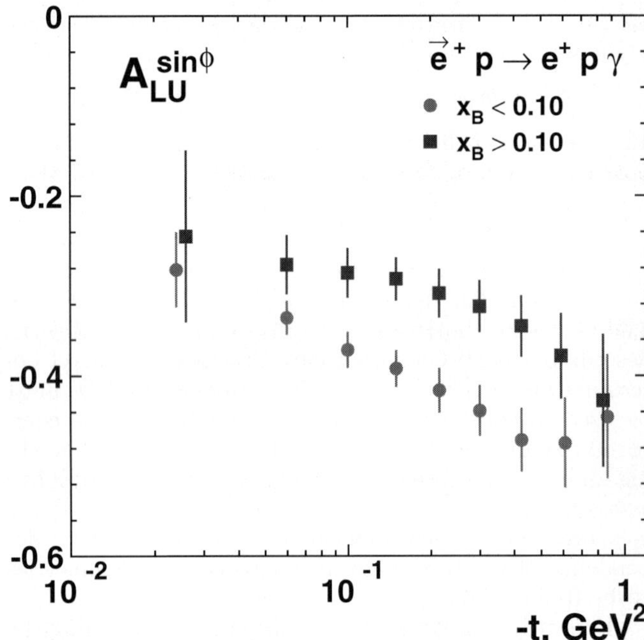

Fig. 9. Projections for the lepton helicity asymmetry in DVCS from 2005-06 running (*see text*)

ues of x_B essentially a $1/x_B$-dependence; for more details we refer to [17]. While the bands in the figure represent the expected statistical uncertainty, the shaded areas indicate a possible systematic uncertainty of the extraction method used. It is obvious, as already stated in the beginning, that the extraction of GPD-related information will be a complicated task.

It is concluded that the measurements planned for HERMES Run II will constitute important steps into yet uncharted territory and thus are expected to considerably enhance the knowledge on the partonic structure of the nucleon.

Acknowledgements. I am indebted to U. Elschenbroich for providing Fig. 1 and to Ralf Seidl for providing Fig. 2. Many thanks to V. Korotkov, G. van der Steenhoven, and J. Volmer for a careful reading of the manuscript.

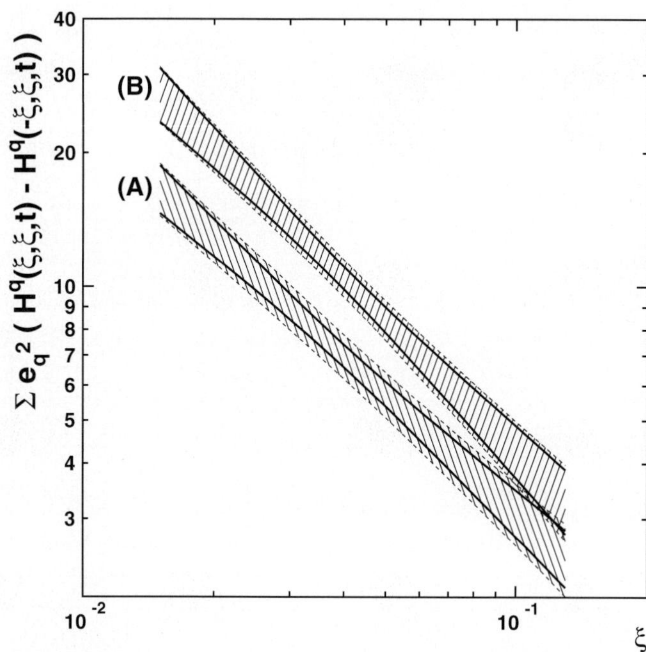

Fig. 10. Projected extraction of *Im H* from measuring DVCS at HERMES with a Recoil Detector (*see text*). The two shaded areas represent two different calculations of this unmeasured quantity

References

1. J. Ashman et al. (EMC): Phys. Lett. B **206**, 364 (1988)
2. J. Ashman et al. (EMC): Nucl. Phys. B **328**, 1 (1989)
3. K. Ackerstaff et al. (HERMES): Nucl. Instr. and Meth. B **330**, 230 (1998)
4. A. Airapetian et al. (HERMES): arXiv:hep-ex/0307064, subm. to Phys. Rev. Lett.
5. M. Burkardt: Phys. Rev. D **62**, 071503 (2000), Erratum-ibid. D **66**, 119903 (2002)
6. M. Diehl: Eur. Phys. J. C **25**, 223 (2002); J.P. Ralston and B. Pire: Phys. Rev. D **66**, 111501 (2002); A.V. Belitsky and D. Muller: Nucl. Phys. A **711**, 118 (2002)
7. X. Ji: arXiv:hep-ph/0304037 (2003)
8. X. Ji: Phys. Rev. Lett. **78**, 610 (1997); Phys. Rev. D **55**, 7114 (1997)
9. W.-D. Nowak: Nucl. Phys. B (Proc. Suppl.) **105**, 171 (2002); W.-D. Nowak: arXiv:hep-ph/0210409 (2002), in 'Spin Structure of the Nucleon', NATO Science Series, II. Mathematics, Physics and Chemistry – Vol. 111, ed. by E. Steffens and R. Shanidze, Kluwer Academic Publishers, p.37 (2003)
10. A.V. Efremov, K. Goeke, and P. Schweitzer: preprint RUB-TPII-05/03 (2003), arXiv:hep-ph/0303062 (2003)
11. D.W. Sivers: Phys. Rev D **41**, 83 (1990); Phys. Rev. D **43**, 261 (1991)
12. J.C. Collins: Phys. Lett. B **536**, 43 2002
13. V.A. Korotkov, W.-D. Nowak, and K.A. Oganessyan: Eur. Phys. J. C **18**, 639 (2001)
14. The HERMES Collaboration: 'A Large Acceptance Recoil Detector for HERMES', DESY PRC 01-01 (April 2001)
15. V.A. Korotkov and W.-D. Nowak: Eur. Phys. J. C **23**, 455 (2002)
16. V.A. Korotkov: talk at '267. WE-HERAEUS-Seminar on Generalized Parton Distributions', Bad Honnef (Germany), Nov. 19 - 21, 2001; see also [14]
17. V.A. Korotkov and W.-D. Nowak: Nucl. Phys. A **711**, 175c (2002)

Eur Phys J A (2004) **19**, s01, 267–273
Digital Object Identifier (DOI) 10.1140/epjad/s2004-03-044-6

EPJ A direct
electronic only

Recent results from JLab on nucleon responses

Kees de Jager

Jefferson Laboratory, Newport News, VA 23606, USA

Received: 9 Sep 2003 / Accepted: 14 Nov 2003 /
Published Online: 6 Feb 2004 – © Società Italiana di Fisica / Springer-Verlag 2004

Abstract. Highlights of recent data from Jefferson Lab on the nucleon response to an electro-magnetic
probe are presented. Recent technological advances in polarized beams and either polarized targets or
nucleon recoil polarimeters have yielded a significant improvement on the precision of the data. An outlook
is presented of planned experiments.

PACS. 24.85.+p – 25.30.-c

1 Electro-magnetic form factors

1.1 Introduction

The nucleon electromagnetic form factors (EMFF) are of
fundamental importance for the understanding of the nucleon's internal structure. In Plane Wave Born Approximation (PWBA) the cross section for elastic electron-
nucleon scattering can be expressed in terms of the Dirac
and Pauli form factors F_1 and F_2, respectively,

$$\frac{d\sigma}{d\Omega} = \sigma_M[(F_1^2+\kappa^2\tau F_2^2)\cos^2(\frac{\theta_e}{2})+2\tau(F_1^2+\kappa F_2^2)^2\sin^2(\frac{\theta_e}{2})] \tag{1}$$

where $\tau = Q^2/4(m_N^2)$, Q is the four-momentum transfer,
m_N the mass of the nucleon, σ_M the Mott cross section for
scattering off a point-like particle, κ the nucleon anomalous magnetic moment, θ_e the electron scattering angle
and E_e the electron energy. F_1 and F_2 can be expressed
in the electric and magnetic Sachs form factors G_E and
G_M, respectively,

$$G_E = F_1 - \tau\kappa F_2$$
$$G_M = F_1 + \kappa F_2 \tag{2}$$

leading to the so-called Rosenbluth [1] formula

$$\frac{d\sigma}{d\Omega} = \sigma_M[\frac{G_E^2 + \tau G_M^2}{1 + \tau} + 2\tau G_M^2 \tan^2(\frac{\theta_e}{2})] \tag{3}$$

This equation illustrates that G_E and G_M can be determined separately by performing cross-section measurements at fixed Q^2 over a range of (θ_e, E_e) combinations (Rosenbluth separation). In the non-relativistic Breit
frame the Sachs form factors can be identified with the
Fourier transform of the nucleon charge and magnetization density distributions.

Through the middle of the previous decade practically
all available proton EMFF data had been collected using
the Rosenbluth separation technique. This experimental
procedure requires an accurate knowledge of the electron
energy and the total luminosity. In addition, since the contribution to the elastic cross section from the magnetic
form factor is weighted with Q^2, data on G_E^p suffer from
increasing systematic uncertainties at higher Q^2-values.
The then available world data set [2] was compared to the
so-called dipole parametrization G_D, which corresponds
to exponentially decreasing radial charge and magnetization densities:

$$G_D = \left(\frac{\Lambda^2}{\Lambda^2 + Q^2}\right)^2 \text{ with } \Lambda = 0.84 \text{ GeV/c and } Q \text{ in GeV/c} \tag{4}$$

For all four EMFF the available data agreed with the
dipole parametrization to within 20%. Both the G_E^p and
the G_M^p data could be fitted adequately with an identical parametrization. However, the limitation of the Rosenbluth separation was evident from the fact that different
data sets for G_E^p scattered by up to 50% at higher Q^2-
values.

1.2 Polarization instrumentation

Over 20 years ago Akhiezer and Rekalo [3] and Arnold
et al.[4] showed that the accuracy of EMFF measurements
could be increased significantly by scattering polarized
electrons off a polarized target (or equivalently by measuring the polarization of the recoiling nucleon). Technological advances have only recently made possible a large
number of new data with a significantly improved accuracy. Polarized electron beams [5,6] are now reliably available with a polarization close to 80% at currents of up to
100 μA. The beam polarization is measured with either
Møller [7] or Compton [8] polarimeters with an accuracy

approaching 1%. The dynamical polarization technique [9] provides polarized hydrogen or deuterium targets with an average polarization of 80 or 20%, respectively, while polarized helium targets are available with a polarization close to 50%, either through spin [10] or metastability [11] exchange at a density of 10 atm. Finally, the polarization of recoiling or knocked-out reaction products can be measured with focal-plane [12] or neutron [13] polarimeters.

1.3 Theory

A frequently used framework to describe the EMFF is that of Vector Meson Dominance (VMD) [14], in which one assumes that the virtual photon - after having become a quark-antiquark pair - couples to the nucleon as a vector meson. The EMFF can then be expressed in terms of coupling strengths between the virtual photon and the vector meson and between the vector meson and the nucleon, summing over all possible vector mesons. A common restriction of VMD models is that they do not predict a correct behaviour of the EMFF at high Q^2-values. Lomon [15] constructed a hybrid model which couples the low Q^2-behaviour of the VMD model to the asymptotic behaviour predicted by pQCD.

Several models [16,17,18] have incorporated constituent quarks on the light front. In doing this the pQCD condition that the transverse momentum is zero, is freed and an orbital angular momentum component is introduced in the wavefunction of the proton, thus giving up helicity conservation. All these models predict a linear dropoff of G_E^p/G_M^p with Q^2, or equivalently $F_2/F_1 \propto 1/Q$.

However, all presently available theories are at least to some extent effective (or parametrizations). Only lattice gauge theory can provide a truely ab initio calculation, but accurate lattice QCD results for the EMFF are still several years away.

1.4 Neutron magnetic form factor

Significant progress has been made in measurements of G_M^n at low Q^2-values by measuring the ratio of quasi-elastic neutron and proton knock-out from a deuterium target. This method is insensitive to nuclear binding effects and to fluctuations in the luminosity and detector acceptance. The basic set-up used in all such measurements is very similar: the electron is detected in a magnetic spectrometer with coincident neutron/proton detection in a large scintillator array. The main technical difficulty in such a ratio measurement is the absolute determination of the neutron detection efficiency. Recently, inclusive quasi-elastic scattering of polarized electrons off a polarized ^{3}He target was measured [27] in Hall A at JLab in a Q^2-range from 0.1 to 0.6 $(\mathrm{GeV/c})^2$. This experiment has provided an independent accurate measurement of G_M^n at Q^2-values of 0.1 and 0.2 $(\mathrm{GeV/c})^2$, in excellent agreement with the Mainz data. At the higher Q^2-values G_M^n was extracted [28] in PWIA, since final-state-interaction effects are expected to decrease with increasing Q^2. A study of G_M^n at

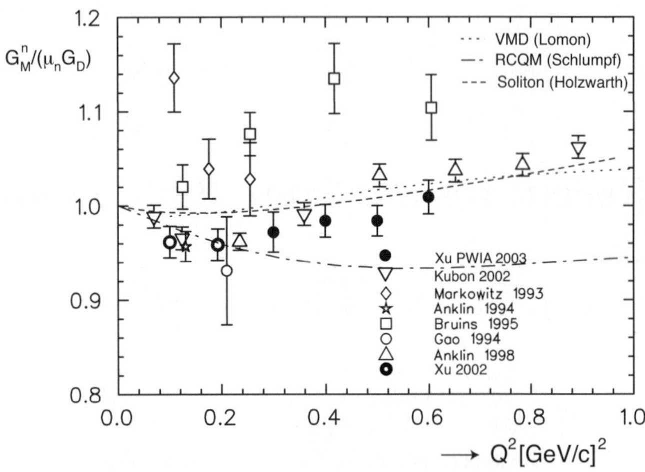

Fig. 1. The neutron magnetic form factor G_M^n, in units of μG_D, as a function of Q^2. The data are from [21,22,23,24,25, 26,27,28]. The theory curves are: *dotted* [15], *dot-dashed* [16] and *short-dashed* [20]

Q^2-values up to 5 $(\mathrm{GeV/c})^2$ has recently been completed in Hall B by measuring the neutron/proton quasi-elastic cross-section ratio using the CLAS detector [29].

Figure 1 shows the results of all completed G_M^n experiments. The Mainz G_M^n data are 8-10% lower than those from ELSA, at variance with the quoted uncertainty of appr. 2%. This discrepancy would require a 16-20% error in the detector efficiency. Two of the theoretical predictions shown [15,20] provide an accurate description of the data.

1.5 Neutron electric form factor

In the last decade a series of spin-dependent measurements have provided accurate data on G_E^n by utilizing the fact that the ratio of the beam-target asymmetry with the target polarization perpendicular and parallel to the momentum transfer is directly proportional to the ratio of the electric and magnetic form factors. A similar result is obtained with the reaction ^{2}H$(\mathbf{e}, e'\mathbf{n})$ when one measures the polarization of the recoiling neutron directly and after having precessed the neutron spin over 90° with a dipole magnet.

Figure 2 shows results obtained through the reactions ^{2}H$(\mathbf{e}, e'n)$[32], ^{2}H$(\mathbf{e}, e'\mathbf{n})$[13,33] and $^3\overrightarrow{\mathrm{He}}(\mathbf{e}, e'n)$[34, 35]. At low Q^2-values corrections for nuclear medium and rescattering effects can be sizeable: 65% for ^{2}H at 0.15 $(\mathrm{GeV/c})^2$ and 50% for $^3\overrightarrow{\mathrm{He}}$ at 0.35 $(\mathrm{GeV/c})^2$. These corrections are expected to decrease significantly with increasing Q, although no reliable calculations are presently available for $^3\overrightarrow{\mathrm{He}}$ above 0.5 $(\mathrm{GeV/c})^2$. Thus, there are now data from a variety of reaction channels available in a Q^2-range up to 0.6 $(\mathrm{GeV/c})^2$ with an overall accuracy of appr. 20%, which are in mutual agreement. The prediction by Miller *et al.* [17], using a constituent quark model on the light front, is in good agreement with the sofar available data, with the pion cloud dominating at low Q^2-values.

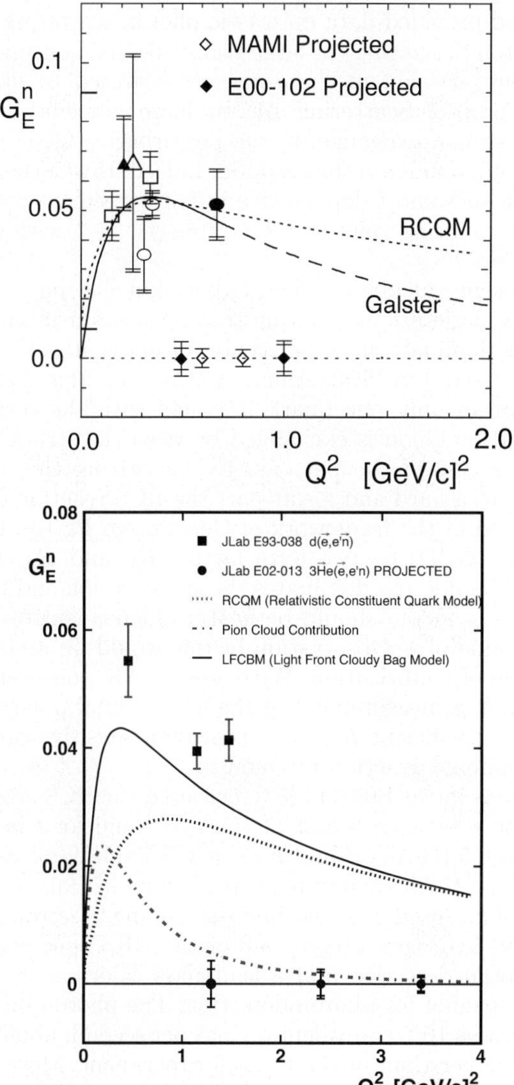

Fig. 2. The neutron electric form factor G_E^n as a function of Q^2. The data are indicated by the following symbols: *solid: triangle* [32], *squares* [39], *diamonds* [38], and *circle* [35]; *open: circles* [31,34], *squares* [13,33], *triangle* [30] and *diamonds* [37]. The curves represent Miller's RCQM prediction [17] (*dotted*) and the Galster parametrization [36] (*full/dashed*). The bottom plot shows the projected accuracy of E02-013 [41] and the separate contributions to the prediction of Miller. Symbols on the zero axis indicate the Q^2-value and the size of the error bars of ongoing experiments

The 30 years old Galster parametrization [36] provides a fortuitously good description of the data.

Shown on the horizontal axis in Fig. 2 are results expected in the near future, from the ^{2}H$(\mathbf{e}, e'n)$[38] channel at JLab and from the ^{2}H$(\mathbf{e}, e'\mathbf{n})$[37] channel at Mainz. In the bottom plot the results for the ^{2}H$(\mathbf{e}, e'\mathbf{n})$channel [39] are shown which extend the existing data set to 1.5 $(\text{GeV/c})^2$ with great accuracy. Also shown is the pion-cloud contribution which Miller added phenomenologically to his RCQM model in order to obtain agreement with the data. Recently an experiment [41] has been ap-

proved at JLab to measure G_E^n up to a Q^2-value of 3.4 $(\text{GeV/c})^2$ using the $^3\overrightarrow{\text{He}}(\mathbf{e}, e'n)$ reaction. In addition, the BLAST facility [40] at MIT is expected to provide accurate data on G_E^n in a Q^2-range from 0.1 to 0.8 $(\text{GeV/c})^2$. Thus, within a couple of years G_E^n data with an accuracy of 10% or better will be available up to a Q^2-value of 3.4 $(\text{GeV/c})^2$.

1.6 Proton electric form factor

The polarization-transfer technique has recently been used in two experiments [42,43] in Hall A at JLab, to measure the ratio G_E^p/G_M^p in a Q^2-range from 0.5 to 5.6 $(\text{GeV/c})^2$. Elastic ep events were selected by detecting electrons and protons in coincidence in the two identical HRS spectrometers. At the four highest Q^2-values a lead-glass calorimeter was used to detect the scattered electrons. The polarization of the recoiling proton was determined with a Focal Plane Polarimeter (FPP) in the hadron HRS, consisting of two pairs of straw chambers with a carbon or polyethylene analyzer in between. The data were analyzed in bins of each of the target coordinates. No dependence on any of these variables was observed. The results for the ratio G_E^p/G_M^p are shown in Fig. 3. The most striking feature of the data is the sharp decline as Q^2 increases. Since it is known that G_M^p closely follows the dipole parametrization, it follows that G_E^p falls more rapidly with Q^2 than G_D. A comparison with the older data from SLAC confirms the expected improvement in accuracy of such a spin-dependent measurement. All theoretical calculations shown in Fig. 3 predict a gradual decrease of G_E^p. At the highest Q^2-values the data do not follow pQCD scaling [45], but rather the $1/Q$ behaviour in F_2/F_1, predicted by RCQM models. If one assumes the linear decrease in $G_E^p/G_M^p$to continue, G_E^p would cross zero at $Q^2 \approx 7.7$ $(\text{GeV/c})^2$. To investigate this possibility an extension [46] of this experiment to a Q^2-value of 9.6 $(\text{GeV/c})^2$has been approved to run in Hall C.

Last year an experiment [47] was completed which used the Rosenbluth technique with a number of improvements. In stead of detecting the scattered electron, the recoil proton was detected and both HRS spectrometers in Hall A were used. Significantly improved systematic errors are expected at Q^2-values up to 4 $(\text{GeV/c})^2$. Recently several studies have investigated the two-photon contribution to the Rosenbluth technique. Blunden *et al.* [48] calculated the elastic contribution to the so-called box diagram in the radiative correction and showed that this accounted for half the discrepancy between the results from the Rosenbluth technique and those from polarization transfer. The inelastic contribution is much harder to calculate but an estimate might be obtained from the transverse beam asymmetry in elastic-proton scattering. Guichon and Vanderhaeghen [49] showed in general that a two-photon exchange contribution needed to explain the discrepancy is of the order of a few percent. However, Rekalo and Tomasi-Gustafsson [50] showed, also on general principles, that the contribution from two-photon exchange should be of order α^2 and thus very small.

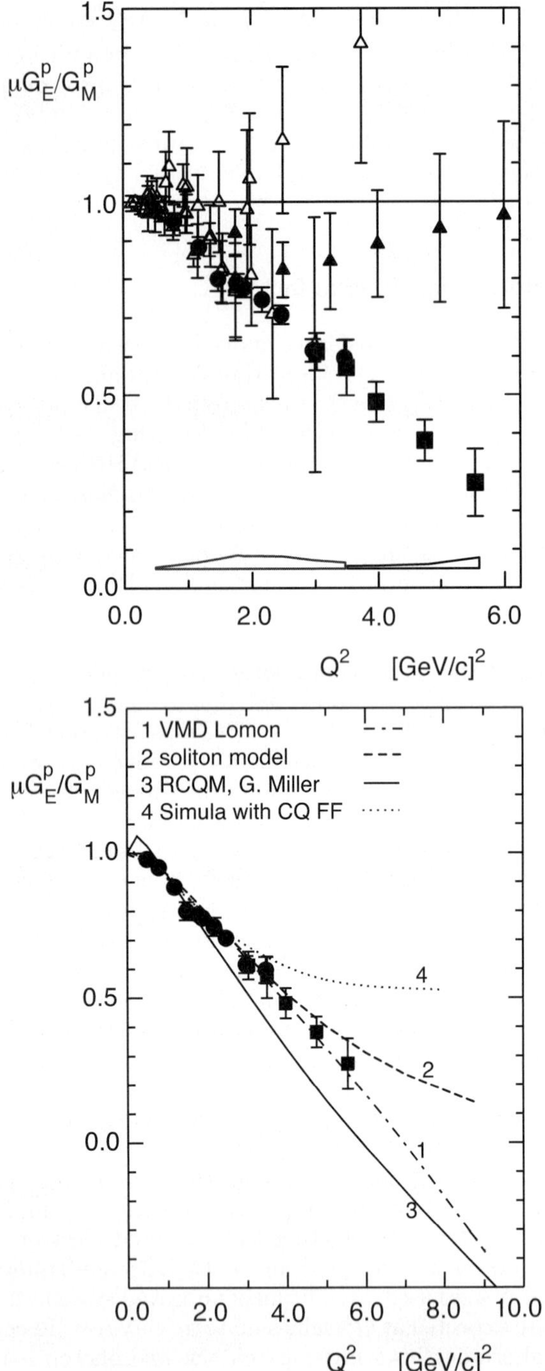

Fig. 3. The ratio $\mu G_E^p / G_M^p$ from [42] (*solid circles*) and [43] (*solid squares*), compared to the older SLAC data (*solid triangles* [44]) (*top*) and various theoretical calculations (*bottom*): *dash-dotted* [15], *dashed* [20], *solid* [17] and *dotted* [19]. The systematic errors for both recent experiments are shown as a band at the bottom of the left figure

2 Wide-angle compton scattering

Until fairly recently, little was known about the response of a nucleon to real photons in the regime where its response is dominated by the internal structure of the nucleon. Only one twenty-year old experiment from Cornell

[51] had provided data on elastic photon scattering (Real Compton Scattering) at large transverse momentum. Two different reaction mechanisms are expected to describe Real Compton Scattering. At very large momentum transfer two-gluon exchange, in the perturbative QCD framework, will dominate the reaction. Indeed, the Cornell data exhibit an s and t dependence in fair agreement with the pQCD counting rule: $d\sigma/dt = f(\theta_{CM}) \cdot s^{-n}$ with $n = 6$ for RCS.

Recent calculations by Radyushkin [52] have shown that at moderate momentum transfer a softer mechanism should dominate at large center-of-mass angles (Wide-Angle Compton Scattering, WACS). In this handbag mechanism only one quark interacts with the real photon and no gluon is exchanged between the struck quark and the rest of the partons. By factorizing the mechanism into a hard and a soft part the RCS reaction can be described in the framework of Generalized Parton Distributions (GPD) by two form factors R_V and R_A. In the WACS regime R_V dominates the cross section and the ratio $d\sigma_{RCS}/d\sigma_{KN}$ should be nearly independent of s at a fixed value of t. Observation of this would be a clear indication of factorization. With a circularly polarized photon beam a measurement of the longitudinal polarization transfer coefficient K_{LL} can provide a sensitive probe of the dominant reaction mechanism.

Experiment E99-114 [53] measured the RCS cross section for s between 5 and 11 $(\text{GeV/c})^2$ and for t between 1.5 and 6.5 $(\text{GeV/c})^2$ and K_{LL} at $s = 7$ $(\text{GeV/c})^2$ and $t = 4$ $(\text{GeV/c})^2$. The experiment, performed in Hall A at Jefferson Lab, used a mixed bremsstrahlung/electron beam, a liquid hydrogen target, one of the HRS spectrometers for proton detection and a lead-glass calorimeter with a sweep magnet for photon detection. The photon beam intensity was 10^{13} equivalent quanta per second, about 1000 times higher than in the Cornell experiment. Most of the background was eliminated by requiring an photon-proton coincidence and RCS events were separated from pion photo-production and elastic electron scattering through kinematical correlations between the energy and the angle of the detected photon and the recoiling proton.

The preliminary value for K_{LL} is shown in Fig. 4 from which a clear preference for the hand-bag mechanism is evident at the kinematics of the experiment. Similarly, preliminary data for the cross section follow the scaling power predicted by Radyushkin. The data of E99-114 have demonstrated that accurate RCS data can be extracted with modern photon facilities which provide sensitive information on the internal structure of the nucleon.

3 Neutron spin structure

3.1 The Q^2-evolution of the Gerasimov-Drell-Hearn sum rule

Over the past twenty years measurements of the nucleon spin structure have shown that only a small fraction (less than 20%) of the nucleon spin can be accounted for by the spin of the quarks. Also the Bjorken sum rule has

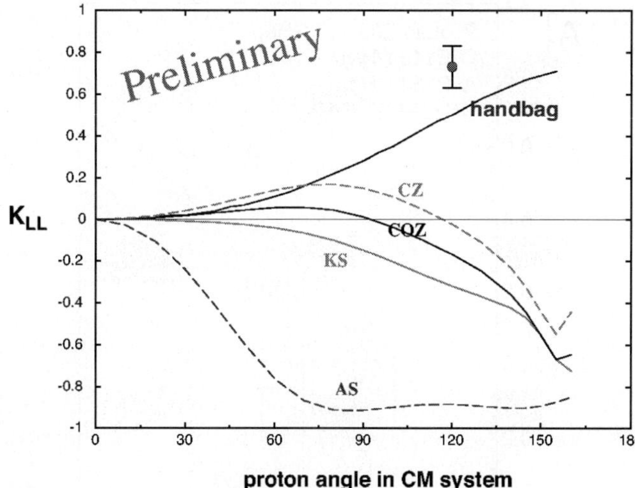

Fig. 4. The preliminary result for the polarization transfer coefficient K_{LL}, compared to a variety of predictions: handbag [52] and pQCD (AS asymptotic, KS King-Sachrajda, COZ Chernyak-Ogloblin-Zhitnitsky and CZ Chernyak-Zhinitsky).

been validated to within 10%. Recently, the Bjorken and the Gerasimov-Drell-Hearn (GDH) sum rules have been generalized [54] over the full range of momentum transfer from the non-perturbative $Q^2 = 0$ to the perturbative $Q^2 = \inf$ limit.

This generalized GDH sum rule $I(Q^2)$ for the neutron has recently been studied [55] from Q^2 is 0.1 to 0.9 $(\text{GeV/c})^2$ by measuring inclusive scattering of longitudinally polarized electrons from a polarized ^{3}He target in Hall A. The polarized ^{3}He target used the principles of optical pumping of a small admixture of Rb and spin exchange between the Rb and the ^{3}He atoms. The scattered electrons were detected in either of the two Hall A HRS spectrometers. A CO_2 gas Čerenkov detector and a lead-glass shower counter were used to separate electrons from the pion background. The results were corrected for the fact that the neutron was embedded in a ^{3}He nucleus by using a calculation by Ciofi degli Atti and Scopetta [56]. The results show a smooth variation of $I(Q^2)$ to increasing negative values as Q^2 drops from 0.9 to 0.1 $(\text{GeV/c})^2$. A recently completed new experiment [57] will extend the data down to 0.02 $(\text{GeV/c})^2$ and establish whether $I(Q^2)$ indeed turns upwards towards the value of the GDH integral.

In both of these experiments measurements were made with the ^{3}He target polarization oriented longitudinal and transverse to the momentum transfer, which allowed the extraction of both spin structure functions g_1 and g_2. The results from the first experiment [55] were used to study two further spin structure moments, d_2 and the Burkhardt-Cottingham sum rule [59]:

$$d_2(Q^2) = \int_0^1 x^2[2g_1(x,Q^2) + 3g_2(x,Q^2)]dx \quad (5)$$

$$\Gamma_2(Q^2) = \int_0^1 g_2(x,Q^2)dx = 0 \quad (6)$$

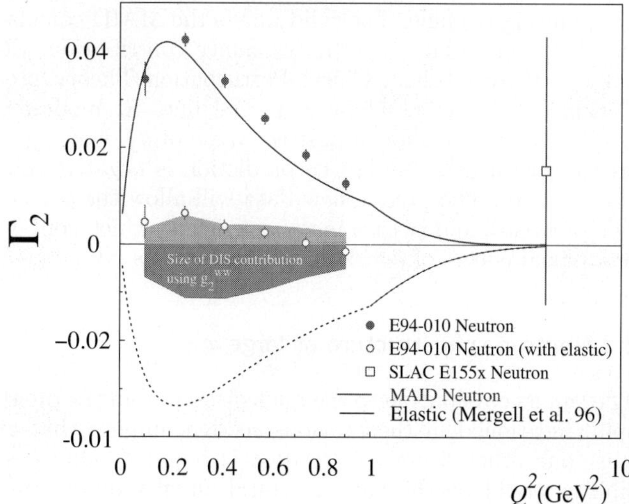

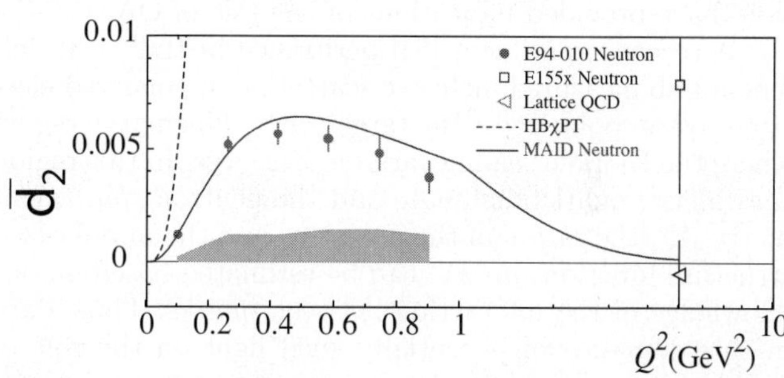

Fig. 5. Results for the Burkhardt-Cottingham sum rule Γ_2^n (*top panel*) and for $\bar{d}_2^n$ (*bottom panel*) along with world data from DIS and theoretical predictions (*see text*)

The values of $g_2(x,Q^2)$ were evaluated from the experimental results, measured at fixed incident beam energies and scattering angles, by interpolation and (for a few points) extrapolation for each of six values of Q^2. The integrals were computed using limits of integration extending from the nucleon pion threshold to a value of x corresponding to an invariant mass $W = 2.0$ GeV. Effects of the nuclear medium were again corrected for using the prescription of Ciofi degli Atti and Scopetta. The final results [58] are shown in Fig. 5.

In the top panel Γ_2^n is indicated by solid circles, the results after addition of the elastic contribution by open circles. The positive light band is an indication of the systematic error and the dark negative band of the estimated DIS contribution. The results clearly indicate that the Burkhardt-Cottingham sum rule is verified within uncertainties over the Q^2-range of the experiment. The open square shows the results of SLAC experiment E155x which had a significantly larger error bar. The solid curve represents the resonance contribution evaluated with the MAID code.

In the bottom panel the results for $\bar{d}_2(Q^2) = d_2(Q^2) - d_2^{elastic}(Q^2)$ are shown as solid circles at the six Q^2-values. The grey band represents their systematic uncertainty. The SLAC E155x result [60] is again shown as an open square, the prediction from Lattice QCD [61] by the open left-pointing triangle. The solid line is the MAID calculation [62] containing only the resonance contribution. Finally, a Heavy-Baryon Chiral Perturbation Theory prediction [63] is indicated by a dashed line. At moderate Q^2 the JLab data show a positive value of $\bar{d}_2^n$, slowly decreasing with Q^2. The lattice prediction is negative, but close to zero. The precise new data will allow the extraction of twist-3 and twist-4 matrix elements, which contain non-trivial effects of quark-gluon interactions.

3.2 Neutron spin structure at large x_B

The small contribution to the nucleon spin from the quark spin, mentioned in the previous section, implies that at least one other degree of freedom is required. One possibility would be the parton orbital angular momentum (OAM). Several recent experimental results, such as on elastic electron scattering and deuteron photo-disintegration, have provided indications of the role of OAM.

A recent experiment [64] performed in Hall A at Jefferson Lab measured inclusive scattering of polarized electrons off a polarized ^{3}He target in a kinematic region where the Bjorken scaling variable x is large. In this region the valence quarks dominate and the neutron spin asymmetry A_1^n (the ration of the polarized and the unpolarized structure functions g_1/F_1) can be estimated based on our knowledge of the interaction between quarks. Thus, data in this domain can potentially shed light on the role of OAM of the valence quarks. This asymmetry is predicted to approach 1 as $x \to 1$. Sofar, all existent data on A_1^n are either negative or consistent with zero.

Data were collected at three x-values, 0.33, 0.47 and 0.60 at Q^2-values ranging from 2.7 to 4.8 $(\mathrm{GeV}/\mathrm{c})^2$. The invariant mass was well above the resonance region. The beam asymmetry was measured with the target polarization oriented both parallel and transverse to the beam polarization. A model [65] for ^{3}He which includes S, S' and D states and a pre-existing $\Delta(1232)$ component, was used to extract A_1^n from $A_1^{^3He}$.

The results for A_1^n are shown in Fig. 6. The new datum at $x = 0.33$ is in good agreement with the world data, while for $x > 0.4$ the precision of the new data shows an order of magnitude improvement. The new data for the first time prove that A_1^n becomes positive at large x and their trend is consistent with models which predict that A_1^n approaches unity as $x \to 1$. None of the new data agrees with calculations which assume the BBS [66] parametrization. In this pQCD parametrization zero quark OAM and helicity conservation is assumed, which implies that a quark with $x \to 1$ must have the same helicity as the nucleon itself (hadron helicity conservation, HHC). On the other hand, the data do agree with predictions which do not assume zero OAM.

One can extract the polarized valence quark distributions $(\Delta u + \Delta \bar{u})/(u + \bar{u})$ and $(\Delta d + \Delta \bar{d})/(d + \bar{d})$ from A_1^n by using the world data set on the unpolarized u and d distributions, if one assumes that in the region $x > 0.3$ valence quarks truly dominate so that all strange quark distributions can be neglected. The results from the Hall A ex-

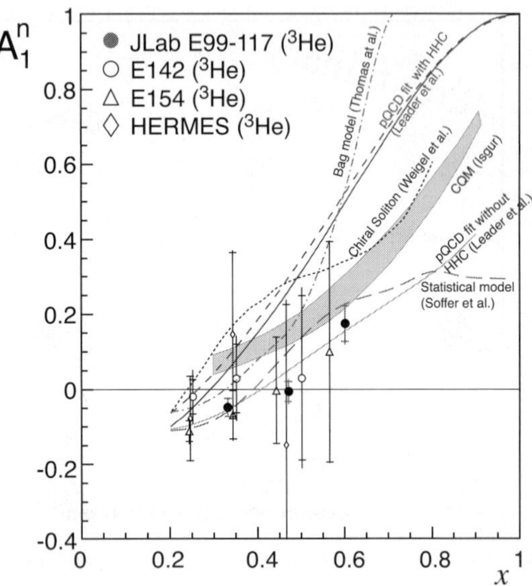

Fig. 6. Results for A_1^n compared to a variety of theoretical predictions and existing data obtained with a polarized ^{3}He target

periment extend to higher x and improve on the accuracy of available HERMES results [67]. Again, the new data agree with e.g. the hyperfine-perturbed CQM prediction by Isgur [68], but not with that from pQCD-based HHC, which suggests that effects beyond leading-order pQCD, such as OAM, could play an important role in this kinematic region.

4 Summary

Recent advances in polarized electron sources, polarized nucleon targets and nucleon recoil polarimeters have made it possible to accurately probe the response of the nucleon to a variety of electromagnetic probes. New data on nucleon electro-magnetic form factors with an unprecedented precision have (and will continue to) become available in an ever increasing Q^2-domain. These data will form tight constraints on models of nucleon structure and will hopefully incite new theoretical efforts. In addition they will significantly improve the accuracy of the extraction of strange form factors from parity-violating experiments.

Acknowledgements. The author expresses his gratitude to Gerry Miller, Zein-Eddine Meziani and Bogdan Wojtsejkhowski for fruitful discussions and for receiving their results prior to publication. This work was supported by DOE contract DE-AC05-84ER40150, under which the Southeastern Universities Research Association (SURA) operates the Thomas Jefferson National Accelerator Facility for the United States Department of Energy.

References

1. M.N. Rosenbluth: Phys. Rev. **79**, 615 (1950)
2. P. Bosted et al.: Phys. Rev. C **51**, 409 (1995)

3. A.I. Akhiezer and M.P. Rekalo: Sov. J. Part. Nucl. **3**, 277 (1974); based on J.H. Scofield: Phys. Rev. **141**, 1352 (1966)
4. R. Arnold, C. Carlson and F. Gross, Phys. Rev. C **23**, 363 (1981)
5. K. Aulenbacher et al.: Nucl. Instrum. Methods Phys. Res. A **391**, 498 (1997)
6. M. Poelker et al.: Proc. of the 14th Int. Symp. on High-Energy Spin Physics, Osaka (2000), p. 943
7. M. Hauger et al.: Nucl. Instrum. Methods Phys. Res. A **462**, 382 (2001)
8. M. Baylac et al.: Phys. Lett. B **459**, 412 (2001)
9. T.D. Averett et al.: Nucl. Instrum. Methods Phys. Res. A **427**, 440 (1999)
10. P.L. Anthony et al.: Phys. Rev. Lett. **71**, 959 (1993); Phys. Rev. D **54**, 6620 (1996); J.S. Jensen and Ph.D. Thesis: CalTech, 2000 (unpublished)
11. R. Surkau et al.: Nucl. Instrum. Methods Phys. Res. A **384**, 444 (1997)
12. L. Bimbot et al.: submitted to Nucl. Instrum. Methods Phys. Res. A
13. M. Ostrick et al.: Phys. Rev. Lett. **83**, 276 (1999)
14. G. Höhler et al.: Nucl. Phys. B **114**, 505 (1976)
15. E.L. Lomon: Phys. Rev. C **64**, 035204 (2001)
16. F. Schlumpf: J. Phys. G **20**, 237 (1994)
17. M.R. Franck, B.K. Jennings, and G.A. Miller: Phys. Rev. C **54**, 920 (1995)
18. J. Ralston et al.: Proc. 7th CIPANP, Quebec City (2000), p. 302
19. F. Cardarelli and S. Simula: Phys. Rev. C **62**, 065201 (2000)
20. G. Holzwarth: Z. Phys. A **356**, 339 (1996)
21. H. Gao et al.: Phys. Rev. C **50**, R546 (1994)
22. P. Markowitz et al.: Phys. Rev. C **48**, R5 (1993)
23. E.E.W. Bruins et al.: Phys. Rev. Lett. **75**, 21 (1995)
24. H. Anklin et al.: Phys. Lett. B **336**, 313 (1994)
25. H. Anklin et al.: Phys. Lett. B **428**, 248 (1998)
26. G. Kubon et al.: Phys. Lett. B **524**, 26 (2002)
27. W. Xu et al.: Phys. Rev. Lett. **85**, 2900 (2000)
28. W. Xu et al.: Phys. Rev. C **67**, 012201 (2003)
29. W. Brooks and M.F. Vineyard: JLab experiment E94-017
30. T. Eden et al.: Phys. Rev. C **50**, R1749 (1994)
31. M. Meyerhoff et al.: Phys. Lett. B **327**, 201 (1994)
32. I. Passchier et al.: Phys. Rev. Lett. **82**, 4988 (1999)
33. C. Herberg et al.: Eur. Phys. Jour. A **5**, 131 (1999)
34. J. Becker et al.: Eur. Phys. Jour. A **6**, 329 (1999)
35. D. Rohe et al.: Phys. Rev. Lett. **83**, 4257 (1999)
36. S. Galster et al.: Nucl. Phys. B **32**, 221 (1971)
37. H. Schmieden et al.: MAMI proposal A1/2-99
38. H. Zhu et al.: Phys. Rev. Lett. **87**, 081801 (2001)
39. R. Madey et al.: Phys. Rev. Lett. **91**, 122002 (2003)
40. Bates Large Acceptance Spectrometer Toroid: http://mitbates.mit.edu /blast
41. B. Wojtsekhowski et al.: JLab experiment E02-013
42. M.K. Jones et al.: Phys. Rev. Lett. **84**, 1398 (2000)
43. O. Gayou et al.: Phys. Rev. Lett. **88**, 092301 (2002)
44. L. Andivahis: Phys. Rev. D **50**, 5491 (1994) and references therein
45. S.J. Brodsky and G. Farrar: Phys. Rev. D **11**, 1309 (1975)
46. C.F. Perdrisat et al.: JLab experiment E01-109
47. J. Arrington and R. Segel: JLab experiment E01-001
48. P.G. Blunden, W. Melnitchouk, and J.A. Tjon: nucl-th/0306076
49. P.A.M. Guichon and M. Vanderhaeghen: hep-ph/0306007
50. M.P. Rekalo and E. Tomasi-Gustafsson: nucl-th/0307066
51. M.A. Shupe et al.: Phys. Rev. D **19**, 1929 (1979)
52. A.V. Radyushkin: Phys. Rev. D **56**, 5524 (1997)
53. C. Hyde-Wright, A. Nathan, and B. Wojtsekhowski: Jefferson experiment E99-114
54. X. Ji and J. Osborne: J. Phys. G **27**, 127 (2001)
55. M. Amarian et al.: Phys. Rev. Lett. **89**, 241301 (2002)
56. C. Ciofi degli Atti and S. Scopetta: Phys. Lett. B **404**, 223 (1997)
57. J.-P. Chen, A. Deur, and F. Garibaldi: Jefferson experiment E97-110
58. M. Amarian et al.: submitted to Phys. Rev. Lett.
59. H. Burkhardt and W.N. Cottingham: Ann. Phys. **56**, 453 (1970)
60. P.L. Anthony et al.: Phys. Lett. B **553**, 18 (2003)
61. M. Göckeler et al.: Phys. Rev. D **63**, 074506 (2001)
62. D. Drechsel, S. Kamalov, and L. Tiator: Phys. Rev. D **63**, 114010 (2001)
63. C.W. Kao, T. Spitzenberg, and M. Vanderhaeghen: Phys. Rev. D **89**, 016001 (2003)
64. X. Zheng et al.: submitted to Phys. Rev. Lett.
65. F. Bissey et al.: Phys. Rev. C **65**, 064317 (2002)
66. S.J. Brodsky, M. Burkardt, and I. Schmidt: Nucl. Phys. B **441**, 197 (1995)
67. K. Ackerstaff et al.: Phys. Lett. B **464**, 123 (1999)
68. N. Isgur: Phys. Rev. D **59**, 034013 (1999)

Eur Phys J A (2004) **19**, s01, 275–278
Digital Object Identifier (DOI) 10.1140/epjad/s2004-03-045-5

A pair polarimeter for linearly polarized high energy photons

C. de Jager[1], B. Wojtsekhowski[1], D. Tedeschi[2], B. Vlahovic[1,3], D. Abbott[1], J. Asai[4], G. Feldman[5], T. Hotta[6], M. Khadaker[7], H. Kohri[6], T. Matsumara[6], T. Mibe[6], T. Nakano[6], V. Nelyubin[1,2], G. Orielly[5], A. Rudge[8], P. Weilhammer[8], M. Wood[2], T. Yorita[6], and R. Zegers[9]

[1] Thomas Jefferson National Accelerator Facility, Newport News, VA 23606, USA
[2] University of South Carolina, Columbia, SC 29208, USA
[3] North Carolina Central University, Durham, NC 27707, USA
[4] Canadian Light Source, University of Saskatchewan, Saskatoon, Canada
[5] The George Washington University, Washington, DC 20064, USA
[6] Research Center for Nuclear Physics, Osaka University, Ibaraki, Osaka 567-0047, Japan
[7] Norfolk State University, Norfolk, VA 23504, USA
[8] CERN EP Division, 121 Geneva 23, Switzerland
[9] Japan Atomic Energy Research Institute, Mikazuki, Hyogo 679-5148, Japan

Received: 1 August / Accepted: 14 Nov 2003 /
Published Online: 6 Feb 2004 – © Società Italiana di Fisica / Springer-Verlag 2004

Abstract. A high quality beam of linearly polarized photons of several GeV will become available with the coherent bremsstrahlung technique at JLab. We have developed a polarimeter which requires about two meters of the beam line, has an analyzing power of 20% and an efficiency of 0.02%. The layout and first results of a polarimeter test on the laser back-scattering photon beam at SPring-8/LEPS are presented.

PACS. 07.60.F polarimeters – 29.40.Wk solid state detectors

1 Introduction

Electron-positron pair photo-production from amorphous matter can be calculated exactly in the framework of QED [1]. The azimuthal distribution in this reaction can be presented as $\sigma = \sigma_{unpol}\,[1 + P_l \cdot A \cdot cos(2\phi)]$, where P_l is degree of photon linear polarization, A the analyzing power of the process and ϕ the azimuthal angle between the polarization plane and the detected particle(s). The value of A ranges from 0.1 to 0.3 depending on the reaction kinematics. The use of the γ, e^+e^- reaction in polarimetry was suggested in 1950 [2]. The main experimental challenge in such a polarimeter is the small angle between the pair components. This was resolved by using a magnetic field to separate the electron and positron [3] at the price of a reduced analyzing power. For photons in the few GeV energy range a different solution was proposed [5] based on a silicon microstrip detector, of which the unique high track resolution allows the observation a e^+e^- pair with a separation as small as 100 μm.

2 Kinematics

Figure 1 illustrates the kinematics of electron-positron pair production and its relevant variables. The use of the angle $\omega_\pm$, which is the azimuthal angle of the e^+e^- plane, yields a value of $A \sim 0.24$ for the case of a very thin converter and equal energies of the electron and positron.

In the determination of the asymmetry we used a vector n_ω constructed from the momenta of the electron($\mathbf{p}_+$), the positron($\mathbf{p}_-$), and the photon($\mathbf{k}$) as

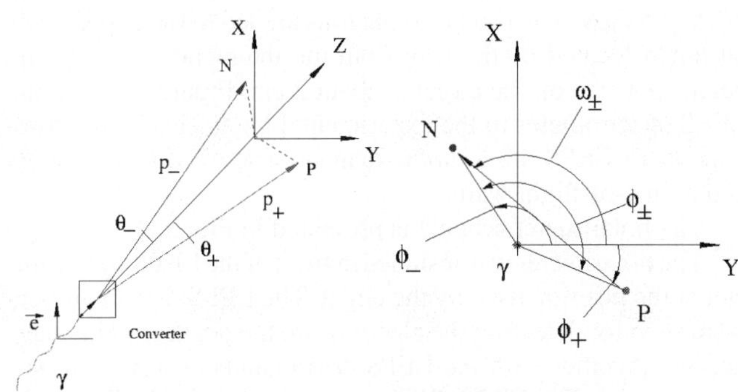

Fig. 1. The kinematics of e^+e^- pair photo production (*left picture*) and the azimuthal angles in the detector plane. The photon momentum is directed along the Z axis. The photon polarization vector **e** is parallel to the X axis. The angle $\phi_+(\phi_-)$ is the angle between the photon polarization plane and the plane constructed by the momentum of the photon and the momentum of the positron (*the electron*). The angle $\phi_\pm$ is called the coplanarity angle. The labels P and N indicate the positions of the crossings of the detector plane by the positron and the electron. The angle $\omega_\pm$ between the polarization plane and the vector $\overline{PN}$ is a directly measurable parameter

$$\mathbf{n}_\omega = \frac{\mathbf{p}_{+\perp}/|\mathbf{p}_{+\|}| - \mathbf{p}_{-\perp}/|\mathbf{p}_{-\|}|}{|\mathbf{p}_{+\perp}/|\mathbf{p}_{+\|}| - \mathbf{p}_{-\perp}/|\mathbf{p}_{-\|}||}$$

where $\mathbf{p}_{\pm\perp} = \mathbf{p}_\pm - \mathbf{p}_{\pm\|}$, $\mathbf{p}_{\pm\|} = \mathbf{p}_\pm \cdot \mathbf{k}/k$. This vector lies in the plane perpendicular to the photon momentum $\mathbf{k}$. The direction of the $\mathbf{n}_\omega$ vector is determind from the positron and the electron tracks in the detector, even the position of the pair creation vertex is not detected.

From the azimuthal dependence of asymmetry we find A_{exp}

$$\frac{N(\omega, \|) - N(\omega + \frac{\pi}{2}, \perp)}{N(\omega, \|) + N(\omega + \frac{\pi}{2}, \perp)} = A_{exp}\, cos(2\omega + \Delta), \quad (1$$

where $N(\omega, \|)$ is the number of e^+e^- pairs observed at the angle ω for the photon polarization $\mathbf{e}$ parallel to the vertical and $N(\omega + \frac{\pi}{2}, \perp)$ is the number of pairs observed at the angle $\omega + \frac{\pi}{2}$ for the horizontal photon polarization, Δ is the tilt of the detector axis.

3 Experiment

3.1 Layout

The experiment was performed at the SPring-8 facility in Japan. SPring-8 is a third-generation synchrotron radiation facility with 62 beam lines for various applications. One of the beam lines with high-energy photons is used for high-energy nuclear physics. The photons are produced by backward Compton scattering laser light off the electrons circulating in the storage ring. The beam line is called LEPS (Laser Electron Photon beam line at SPring-8) [4]. The laser system focuses the light at the intersection region located between two bending magnets in the storage ring. An Argon laser provides the laser beam with a wave length of 351 nm for which the maximum energy of the produced photons is 2.4 GeV. The tagging system employs the bending magnet located down stream of the interaction region for analyzing the momentum of the electrons. The energy resolution of the tagging system is about 15 MeV in the region from 1.5 to 2.5 GeV. The produced photons are led to the experimental hutch located 69 m away from the interaction region. The beam spot size on the target is about 2 cm. Figure 2 shows the LEPS spectrometer in the experimental hutch. The main components of LEPS are a dipole magnet, a set of drift chambers and a time-of-flight wall.

The polarimeter scheme is presented in Fig. 3.

The polarimeter was installed in front of the LEPS spectrometer at the position used by the target. The LEPS drift chambers were used for detecting the electron and the positron. The magnet SM together with the LEPS drift chambers formed a pair spectrometer. The LEPS dipole magnet was turned off.

3.2 Photon beam

Figure 4 shows the photon energy spectrum detected by the tagger. The calculated photon polarization vs photon energy is shown in Fig. 5. It has a maximum value of 93% at the Compton edge of 2.4 GeV. The energy dependence of the polarization was taken into account in the extraction of the analyzing power (see Sect. 4.2).

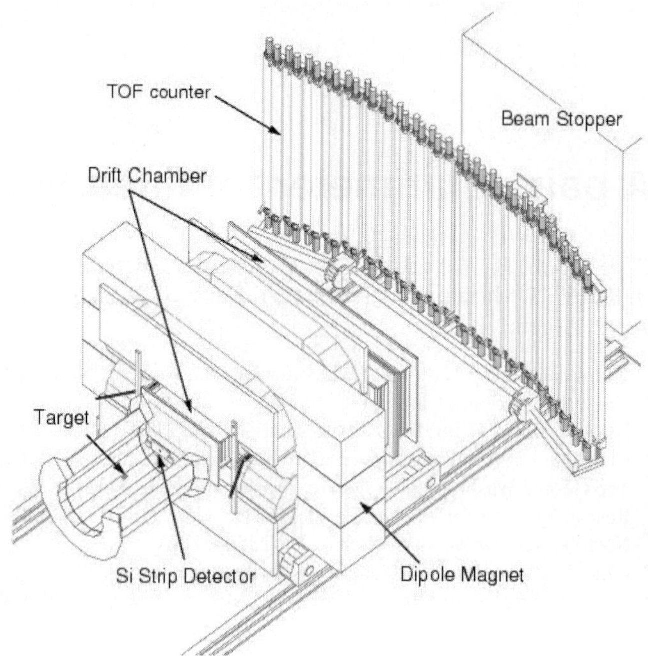

Fig. 2. The LEPS spectrometer

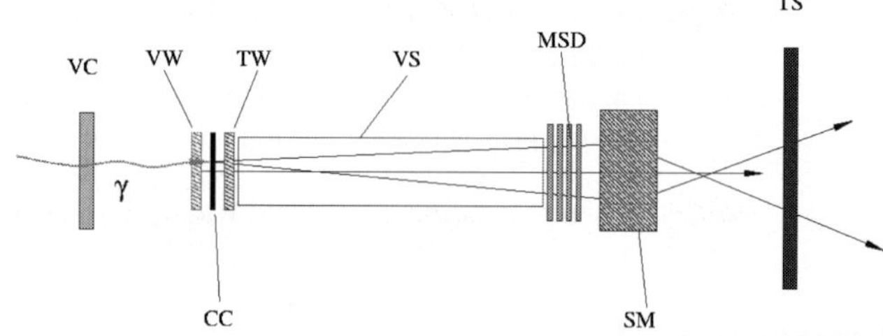

Fig. 3. The layout of the photon polarimeter. The photon arrives from the left. The veto detector is marked as VC, the veto wire chamber as VW, the converter as CC, the trigger wire chamber as TW, the vacuum straight section as VS, the set of micro-strip detectors as MSD, the separation magnet as SM, and the trigger scintillator counter as TS

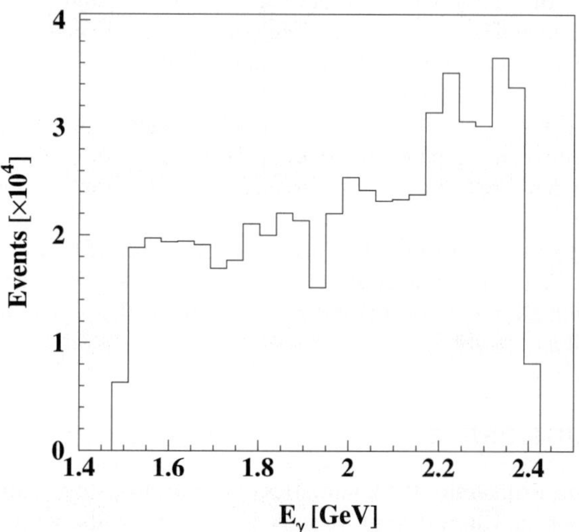

Fig. 4. The tagged photon energy spectrum

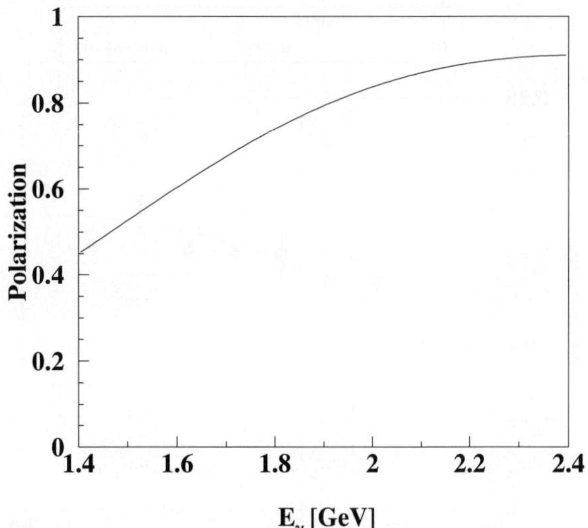

Fig. 5. The photon polarization as a function of the photon energy

3.3 Trigger logic

The signals from VC, WV, TW and TS are used to form a polarimeter trigger signal (PT). The coincidence between PT and a tagger trigger signal was used to initiate a readout cycle of both the polarimeter and the LEPS electronics. The amplitude from the trigger wire chamber TW was recorded in both DAQs and used to cross check the synchronization of information.

3.4 Data collection

Measurements were done at two orientations of the laser polarizations – vertical and horizontal. The orientations were interchanged typically every 2 hours. The degree of the laser photon polarization was 99%. The typical tagger rate was 150 kHz, the wire chamber rates 20 kHz and the event rate 150 Hz. In the trigger rate the contribution from the unpolarized background of photons produced on the residual gas in the storage ring was below 2%.

4 Analysis and results

4.1 Data analysis

Only events with one hit in the tagger were considered. For reconstruction of the tracks in the MSD and the drift chambers first clusters were searched. After a cluster was recognized and its position determined, the hit pattern was used for a track search. We analyzed all combinations and selected two with best matching of the hit positions. The distribution of the residuals for the best combination has a peak, of which width allows to estimate the MSD spatial resolution to be 15 μm. Events with two tracks in the MSD were used in the analysis of the pair angular and position distributions. The drift chamber and tagger information were used to calibrate the deflection parameter of the pair spectrometer. The energy of the electron and the positron were used to select pairs with $E_+ \sim E_-$. The separation between

the electron and the positron tracks in the MSD (Δr) and the photon energy E_γ were used to calculate the reduced open angle $\Theta = \Delta r/L \cdot E_\gamma/m_e$, where L is the distance between the converter and MSD, m_e the electron mass.

4.2 Results

The beam polarization effect in the azimuthal distribution (see 1) of the pair plane is shown in Fig. 6 for events with a photon energy in the range 1.5-2.4 GeV.

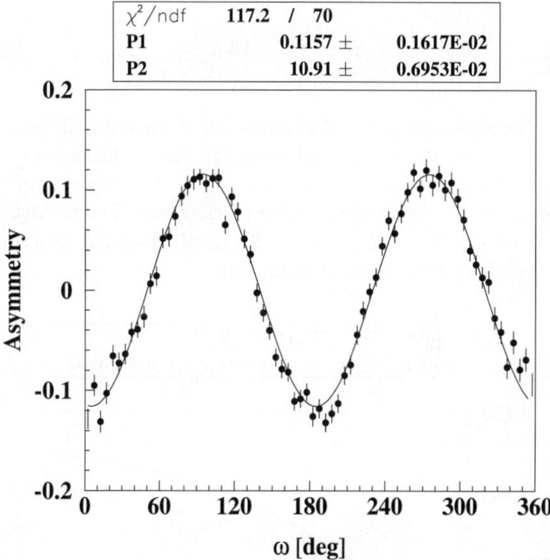

Fig. 6. The beam polarization effect in the azimuthal distribution of e^+e^- pairs. ω is the angle between the e^+e^- plane and the photon polarization plane. Parameters P1 and P2 of the fit reflect the asymmetry A_{exp} and Δ in 1

The analyzing power of the reaction has a strong dependence on the open angle between the pair components. Figure 7 shows a corresponding experimental result. Two cases of the averaged analyzing power are also presented in this plot. These running averagies show how to select the low and the upper limits on the reduced open angle to minimize the sensitivity of the analyzing power A to these cuts.

For the next step in the analysis we used only events with a reduced open angle in the range from 4 to 20. The asymmetry as a function of the photon energy is plotted in Fig. 8. The curve follows the dependence of the photon beam polarization as a function of the photon energy.

The selection of the pairs with almost equal e^+ and e^- energies leads to a higher asymmetry as is shown in Fig. 9. The values of the analyzing power for these two cases are 0.119 and 0.192. They agree within 5% with calculations [6] for the thickness of the carbon converter used in our experiment (0.1mm).

5 Conclusion

Here, we described the results of a polarimeter test at SPring-8/LEPS. For the first time the polarization effect in the pair

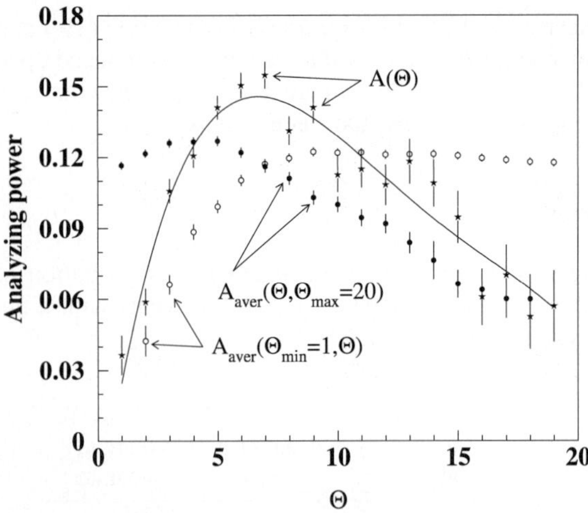

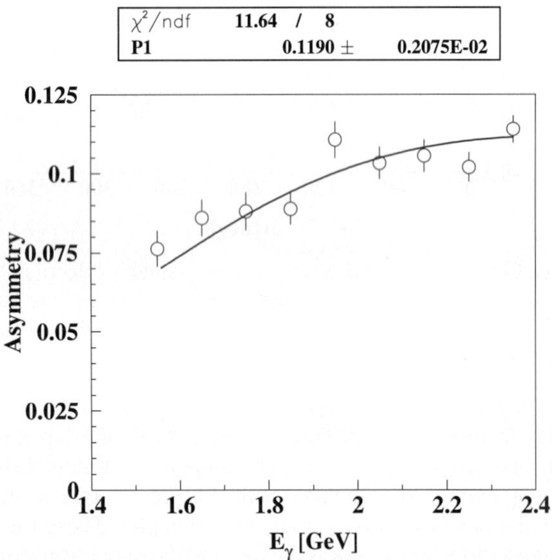

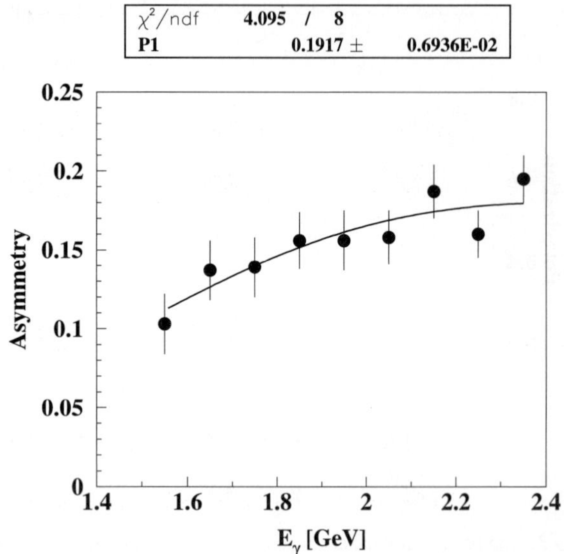

Fig. 7. The analyzing power A as a function of the reduced open angle Θ. The *solid line* shows the trend of the experimental data (*stars*). The average value (*open circles*) $A_{aver}(\Theta_{min} = 1, \Theta)$ shows the effect of a running upper limit for a fixed value of a low limit. The average value (*closed circles*) $A_{aver}(\Theta_{min} = \Theta, 20)$ shows the effect of a running lower limit for a fixed value of upper limit

Fig. 8. The asymmetry A_{exp} as a function of the photon energy for all pairs. The shape of the curve follows the polarization of the photon beam

Fig. 9. The asymmetry A_{exp} as a function of the photon energy for the pairs with a condition $0.8 < E_+/E_- < 1.2$

photo-production from amorphous matter was observed in the GeV energy range. We demonstrated that a compact polarimeter for photon energies of several GeV can be constructed by using silicon micro-strip detectors. An analyzing power was observed of 0.192 for 0.1 mm carbon converter and near equal energies of the electron and the positron, which is within 5% of the expected value.

We acknowledge the crucial support of W. Briscoe, L. Cardman, and B. Mecking.

This work was supported in part by the National Science Foundation in grants PHY-0099487 for the North Carolina Central University and PHY-0072361 for the University of South Carolina and by DoE contract DE-AC05-84ER40150 under which the Southeastern Universities Research Association (SURA) operates the Thomas Jefferson National Accelerator Facility for the United States Department of Energy.

References

1. H. Olsen and L.C. Maximon: Phys. Rev. **114**, 887 (1959); L.C. Maximon and H. Olsen: Phys. Rev. **126**, 310-319 (1962)
2. C.N. Yang: Phys. Rev. **77**, 722 (1950); J.H. Berlin and L. Madansky: Phys. Rev. **78**, 623 (1950)
3. G. Barbiellin et al.: Phys. Rev. Lett. **9**, 396 (1962); H. Sasaki et al.: Nucl. Instr. and Meth. **62**, 45 (1968); M. Kobayashi and K. Kondo: J. Phys. Soc. Japan **28**, 277 (1970); M. Kobayashi and K. Kondo: Nucl. Instr. and Meth. **104**, 101 (1972)
4. T. Nakano et al.: Nucl. Phys. A **684**, 71c (2001)
5. B. Wojtsekhowski et al.: Polarimeter for High Energy Photons, JLab Technical Note 98-039, 1998
6. B. Wojtsekhowski, D. Tedeschi, and B. Vlahovic: accepted for publication in Nucl. Instr. and Meth.

Eur Phys J A (2004) **19**, s01, 279–282
Digital Object Identifier (DOI) 10.1140/epjad/s2004-03-046-4

EPJ A direct
electronic only

Future experiments at MAMI

Reinhard Beck[1] and Alexander Starostin[2]

[1] Institut für Kernphysik, Johannes Gutenberg-Universität, D-55099 Mainz, Germany
[2] Dept. of Physics and Astronomy, UCLA, box 1547, Los Angeles, CA, 90025-1547, USA

Received: 15 Oct 2003 / Accepted: 14 Nov 2003 /
Published Online: 6 Feb 2004 – © Società Italiana di Fisica / Springer-Verlag 2004

Abstract. A new experimental program is about to get underway at the Mainz Microtron (MAMI) Facility. The experimental apparatus consists of the famous Crystal Ball together with TAPS detector as a forward wall, and a central tracker. This configuration provides a geometrical acceptance close to 4π combined with good energy and angular resolution in particular for neutral final states. After the energy upgrade of the Mainz Microtron (MAMI-C), the Crystal Ball and TAPS setup will be equipped with a frozen-spin polarized target filled with ^{1}H, or ^{2}H to perform new high precision, high statistics measurements of neutral meson production on the nucleon up to 1.5 GeV incident photon energies. In particular it provides a unique opportunity to investigate the GDH sum rule on a neutron target in the contributions by reactions $\gamma\mathbf{n} \to \pi^0 n$, $\gamma\mathbf{n} \to \pi^0\pi^0 n$ and $\gamma\mathbf{n} \to \eta n$.

1 Introduction

After the successful completion of a rich experimental program at Brookhaven National Laboratory the Crystal Ball, CB, photon spectrometer has been moved beginning of November 2003 to the Mainz Microtron Facility for a series of high precision experiments requiring high intensity, high quality, ultra-fine tagged, linearly and circularly polarized photon beams. The TAPS BaF$_2$ calorimeter [1] will be used in the measurements as a forward wall. The high granularity and large acceptance of the experimental setup allows the data to be simultaneously obtained for different final states over the full range of angles with good energy and angular resolution and high statistical accuracy.

Over the past years (1997-2002) the Crystal Ball detector has been used successfully in medium energy beams for nucleon and hyperon spectroscopy. The experiments have provided novel data on K^- and π^- induced reactions for kaon beams of 500-750 MeV/c and pion beams of 150 to 750 MeV/c as well as produced new upper limits for various η decay modes, see for example [2,3,4,5,6]. Here we use some of the K^- and π^- results to illustrate the extraordinary capabilities of the Crystal Ball.

2 MAMI-B

The Mainz Microtron is a unique facility which satisfies our beam demands. The maximum photon energy available with MAMI-B is E_γ^{max} =855 MeV. The energy upgraded machine (MAMI-C) will provide E_γ^{max} =1.5 GeV, which corresponds to $\sqrt{s} = 1.92$ GeV. The resolution of the MAMI/Glasgow tagging facility [7] for bremsstrahlung photons is $\Delta E_\gamma = 2$ MeV for an incoming electron beam of 855 MeV. The resolution can be improved at least by factor of three using the tagger microscope. The microscope covers approximately 80 MeV of the photon bremsstrahlung spectrum. The maximum photon flux is $N_\gamma^{max} = 5 \times 10^5$ [s^{-1} MeV^{-1}], that is much higher than most other facilities in the world. The degree of photon polarization is 70% for linearly and 85% for circularly polarized photons.

3 MAMI-C

A Harmonic Double Sided Microtron (HDSM) is presently under construction to increase the end energy of the three staged cw Race-Track Microtron (RTM) cascade MAMI from 0.855 to 1.5 GeV. This new accelerator, scheduled to come into operation in 2004, consists mainly of two pairs of 90° bending magnets and two linear accelerators (see Fig. 1). Special features of the HDSM are the operation of the two linacs at different frequencies, 2.45 GHz and 4.90 GHz, for higher longitudinal stability, and a relatively strong field gradient in the bending magnets for the compensation of vertical edge defocusing. In this section the main design considerations and a short report of the status for construction is presented.

In order to extend the experimental possibilities at the Institute of Nuclear Physics at the University of Mainz the end energy of MAMI will be increased by a fourth stage to 1.5 GeV. The main prerequisites were that there should be no substantial degradation of the excellent beam quality and operational reliability of the RTM cascade (see Ta-

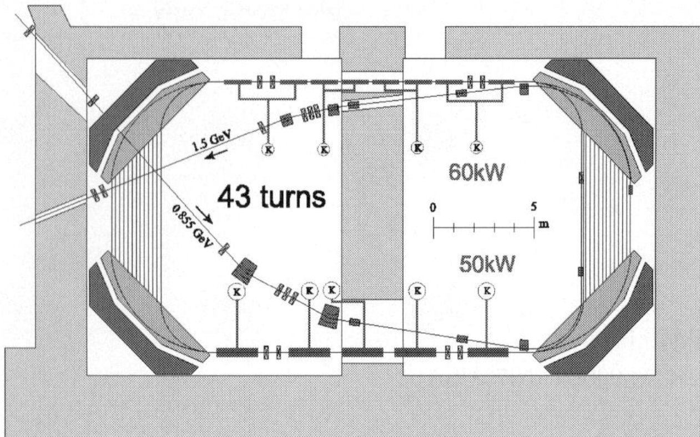

Fig. 1. General layout of HDSM

Table 1. Main parameters of RTM3 and HDSM

		RTM3	HDSM
General			
$E_{Inj.}$	MeV	180.2	855.5
$E_{Extr.}$	MeV	854.4	1507
turns	#	90	43
power consumption	kW	650	1500
RF system			Linac: 1 / 2
total energy gain	MeV	7.49	16.63 - 13.93
frequency	GHZ	2.449532	4.899 / 2.449
sections/klystrons	#	5/5	8/4 / 5/5
linac length	m	8.870	8.59 / 10.10
shunt-impedance	$M\Omega/m$	67	78 / 68
dissipated power	kW/m	11.5	14.2 / 12.4
beam power	kW	67.4	65.2
Magnet system			
min./max. field	T	1.2842	0.95 - 1.53
min./max. gap	mm	100	85 - 138
min./max. radius	m	0.47 - 2.22	2.23 - 4.60
weight	to	900	1000
Beam parameters			
energy width	keV	13	110
hor. emittance	nm rad	7.8	9.2
vert. emittance	nm rad	0.5	0.4

ble 1), and that an existing experimental hall (29×15 m^2) should be used for the installation of the accelerator. At the beginning two different solutions were studied: a) a superconducting isochronous recirculator with only a few turns and b) a double sided microtron (DSM) with smaller energy gain and greater number of turns. Detailed calculations for solution b) showed that the emittance increase by quantum fluctuations from synchrotron radiation was only about a factor of 1.5 in both horizontal and longitudinal directions [8]. Therefore, and because of higher investment costs of solution a) it was decided to build a DSM. For more details on the design and status of the 1.5 GeV double sided microtron see [9].

4 Physical program

The first stage of our physical program at MAMI (MAMI B) is centered on the first measurement of the magnetic dipole moment of the $\Delta^+(1232)$ resonance. The magnetic dipole moment, μ_b, provides us with a simple way for testing the validity of the theoretical hadron description in the non-perturbative sector of QCD. This includes quark soliton models, the standard quark models, various effective Lagrangians and lattice QCD calculations. Our experimental technique takes advantage of the very short Δ lifetime by having the Δ radiatively decay to itself. This method has been successfully pioneered for the Δ^{++} using the reaction $\pi^+ p \to \gamma'\Delta^{++} \to \gamma'\pi^+ p$ [10]. We propose to determine $\mu_b[\Delta^+(1232)]$ using radiative π^0 photoproduction: $\gamma p \to \Delta^+ \to \gamma'\Delta^+ \to \gamma'\pi^0 p$. A first pilot experiment $\gamma p \to \gamma'\pi^0 p$ has been performed with the TAPS calorimeter at MAMI for energies $\sqrt{s} = 1221 - 1331$ MeV. Angular and energy differential cross section have been determined for all particles in the final state in three bins of the excitation energy [11]. The theoretical aspects have been dealt with in detail already by the theory groups at MAMI [12] and Tuebingen [13]. μ_b can be determined from the differential cross section $d\sigma^5/d\Omega_\gamma d\Omega_\pi dE_\gamma$ and from the asymmetry, Σ, for linearly polarized photons.

The broad spectrum of MAMI bremsstrahlung photons from $E_\gamma^{min} \approx 100$ MeV to $E_\gamma^{max} \approx 1500$ MeV together with the 4π acceptance of the experimental apparatus allows the simultaneous survey of π^0, $2\pi^0$, $3\pi^0$ and η production at all energies and for the full angular range. Such measurements will be perform with LH$_2$ and LD$_2$ targets using linearly and circularly polarized photon beams. A unique frozen spin target filled with ^{1}H, or ^{2}H will be used in the second stage of the experiment (MAMI-C). The target makes possible new high precision, high statistics measurements of the cross sections for the $\gamma\mathbf{N} \to \pi^0 N$ and $\gamma\mathbf{N} \to \pi^0\pi^0 N$ processes at incident photon energies up to 1.5 GeV. In particular it provides a unique opportunity to measure the partial contributions to the GDH sum rule on a neutron target in the reactions $\gamma\mathbf{n} \to \pi^0 n$ and $\gamma\mathbf{n} \to \pi^0\pi^0 n$. Our measurements will also provide new information on the photon coupling of low-mass baryon and hyperon resonances.

The production of $2\pi^0$'s on nuclei is interesting because of predicted medium modification effects. They have recently received much attention both, experimental and theoretical, see for example [6,14,15,16]. We plan to make new high statistics, high precision measurements of $2\pi^0$ production on complex nuclei with the Crystal Ball at Mainz. The data will be obtained simultaneously over the full range of incident photons energies up to 1.5 GeV. That will allow us to investigate in better details the possibility of chiral restoration in normal nuclear matter. It also enable us to make better estimate of the corrections due to the pion rescattering in nuclei.

An incomplete list of other possible measurements includes: (i) threshold photoproduction of π^0 and η at MAMI-B as well as η', ω and K_s^0 at MAMI-C with polarized and unpolarized beams and targets; (ii) measure-

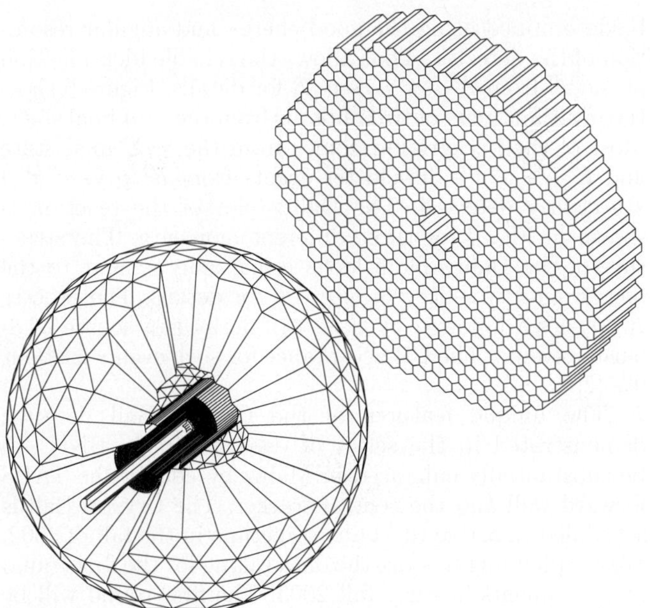

Fig. 2. The experimental apparatus proposed for the experimental program at MAMI. Some of the Crystal Ball crystals have been omitted in order to show the position of the cylindrical wire chamber and the target inside the Crystal Ball. The central detector, the Crystal Ball, TAPS as forward wall, the cylindrical wire chamber, and the liquid hydrogen target are shown

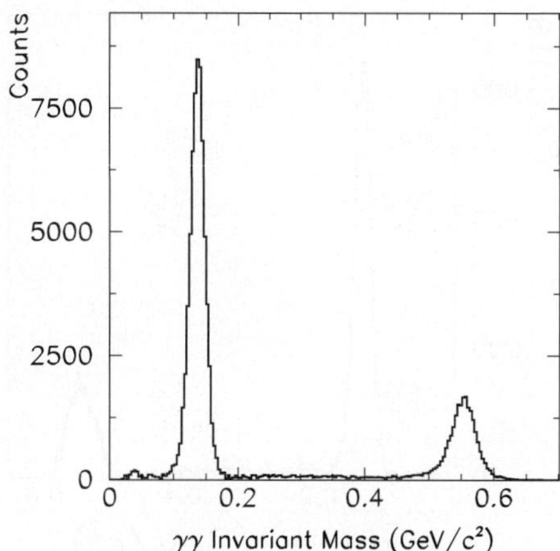

Fig. 3. Invariant mass of two photons obtained for a 1.8–cm–thick CH_2 target in a 750 MeV/c π^- beam. The normalized carbon and empty target spectra have been subtracted

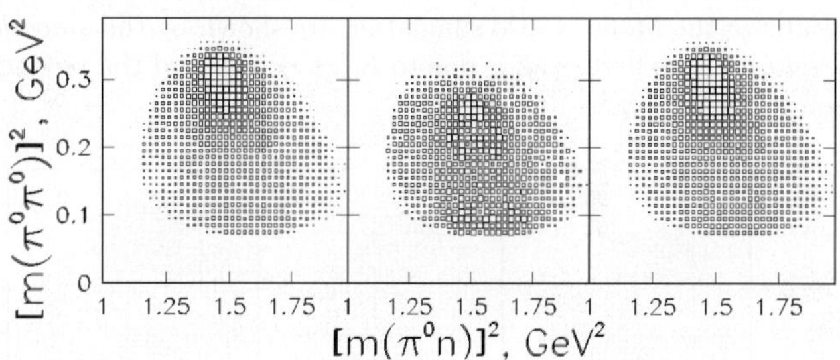

Fig. 4. Dalitz plots measured for $\pi^- p \to \pi^0 \pi^0 n$. There are two entries for every event. **Left:** Only 4-cluster (*four photon*) events are used. **Center:** Only 5-cluster (*four photon and the neutron*) events are used. **Right:** Sum of 4- and 5-cluster events

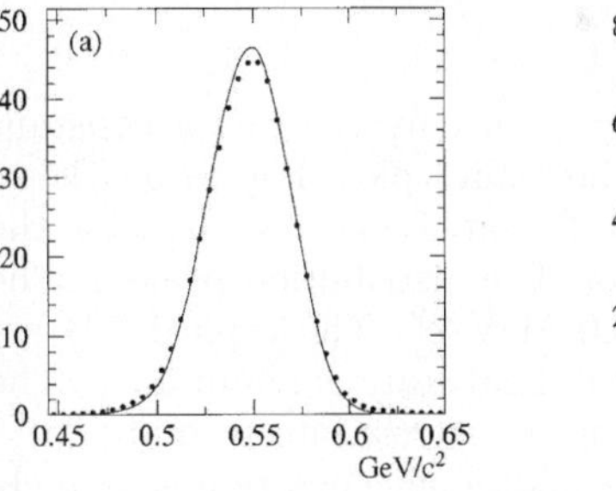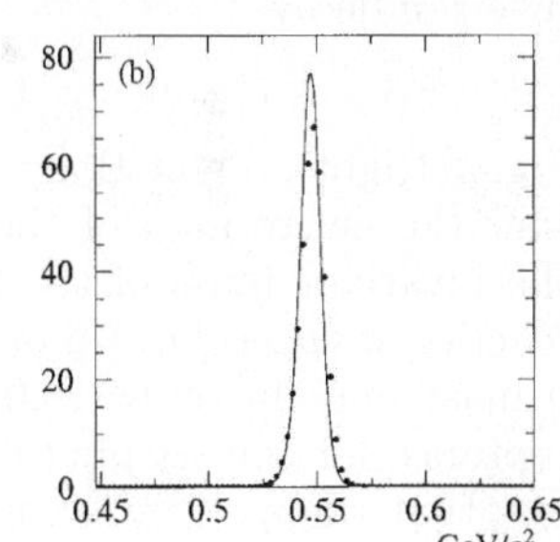

Fig. 5. a Invariant mass spectrum for the six clusters in $\pi^- p$ interactions at 720 MeV/c. **b** Invariant mass spectrum for the $3\pi^0$ for a constrained fit to the process $\pi^- p \to \eta n \to 3\pi^0 n \to 6\gamma n$, a 4C fit. The *dots* are our data and the *line* is the Monte Carlo showing excellent agreement

ments of the $N^*(1535)$ magnetic dipole moment using $\gamma p \to \gamma' \eta p$; (iii) a new measurement of the η mass.

The proposed experimental apparatus is shown in Fig. 2. The Crystal Ball with TAPS as the forward wall will be used for detection of photons and nucleons. In addition the polar and azimuthal angles of the outgoing proton for $\Theta_{lab} > 20°$ will be measured by the central tracker which is based on the DAPHNE cylindrical multiwire proportional chamber. The chamber will be inserted into the Crystal Ball beam cavity.

The Crystal Ball was build at SLAC and used in J/ψ measurements at SPEAR and b–quark physics at DESY [17]. The CB is constructed of 672 optically isolated NaI(Tl) crystals, 15.7 radiation lengths thick. The counters are arranged in a spherical shell with an inner radius of 25.3 cm and an outer radius of 66.0 cm. The hygroscopic NaI is housed in two hermetically sealed evacuated hemispheres. Each crystal is shaped like a truncated triangular pyramid, 40.6 cm high, pointing towards the center of the Ball. The sides on the inner end are 5.1 cm long and 12.7 cm on the far end. Electromagnetic showers in the spectrometer are measured with an energy resolution

$$\sigma_E/E \sim 1.7\%/(E~(\mathrm{GeV}))^{0.4} \, ;$$

the angular resolution for photon showers at energies of 0.05–0.5 GeV is $\sigma_\theta = 2°$–$3°$ in the polar angle and $\sigma_\phi = 2°/\sin\theta$ in the azimuthal angle.

High granularity and a large acceptance make the Crystal Ball a unique instrument for measuring reactions with multiphoton final states. The CB detects neutrons

with an efficiency of $\approx 35\%$ at $E_n = 150$ MeV [4]. Figures 3, 4, and 5 show examples of our pion data for 2-cluster, 4-cluster, 5-cluster and 6-cluster events. The invariant mass of two photons exhibits two narrow peaks from $\pi^0 \to \gamma\gamma$ and $\eta \to \gamma\gamma$ events, see Fig. 3. The small background under the peaks ($< 2\%$) comes from photon-neutron misidentification. Figures 4(Left) and 4(Center) show the Dalitz plots of the $2\pi^0 n$ final state from 4-cluster (four photon) and 5-cluster (four photon and a neutron) events. The sum of 4- and 5-cluster events is shown in

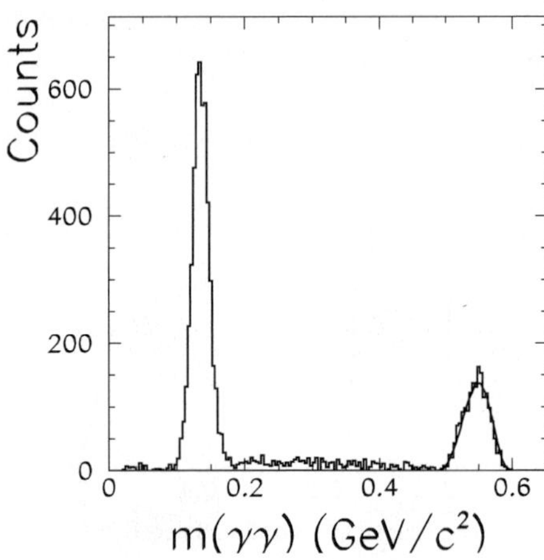

Fig. 6. The invariant mass of two photons in $K^-p \to \gamma\gamma\Lambda$ obtained with 750 MeV/c K^- beam. We have identified the π^0 that comes from Λ decay by its unique kinematics. The normalized empty target spectrum has been subtracted. The results of the Monte Carlo simulation are shown by the *smooth solid line*. The first peak is due to $K^-p \to \pi^0\Lambda$ and the second one to $K^-p \to \eta\Lambda$

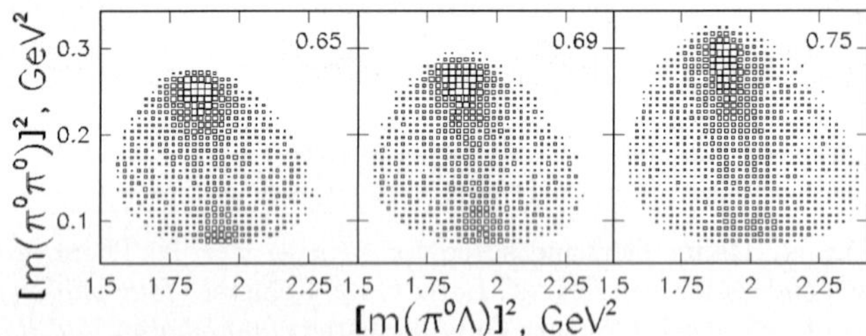

Fig. 7. Dalitz plots for $K^-p \to \pi^0\pi^0\Lambda$. Λ is detected via $\Lambda \to \pi^0 n$ decay

Fig. 4(Right). The CB acceptance for the sum is constant over the entire area of the Dalitz plot. Figure 5a shows the invariant mass of the 6γ satisfying a χ^2 test for the reaction $\pi^-p \to \eta n \to 6\gamma n$. The distribution peaks at the η mass and has $\sigma \approx 18.0$ MeV/c^2. The typical CB acceptance for the six photon final state is about 15%. The fact that the $\eta \to 6\gamma$ decay occurs via intermediate $3\pi^0$'s allows us to apply three additional constraints, namely the masses of the $3\pi^0$'s. Such constraints are useful to reduce the background and they improve the experimental resolution. The improved mass resolution when applying a constrained fit to the reaction $\pi^-p \to \eta n \to 3\pi^0 n$ is shown in Fig. 5b. The width of the η peak is $\sigma \approx 5.0$ MeV/c^2. The background under the peak is less than 1%.

The CB has been successfully used for extensive studies of hyperon production: the Λ is measured via $\Lambda \to \pi^0 n$, Σ^0 via $\Sigma^0 \to \Lambda\gamma$ and K_s^0 via $K_s^0 \to 2\pi^0$ decay. Λ and K_s^0 travel a few centimeters before decay. The distance between primary (production) vertex and decay vertex can

be determined using the good energy and angular resolution of the detector. That allows the reliable identification of the π^0 from $\Lambda \to \pi^0 n$, see [2] for details. Figure 6 show the invariant mass of two photons from the $\gamma\gamma\Lambda$ final state. Most of the background comes from the $\gamma\gamma\Sigma$ final state and $K_s^0 \to \pi^0\pi^0$. Six-cluster events from $K^-p \to \pi^0\pi^0\Lambda$ are clearly detected. The Dalitz plot of the reaction is shown in Fig. 7 for three different momenta. The structure of the Dalitz plots looks remarkably similar to the ones from $\pi^-p \to \pi^0\pi^0 n$, see [18] for details. The process $K^-p \to \eta\Lambda$ followed by $\eta \to 3\pi^0$ decay is a source of 8-cluster events. The CB acceptance for such events is about 9% [2].

The unique features of the Crystal Ball detector demonstrated in the series of experiments in BNL will be substantially enhanced in Mainz by adding the TAPS forward wall and the central tracker. The Crystal Ball is scheduled to arrive to Mainz sometime in the fall of 2002. We are planning to start the data taking for the first round of experiments in early fall 2003. The first round will be finished by spring 2004 followed by the upgrade of MAMI and the tagger. The Crystal Ball collaboration is looking forward for the first experiment on the upgraded MAMI-C with the new tagger. The anticipated beginning date for the second stage of the experimental program is spring 2005.

References

1. R. Novotny: IEEE Trans. Nucl. Sci. **38**, 379 (1991)
2. A. Starostin et al.: Phys. Rev. C **64**, 055205 (2001)
3. W.B. Tippens et al.: Phys. Rev. Lett. **87**, 192001 (2001)
4. T.D. Stanislaus et al.: Nucl. Instrum. Methods A **462**, (2001)
5. S. Prakhov et al.: Phys. Rev. Lett. **84**, 4802 (2000)
6. A. Starostin et al.: Phys. Rev. Lett. **85**, 5539 (2000)
7. I. Anthony et al.: Nucl. Instrum. Methods A **301**, 230 (1991)
8. K.-H. Kaiser et al.: Proc. PAC'99, p. 2915
9. A. Jankowiak et al.: Proc. EPAC2002, Paris, (2002)
10. B. Nefkens et al.: Phys. Rev. D**18**, 3911 (1978)
11. M. Kotulla et al.: Phys. Rev. Lett. **89**, 272001 (2002)
12. D. Drechsel and M. Vanderhaeghen: Phys. Rev. C **64**, 065202 (2001)
13. A.I. Machavariani and Amand Faessler: arXiv:nucl-th/0202060 (2002)
14. R. Rapp et al.: Phys. Rev. C **59**, R1237 (1999)
15. M.J. Vicente Vacas and E. Oset: Phys. Rev. C **60**, 064621 (1999)
16. F. Bonutti et al.: Phys. Rev. Lett. **77**, 603 (1996)
17. E.D. Bloom and C.W. Peck: Ann. Rev. Nucl. Sci. **33**, 143 (1983)
18. B.M.K. Nefkens, S. Prakhov, and A. Starostin: nucl-ex/0202007

Eur Phys J A (2004) **19**, s01, 283–287
Digital Object Identifier (DOI) 10.1140/epjad/s2004-03-047-3

EPJ A direct
electronic only

Status of the BLAST experiment

D.K. Hasell, for the BLAST Collaboration

Massachusetts Institute of Technology, Cambridge, MA 02139, USA

Received: 10 Oct 2003 / Accepted: 14 Nov 2003 /
Published Online: 6 Feb 2004 – © Società Italiana di Fisica / Springer-Verlag 2004

Abstract. The BLAST experiment is beginning operation at the MIT-Bates Linear Accelerator Laboratory. The experiment will study the spin dependent electro-magnetic interaction in few nucleon systems at momentum transfers between 0.1 and 1.0 GeV^2. This will provide improved measurements of the nucleon form factors, particularly G_E^n, as well as study the structure of D and ^{3}He. Other reaction channels such as pion production and inclusive scattering will also be studied. The experiment, physics goals, and current status are described briefly.

PACS. 25.30.-c – 14.20.Dh – 13.40.Gp – 21.10.Ft

1 Introduction

BLAST, Bates Large Acceptance Spectrometer Toroid, has been designed to study in a systematic manner the spin dependent, electro-magnetic interaction of few nucleon systems. Utilising a polarised electron beam with polarised targets of pure H, D, or ^{3}He BLAST will provide improved measurements of the nucleon form factors and the spin structure of D and ^{3}He. In addition, the general purpose, symmetric detector will simultaneously measure quasi-elastic, inclusive, and production channels. The strength of the BLAST program rests on the ability to measure all these quantities with a single detector while flipping the beam and target spins to reduce systematic errors.

The following sections give a brief description of the experiment, the physics motivation for BLAST, and describe the current status of the experiment with some preliminary results.

2 BLAST experiment

The BLAST experiment is situated at the MIT-Bates Linear Accelerator Laboratory[1]. The accelerator provides polarised electrons with energies up to 1 GeV. The BLAST detector, situated on the South Hall storage ring, has a symmetric design with a toroidal magnetic field and an array of detectors to provide particle tracking and identification. A variety of internal gas targets are available providing isotopically pure, polarised H, D (vector and tensor), or ^{3}He.

2.1 MIT-Bates linear accelerator

The MIT-Bates linear accelerator consists of a 500 MeV linac with a recirculator providing electron energies between 0.25 and 1 GeV. Laser induced photo-adsorption on a strained gallium-arsenide crystal is used to produce polarised electrons which are then accelerated and directed into one of two experimental halls: north or south. The initial electron beam helicity can be switched by changing the polarisation of the laser at the ion source.

BLAST is situated on the South Hall Storage Ring, SHR. Typically electron beam currents averaging 80 mA with lifetimes of ∼ 25 minutes and beam polarisations of 65–70% are available. Siberian snakes are used to maintain the longitudinal beam helicity and a spin flipper can be used to reverse the helicity of the stored beam during data acquisition to reduce systematic errors. Tests with the spin flipper show ∼ 98% conservation of polarisation for each spin reversal.

A Compton polarimeter[2] is situated on the SHR to measure the beam polarisation. The polarimeter uses circularly polarised laser light focused on the on-coming electron beam and measures the back-scattered photons in a CsI detector. The polarisation of the laser light can be switched in a Pockels cell and a chopper wheel allows simultaneous measurements with no laser beam for background corrections. The combined systematic and statistical error in measured beam polarisation is about 3%.

2.2 BLAST detector

The BLAST detector (see Fig. 1) is based around eight, water cooled copper coils. Each coil has 26 turns and normally carries 6730 A of current. This results in a toroidal

---

¹ Operated by MIT for the US Department of Energy

² contribution from MIT

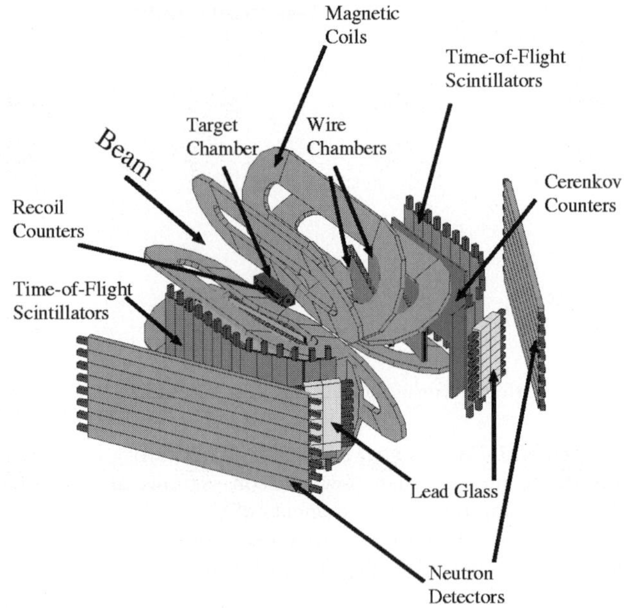

Fig. 1. Schematic, isometric view of the BLAST detector showing the main detector elements

magnetic field of up to 3500 G between the coils but less than 30 G in the region of the target. The magnetic field was mapped with a 3D Hall probe and the results agreed with calculated values to within 1%.

The detectors for BLAST are designed to detect particles scattered into the two, left and right, horizontal sectors, nominally subtending 20–80° in polar angle and ±15° in azimuthal angle though plans are underway to extend the polar coverage. The top sector is used for the internal target and the bottom sector has vacuum pumps. The remaining four sectors are not used at present.

The instrumented horizontal sectors (see Fig. 1) are symmetric; each consisting of wire chambers for particle tracking, Čerenkov detectors for electron identification, time of flight scintillator bars to provide a trigger and relative timing, and neutron wall to detect neutrons. In addition there are plans to install silicon strip, recoil detectors inside the target chamber to detect D and ^{3}He nuclei. The lead glass calorimeters shown in the forward region are available but not installed. Not shown in this figure are some addition neutron detectors obtained from the LADS experiment which will be used to augment the neutron detection efficiency.

In each sector there are three wire chambers[3] combined into a single gas volume to reduce multiple scattering. Each chamber has two super-layers of jet-style drift cells with 3 planes of sense wires. The super-layers are inclined ±5° to the vertical to allow reconstruction in three dimensions. This results in 18 layers of sense wires with which to track charged particles to determine momentum, origin in the target, and scattering angles.

The Čerenkov detectors[4] use 1 cm thick Aerogel tiles inside boxes painted with a white reflective paint. PMT's (5″ diameter) at the top and bottom of the boxes detect the light produced as electrons pass through the Aerogel which has a refractive index of 1.02–1.03. There are four boxes in each sector with 6, 8, 12, and 12 PMT's for readout.

Immediately behind the Čerenkov detector are sixteen vertical bars of 1″ thick scintillator. Each bar is 8″ wide. These form the time of flight, TOF, detector[5] and are used to measure the relative timing between particles striking the TOF. Each bar is readout through PMT's at the top and bottom. The timing for each PMT is adjusted so that a relativistic particle from the target produces a signal with the same timing regardless of which TOF fires. This is necessary as the TOF also determines the timing of the trigger for the data acquisition and provides the common stop for the wire chamber TDC's.

Beyond the TOF detector is a wall of 8 horizontal scintillator bars 10 cm thick, 22.5 cm wide and 400 cm long used to detect neutrons. The neutron wall[6] bars are readout at both ends by PMT's. Additional neutron detectors are being installed using scintillator bars from the LADS experiment. These are in two varieties 15 and 20 cm thick and will double or triple the neutron detection efficiency.

The BLAST data acquisition system is based on the CODA system from TJLAB and the trigger is the same as used in Hall A of TJLAB. This allows a vary flexible system with multiple, simultaneous triggers so data can be accumulated for elastic, quasi-elastic, inclusive, and production reactions at the same time.

2.3 Internal targets

Three different internal targets[7] have been developed for the BLAST experiment: an atomic beam source (ABS), a laser driven target (LDT), and a ^{3}He target.

The ABS consists of a RF dissociator to dissociate hydrogen or deuterium molecules to atoms, a cooled nozzle for atomic beam formation, two sets of sextupoles for Stern-Gerlach selection of the desired hyperfine states and to focus them into the target cell, and RF transitions units to interchange hyperfine states as required to maximise the number of atoms with the desire polarisation. This provides polarised H and both vector and tensor D for the experiment and allows the polarisation to be changed quickly during data acquisition to minimise systematic errors. The ABS is expected to provide target densities of 5×10^{13} atoms/cm^2 with polarisations up to 80%.

The laser driven target uses circularly polarised laser light to optically pump a trace amount of K in a heated volume of dissociated H. The hydrogen atoms become polarised through spin exchange with the potassium atoms. The advantage of the LDT is the 25-30 times higher densities possible.

---

[3] contribution from MIT

[4] contribution from Arizona State University
[5] contribution from University of New Hampshire
[6] contribution from Ohio University
[7] contribution from MIT

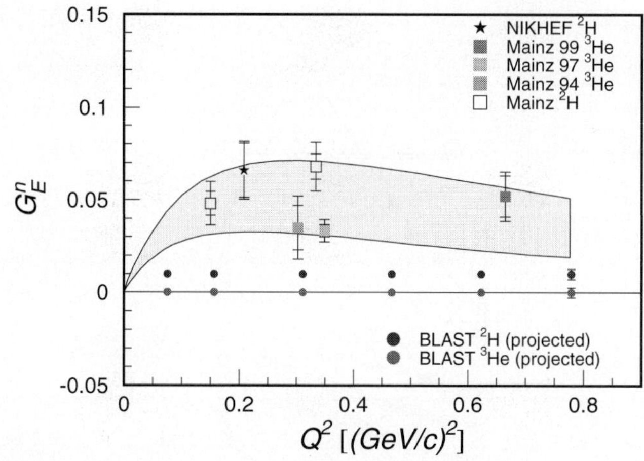

Fig. 2. The existing data for the neutron charge form factor, G_E^n, is shown. The precision and range of data expected from BLAST are shown for measurements from both D and ^{3}He targets

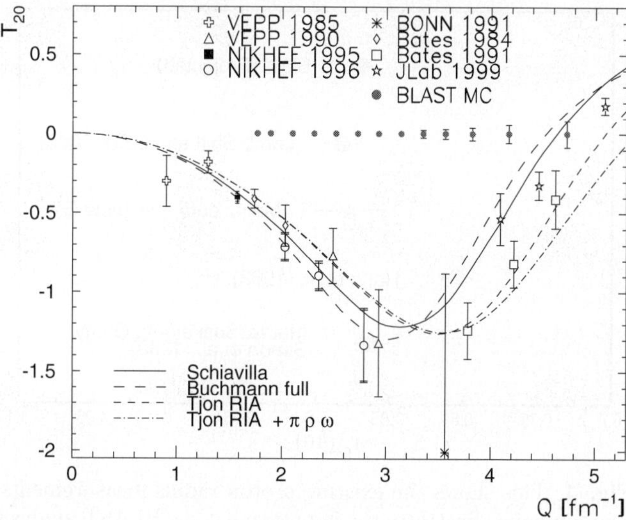

Fig. 3. The present data for T_{20} is shown together with the expected precision of the BLAST measurements

The polarised ^{3}He target provides BLAST with an effective polarised neutron target and allows a independent measurement of G_E^n from that obtained using deuterium as a target.

The polarised ^{3}He target is effectively a polarised neutron target and again allows a measurement of quasi-elastic *en* scattering which can be compared with the deuterium results.

3 Physics goals

The design of the BLAST experiment allows many measurements to be made simultaneously. Flipping beam and target polarisations during running and the symmetric detector design reduce systematic errors. Thus BLAST will provide new and improved data for a number of elastic, quasi-elastic, inclusive, and production reactions. A few of the specific studies planned for BLAST are outlined below.

3.1 Nucleon form factors

One of the important contributions BLAST will make is in the area of nucleon form factors. The proton form factors G_E^p and G_M^p and the neutron form factor G_M^n are already reasonably well measured. BLAST will add to these measurements but the main contribution will be in measuring the neutron charge form factor G_E^n. Figure 2 shows the existing data for G_E^n which are rather sparse and have considerable uncertainty. The expected precision and range of measurements from BLAST are indicated near the bottom of the figure. BLAST will measure G_E^n by two independent methods: using polarised D or ^{3}He as targets, which will serve as a cross check of the results. With deuterium the neutron form factor is derived from studying the quasi-elastic *en* scattering. Corrections must be made for the presence of the proton but BLAST can check these though the simultaneously measurement of quasi-elastic *ep* scattering which can be compared with the direct measurement BLAST makes using a polarised hydrogen target.

3.2 Deuteron form factors

Deuterium has three elastic form factors: G_C, G_M, and G_Q. Previous experiments, which did not enjoy polarised beams plus polarised targets, were unable to fully unfold the contributions from the three form factors or had to perform difficult double scattering experiments to measure the polarisation of the scattered deuteron. With a pure, polarised D target available in both vector and tensor polarisations BLAST can resolve all terms[1]. Existing data for T_{20} is shown in Fig. 3 together with the expected precision and extent of the BLAST measurements. The other spin asymmetries: T_{22}, T_{11} and T_{10} will also be measured and used to unfold G_C, G_M, and G_Q.

3.3 Proton radius

The proton charge and current radii are important fundamental quantities in physics. Precise determination of the proton charge radius is extremely important to the understanding of the proton structure in terms of quark and gluon degrees of freedom in QCD. It is also essential for high-precision tests of QED in the hydrogen Lamb shift measurements. The existing proton charge radius data is shown in Fig. 4 together with the expected precision of the BLAST measurement. The BLAST measurement involves measuring the elastic *ep* cross section at low momentum transfers and then determining the slope dG_E^p/dQ^2 as $Q^2 \to 0$ which is proportional to the proton radius.

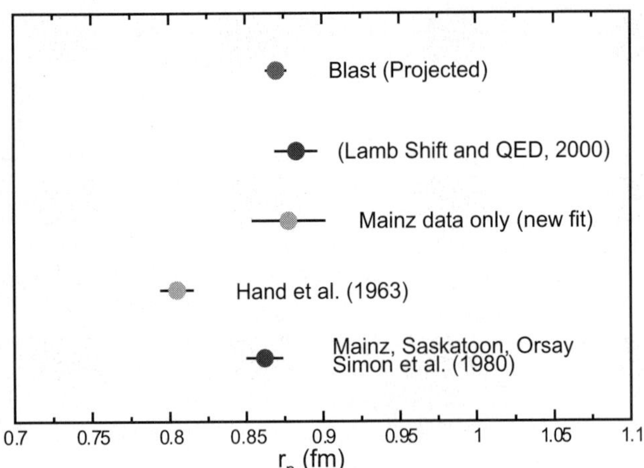

Fig. 4. This shows the existing proton radius measurements. The top point illustrates the improvement the BLAST analysis is expected to achieve

3.4 Ratio of proton form factors

There has been some recent interest[2,3,4] in the ratio of proton form factors: $\mu_p G_E^p / G_M^p$. Differing experimental results show the ratio either remaining flat and close to 1 or decreasing quite rapidly with increasing momentum transfer. The planned BLAST measurements, while quite accurate, unfortunately will be below 1 GeV^2 in momentum transfer and most likely unable to strongly support one case or the other. However, relatively simple upgrades to the South Hall Ring would permit momentum transfers up to 1.8 GeV^2 which could distinguish between the two trends.

4 Current status

The BLAST detector has been fully assembled and has been undergoing commissioning studies since May, 2003. A brief summary of the performance of main detector components is given below.

The wire chambers showed clear tracks but initially suffered from a space charge effect which caused excessive noise. This was solved by changing the gas mixture to a 70:30 mixture of He:iso-butane and adding a trace amount of water vapour to the gas system. The chambers now run very quietly as can be seen in Fig. 5 which shows the wire chamber hits along the two tracks and no random or noise hits elsewhere in the chamber. Another significant factor in the chamber operation was found to be a careful tune of the storage ring beam. Four scintillators placed downstream of the target are used to optimise the beam transport through the target cell. Determining the calibration parameters for the wire chamber is on-going. Currently the average resolution for individual wires is better than 200 μm but overall track reconstruction shows a momentum resolution of $\sim$5%. Improved calibration and reconstruction including energy loss corrections and kinematic fitting should improve the momentum resolution to better than 2%.

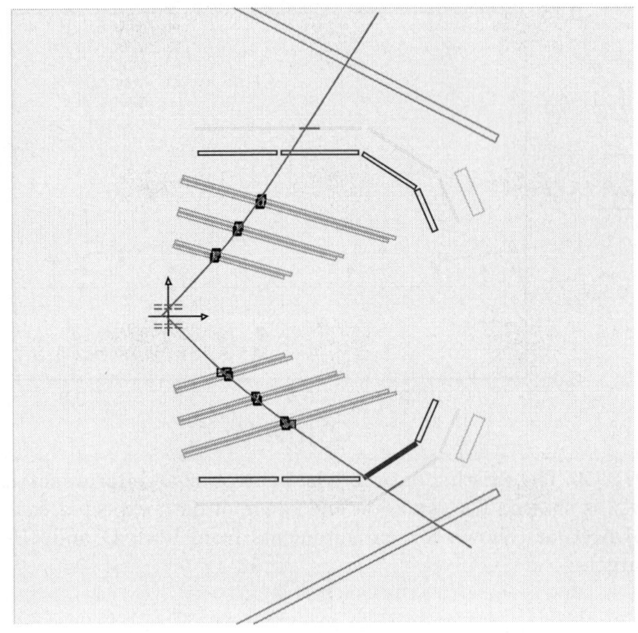

Fig. 5. A typical elastic ep scattering event with reconstructed tracks. The lower track shows an electron which fires the Čerenkov detector and TOF while the top track is the proton with only a recorded hit in the TOF

The PMT's for the Čerenkov detectors initially showed a 50% decrease in efficiency when the toroidal magnetic field was on. The 5″ PMT's used are sensitive to stray fields of only 0.5 G and additional shielding was required to eliminate such fields. Now electron identification efficiency is 85–90% and can be used to significantly clean-up event selection.

The PMT's on the TOF detector also changed slightly when the toroid was on so an extra layer of μmetal was necessary. The meantime of the TOF signals were timed so the trigger (and hence COMMON STOP for the wire

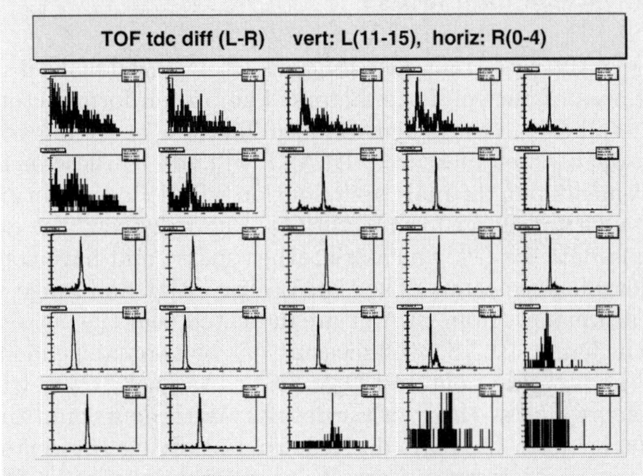

Fig. 6. TDC distributions for the back angle TOF's in the *left sector* in coincidence with forward angle TOF's of the *right sector*. The strong peak seen in the diagonal elements correspond to elastic *ep* scattering

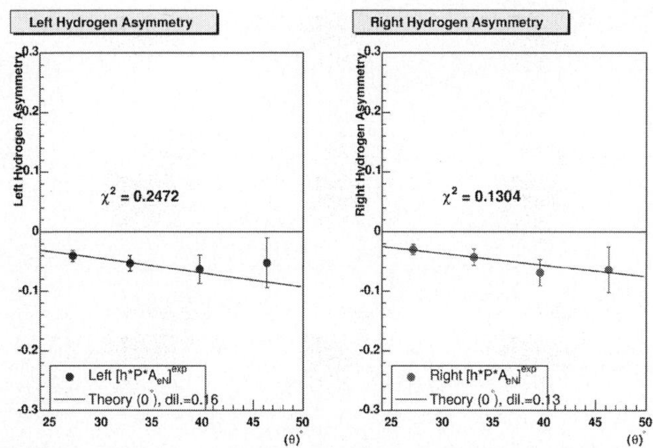

Fig. 7. Asymmetry measured in *left* and *right sectors* for *ep* elastic scattering. The *solid lines* indicate the expected asymmetries assuming a product of beam and target polarisations of 0.13–0.16

chambers) is independent of which TOF is struck. Using just the relative timing of the TOF detector it is possible to select very clean elastic scattering events as illustrated in Fig. 6 which shows a strong, correlated timing peak between the left and right sector TOF's corresponding to elastic *ep* scattering. A timing resolution of better than 600 ps has been achieved for the TOF detector.

The atomic beam source, ABS, was the first target used at BLAST and several initial problems had to be overcome. Firstly, the toroidal magnetic field caused a precession of the atomic spin during transport which decreased the intensity focused into the target cell. Adding iron shielding solved this problem. However, the combination of toroidal field and additional iron caused hysteresis affects when the magnetic fields of the ABS transition units were changed. Since the controllers monitored cur-

rent rather than produced field the wrong transition field was often set which reduced he polarisation. Studies with the ion polarimeter to measure atomic fraction and careful tuning of the transitions for a fixed state did enable polarised running but without the rapid spin reversal of the target. Figure 7 shows the measured asymmetries in the left and right sectors as a function of the scattering angle. With an average beam polarisation of $\sim 65\%$ the measured asymmetry is consistent with a target polarisation of $\sim 22\%$.

The ABS is currently removed to improve the magnetic shielding, install Hall probes to monitor the field of the transition units, improve the pumping, and to upgrade the RF control system. These improvements should enable the ABS to be operated as intended.

5 Conclusion

The BLAST experiment expects to begin production running in the Fall, 2003. Problems identified during the commissioning studies have been solved. With the polarised electron beam from the MIT-Bates linear accelerator, high polarisation, isotopically pure internals targets of H, D, or ^{3}He, and a symmetric, general purpose detector system BLAST will make significant measurements of a number of important physics quantities.

References

1. T.W. Donnelly and A.S. Raskin: Ann. Phys. **169**, 247 (1986)
2. M.K. Jones et al.: Phys. Rev. Lett. **84**, 1398 (2000)
3. O. Gayou et al.: Phys. Rev. C **64**, 038202 (2001)
4. O. Gayou et al.: Phys. Rev. Lett. **88**, 092301 (2002)

Eur Phys J A (2004) **19**, s01, 289–290
Digital Object Identifier (DOI) 10.1140/epjad/s2004-03-048-2

eRHIC: The electron ion collider at BNL

A. Deshpande

RIKEN-BNL Research Center, Brookhaven National Laboratory, Upton, NY 11973-5000

Received: 31 Jul 2003 / Accepted: 14 Nov 2003 /
Published Online: 6 Feb 2004 – © Società Italiana di Fisica / Springer-Verlag 2004

Abstract. Addition of a high intensity & high energy polarized electron/positron beam facility to the already existing Relativistic Heavy Ion Collider (RHIC) complex so that electrons/positron from that facility could collide with the RHIC hadron beams, would significantly enhance RHIC's ability to explore fundamental and universal aspects of QCD. We present here the high lights of the physics program that could result as a result of such a RHIC upgrade.

PACS. 14.20.Dh – 13.40.Gp – 21.10.Ft

1 Introduction

In the past few years considerable interest has developed in the experimental and theoretical nuclear physics community to study the scope of a physics program that could be pursued if a high energy electron beam facility is built at BNL and that beam is made to collide with the already existing hadron beams from RHIC. Such a facility is called eRHIC [1]. If the electron beam is polarized then not only could one study unpolarized deep inelastic scattering (DIS) using the unpolarized protons (e-p) or heavy ions (e-A) at RHIC, but also polarized DIS ($\mathbf{e} \cdot \mathbf{p}$) using the polarized protons from RHIC. No other experimental facility planned in near or distant future would have possibility of such a wide scope of experimental program.

2 What is eRHIC?

The nominal electron beam energy for this future facility under consideration is about 10 GeV, but it could be varied between 5 and 10 GeV. The variation of the hadron beam energy is already possible in the RHIC accelerators today. Using the highest possible value of the proton beam energy possible in the present RHIC of 250 GeV and a 10 GeV electrons from a new facility one could envision DIS with $\sqrt{s} \sim 100$ GeV. For polarized DIS which till now has only been pursued in fixed target mode this would be a factor of ~ 3 to ~ 20 increase in the value of $\sqrt{s}$. The facility will deliver at least 10^{33} cm^{-2}sec^{-1} luminosity. For unpolarized DIS off of heavy ions, it is expected that RHIC would provide beams of 100 GeV/nucleon. This implies that using a 10 GeV electron beam one could attain $\sqrt{s} \sim 60$ GeV for e-A collisions. Using the electron, proton beam energies mentioned above and the the simple relation between the kinematic variables it is obvious that with eRHIC one can reach x as low as few $\times 10^{-4}$ and as high as 0.8 for $Q^2 \geq 1$ GeV^2. If the requirement of $Q^2 > 1$ is removed, eRHIC could access (in photo-production region, for example) $x \sim 10^{-5}$. For a nuclear beam one could reach $x \sim 10^{-3}$ values for $Q^2 > 1$ GeV^2. Figure 1 shows the $x - Q^2$ kinematic reach of the eRHIC (e-p) compared to the fixed target experiments of the past and present.

3 The polarized e-p physics at eRHIC

In the case of polarized DIS off of proton beams the program for such a collider facility is well defined [2,3,4]: eRHIC will enable accucrate measurement of the spin struc-

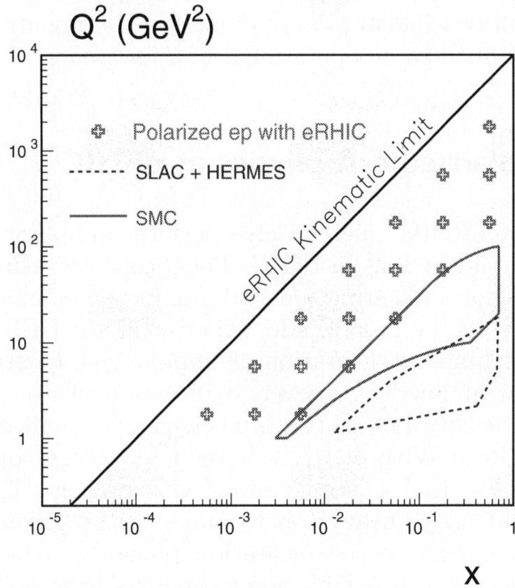

Fig. 1. The $x - Q^2$ kinematic coverage of the eRHIC compared with the present and past fixed target polarized DIS experiments

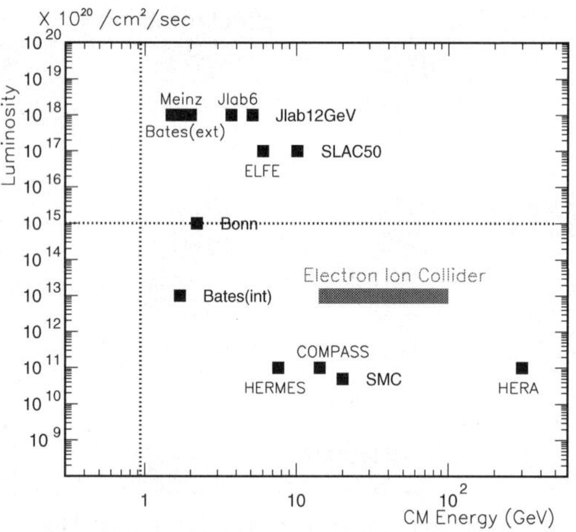

Fig. 2. The center of mass energy vs the e-p luminosity plot for past and some future DIS facilities

ture function of the proton and of neutron (if $^3\mathrm{He}^{+2}$ can be stored in the RHIC ring) in the low-x region which is the source of the largest uncertainty in the first moments $\Gamma_1^{p/n}$ and the Bjorken sum rule. The spin carried by the polarized gluon inside a nucleon is a yet unmeasured quantity. eRHIC would measure that unambiguously and using theoretically and experimentally clean techniques of pQCD analysis at NLO, the photon-gluon-fusion process resulting in di-jet or 2 high p_T oppositely charged tracks, and using photoproduction of di-jets and high p_T tracks. Further eRHIC facility would be unique in its ability to study and resolve the polarized parton distribution inside a polarized photon. Investigations based on charged current e-p scattering would allow a first and unique measurement of the parity violating structure function $g_5^{\pm}$. There would also be other measurements possible with eRHIC that would be semi-inclusive and exclusive in nature, which will help resolve the spin structure of the nucleon along with the above mentioned measurements with much better accuracy than possible in present day DIS experiments and by those planned to be operational in a few years.

4 Unpolarized e-A physics at eRHIC

With the eRHIC, one can also perform unpolarized DIS using the heavy ions in RHIC. The most interesting kinematic region indicating unusual nuclear phenomena is at low x [5]. In the last decade, experiments at HERA have provided hints of saturation of unpolarized F_2 structure functions at low x. However, exploration of even higher gluonic density kinematics is necessary to confirm these observations. With eRHIC's lower CM energy compared to HERA's, this is not possible with protons, however, one could use a heavy ion to enhance the gluon densities and expect to see saturation phenomena using the RHIC beams in e-A DIS. Some theorists have boldly described this saturation phenomenon as indicate formation of a new and unique state of matter, and called it the Color Glass Condensate (CGC). Indirect evidence for its

existance seen from HERA data could be confirmed with direct measurements of e-A at eRHIC. Properties of CGC can be probed by measuring both the inclusive as well as exclusive processes in e-A scattering. Understanding the CGC would be critical in the investigations of QGP underway at RHIC and at LHC in future. Experimental signatures for this kind of unusual phenomena have been predicted in various recent publications [5]. The eRHIC will be a unique place to study this physics in detail as a function of A (e.g. the atomic number) of the beam. Also, the variable energies for each nuclear beam will allow a study of the onset of saturation physics. Other unpolarized DIS phenomena which have been studied in the past in the fixed target environment have been mainly related to color transparency, EMC effect and other intermediate low-x related phenomena. These could also now be studied at eRHIC with ease.

5 Outlook

Various subgroups are now being formed to realize the EIC project and its physics [1]. The activities of the accelerator working group formed principally by the collaboration of MIT/Bates and BNL have resulted in a preliminary design for the electron ring that could realize the EIC at BNL. A small group of people met recently at BNL to discuss the electron beam polarimetry and its into the electron ring accelerator design. Physics groups have continued their activities in smaller subgroups and are expected to present their investigations at regular intervals. EIC collaborations meetings are presently planned at a frequency of once a year. The detector development activity is about to start. Integration of the detector with the accelerator lattice is one of the issues of particular interest to the EIC community because of the particular interest in low x physics and the demanding particle ID requirements posed by exclusive physics interests [1,3].

Acknowledgements. I would like to thank the organizers and the ICTP Trieste for the organization of this meeting and inviting me for it.

This work was supported by the RIKEN-BNL Research Center at Brookhaven National Laboratory.

References

1. Home page of eRHIC/EIC: `http://www.bnl.gov/eic`
2. Proceedings of eRHIC Workshop at Yale, BNL-Report-No. Editors A. Deshpande, and V.W. Hughes, 2001
3. A whitepaper on *The Electron Ion Collider, A high luminosity probe of the partonic substructure of nucleons and nuclei*, Editors A. Deshpande, R.G. Milner, and R. Venugopalan, BNL-Report-No.68933-02/07 Rev, February 2002
4. A. Deshpande: *The EIC Project, physics prospects and present status*, Nucl. Phys. **B** (Proc. Supp.) 105 (2002) February 2002, Editors S.D. Bass, A. De Roeck, and A. Deshpande
5. E. Iancu, A. Leonidov, and L. McLerran: *The color glass condensate: An introduction*, `hep-ph/0202270`

Eur Phys J A (2004) **19**, s01, 291–296

Digital Object Identifier (DOI) 10.1140/epjad/s2004-03-049-1

EPJ A direct

electronic only

The COMPASS experiment

A. Bressan, on behalf of the COMPASS Collaboration

University of Trieste and INFN, Sezione di Trieste, Trieste, Italy

Received: 19 Sep 2003 / Accepted: 14 Nov 2003 /
Published Online: 6 Feb 2004 – © Società Italiana di Fisica / Springer-Verlag 2004

Abstract. The COMPASS Experiment at the CERN SPS has a broad physics program focused on the study of the spin structure of the nucleon and on hadron spectroscopy. Key measurements for the spin program are the gluon contribution to the spin of the nucleon, semi-inclusive measurements, and the first measurement of the transverse structure function $\Delta_T q(x)$. Its state-of-the-art apparatus consists of a two-stage large acceptance spectrometer designed for high data rates and equipped with high-resolution tracking, particle identification and electromagnetic and hadronic calorimetry.

The first year of physics run (2002) was devoted to the spin programme, using a polarised μ^+ beam at 160 GeV/c and a polarised ^{6}LiD target.

PACS. 13.60.-r – 13.85.Hd – 14.40.Lb – 25.30.Mr

1 Introduction

The primary physics goals of the CERN COMPASS [1] collaboration (about 270 physicists from 27 institutes and 11 countries) are twofold:

- A rich spin program with a polarised muon beam and a polarised target. It is aimed to provide a direct measurement of the gluon polarisation $\Delta G/G$ by measuring the spin dependent asymmetry of the photon-gluon process, that can be accessed by detecting open-charm events and high-p_T correlated hadron pairs. It is also aims to determine for the first time the transversity structure function $\Delta_T q(x)$. Accurate measurements of the flavour decomposition of the quark helicity distributions, and the measurement of the polarised fragmentation function complete the spin program.
- A wide program of hadron spectroscopy will be performed by using hadronic beams. COMPASS will study π and K polarisabilities (Primakoff reactions), allowing to test predictions of the chiral perturbation theory; extensive meson spectroscopy to investigate the presence of exotics states; collections of large samples of semileptonic decays of charmed mesons and baryons to determine form factors and probe predictions from the Heavy Quark Effective Theory.

The experiment was constructed in 1998–2000 and commissioned in 2001. 2002 run was the first year of physics, devoted to the spin program, in which a total of 260 TB of data have been collected over the 80 days of the run.

2 The COMPASS apparatus

The COMPASS apparatus consists a two stage spectrometer (Fig. 1): the Large Angle Spectrometer (LAS) covers an angular region up to ±180 mrad, just downstream the polarised target; the small angle spectrometer (SAS) subsequentially detects particles within ±30 mrad. The two conventional dipole magnets (SM1 and SM2) provide respectively field integrals of 1 and 4.4 T · m for the two sections of the spectrometer. Different tracking devices are used to cope with different fluxes and to fit the needed resolution. Tracking in the beam region is provided by scintillating fibre (SciFi) and silicon detectors. For the region up to 20 cm from the beam new micropattern detectors, i.e. $\mu\Omega$ [3] (between the target and SM1) and GEMs [4] (in the rest of the spectrometer) are used. The large area outer tracking utilizes multi-wire proportional chambers (MWPC), drift chambers SDC, straw tubes, and large size drift chambers (W45). Limited streamer Iarocci-type tubes (MW1) and drift tubes track the muon traversing the hadron absorbers. Both LAS and SAS include hadronic calorimeters, also used at trigger level. In the LAS a ring imaging Cherenkov detector (RICH-1 in the figure) allows to separate pions and kaons up to 40 GeV.

Some of the key points of the experiment will be reviewed in more details in the following.

2.1 Beam definition

The COMPASS muon beam is produced from the decays of the hadronic particles (manly pions) generated by the 450 GeV SPS proton beam impinging on Beryllium target. Muons are produced in a decay path of 300 m, and selected

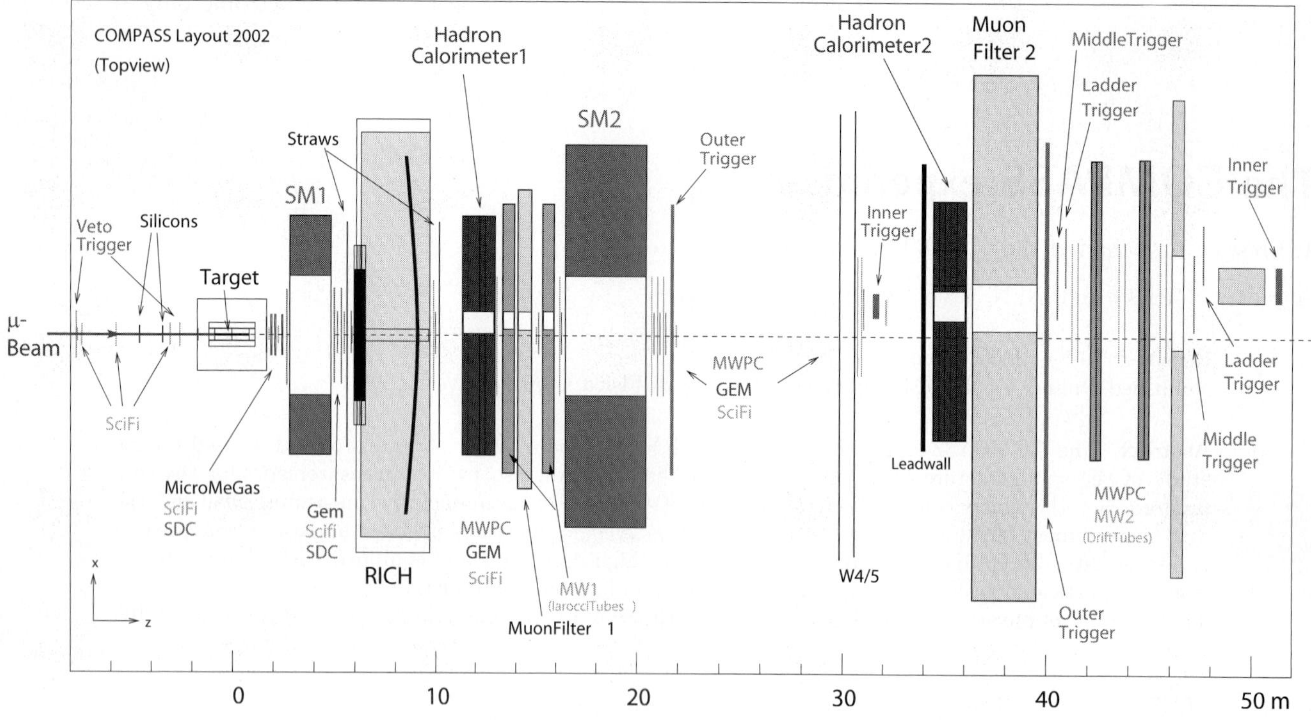

Fig. 1. The COMPASS apparatus for the 2002 run

in momentum by the acceptance of the transport beam line. For 2002 a 160 ± 3.5 GeV/c μ^+ beam was used, with a -80% average polarisation [2] and an intensity of $2 \times 10^8 \mu$/spill.

Four scintillator hodoscopes upstream and downstream of the last bending magnet measure the muon momentum with an uncertainty of 0.5%. The trajectory of the muon upstream and downstream of the target is reconstructed by 12 scintillating fibre hodoscopes; spatial resolution of $130 \div 250$ μm, time resolution of $350 \div 500$ ps for an efficiency around 99% are the characteristics of these devices. Four double-sided silicon detectors with $50 \mu m$ pitch, before and after the target, increase the resolution on the muon tracks.

2.2 The target

COMPASS uses the target system of the Spin Muon Collaboration SMC [5].

The SMC superconducting solenoid produces a 2.5 T field, used to keep the target polarisation in a longitudinal state. An additional dipole provides a transverse field and is used to run the experiment with the target transversely polarised with respect to the longitudinal muon beam polarisation.

The target system is shown in Fig. 2; the beam enters from the left of the figure and passes through the two 60 cm long, oppositely polarised target cells. The average polarisation of the ^{6}LiD target during 2002 was about +53% and -48%. To keep the systematics under control, the spin orientations are inverted every 8 h. During the transverse spin runs the polarisation is held by the 0.5 T dipole field.

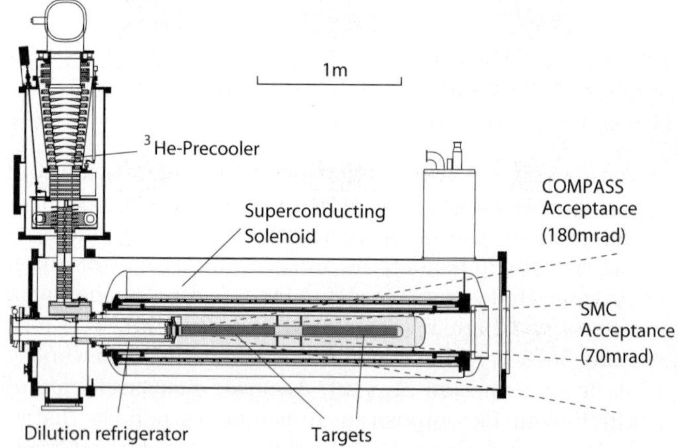

Fig. 2. The target system for the 2002 run; also shown are the acceptance for the SMC magnet and the new COMPASS expected acceptance

2.3 Calorimetry and particle identification

The hadron calorimeters of LAT (HCAL1) and SAT (HCAL2) are Fe scintillators sandwiches with planar WLS. The resolution is $\sigma/\mathrm{E} \simeq 6\% \oplus 60\%/\sqrt{\mathrm{E}}$ for pions and $\sigma/\mathrm{E} \simeq 0.6\% \oplus 24\%/\sqrt{\mathrm{E}}$ for electrons.

Muons are very efficiently identified by large detector planes placed before and after a 60 cm thick hadron absorber which filters the other particles. Aluminium Iarocci-type limited streamer tubes were used for the LAS, while drift tube planes equipped the SAS.

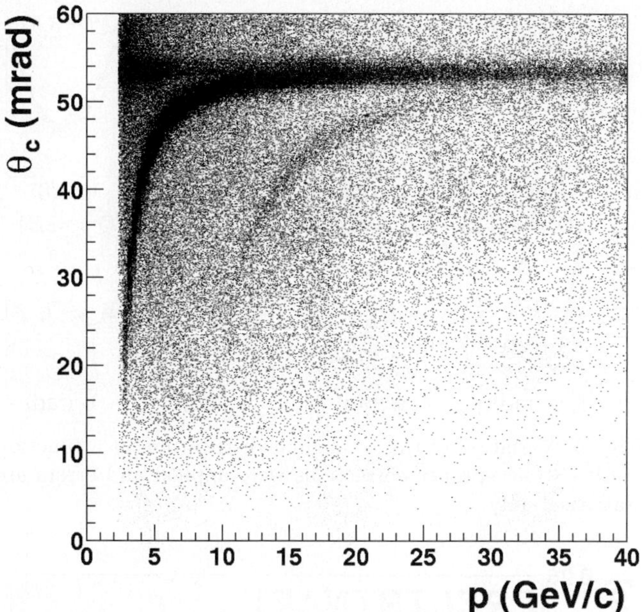

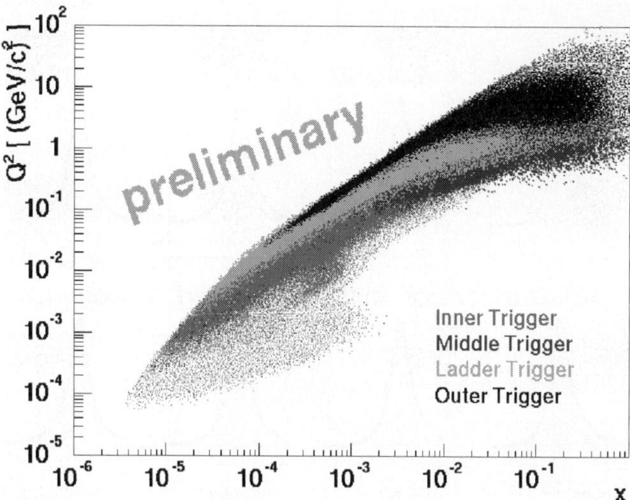

Fig. 4. The kinematical range covered by the different triggers during the 2002 run

Fig. 3. Reconstructed Cherenkov angle θ as a function of the momentum; the main band is given by π, the smaller band are K, the spot at saturated Cherenkov angle is given by electrons

Hadron identification in the LAS is provided by RICH-1 [6], designed to separate π and K, over the whole LAS angular acceptance up to 60 GeV.

RICH-1 consists of a 3 m long C_4F_{10} radiator at atmospheric pressure, a wall of 116 spherical mirrors (3.3 focal length) covering an area of >20 m^2 and two sets of far UV photon detectors placed above and below the acceptance region. The Cherenkov photons are detected by MWPCs equipped with CsI photocathodes [7] covering a surface of 5.3 m^2. To guarantee a transmission higher than 70% for 165 nm photons over the typical 4.5 m path the radiator gas is carefully pre-cleaned and constantly filtered during operation [9]. Photocathodes are segmented in pads of 8×8 mm^2 (83000 total elecctronic channels) read-out by a system of front-end boards [8] with local intelligence Typical noise level of the front-end electronics is $\sigma_n \simeq 1100$ electrons equivalent. The occupancy is around 3% and the dead time per event is 500 ns. Figure 3 shows the reconstructed Cherenkov angle θ_c as a function of the momentum of the track: θ_c saturates at $\simeq 52$ mrad corresponding to a radiator refrective index of 1.0049; pion an K bands are clearly visible. The spot of particles saturating at very low momentum are electrons. In the reconstructed mass spectra the proton peak is also visible.

2.4 Trigger and data acquisition

The trigger is based on the scattered muon. 2 ns wide coincidences between more than 500 elements, composing various hodoscope planes at different positions along the beam, select the scattered muons in the kinematical region of interest, on the bases of target pointing and energy release. The different kinematic ranges covered by the various triggers for the 2002 run is shown in Fig. 4. An hadron shower in the hadronic calorimeter provides more selective triggering. The overall typical trigger rate was 5 kHz with a dead time of about 7%.

DAQ uses a parallel read-out electronics with local pre-event buildings of the ~ 190 k channels [10]. A pipeline acquisitions system transfers the data via S-link to 16 PCs, where the data is buffered during spills. Giga Ethernet lines are then used to transfer the data through network switches to 12 Event Builder PCs. The typical event size is 44 kB, with a dato flow of 220 MB/s during the spill. The data is transferred to the CERN Central Data Recording (CDR); speeds up to 50 MB/s have been reached, for a total of 3 TB/d recorded data.

3 The analysis

For both data recording and reconstruction COMPASS used a farm of 100 dual processor PCs at CERN and the CASTOR file system, on which the 260 TB of data were stored in about 260000 files.

The analysis is performed with a fully object oriented, with a modular architecture, C++ program, entirely written within the collaboration. In 2002 a considerable amount of work was devoted to update the tracking and PID packages in this program (called COmpass Reconstruction and AnaLysis or CORAL).

The data analysis have not been finalized yet and the present status allows only to derive expectations on the physics results that can be obtained by the 2002 run. We list in the following the different items that have been attached by the analysis.

1) Low-Q^2 vector mesons ρ^0 and ϕ production. Elastic (or quasi elastic) production of the vector mesons (VM) ρ^0 and ϕ proceed via coherent or incoherent muon scattering over the nucleus of bound or free nucleons. The VM signal is selected by looking at exclusive events

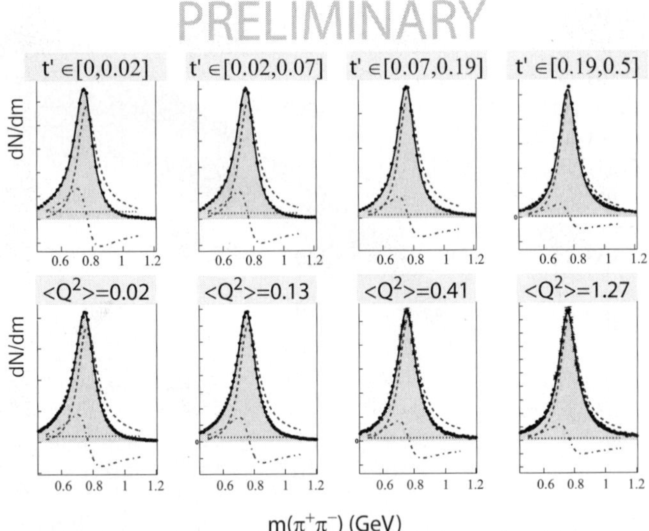

Fig. 5. Fits of the $\pi^+\pi^-$ invariant mass according to the Söding parametrizatin. Dashed are relativistics p-wave Breit-Wigner Contributions, *dashed-dotted lines* correspond to the interference term and *dotted lines* are the non-resonant part

(no missing energy) with only two hadrons in the final state. ρ^0 and ϕ productions are identified by selecting the correct mass interval in the invariant mass distribution for the $\pi^+\pi^-$ and for the K^+K^- hypothesis. Figure 5 shows the fit of the two-pions mass spectrum by considering the coherent sum of resonant and non-resonant $\pi^+\pi^-$ production [11]. The enhancement of events at low masses and the depletion at large can be described by the Söding model [12]. Looking at different t' and Q^2 ranges, the distortion due to the interference term decreases with $-t'$ increasing. The non-resonant di-pion production is also reduced with Q^2 increasing.

The angular distributions (Fig. 6) give a direct indication of the polarisation of the VM. Both the polar and azimuthal distributions are shown.

Provided that a precise modelling of the acceptance can be obtained, the large statistics available allows high precision measurements and tests. As an example one sees that the S-channel helicity conservation (SCHC) hypothesis for the production of the ρ^0 and ϕ mesons is fully compatible with our results.

2) Λ and $\bar{\Lambda}$ hyperon polarisations.

Λ^0, $\bar{\Lambda}^0$ and neutral kaons K^0 are clearly reconstructed and identified by the spectrometer as shown by the Armenteros-Podolanski plot of Fig. 7. The angular distributions of the Λ decay products allow to extract both the transverse and longitudinal components of the Λ polarisation. Figure 8 shows a comparison between data and MonteCarlo simulations for the x_F and Q^2 distribution of Λ^0, $\bar{\Lambda}^0$ and neutral kaons. The agreement is qualitatively good over most of the kinematic range.

The analysis of the full 2002 data sample will increase the statistics on the Λ by a factor of 6 with respect of what shown here, and the same amount will come from

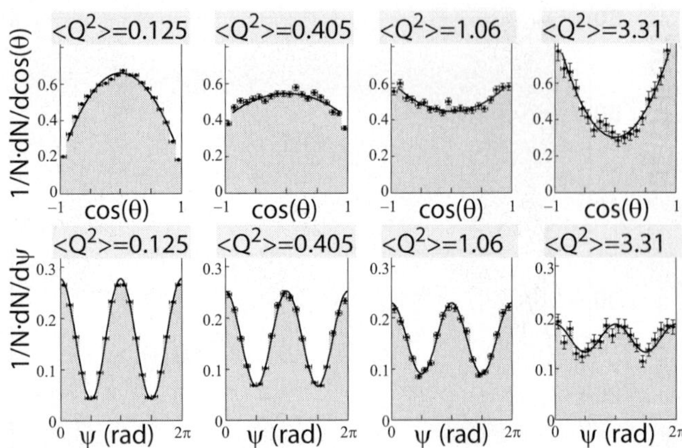

Fig. 6. Variation of the ρ^0 angular distribution as a function of Q^2. No acceptance corrections were applied and errors are statistical only

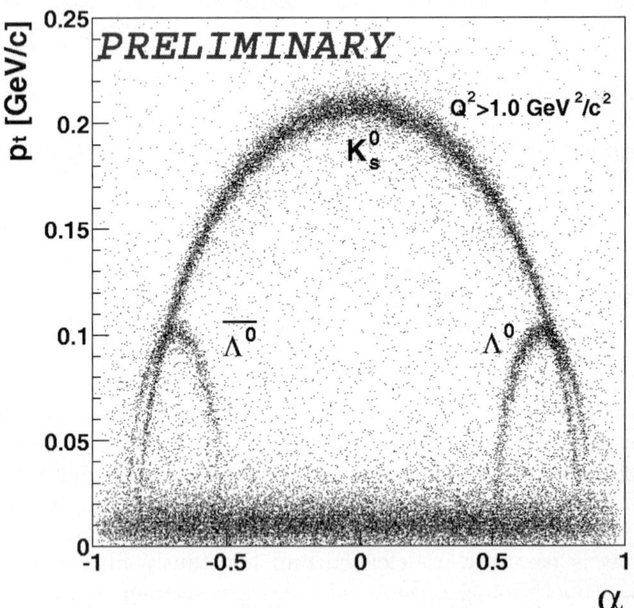

Fig. 7. K and Λ identified using the Armenteros-Podolanski plot. The definition of the α variable is the following: $\alpha = (p_L^+ - p_L^-)/(p_L^+ + p_L^-)$

the 2003 run. Therefore high precision Λ and $\bar{\Lambda}$ hyperons polarisation can be expected.

3) $\Delta G/G$ from charm photo-production and from high-p_T hadron pairs.

In COMPASS the gluon polarisation $\Delta G/G$ will be studied by identifying the photon-gluon fusion process, tagged either by open-charm production or by high-p_T hadron pair production. Open-charm events are identified by reconstructing D^0 and D^* mesons from they decay products, i.e. $D^0 \to K\pi$ and $D^* \to D^0\pi^0 \to K\pi\pi^0$. In the first case, cuts on the K direction in the D^0 rest frame ($|\cos(\theta_K^*)| < 0.5$) and on the D^0 energy fraction ($z_D = E_D/E_{\gamma^*} > 0.25$) are needed to reduce the background contamination. The second case is much cleaner given the unique cinematics.

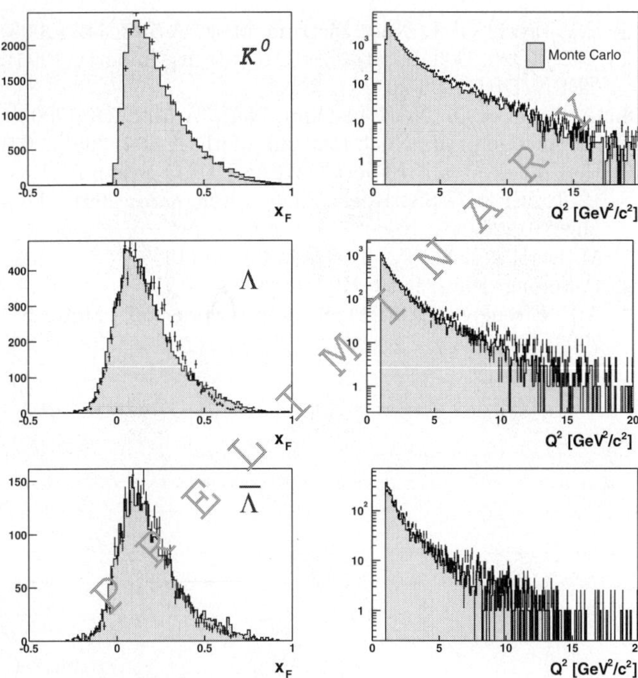

Fig. 8. Comparison between MC (*histograms*) and data (*points*) for Q^2 and x_F distributions for K^0, Λ and $\bar{\Lambda}$

The gluon polarisation will be determined from the measured charm production asymmetry:

$$A^{meas.} = (N_{c\bar{c}}^{\uparrow\downarrow} - N_{c\bar{c}}^{\uparrow\uparrow})/(N_{c\bar{c}}^{\uparrow\downarrow} + N_{c\bar{c}}^{\uparrow\uparrow}) = P_B P_T f D A_{c\bar{c}}^{\gamma^* N}$$

where $N_{c\bar{c}}^{\uparrow\uparrow}(N_{c\bar{c}}^{\uparrow\downarrow})$ represents the number of charm events with target spin parallel (anti-parallel) to the μ helicity and D is the γ^* depolarisation factor, with $< D > \approx 0.66$ for COMPASS.

Preliminary signals of the D meson are shown on Fig. 9. Kaon-pion pairs are selected by asking: $z_D > 0.2$; $|\cos(\theta_K^*)| < 0.85$; $10 < p_K < 35$ GeV in order to be in the RICH K identification region. A soft pion (< 10 GeV) is also required.

4) Collins asymmetry and transversity. Semi-Inclusive Deep Inelastic Scattering (SDIS) provides the possibility to measure the transverse polarised parton distribution function $\Delta_T q(x)$ via the azimuthal dependence (Collins angle ϕ_C) of the leading hadron. This measurement requires to operate with a transversely polarised target: about 3 weeks of 2002 run have been devoted to this measurement. Figure 10 shows the expected statistical error on the Collins asymmetry from the 2002 transverse polarisation data when a positive or negative leading hadron is selected, compared to the Efremov [13] calculation of asymmetry A_{UT}, which include the transversity structure function $\Delta_T q(x)$ by the following linear combination over the quark flavours:

$$A_{UT} = \frac{\sum_q e_q^2 \Delta_T q(x) \Delta_t D_q^h(z)}{e_q^2 \Delta q(x) D_q^h(z)}$$

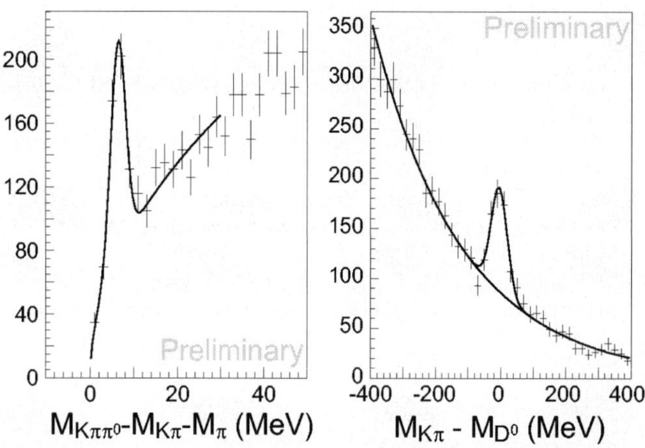

Fig. 9. D^* produced by requiring the invariant mass of the $K\pi$ pair to be in the 60 MeV window around the D^0 peak, together with a detected soft pion. Using this D^* cut, the D^0 peak in the invariant mass spectrum of $K\pi$ is very clear

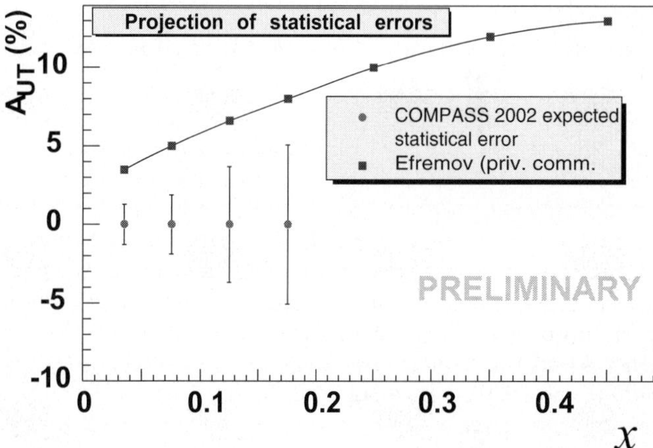

Fig. 10. Expected statistic error compared to Efremov calculation of asymmetry A_{UT} in case of deuterium target. The statistic error is extrapolated to the whole statistics available from year 2002 run and to the combined (*positive and negative charged hadrons*) signal

4 Conclusion

The 2002 run, first year of COMPASS data taking, was successfully; novel detector techniques were integrated in the experiment and were performing according to expectations.

A large amount of interesting data have been collected and is been analysed. The hints from the analysis of VMs and Λ's look very promizing and the large statistics available in COMPASS will allow precise mesurements of polarisations and tests of the current theories.

A long work for extracting the gluon polarisation from the data is foreseen but the D's signals looks very encouraging for the future.

We look forward for the first physics results.

References

1. G. Baum et al.: COMPASS proposal, CERN-SPSLC-96-14 (1996)
2. D. Adams et al.: [Spin Muon collaboration], Nucl. Instrum. Meth. A **443**, 1 (2000)
3. Y. Giomataris et al.: Nucl. Instrum. Meth. A **376**, 29 (1996); D. Thers et al.: Nucl. Instrum. Meth. A **469**, 133 (2001)
4. F. Sauli et al.: Nucl. Instrum. Meth. A **386**, 531 (1997)
5. D. Adams et al.: Nucl. Instrum. Meth. A **437**, 23 (1999)
6. E. Albrecht et al.: Nucl. Instrum. Meth. A **502**, 112 (2003)
7. The RD26 Collaboration: RD26 Status Report (1996), CERN/DRDC 96-20
8. G. Baum et al.: Nucl. Instrum. Meth. A **502**, 246 (2003)
9. E. Albrecht et al.: Nucl. Instrum. Meth. A **502**, 266 (2003)
10. H. Fischer et al.: "The COMPASS DAQ System", Proc $12^{t}h$ IEEE-NPSS Real Time Conference 2001, hep-ph/0101225.
11. M. Derrick et al.: Z. Physiks C **69**, 39 (1995)
12. P. Söding: Phys. Lett. **19**, 702 (1966)
13. A.V. Efremov: "COMPASS analysis workshop", March 3rd 2003, Dubna

Printing and Binding: Strauss GmbH, Mörlenbach